RENEWABLE ENERGY POWERED DESALINATION HANDBOOK

RENEWABLE ENERGY POWERED DESALINATION HANDBOOK

APPLICATION AND THERMODYNAMICS

VEERA GNANESWAR GUDE

ELSEVIER

Butterworth-Heinemann
An imprint of Elsevier

Butterworth-Heinemann is an imprint of Elsevier
The Boulevard, Langford Lane, Kidlington, Oxford OX5 1GB, United Kingdom
50 Hampshire Street, 5th Floor, Cambridge, MA 02139, United States

Library of Congress Cataloging-in-Publication Data
A catalog record for this book is available from the Library of Congress

British Library Cataloguing-in-Publication Data
A catalogue record for this book is available from the British Library

ISBN: 978-0-12-815244-7

Publisher: Mathew Deans
Acquisition Editor: Ken McCombs
Editorial Project Manager: Peter Jardim
Production Project Manager: Surya Narayanan Jayachandran
Cover Designer: Matthew Limbert

Typeset by SPi Global, India

DEDICATION

To my late father Gude Bhaskar Rao, to my mother Gude Fathima, to my sister Kalyani, and to my brothers Kiran and Karunakar, and to Preeti Muire.

CONTENTS

CONTRIBUTORS

Nassir Al Arifi
King Saud University, Riyadh, Saudi Arabia

Ahmed Al Ghamdi
Yanbu Industrial College, Yanbu, Saudi Arabia

Abdulaziz M. Al-Bassam
King Saud University, Riyadh, Saudi Arabia

Anne-Perrine Avrin
University of California, Berkeley, CA, United States

Dmitrii Bogdanov
Lappeenranta University of Technology, Lappeenranta, Finland

Rolando Bosleman
Energy Recovery Inc., San Leandro, CA, United States

Christian Breyer
Lappeenranta University of Technology, Lappeenranta, Finland

Upeksha Caldera
Lappeenranta University of Technology, Lappeenranta, Finland

Pietro Catrini
University of Palermo, Palermo, Italy

Dornadula Chandrasekharam
India Institute of Technology Hyderabad, Hyderabad, India

Andrea Cipollina
University of Palermo, Palermo, Italy

Mário Costa
Instituto Superior Técnico, Lisboa, Portugal

Syed Noman Danish
King Saud University, Riyadh, Saudi Arabia

Neven Duić
University of Zagreb, Zagreb, Croatia

Francesco Giacalone
University of Palermo, Palermo, Italy

Veera Gnaneswar Gude
Mississippi State University, Mississippi State, MS, United States

Sajjad Haider
King Saud University, Riyadh, Saudi Arabia

Gang He
Stony Brook University, Stony Brook, NY, United States

Eric Kadaj
Energy Recovery Inc., San Leandro, CA, United States

Soteris A. Kalogirou
Cyprus University of Technology, Limassol, Cyprus

Daniel M. Kammen
University of California, Berkeley, CA, United States

Salah Ud-Din Khan
King Saud University, Riyadh, Saudi Arabia

Shahab Ud-Din Khan
COMSATS Institute of Information Technology, Attock, Pakistan

Arif Lashin
King Saud University, Riyadh, Saudi Arabia

Giorgio Micale
University of Palermo, Palermo, Italy

Jayanta D. Mondol
Ulster University, Newtownabbey, Northern Ireland, United Kingdom

Ibrahim Mustafa
Yanbu Industrial College, Yanbu, Saudi Arabia

Jamel Orfi
King Saud University, Riyadh, Saudi Arabia

Antonio Piacentino
University of Palermo, Palermo, Italy

Adrian Pugsley
Ulster University, Newtownabbey, Northern Ireland, United Kingdom

Usman Ali Rana
King Saud University, Riyadh, Saudi Arabia

Lovedeep Sahota
Indian Institute of Technology (I.I.T.) Delhi, New Delhi, India

Raquel Segurado
Instituto Superior Técnico, Lisboa, Portugal

Mervyn Smyth
Ulster University, Newtownabbey, Northern Ireland, United Kingdom

Alessandro Tamburini
University of Palermo, Palermo, Italy

Gopal N. Tiwari
Bag Energy Research Society, Varanasi, India

Eftihia Tzen
Centre for Renewable Energy Sources & Saving, CRES, Pikermi, Greece

Aggelos Zacharopoulos
Ulster University, Newtownabbey, Northern Ireland, United Kingdom

BIOGRAPHY

Veera Gnaneswar Gude is an associate professor of Civil and Environmental Engineering at Mississippi State University (MSU). He has more than 15 years of academic, research and industrial experience in desalination, water-wastewater treatment, biofuel scientific, and technological areas. He received a BS degree in chemical engineering technology from Osmania University in 2000 and worked for Du Pont Singapore after his graduation from 2000 to 2004. He received an MS degree in environmental engineering from the National University of Singapore in 2004 and a PhD degree in environmental engineering from the New Mexico State University in 2007, under the direction of Prof. Nagamany Nirmalakhandan, for his research in low-temperature thermal desalination. He has expanded his research interests into water and wastewater treatment, microbial desalination, biofuel synthesis using sustainable chemistry principles and process intensification topics during his postdoctoral research, industrial and academic appointments. He is a licensed professional engineer and a board certified environmental engineer (also known as diplomate of environmental engineering) by the American Academy of Environmental Engineers and Scientists (AAEES).

Dr. Gude has published more than 70 scientific research articles on desalination (thermal, membrane, hybrid, and microbial), water-wastewater treatment, and on biofuels research in well-regarded journals. His research work is well utilized in the literature with more than 2500 citations by 2017. He has published 3 books on desalination research (Elsevier) and 2 books on biofuels research (CRC Press), 15 invited book chapters, 50 conference proceedings papers, 15 technical reports, several popular press articles and media releases, and 2 patents on low-temperature desalination and microalgae biofuels technologies. He has delivered 35 invited lectures including 6 plenary/keynote lectures and more than 140 scientific research and educational presentations. He has organized many workshops on water-energy-environment nexus topics at national and international conferences. He was the chair and board representative for clean energy and water division of ASES between 2011 and 2016. He serves on numerous scientific advisory boards and task committees across the world including ASCE-EWRI, ASEE, ASES, and AWWA. He is a member of several editorial boards and editor for many scientific journals including *ASCE Journal of*

Environmental Engineering, Heliyon, Nature npj Clean Water, Renewable Energy, Resource-Efficient Technologies, and *Water Environment Research Journal.* His research is supported by NSF, USEPA, USGS, USDA, and by many industrial and international agencies. He has received much recognition for his research, teaching, and service activities at regional, national, and international conferences and by professional societies (ASCE, ASEE, ASES, and Chi Epsilon National Civil Engineering Honor Society) including the 2017 Albert Nelson Marquis Lifetime Achievement Award and 2017 ASCE MS Engineer of the Year Award.

PREFACE

Desalination processes, both membrane and thermal systems, require significant quantities of energy either in the form of heat or electricity. Currently, the energy demands are being met by conventional energy sources derived from fossil fuels which is an unsustainable approach, especially when considering the expected growth in global needs for desalinated freshwater across the world. Renewable energy sources, that is, solar, wind, geothermal, and tidal energy sources are identified as ideal alternatives for replacing the conventional energy demands for desalination.

Renewable energy technologies and their integration with desalination processes have advanced in recent years. The desalination costs have also reduced drastically owing to improvements in the two technological areas. However, water supply alternative using renewable energy powered desalination approach is not yet affordable in all locations and at all capacities. There is a growing need for understanding the energy and water production technologies and their integration for maximizing the process benefits while minimizing the environmental impacts toward the goal of achieving sustainable desalination.

This book is intended to serve as a resource and guidance manual for understanding the principles, mechanisms, design, and successful implementation of desalination technologies for undergraduate and graduate researchers, science and engineering professionals, and industrial practitioners. This book is a critical resource for researchers from chemical engineering and science, environmental science and engineering, mechanical and electrical engineering, and industrial and process engineering and management disciplines with a desire to understand the science, engineering, and process management related to sustainable desalination.

The handbook has a two-part coverage which begins with an overview of the different renewable technologies currently used in the field. This coverage includes: solar, wind, geothermal, and nuclear desalination. The cost of desalination is largely dominated by the energy cost. Therefore, an evaluation of costs is also included. This coverage is followed by equally important and rigorous discussion of energy recovery and thermodynamics of desalination processes. Some of the highlights of this contribution are as follows.

- Covers clean and renewable energy sources such as nuclear, geothermal, solar, and wind powered desalination and energy storage and optimization
- Includes energy recovery schemes, optimization, and process controls

- Elaborates the principles of thermodynamics and second law efficiencies to improve process performance including solar desalination
- Explains global applicability of solar, wind, geothermal, and nuclear energy sources with case studies
- Discusses renewable energy–desalinated water optimization schemes for island communities

This book is presented in two parts and 14 chapters. The first part presents the potential renewable energy source applications in desalination plants, while the second part discusses the energy consumption, recovery schemes and storage options for desalination plants and presents studies with elaborated thermodynamic analysis of MED, MSF and solar desalination systems. Details of the chapters are presented below.

Chapter No.	Title
1	Introduction to renewable energy powered desalination
2	Solar desalination potential around the world
3	Wind-powered desalination—Principles, configurations, design, and implementation
4	Geothermal source for water desalination—Challenges and opportunities
5	Desalination of seawater using geothermal energy for food and water security: Arab and sub-Saharan countries
6	Nuclear energy powered seawater desalination
7	Relevance of nuclear desalination as an alternative to water transfer geoengineering projects: Example of China
8	Desalination costs using renewable energy technologies
9	Integrated planning of energy and water supply in islands
10	Energy storage for desalination
11	Energy recovery devices in membrane desalination processes
12	Thermodynamic, exergy, and thermoeconomic analysis of multiple effect distillation processes
13	Exergy analysis of thermal seawater desalination—A case study
14	Exergy and technoeconomic analysis of solar thermal desalination

PART 1—RENEWABLE ENERGY APPLICATIONS

Chapter 1 provides an overview of various desalination technologies and their integration with renewable energy sources. Phase change and non-phase change and hybrid desalination processes are presented. The chapter concludes with some general guidelines which can be used for the selection of desalination and renewable energy systems, their combination to produce freshwater in an environmentally friendly way, and the parameters that need to be considered.

Chapter 2 examines the spatial relationships between freshwater scarcity and stress, population density, solar insolation, and saline water availability from a global perspective, using the objective data to identify regions suitable for solar desalination technology deployments. The potential for Middle Eastern and North African countries and other low insolation, high population regions in utilizing the solar energy source for desalination is evaluated.

Chapter 3 highlights the history of technological advancements in wind turbines. A state-of-the-art review on the wind energy production principles, operation and maintenance and installed capacities around the world are provided. Technical feasibility of on-grid and off-grid wind energy applications for desalination is discussed. The selection of the optimum wind desalination combination is discussed, which include wind energy mechanical vapor compression, wind energy electrodialysis, and wind energy reverse osmosis technologies. The current status of the wind and desalination technologies, economic issues, as well as basic guidelines for the design and implementation of a wind desalination project along with future directions of research and development are discussed.

Chapter 4 provides an elaborated view of the potential for geothermal sources to support desalination processes. Direct use and power generation based on geothermal sources is considered for thermal and membrane desalination applications. The benefits of geothermal source utilization for desalination over other renewable energy sources are discussed in detail. Cogeneration and polygeneration schemes are discussed. Selection of the desalination processes and considerations for geothermal-driven desalination and other renewable energy integrated combinations are presented in detail.

Chapter 5 presents the water resource scenario for Gulf Cooperation Council, Middle Eastern and North African and sub-Saharan countries. The role of desalination technologies in the water portfolio for these communities has been discussed. Geothermal energy source availability and its potential role in sustainable desalination of brackish and seater sources is presented with case studies in these regions.

Chapter 6 deals with the state-of-the-art and recent desalination technologies in-line with their application in renewable energy as a comparative approach to nuclear energy. Most up-to-date and available data of all operating and planned nuclear desalination units have been discussed. The option of small modular reactors as an energy source for desalination technology has been included along with the economic values. Theoretical and computational models needed for evaluating the performance of the coupled nuclear and desalination units have been included. Computational and experimental facilities required for the desalination process have been included and discussed. Some case studies from the countries of the Gulf Cooperation Council for the implementation of nuclear power and its possibilities and obstacles have been discussed based on scientific and policy-making issues along with a technoeconomic assessment of nuclear powered desalination units.

Chapter 7 includes a critical assessment of the megaprojects where resource demands exceed environmental capacity. Using the Desalination Economic Evaluation Program developed by the International Atomic Energy Agency, the nuclear desalination capacity for China was evaluated as an alternative over the mega-water diversion option. Affordability of nuclear desalination and comparison of coal and nuclear desalination with the currently planned South-North Water Transfer megaproject was analyzed in terms of costs, critical vulnerabilities, and future resource demands in coastal regions.

Chapter 8 discusses the cost of desalination powered by 100% renewable energy power plants and demonstrates how and why renewable energy powered desalination is a lucrative option to meet the future global water demand. The global desalination demand by 2030, based solely on water stress and water demand is estimated. Using the respective 2030 costs of renewable energy power plants, storage and seawater reverse osmosis plants, the levelized cost of electricity and water was derived, the details of which are elaborated in this chapter.

Chapter 9 deals with the issues at the nexus of energy and water sources for small island communities. To overcome the intermittent nature of the renewable energy sources, energy storage is considered as a solution to provide reliable power supplies for desalination. Teo options, i.e., Pumped hydro when the topography is suitable and excess wind power for desalination and pumping are considered as viable options. An integrated approach to power and water supply systems' planning, with the purpose of increasing the integration of intermittent RES and minimizing the electricity and water production costs is presented with illustrations of the scenarios for the Island of S. Vicente, in Cape Verde.

PART 2—THERMODYNAMICS

Chapter 10 considers the mismatch between the source of supply and demand and intermittent nature of these energy resources and proposes energy storage as a must for reliable and continuous operation of desalination facilities. The key components of energy storage, that is, accumulation, storage, and supply are elaborated in detail for thermal and membrane desalination processes. A detailed discussion of current energy storage options for different desalination technologies using various renewable energy (solar collectors, photovoltaic arrays, geothermal, and wind and tidal energy sources) and waste heat sources with a focus on thermal energy storage and battery energy storage systems is presented. The principles of energy storage (thermal and electrical energy) are discussed with details on the design and sizing for desalination process applications.

Chapter 11 presents a detailed account of the recent developments in energy recovery options and devices for membrane desalination processes. Energy recovery devices for minimizing the specific energy consumption and water desalination costs are discussed with the process design, configurations, and implementation issues and case studies.

Chapter 12 presents the various opportunities to improve the performance of multieffect distillation (MED) desalination technologies. The fundamentals of thermodynamic and exergy analysis for MED process optimization are presented. "Exergy costing" is also discussed and proposed as an innovative method capable of reflecting how the thermal inefficiencies contribute to the gradual increase in the economic value (to be intended as a cost of generation) of material streams along the process. To achieve a more comprehensive view, the aforementioned analyses are carried out for a reference MED plant assumed as a case study.

Chapter 13 deals with exergy analysis of a thermal desalination plant (multistage flash distillation process) located in Yanbu, Saudi Arabia. Exergy flowrates are calculated and exergy destruction around each of the equipment is estimated. The second law efficiency of the plant for normal and peak loads are reported. Two options for improving the exergy efficiency, that is, by decreasing the inlet exergy to the plant and controlling exergy losses during process operation are presented by considering two variations in MSF designs which are once-through and brine mixing.

Chapter 14 discusses the performance of renewable energy powered desalination systems, especially solar desalination, through energy and exergy analysis. Exergoeconomic analysis and lifecycle cost analysis of conventional single-slope and double-slope solar still; and active solar still integrated with partially covered N-photovoltaic thermal flat plate collectors (N-PVT-FPC) are presented. The modeling of fundamental equations of exergoeconomic analysis for these systems have been studied including the case study of glazed, hybrid PVT module air collector. In addition, the ratio of thermodynamic loss rate to capital cost has been discussed on the basis of energy and exergy for different devices of coal-fired electricity generation stations.

ACKNOWLEDGMENTS

I would like to express my appreciation to my colleagues in the Civil and Environmental Engineering Department, the Bagley College of Engineering, and the Office of Research and Economic Development at Mississippi State University (MSU) for their support. My sincere appreciation and gratitude are extended to my doctoral and postdoctoral advisors Dr. Nagamany Nirmalakhandan, Dr. Ricardo Jacquez, Dr. Shuguang Deng, and Dr. Adrian Hanson of New Mexico State University (NMSU) for their support in my professional career. Prof. Khandan is gratefully acknowledged for introducing me to desalination research during my doctoral research studies at NMSU. His encouragement and guidance had a tremendous impact on my professional development. Colleagues at the South-West Technology Development Institute of NMSU, especially Corey Asbill and Andrew Rosenthal and Robert Keith Payne of Engineering Technology Department are thanked for their support in my research while at NMSU. Dr. Craig Ricketts is thanked for sharing his workplace at the engineering research facility. Industrial collaborations with Thomas Leggiore, Charles Oborne, and George Forbes are also acknowledged. My graduate research students at Mississippi State University and New Mexico State University are appreciated for their diligence and energy in work environment and research contributions, especially Dr. Bahareh Kokabian and Dr. Edith Martinez-Guerra of MSU for their work on microbial desalination and solar desalination research projects, respectively. Finally, I thank Ms. Preeti Muire for her support and love.

Veera Gnaneswar Gude

PART 1

Renewable Energy Applications

CHAPTER 1

Introduction to Renewable Energy Powered Desalination

Soteris A. Kalogirou
Cyprus University of Technology, Limassol, Cyprus

Abbreviations

ED	electrodialysis
ER-RO	RO with energy recovery
LCZ	lower convecting zone
MD	membrane distillation
MED	multiple effect distillation
MES	multiple effect stack
MSF	multistage flash
MVC	mechanical vapor compression
NCZ	nonconvecting zone
ppm	parts per million
PV	photovoltaics
RES	renewable energy systems
RO	reverse osmosis
TDS	total dissolved solids
TVC	thermal vapor compression
UCZ	upper convecting zone
VC	vapor compression

1.1 INTRODUCTION

The provision of good-quality freshwater is becoming an important and difficult to solve problem in many parts of the earth. It is well known today that in arid areas of the earth, which have high temperatures and high amounts of solar radiation, potable water is very scarce and the continuation of habitation of these areas strongly depends on the ability to produce adequate quantities of freshwater.

The existence of life on earth depends on water. This is one of the most plentiful resources on the planet, covering three-fourths of its surface. About 97%, however, of this water is saltwater contained in the oceans and a small 3% (about 36 million km^3) is freshwater found in lakes and rivers, the poles (in the form of ice), and groundwater, which supply most of the human and

Renewable Energy Powered Desalination Handbook
https://doi.org/10.1016/B978-0-12-815244-7.00001-5

animal needs. The biggest part of this tiny 3% of the world's freshwater, nearly 70%, is frozen in glaciers, permanent snow cover, ice, and permafrost. The rest 30% of all freshwater is found underground, mostly in deep aquifers. The quantity contained in lakes and rivers together is around 0.25% of all freshwater and of the two, lakes contain most of the quantity [1].

1.1.1 Water and Energy

The history of mankind proves that water and civilization are two very related and often inseparable concepts because they are able to govern the lives of humanity and promote the flourishing of civilization. This is proved by many examples in the history of humanity as all great civilizations were developed and prospered near large sources of water. Rivers, lakes, and oases have attracted mankind to their proximity because water is required every day to sustain life. Maybe the most noteworthy example of this influence is the Nile River in Egypt, which provided water for irrigation and mud full of nutrients for agriculture. During that period, ancient Egypt became the main wheat exporting country in the whole Mediterranean [2]. Owing to the flourishing of civilization at that time, various sciences such as mathematics, law, and astronomy, and organization of the society such as currency and police were created for the first time.

Moreover, energy is as important as water for the development of good standards of life because it is the motive power of all human activities. Under certain situations, water is also a power generating force, as for example the harnessing of waterpower to operate mills to grind grain, which is reported to have occurred more than 2000 years ago [3].

The Greek philosophers were the first to express concepts about the nature of water and energy. One of the seven wise men of antiquity, Thales of Miletus (640–546 BC), wrote about water [4,5] that it is fertile and moulded (can take the shape of its container) and considered the immense sea that surrounds the earth to be the primary mother of all life. Another philosopher, Embedokles (495–435 BC), a few years later developed the theory of the elements [4], describing that the world consists of four primary elements: fire, air, water, and earth. With today's knowledge and understanding, these may be considered as the energy, atmosphere, water, and soil, which are the four basic constituents that affect the quality of human life [6].

One of the greatest philosophers and scientists of antiquity, Aristotle (384–322), described in an astonishingly precise way the origin and properties of natural, brackish, and seawater. He wrote about the water cycle in nature [7]: "Now the sun moving, as it does, sets up processes of change and becoming and decay, and by its agency the finest and sweetest water is every day carried out and is dissolved into vapor and rises to the upper regions, where it is condensed again by the cold and so returns to the earth."

Even today, no better explanation can be given for the water cycle as occurring in nature. In fact, the sea and the atmosphere is a huge solar energy open distiller in a perpetual operational cycle.

Moreover, about seawater, Aristotle wrote [8]: "Salt water when it turns into vapor becomes sweet, and the vapor does not form salt water when it condenses again. This is known by experiment."

1.1.2 Water Demand and Consumption

As was seen above, humans have been dependent on rivers, lakes, and underground water reservoirs for freshwater necessities in domestic life, agriculture, and industry. In recent decades, the rapid worldwide population growth and industrial growth have caused a large escalation of freshwater demand, both for the household needs and for crops to produce adequate quantities of food. In addition to the increased quantities required, the reserves of freshwater are reduced because of the problem of pollution of rivers and lakes by industrial wastes and the large amounts of sewage discharged. An additional problem is the uneven distribution as countries such as Canada have a tenth of the world's surface freshwater but less than 1% of its population, whereas other countries with much bigger populations do not have adequate quantities of freshwater [1].

Worldwide consumption figures suggest that about 70% of the total water consumption is used by agriculture, 20% is used by the industry, and only 10% is used for the household needs. As desalination is generally expensive before considering this method, water conservation measures should be developed and applied. In line with the slogan that "the greener kWh is the kWh not consumed," here the "the cheaper water is the water not spent." For example, systems exist nowadays such as drip irrigation which uses 30%–70% less water than traditional methods and increases crop yield. This is using perforated plastic pipes to deliver the water to crops directly. Although this system was developed more than 50 years ago, until today it is used in less than 1% of the irrigated land. Moreover, to sustain food production in many countries, governments heavily subsidize irrigation water and therefore farmers have no incentive to invest in water saving methods [1].

1.1.3 Desalination and Energy

As was mentioned before, the only nearly inexhaustible sources of water are the oceans, but to use this water we have to reduce its salinity considerably. Therefore, since prehistoric times, it was considered attractive to produce freshwater with desalination of seawater. Desalinize in general means the removal of salts from seawater or saline water down to health acceptable limits. The World Health Organization (WHO) recommends a permissible limit of salinity in water of 500 parts per million (ppm) and, in special cases, up to 1000 ppm. The amount of salt reduction is evident by considering that most of the water available on earth has a salinity of up to 10,000 ppm, while seawater normally has a salinity in the range of 35,000–45,000 ppm, both in the form of total dissolved salts [9]. As can be understood, the purpose of any desalination system is to remove salts and produce water with total dissolved solids (TDS) of 500 ppm or less.

Desalination processes require substantial quantities of energy to achieve the removal of salts from seawater. This is a very important parameter as energy is an expensive commodity and this is a permanent running cost, which varies with fuel prices, and which few of the water-short areas of the world can afford. Some countries, mainly in the Middle East, who suffer from water shortage problems, have a large amount of oil to power desalination processes and, because of the oil income, have enough money to invest in desalination equipment. Many other areas of the world do not have the resources to develop in a similar manner.

A possible substantial increase in desalinated water supply will also create a series of problems related to energy consumption and its effect on the environment. Therefore, even if oil was much more widely available, it is doubtful if it can be burned at the scale needed to provide all humanity with freshwater. Furthermore, sources of energy that are not polluting will have to be utilized for desalination. Fortunately, many arid parts of the world that face water shortage problems have usable renewable sources of energy, mainly solar radiation, that can be used to drive desalination processes. These systems are called solar desalination systems. Therefore, solar desalination of seawater is one of the ways of meeting water demand.

It should be noted that, on a global scale, solar desalination is a natural phenomenon which produces rainfall and snowfall, which are the main source of freshwater supply worldwide. The process is initiated from the solar radiation falling on the surface of the sea, which is absorbed as heat and causes evaporation of the water. The vapor rises above the surface and is moved by winds. When this vapor cools down to its dew point, condensation occurs and freshwater precipitates as rain or, if the temperature is low, as snow. All man-made distillation systems are small-scale replications of this natural process.

Renewable energy systems (RES) produce energy from resources that are "freely" available in nature and do not deplete with consumption. Above all, their collection and use is done in a pollution-free way friendly to the environment. Therefore, freshwater production using renewable energy powered desalination technologies is considered to be a viable solution to water scarcity problems in areas that are characterized by lack of freshwater and conventional energy supply, such as heat or electricity grid. Several renewable energy desalination pilot plants have been installed in many parts of the world. Most of them have been operated successfully for a number of years, whereas others were just experimental and were set up to prove their suitability and viability. All of these are custom-designed for specific application of solar, wind, or geothermal energy to produce freshwater. It is anticipated that operational data and experience obtained from these plants can be valuable to achieve better designs in the future with higher reliability and cost reduction. It is a well-known fact that today renewable energy powered desalination systems cannot compete with conventional systems in terms of the cost of water produced. They are, however, applicable in areas with abundant renewable energy resources and are expected to become economically viable options soon.

After the presentation of this general relationship between energy and water and the possibilities to use RES, this chapter aims to present a short description of the various methods used for seawater desalination and particularly their integration with RES. More details on solar thermal desalination can be found in a recently published book of the author [10].

1.2 DESALINATION PROCESSES

Desalination can be achieved by using a number of techniques which fall into two main categories: the ones which use either phase change or involve semipermeable membranes to separate the solvent or some solutes. They are:

(i) phase-change or thermal processes and

(ii) single phase or membrane processes.

Table 1.1 presents the most important technologies in use today. The motive power in the phase-change or thermal processes is a thermal energy source which may be obtained from fossil fuels, nuclear energy, or from a

Table 1.1 Desalination processes

Phase-change processes	Membrane processes
1. Multistage flash (MSF) **2.** Multiple effect distillation (MED) **3.** Vapor compression (VC) – Thermal vapor compression (TVC) – Mechanical vapor compression (MVC) **4.** Membrane distillation (MD) **5.** Freezing **6.** Humidification/dehumidification (H/D) **7.** Solar distillers (SD) – Conventional solar stills – Special solar stills – Cascaded type solar stills – Wick-type solar stills – Multiple-wick-type solar stills	**1.** Reverse osmosis (RO) – RO without energy recovery – RO with energy recovery (ER-RO) **2.** Electrodialysis (ED)

nonconventional solar energy source or geothermal energy. In the membrane or single-phase processes, electricity is used, obtained from the grid, photovoltaics (PV), or wind power. An exception is membrane distillation (MD), which is classified as a phase-change process but needs membranes to operate.

All processes presented in Table 1.1 require chemical pretreatment of seawater to avoid foaming, fouling, scaling, and corrosion, as well as chemical posttreatment, mainly for disinfection.

Thermal processes are also distinguished into direct and indirect collection systems. In a direct collection system, the same equipment is used to collect and desalinate seawater, whereas in indirect collection systems two separate systems are used, one for energy collection and the other for desalination. It should be noted that the conventional desalination systems are similar to solar systems since the same type of equipment is utilized. The main difference is that in the conventional systems, either a boiler is used to provide the required heat or mains electricity to provide the required electric power, whereas in the solar ones, solar radiation is employed [1].

A representative example of direct collection systems is the solar still. This uses the greenhouse effect to evaporate salty water and consists of a basin, in which a constant amount of seawater is enclosed in an inverted "V"-shaped glass envelope as shown in Fig. 1.1. The sun's rays passing

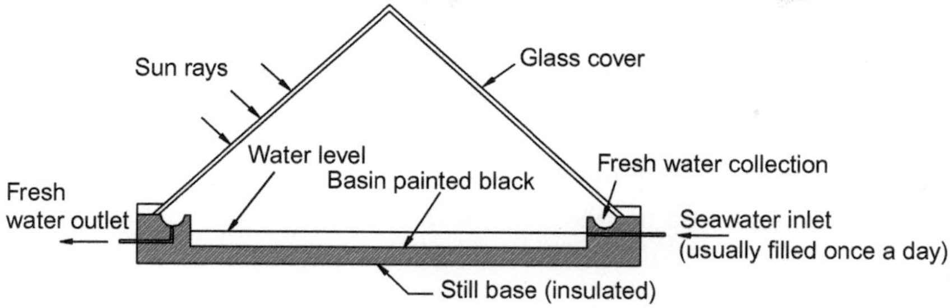

Fig. 1.1 Basic design features of a solar still.

through the glass roof are absorbed by the basin, which is black in color. As the water is heated, its vapor pressure is increased and water vapor is produced, which is condensed on the underside of the roof. This condensate, which is water, is free from any salts and runs down into troughs, which collect the distilled water and transfer it to a reservoir. The still operates as a greenhouse, trapping the incoming sunlight passing through the transparent roof, but it is opaque to the infrared radiation emitted by the hot water. The glass roof should avoid vapor leaks and prevent the wind from reaching and cooling the salty water. This is also known as a roof type or greenhouse type solar still. The basic parts of a solar still are the airtight basin, usually constructed from concrete/cement, a galvanized iron sheet or various plastic materials, and a top cover of transparent material such as glass or plastic [1].

Many other designs of solar stills have been developed; some typical ones are shown in Fig. 1.2. Solar stills require frequent flushing, which is usually done during the night, to prevent salt precipitation [11]. A typical still efficiency, defined as the ratio of the energy utilized in evaporating the water in the still to the solar energy incident on the glass cover, is 35% (maximum) and daily still production is about 3–4 L/m^2 [12]. Although the yield of solar stills is quite low, their use may prove to be viable if small freshwater quantities are required and the cost required to supply an arid area with naturally produced freshwater is high.

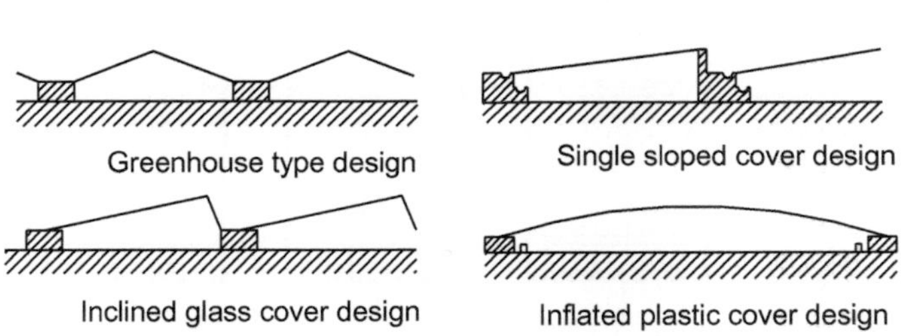

Fig. 1.2 Various other designs of solar stills, usually combined with rainwater catching.

Many good review papers are published on solar stills by various researchers. Talbert et al. [13] give an excellent historical review of solar distillation. Delyannis and Delyannis [14] reviewed the major solar distillation plants erected around the world. Kalogirou [15] also reviewed various types of solar stills. Malik et al. [16] reviewed the work on solar distillation systems until 1982, updated up to 1992 by Tiwari [17], which also included active solar distillation, that is, the combination of solar stills with solar thermal collectors to increase more the seawater temperature and thus enhance the evaporation rate of seawater. Fath [18] also reviewed various designs of solar stills and studied the suitability of solar stills for providing potable water.

Indirect collection systems comprise thermal powered phase-change processes and electricity powered membrane processes.

Thermal energy powered, commercial desalination processes are mainly based on multistage flash (MSF) distillation, multiple effect distillation (MED), and vapor compression (VC), which is subdivided into thermal (TVC) or mechanical vapor compression (MVC) and MD. Both MSF and MED processes utilize a number of stages, which are at successively decreasing pressure and temperature. In the MSF process, shown schematically in Fig. 1.3, the seawater is heated in a number of heat exchangers and the generation of vapor from seawater or brine is done due to a sudden pressure reduction (flashing) when seawater enters a chamber, which is under vacuum. The process is repeated stage by stage at successively decreasing pressure. In each stage, condensing steam is used to preheat the seawater feed. By fractionating the overall temperature differential between the warm source and seawater into a large number of stages, the system approaches ideal total latent heat recovery. This process requires an external steam supply, normally at a temperature around 100°C, which can be obtained by fossil fuels or solar thermal collectors. The maximum temperature is limited by

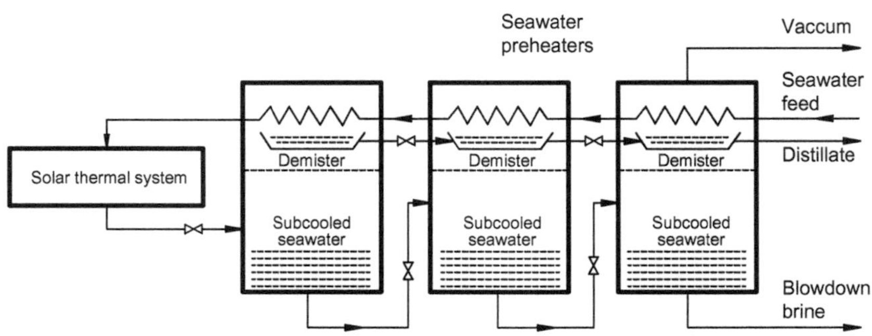

Fig. 1.3 The MSF process schematic.

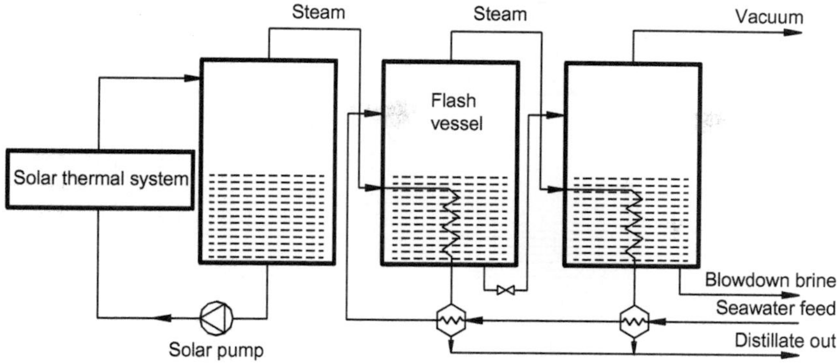

Fig. 1.4 The MED process schematic.

the salt concentration to avoid scaling and this maximum limits the performance of the process. Operation of this system requires pressure gradients in the plant. Current commercial installations are designed with 10–30 stages (2°C temperature drop per stage) [10].

In MED, shown schematically in Fig. 1.4, vapors are generated due to the absorption of thermal energy by the seawater. The steam generated in one effect is used to heat the saline solution in the next effect, as the next effect is at a lower temperature and pressure. The performance of MED and MSF processes is proportional to the number of effects or stages, respectively. An MED plant typically uses steam supplied externally at a temperature of about 70°C. In this process, vapor is produced by flashing and by boiling, but the majority of the distillate is produced by boiling. Many variations of MED plants were used; their difference concerns the combinations of heat-transfer configurations and flow sheet arrangements used. The first plants produced were of the submerged tube design and used only two to three effects. In modern systems, however, the problem of low evaporation rate of pool boiling has been resolved by using thin film evaporator designs. These may have vertical or horizontal tubes. The vertical tube designs are of two variations: the climbing film, natural and forced circulation type or long tube, vertical, straight falling film type. In the latter, the brine boils inside and the steam condenses outside the tubes. In the horizontal tube, usually the steam condenses inside with the brine evaporating on the outside of the tubes [1].

In a VC plant, shown schematically in Fig. 1.5, heat recovery is achieved by raising the pressure of the steam from a stage by means of a compressor. By doing so, the condensation temperature increases, and therefore the

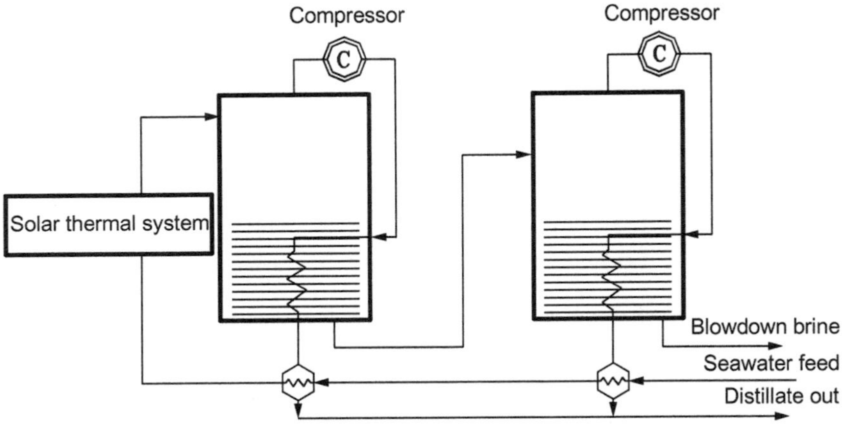

Fig. 1.5 The VC process schematic.

steam can be used to provide energy to the same stage or to other stages [11,19]. Similar to a conventional MED system, vapors produced in the first effect are used as the heat input to the second effect that is at a lower pressure. The vapors produced in the last effect are then passed to the vapor compressor where they are compressed and their saturation temperature is raised before they return to the first effect. In this system, the compressor represents the major energy input to the system. Various cost estimates and process designs showed that this type of plant is not particularly appropriate, unless it is combined with an MED system. Therefore, the VC system can be used in conjunction with an MED system and operated at periods of low solar radiation or overnight.

VC systems are subdivided into two main categories, namely: MVC and thermal vapor compression (TVC). The MVC systems employ a mechanical compressor to compress the vapor, whereas the thermal ones utilize a steam jet compressor. On TVC and MVC, after the initial vapor is generated from the saline solution, this vapor is thermally or mechanically compressed to generate additional production.

Freezing, humidification/dehumidification, and MD processes also involve phase change.

In the freezing process, the conversion of salty water to freshwater is done in a natural way, known for thousands of years. In desalination of water by freezing, freshwater is removed in the form of ice and leaves behind concentrated brine as cool water. This is a separation process based on the solid-liquid phase-change phenomenon. It is based on the fact that when the temperature of saline water is reduced to its freezing point, ice crystals of pure water are formed within the salt solution, mechanically separated from the concentrated solution, and remelted to obtain pure water. As it is understood, the basic energy input for this process is electricity mainly used for the refrigeration system [20].

The humidification/dehumidification process also uses refrigeration but the operating principle is different. It is based on the principle that air can be mixed with large quantities of water vapor and its vapor carrying capability increases with temperature [21]. Therefore, in this process, seawater is "added" into an airstream and increases its humidity. This humid air is then directed to a cool coil which is at a temperature below the dew point; therefore, water vapor contained in the air is condensed on the surface of the coil and collected as freshwater in the same way that condensate is produced in an air-conditioning equipment.

MD is a relatively new and promising desalination technology, suitable for small-to-medium capacities. It operates at atmospheric pressure and temperatures below 100°C and therefore is suitable to be powered with solar thermal energy. MD is a nonisothermal membrane separation process, in which the water vapor molecules are transferred through a microporous hydrophobic membrane. The driving force of the process is the partial pressure difference across the membrane created by the temperature difference between the two sides of the membrane. It gives high-quality distillate and the possibility of operating at low temperatures. Moreover, the capability of utilizing solar thermal energy makes it a cost-effective and environmental-friendly process.

MD is a relative simple hybrid process, which combines thermal distillation and the normal membrane technologies, presented subsequently. It is in fact a thermal membrane separation process in which water vapor molecules are transferred from a hot aqueous solution, usually seawater, through a microporous hydrophobic membrane, because of the partial pressure difference created due to the temperature difference on both sides of the membrane [22].

The MD process is based on the principle of the liquid-vapor equilibrium. In this, the hot aqueous solution flows toward the hot membrane surface, releasing vapor. The hydrophobic nature of the membrane, owing to the surface tension forces, prevents the penetration of the liquid phase into the membrane pores but allows the passage of the vapor molecules through the membrane. Since the surface tension of the aqueous phase is higher than the critical surface tension of the membrane polymeric film, it withholds the liquid solution and does not allow it to enter into the membrane pores. This results in a liquid-vapor interface created at the entrance of the membrane pores; thus, the liquid phase is separated from the gas phase. As the process is nonisothermal, the temperature difference across the membrane creates a partial pressure gradient, which causes the water vapor molecules and the volatile components to evaporate in the liquid-vapor interface at the entrance of the membrane pores, and subsequently the vapor formed to diffuse in the membrane. Since a vapor pressure difference is maintained between the two membrane sides, vapor molecules are transferred through the membrane pores, from the high partial pressure area (hot surface) to the low partial pressure area (cold surface), and finally it is condensed either internally downstream of the membrane or externally outside the membrane module [22–24].

The other major category of industrial desalination processes does not involve phase change but employs membranes to achieve separation. These are reverse osmosis (RO) and electrodialysis (ED). The first one requires electricity or shaft power to drive the pump that increases the pressure of the saline solution to that required. The RO system, shown schematically in Fig. 1.6, depends on the properties of semipermeable membranes, which, when used to separate water from a salt solution, such as seawater, allow freshwater to pass into the brine compartment under the influence of osmotic pressure. If a pressure in excess of this value is applied to the seawater, freshwater will pass from the brine into the water compartment. In principle, the only energy requirement is for pumping the feed seawater at a pressure above the osmotic pressure. In practice, however, higher pressures must be used, typically 50–80 atm, in order to have a sufficient amount of water passing through a unit area of membrane [25]. Part of the seawater feed passes through the membrane where the majority of the dissolved solids are removed. The remainder, together with the remaining salts, which creates a very high salinity water, is rejected at high pressure back to the sea. It is therefore economically viable in larger plants to recover the rejected brine high pressure with a suitable brine turbine, called energy recovery reverse osmosis (ER-RO) systems.

ED also requires electricity for the ionization of water which is cleaned by using suitable membranes located at the two opposite-charged electrodes. A simplified schematic diagram of this process is shown in Fig. 1.7. In this system, salinity is reduced by transferring ions from the feed water compartment, through membranes, under the influence of an electrical potential difference. To do so, the process utilizes a direct current (DC) electric field to remove salt ions in the brackish water. Saline feed water contains dissolved salts separated

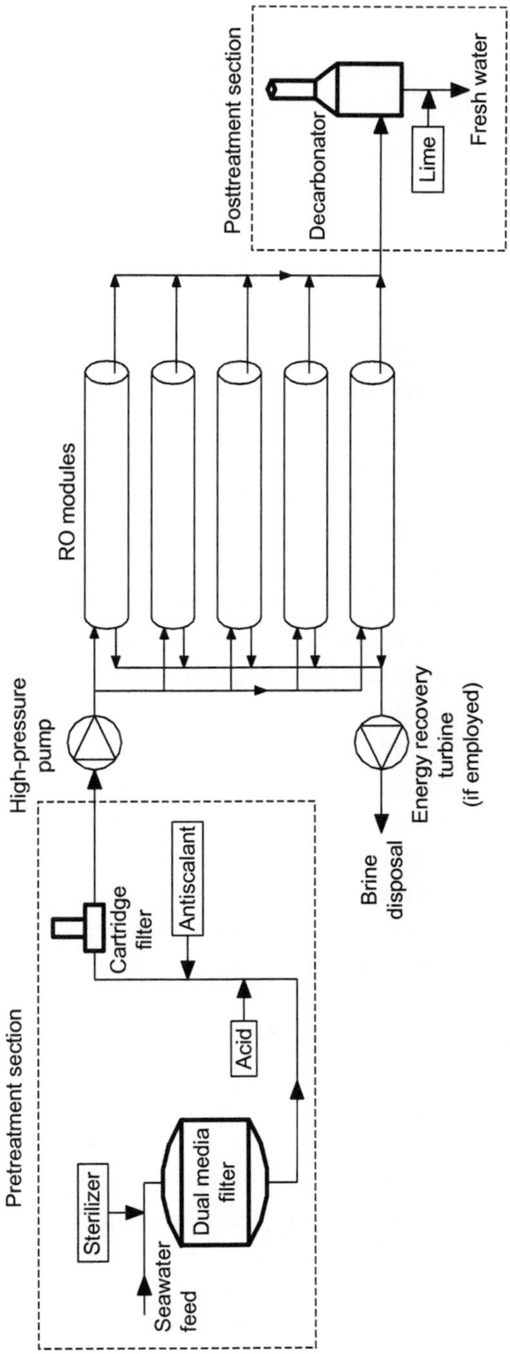

Fig. 1.6 The RO process schematic.

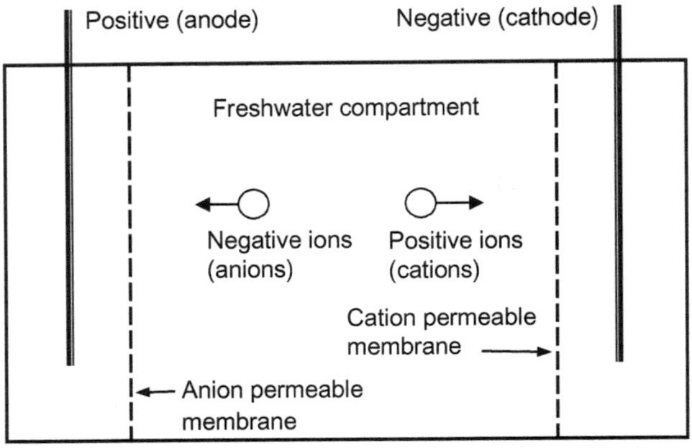

Fig. 1.7 A simplified ED process schematic.

electrically into positively charged sodium and negatively charged chlorine ions. These ions move toward two electrodes immersed in the solution, that is, positive ions (cations) will go to the negative electrode (cathode) and negative ions (anions) to the positive electrode (anode). To obtain freshwater, special membranes, alternatively cation-permeable and anion-permeable, separate the two electrodes; thus, the center space between these two membranes will be cleared from salts [26]. In a real system, a comparatively large number of alternating cation and anion membranes are used, separated by plastic flow-spacers that allow the passage of water. The streams of alternating flow-spacers create a sequence of diluted and concentrated water streams which flow in parallel to each other. To prevent scaling problems, the polarity of the electric field is reversed about every 20 min.

Since the energy requirement of the process is proportional to the water's salinity, the method is more feasible for brackish water with a salinity of not more than about 6000 ppm dissolved solids. Moreover, owing to the low conductivity of very pure water, which increases the energy requirements of the process, this is not suitable for water of less than about 400 ppm of TDS. It is interesting to note that the process operates with dc power; thus, solar energy can be used by directly producing the voltage difference required with PV panels.

Both RO and ED can be used for brackish water desalination, but RO is also suitable for seawater desalination. The dominant desalination processes today are MSF and RO, which account for 44% and 42% of worldwide capacity, respectively [27]. Furthermore, the MSF process represents more than 93% of the thermal process production, while the RO process represents more than 88% of membrane processes production [28].

1.3 RENEWABLE ENERGY POWERED DESALINATION SYSTEMS

RES offer alternative and environmental-friendly solutions to decrease the dependence on fossil fuels for the energy needs to power various processes including desalination. Compared to fossil fueled desalination plants, the total worldwide renewable energy desalination installations are very small and the total capacity is less than 1% [2]. This is mainly due to the high initial capital investment required by renewable energy, making these desalination plants less competitive compared to conventional fossil fuel-powered desalination plants.

Solar energy can be used for seawater desalination either by producing the thermal energy required to drive the phase-change processes or by producing electricity required to drive the membrane processes. The most promising and applicable RES desalination combinations are shown in Table 1.2, obtained mostly from the THERMIE Programme booklet [29].

Over the years, numerous desalination systems utilizing renewable energy have been constructed and operated. Most of these systems have been built as research or demonstration pilot plants of small capacity mainly to prove the principle of operation rather than producing large quantities of freshwater.

Table 1.2 RES desalination combinations

RES technology	Feed water salinity	Desalination technology
Solar thermal	Seawater	Multiple effect distillation (MED)
Geothermal	Seawater	Multistage flash (MSF)
	Seawater	Thermal vapor compression (TVC)
	Seawater	Membrane distillation (MD)
Photovoltaics (PV)	Seawater	Reverse osmosis (RO)
	Brackish water	Reverse osmosis (RO)
	Brackish water	Electrodialysis (ED)
Wind energy (electricity)	Seawater	Reverse osmosis (RO)
	Brackish water	Reverse osmosis (RO)
	Brackish water	Electodialysis
	Seawater	Mechanical vapor compression (MVC)
Wind energy (mechanical energy)	Seawater	Reverse osmosis (RO)
	Seawater	Mechanical vapor compression (MVC)

Most of these pilot plants are not in operation today, but it is hoped that the knowledge obtained is being passed on and is applied to the plants currently being erected. The main renewable energy equipments that can be used in desalination system are briefly presented below. The interested reader can find more information on specific books and papers suggested.

1.3.1 Solar Collectors

Solar energy collectors are a special type of heat exchangers that convert solar radiation into thermal energy of the transport medium. The major component of any solar thermal system is the solar collector, which absorbs the incoming solar radiation, converts it into heat, and transfers this heat to a fluid (usually air, water, or thermal oil) flowing through the collector pipes or ducts. The solar energy collected is carried from the circulating fluid either directly to the desalination equipment, or to a thermal energy storage tank from which it can be drawn to produce desalinated water at night and/or on cloudy days [30].

The two major types of solar collectors are stationary or nonconcentrating and tracking or concentrating, the latter producing higher temperatures. A nonconcentrating collector has the same area for intercepting and for absorbing solar radiation, whereas a sun-tracking concentrating solar collector usually has concave reflecting surfaces to intercept and focus the sun's beam radiation to a smaller receiving area, thereby increasing the radiation flux. A large number of solar collectors are available in the market. A comprehensive list is shown in Table 1.3 [31].

A comprehensive review of the various types of collectors currently available is presented in Ref. [32] and may not be repeated here. More details on the design of such system can also be found in the book Solar Energy Engineering, published by the author [30]. The interested reader can refer to this publication where details of the various collectors available are described together with their possible applications, including desalination.

1.3.2 Solar Ponds

The basic principle of operation of a solar pond is to heat a large amount of water in such a way as to prevent the heat losses to the environment that would occur if less dense heated water is allowed to rise to the surface of the pond and lose energy by convection and radiation. As shown in Fig. 1.8, this is achieved if a stagnant, highly transparent insulating zone is created in the upper part of the pond to contain the hot fluid which is collected

Table 1.3 Solar energy collectors [31]

Motion	Collector type	Absorber type	Concentration ratio	Indicative temperature range (°C)
Stationary	Flat plate collector (FPC)	Flat	1	30–80
	Evacuated tube collector (ETC)	Flat	1	50–200
	Compound parabolic collector (CPC)	Tubular	1–5 5–15	60–240 60–300
Single-axis tracking	Linear Fresnel reflector (LFR)	Tubular	10–40	60–250
	Parabolic trough collector (PTC)	Tubular	15–45	60–300
	Cylindrical trough collector (CTC)	Tubular	10–50	60–300
Two-axes tracking	Parabolic dish reflector (PDR)	Point	100–1000	100–500
	Heliostat field collector (HFC)	Point	100–1500	150–2000

Note: Concentration ratio = aperture area divided by receiver/absorber area of the collector.

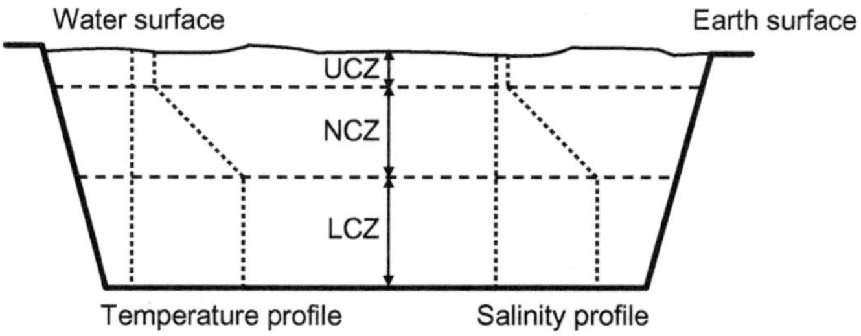

Fig. 1.8 Schematic vertical section through a salt-gradient solar pond.

and stored at the lower part of the pond. Therefore, solar ponds are both solar energy collectors and heat stores. Salt gradient lakes, which demonstrate an increase in temperature with depth, can occur naturally. A salt-gradient nonconvecting solar pond consists of the following three zones [33]:

1. The upper convecting zone (UCZ), which has an almost constant low salinity and is close to ambient temperature.
2. The nonconvecting zone (NCZ), in which both salinity and temperature increase with depth.

3. The lower convecting zone (LCZ) of almost constant, relatively high salinity (typically 20% by weight) at a high temperature. This is the layer that stores the solar energy collected. The high salinity prevents the formation of natural convection currents, which minimizes the thermal losses.

The basic feature of a solar pond is to suppress natural convection in order to allow it to store thermal energy. In solar ponds, this is achieved with salt stratification. As can be seen from Fig. 1.8, salinity increases with depth in the NCZ until the LCZ is reached. Solar radiation heats the high salinity water and, owing to its high relative density, natural convection is suppressed; therefore, hot salty water cannot rise and thus the heat is stored in LCZ. For this purpose, chemically stable salts can be used. Such a salt must be nontoxic, cheap, safe to handle, readily available, and its solubility should be temperature dependent and should not reduce significantly the insolation transmission characteristics of water [33].

Solar ponds can be used to provide thermal energy for a variety of possible applications. Smaller ponds can be used for space and water heating, while the larger ponds are proposed for industrial process heat, electric power generation, and desalination.

The thermal energy from a salt-gradient solar pond can be used to operate a low-temperature distillation unit to desalt seawater. This idea can be applied in desert areas near the oceans which have abundant solar radiation and suffer from water shortage problems. Solar pond-coupled desalination uses the hot brine from the pond as a thermal energy source to evaporate the seawater at low pressure in combination with an MED evaporator.

These, according to Matz and Feist [34], offer a good solution of brine disposal of inland ED plants, as well as a source of thermal energy to heat the feed of an ED plant which can increase substantially its performance.

1.3.3 Photovoltaics

PV cells or panels convert sunlight, which is the most abundant energy source on earth, directly into electricity. They have many advantages including completely silent operation, adaptability into various weather and installation environments, and no moving parts. They also require minimal maintenance and have a long life. Maybe the most important advantage is that they generate electricity without producing emissions of greenhouse or any other gases. A PV cell consists of two or more thin layers of semiconducting material, primarily silicon. When they are exposed to light, electrical charges are generated, conducted away by metal contacts as DC electricity. Generally, the electrical output from a single cell is small, so a number of cells are connected together in series and parallel to produce the required current and voltage, and encapsulated usually with a glass cover and plastic sheet to form a module usually called a "panel" [30].

The basic element of a PV system is the PV panel and any number of panels can be connected together, again in series or parallel, to produce the desired electrical output. Actually, this modular structure is a considerable advantage of the PV system, as more panels can be added to an existing system according to the needs as required.

PV systems can be effectively used in any application powered by electricity, although cost-effective systems include earth and space communications, remote power supply in relay stations, remote monitoring systems, street, parking space and domestic lighting, water pumping, and battery charging, mainly for emergency vehicles such as fire trucks. The global installed capacity of PV at the end of 2015 exceeded, for the first time, the 227 GWp limit [35].

As mentioned above, PV modules are designed for outdoor use and can be used in various harsh situations such as desert, marine, tropic, arctic, and space environments. A number of PV active materials can be used and the choice of the material to use in each case has important effects on the PV system design and performance. These materials differ with respect to the composition of the material and its atomic structure. The atomic structure of a PV cell can be based on one of the three main types; single-crystal (monocrystalline), polycrystalline, or amorphous silicon; the most commonly PV material produced and used is crystalline silicon, either as polycrystalline or as monocrystalline. The greatest advantage of amorphous silicon solar cells is that they can be deposited on a variety of shapes, such as roof-tiles, which opens new horizons in the possible applications of PV [30].

As seen above, a PV panel, also called a PV module, is a collection of PV cells that protects the cells and provides a usable operating voltage. PV cells are delicate components made as thin sheets which are fragile and susceptible to corrosion by humidity and fingerprints. Additionally, the operating voltage of a single PV cell is very low, usually less than 1 V, which is unsuitable for many practical applications. Furthermore, depending on the manufacturer and the type of PV material employed, modules have different appearances (such as bluish or greenish color) and performance characteristics.

1.3.4 Wind Energy

According to Betz, in a one-dimensional wind turbine model, the theoretical maximum aerodynamic conversion efficiency of wind turbines from wind to mechanical power is 59% [30]. The actual properties of even the most efficient modern airfoil sections, used for blades of medium to large size wind turbines, however, limit the peak efficiency achieved to around 48%. Today the need to lower the cost of blades leads to the construction of slender blades for wind turbines, which are rotated fast and have a peak efficiency somewhat below the optimum, at around 45%. It should be noted that the average year-round efficiency of most commonly used turbines is about one-half this number. This is due to the need to protect the wind turbines by shutting them down in low or high winds and to limit the power once the rated level is reached. Furthermore, a reduction of the actual average efficiency is caused by the gearbox and generator losses and because the machines do not always operate at their optimum working point [36].

For some years now, the wind turbines represent a mature technology for power production and they are commercially available on a wide range of nominal power starting from a few Watts to the MW range. More details can be found on the comprehensive review of wind technology presented by Ackermann and Soder [37]. In spite of its maturity, current research is focussed on new control strategies and energy storage systems in an attempt to increase the energy production from wind turbines.

Although wind generators are an accepted technology, the growth of the installed wind power is hindered by a number of barriers. The main barriers are public acceptance, visual impact, noise problems, telecommunication interference, and various influences on natural habitat and wild life. Most of these problems, however, are not scientifically proved and anyhow solved by the installation of offshore wind parks at distances that are not visible from the shoreline. Irrespective of these barriers, installed wind capacity increases from year to year and today amounts to 434.9 GW [38].

The technology of the wind turbine generators, which is currently in use, is only 30 years old. As seen above, the worldwide installed wind capacity at the end of 2015 was 434.9 GW_e with an increase of about 30% per year. Although a lot of variations have been tried, nearly all the wind turbines manufactured by the industry are of the horizontal axis type and most of them have three blades.

1.3.5 Geothermal Energy

Geothermal energy is due to the heat interactions taking place at the center of the earth. Actual measurements, however, show that the ground temperature below a certain depth remains fairly constant throughout the year. This is because the temperature fluctuations at the surface of the ground are reduced as the depth of the ground increases due to the high thermal inertia of the soil. This applies to a depth of several hundreds of meters, depending on the location, before the temperature is affected from the earth core temperature. For example, Popiel et al. [39] present the temperature distributions measured in the ground for the period between summer 1999 and spring 2001 in Poland and found that from the point of view of the temperature distribution three ground zones can be distinguished:

1. *Surface zone* reaching a depth of about 1 m. In this zone, the ground temperature is very sensitive to short time changes of weather conditions, such as sunshine and precipitation.
2. *Shallow zone* extending from a depth of about 1–8 m (for dry light soils) or 20 m (for moist heavy sandy soils). In this zone, the ground temperature is almost constant and close to the average annual air temperature. Here the ground temperature distributions depend mainly on the seasonal cycle weather conditions with a large time lack.
3. *Deep zone* below the depth of the shallow zone. In this zone, the ground temperature is practically constant and rising very slowly with depth according to the geothermal gradient. A noticeable increase in temperature is observed at a depth of more than several hundreds of meters.

Depths up to about 200 m are usually employed for shallow geothermal systems in combination with heat pumps (HPs) to increase their COP. Energy from the earth is usually extracted with ground heat exchangers. These are made of a material that is extremely durable but allows heat to pass through efficiently. Manufacturers of ground heat exchangers typically use high-density polyethylene, which is a tough plastic, with heat fuse joints. This plastic is usually warranted for as long as 50 years. The fluid in the loop is usually water or an environmentally safe antifreeze solution. The length of the loop depends on a number of factors such as the loop configuration, the thermal load served, the soil conditions (including the presence of gravel or running water), local climate, and many others. A review of ground heat exchangers is given in Ref. [40].

For installations in the deep geothermal zone, there are different geothermal energy sources which can be utilized and may be classified in terms of the measured temperature as low (<100°C), medium (100–150°C), and high (>150°C) temperature. The thermal gradient in the Earth varies between 15°C and 75°C per km depth but the heat flux is irregular in different continental areas. Besides, there are local centres of heat, located between 6 and 10 km deep, which are created because of the disintegration of radioactive elements. A complete overview of geothermal energy technology is presented by Barbier [41,42]. Baldacci et al. [43] presented a report which shows that the cost of electrical energy is generally competitive, at a price of 0.6–2.8 US cents/MJ (2–10 US cents/kWh), and that a small 0.3%, which corresponds to 177.5 billion MJ/year (49.3 billion kWh/year) of the world total electrical energy, is generated in the year 2000 from geothermal resources. As will be shown in the next section, geothermal energy can be used as a power input for desalination evaporators.

1.4 INTEGRATION OF RE AND CONNECTION WITH DESALINATION

1.4.1 Solar Thermal Energy

Since most of the distillation technologies are powered by thermal energy, solar thermal energy is one of the most promising applications of renewable energy to seawater desalination. These are applied to indirect solar desalination solar distillation systems which, as was seen before, consist of two separated devices, the solar collector and the conventional distiller. These systems usually employ a commercial desalination plant that is connected to commercial solar thermal collectors of various types. In some designs, corrosion-free collectors are used; details of these collectors are given in Refs. [44,45].Some systems may employ special or application oriented solar thermal collectors, such as a system developed by Rajvanshi [46], who designed such a solar collector to be connected to an MSF distillation plant. The design and testing of a corrosion-free solar collector, used for driving a multieffect humidification process, is presented by Hermann et al. [47]. This collector is employed in a pilot plant installed at PozoIzquierdo in Gran Canarias, Spain [48]. Details of the ACE-20 parabolic trough collector, which is optimized for driving a solar MED plant coupled to a double effect absorption HP installed at the solar research center in Plataforma Solar de Almeria, Spain, are given by Ajona [49].

Furthermore, there are also special designs of distillers which can be coupled to solar thermal collectors. Miyatake et al. [50] presented one such application in which a prototype distiller design uses the steam generated in the desalination process for driving another process.

Some of the most important designs of indirect solar desalination pilot plants implemented at various locations are as follows:

- An MED plant with a capacity of 6000 m^3/day driven by parabolic trough collectors was installed in the Arabian Gulf [51].
- A 16 effect-MED, plant capacity 16 m^3/day; solar collectors, flat plate, installed on Takami Island, Japan [51].
- An autoregulated MSF with a capacity of 100 m^3/day driven by parabolic trough collectors is installed in Kuwait [51].
- A 10 m^3/day-MSF driven by solar collectors is installed in Safat, Kuwait [52].
- An 18-effect MED with a plant capacity of 120 m^3/day driven by evacuated tube collectors is installed in Abu Dhabi, UAE [53].
- A plant consisting of a hybrid distillation system, which includes an MSF of 55 stages and an MSF of 75 stages with a capacity of 500 m^3/day, driven by parabolic trough collectors, is installed in Al-Ain, UAE [54].
- A small experimental pilot plant consisting of a four-stage MSF, driven by solar thermal collectors and PV cells to drive the auxiliary equipment, is installed at Al Azhar University in Gaza. Tests carried out showed a maximum daily production of 0.2 m^3/day [55].
- A parabolic trough solar field connected to a 14 effect-MED-TVC plant with a capacity of 72 m^3/day is installed at the Plataforma Solar de Almeria (PSA), Spain [56]. In a second phase of this project, a double-effect absorption HP was coupled to the solar desalination plant, which improved its performance substantially [57].
- A distillation plant with a capacity range of 0.1–0.35 m^3/day, driven by a solar collector, is set up in Hzag, Tunisia [52].
- A 10 m^3/day-MSF plant driven by low concentration solar collectors is erected in Gran Canarias, Spain [58].
- An MED-14 effects, with capacity of 40 m^3/day driven by evacuated tube collectors is installed on La Desired Island, France [59].
- A small 0.3 m^3/day-MSF plant powered by low concentration solar collectors is tried on Lampedusa Island, Italy [60].
- Finally, a 10-stage MSF with a plant capacity of 10 m^3/day, driven by flat plate and parabolic trough collectors, is installed in La Paz, Mexico [61].

Several researchers [62] have suggested solar pond-powered desalination as being one of the most cost-effective systems. An MED plant driven by a solar pond was developed by Tleimat and Howe [63]. Hoffman [64] also proposed solar pond-driven MED plants as the most cost-effective and competitive to the use of fossil fuel. Another two examples of solar pond-powered MED plants are the plant installed at the University of Ancona, Italy, which comprises a hybrid plant with a capacity of 30 m^3/day [65] and a plant installed near the Dead Sea with a capacity of 3000 m^3/day [52].

Various other systems were developed which couple a solar pond to an MSF process. These include a plant installed in Margarita de Savoya, Italy, with a plant capacity of 50–60 m^3/day [51], a small prototype installed at the laboratoire of thermique Industrielle, Tunisia, where a solar pond of 1500 m^2 drives an MSF system with a capacity of 0.2 m^3/day [66], and finally a plant installed in El Paso, Texas, which includes a plant capacity of 19 m^3/day [67].

Since the MSF process is not able to operate with a variable heat source, a modified MSF system called "Autoflash" was developed by ATLANTIS company with a plant capacity of 300 m^3/day, which can be powered from a solar pond [68] installed on the islands of Cape Verde. Generally, in pilot desalination plants coupled to salinity gradient solar ponds, the seawater or brine pre-heater of the desalination plant passes through the storage zone of the solar pond (LCZ) in the form of a plastic heat exchanger, to avoid scale deposits, where it absorbs the thermal energy collected and stored in this zone [69].

Hybrid or dual purpose solar pond plants were also developed. Thermal energy delivered by a salinity-gradient solar pond has been used in seawater and brackish water RO desalination [70]. A solar pond-powered desalination plant erected in El Paso, United States, was developed by Lu et al. [69] in which the solar pond drives both the thermal and RO plants. Safi and Korchani [71] also analyzed the cost of dual purpose plants connected to a solar pond.

Rheinlander and Lippke [72] examined a cogeneration system in which a MED plant is powered by a solar tower power plant. Glueckstern [73] presented a detailed analysis of dual-purpose solar plants. A hybrid MSF-RO system driven by a dual purpose solar plant was installed in Kuwait [51]. The desalination system consists of two systems: a 25 m^3/day MSF plant and a 20 m^3/day RO plant.

Another possible way of using solar energy is by powering an RO process or an MVC by coupling the shaft of these processes with a solar thermal power system. A prototype of an RO plant of 300 m^3/day was used in this way powered from flat plate collectors which use Freon as the working fluid [74]. Childs et al. [75] also presented an RO plant that can be connected to solar thermal collectors. More research in this domain is required.

1.4.2 Solar Photovoltaics

Since PV can produce electricity directly, this technology can be connected directly to power an RO system. Owing to the recent cost reduction in PV technology, this is a viable process. The cost of the electrical energy produced by PV with respect to that produced with conventional energy depends on the plant capacity and on the salt concentration of the feed, which determines the power required per m^3 of the freshwater produced. All publications examined subsequently refer to the price of PV before the drastic reduction in their price occurred in the last few years.

Both the author [76] and Tzen et al. [77] presented an analysis of the cost of PV-RO desalination systems. A detailed cost analysis of a system installed in Oman is also presented by Al Suleimani and Nair [78]. Thomson and Infield [79] presented the simulation and application of a PV-driven RO for Eritrea. This system is not using battery storage and is able to operate with a variable flow that is unavoidably produced by the PV system. The production capacity of the system was 3 m^3/day and the PV array size was 2.4 kWp. Finally, the Canary Islands Technological Institute (ITC) developed a stand-alone system called DESSOL which has a capacity of 1–5 m^3/day at nominal output [80].

One of the largest PV-RO brackish water desalination plant with a capacity of 15 m^3/day was installed in Riyadh, Saudi Arabia. This is connected to solar stills which can produce 5 m^3/day. In fact, the feed water of the solar stills is the outlet brine from the RO unit which produces another 10 m^3/day freshwater [81].

The DC produced from a PV system can be used to power an ED process. For brackish water desalination in remote areas, ED is a more suitable process than RO. Until now, a number of ED pilot plants have been connected to PV panels, usually employing batteries for storage. An analysis performed by Gomkale [82] on solar desalination for Indian villages concluded that solar PV operated ED is more cost-effective for desalting brackish water than the conventional solar still. A PV-driven ED plant developed by the Bureau of Reclamation, United States, was installed at the Spencer Valley in New Mexico [52,83]. The University of Bahrain [84] and the Laboratory for Water Research at the University of Miami, Florida, United States [85] also performed experimental research in PV-ED.

Some other pilot PV-ED systems have also been installed in various countries:

- A 0.120 m^3/h, brackish water, powered by a 450 Wp PV system installed in Thar Desert, India [86].
- A 2.8 m^3/day brackish water desalination system was installed in Spencer Valley, New Mexico [52].
- A relatively large 10 m^3/day system, suitable for seawater desalination, was installed on Ohsima Island, Nagasaki, Japan [87].
- Finally, an 8.33 m^3/h system desalting brackish water was erected in Fukue city, Nagasaki, Japan [88].

1.4.3 Wind Power

As will be shown in the next section, RO is the desalination process with the lowest energy requirements, and usually in coastal areas there is a high availability of wind power resource [89]; therefore, wind powered desalination is one of the most promising options of renewable energy desalination. Garcıa-Rodrıguez et al. [90] presented a preliminary cost evaluation of wind powered RO. They examined seawater RO driven by wind power, and the influence of climatic conditions related to the wind resource characteristics and plant capacity on the freshwater product cost. Moreover, they examined the possible evolution of freshwater cost due to possible future changes in wind power and RO technologies. Finally, they evaluated the influence on the competitiveness of wind-powered RO against conventional RO plants due to the modifications of the financial parameters and conventional energy cost. Some other researchers performed feasibility studies on the use of this type of systems.

A detailed analysis of a wind-powered RO plant by considering a number of different wind turbines and membranes was performed by Kiranoudis et al. [91]. They considered various design parameters selection and operation aspects as well as the use of both seawater and brackish water feed.

An RO system with variable load connected to a wind turbine was reported by Ehmann and Cendagorta [92]. This was installed at the Canary Islands Technological Institute (ITC) at PozoIzquierdo, Gran Canarias, Spain.

A stand-alone system with capacities between 5 and 50 m^3/day, known as AEROGEDESA, was also developed at the Canary Islands Technological Institute [80,93]. For a plant with a nominal output of 13 m^3/day, the average production was reported to be 11 m^3/day for an annual average wind speed of 7 m/s.

Details of a wind driven RO seawater desalination system with a capacity of 56 m^3/day installed at Pajara, Fuerteventura Island, Spain, are given by

Gonzalez [94]. This hybrid diesel-wind system, which provides the water requirements of a village of 300 people, consists of two diesel engines and a 225 kW wind turbine.

A wind-powered RO plant with a number of possible combinations for autonomous diesel operation is reported by Heukelom et al. [95]. They reported that as the system is using a relatively cheap wind energy source, when combined with a diesel generator set, results in a reliable energy system for water desalination.

A study of a wind-powered brackish water RO plant for Jordan is presented by Habali and Saleh [96]. The product cost reported in this study is lower than that obtained by using conventional diesel engines.

Miranda and Infield [97] developed another pilot plant in which the RO system is able to adapt to the variable wind energy. Furthermore, Robinson et al. [98] also described a wind-RO brackish water desalination system designed for small remote communities in Australia.

Some other wind-driven RO systems developed are:

- A 25 m^3/day RO plant powered by a hybrid wind-diesel system was installed in the Middle East [99].
- A wind powered RO system was set in operation in 1995 in Drepanon, Achaia, near Patras, Greece [100].
- A RO system driven by a wind power plant was installed on the Island of the County Split and Dalmatia [101].

Another possibility examined is the direct coupling of a wind energy system and an RO unit, that is, direct rotation of the RO system high-pressure pump by wind. The main research in this field has been performed by the Canary Islands Technological Institute [102]. A brackish water desalination wind-powered RO plant was installed on Coconut Island in Oahu, Hawaii. In this system, the shaft power of a windmill is directly connected to the high-pressure pump of the RO system. It is reported that the system can maintain a constant freshwater production of 13 L/min for a wind speed of 5 m/s [103].

Another possibility considered is the use of wind power directly with an MVC system. Karameldin et al. [104] presented a detailed analysis on the influence of the main parameters of such a system. A pilot plant with a freshwater production of about 0.3–2 m^3/h operating on this principle was erected in Borkum Island in the North Sea [52,105]. Another pilot plant was installed on Ruegen Island, Germany, employing a 300 kW wind energy converter and 120–300 m^3/day production MVC system [52,106].

1.4.4 Hybrid Solar PV-Wind Power

A promising alternative to drive a desalination unit is the use of hybrid wind-solar systems. This is so because there is a complementarity of wind and solar resources, as usually when there is no sun the wind is stronger and vice versa. A useful software for simulation of hybrid wind-PV RO systems is presented by Manolakos et al. [107]. A pilot unit operating on both PV and wind was installed in 1980 in Borj-Cedria, Tunisia [83]. The system comprises a 0.1 m^3/day-compact solar distiller, a 0.25 m^3/h RO plant, and an ED plant for 4000 ppm brackish water. The power supply system includes a 4 kW peak PV field and two wind turbines. Other systems that have been operating based on this principle are two RO desalination plants supplied by a 6 kW wind energy converter and a 2.5 kW solar generator for remote areas [108], a 3 m^3/day-plant for brackish water, installed in Israel [109], and a system with a capacity of 1 m^3/h in Oman [110].

1.4.5 Geothermal Energy

Possibly the first paper published on geothermal energy assisted desalination plants was authored by Awerbuch et al. [111]. According to this paper, in 1972 the Bureau of Reclamation of the US Department of the Interior set up an experimental pilot plant near Holtville, California, United States to investigate the geothermal powered desalination. Boegli et al. [112] from the same department reported the experimental results of geothermal desalination at the East Mesa Test Site. The processes analyzed included an MSF system and a high-temperature ED. They also analyzed various evaporation tubes and membranes. Ophir [113] presented an economic analysis of geothermal desalination in which sources of 110–130°C were considered. He concluded that the price of geothermal desalination is as low as the price of large multieffect dual purpose plants. A technical and economic analysis on the use of geothermal sources between 75°C and 90°C with MED is presented by Karytsas [114,115].

Another two geothermal-powered distillation plants were installed, one in the south of Tunisia [116] and the other in France [117]. Both systems employ condensers and evaporators made from polypropylene and operate at a temperature range between 60°C and 90°C [118]. The performance of a solar still fed by brackish underground geothermal water is presented by Bouchekima [119].

Another possibility is the use of high-pressure geothermal power directly as shaft power for desalination. Furthermore, nowadays there are commercial membranes that withstand temperatures up to 60°C, which allows the direct use of geothermal brines for desalination [120].

1.5 PROCESS SELECTION AND RENEWABLE ENERGY OPTIMIZATION

As in all other systems, during the design phase of a renewable energy powered desalination system, a process suitable for a particular application will need to be selected by the designer. There are many factors that need to be considered for such a selection. The main ones are as follows [121]:

- **(i)** The amount of freshwater required for the particular application in combination with the range of applicability of the various desalination processes.
- **(ii)** The effectiveness of the process with respect to energy consumption, which would allow the use of the smallest possible renewable energy system.
- **(iii)** Suitability of the process for renewable energy application, especially due to its intermitted nature or its suitability to follow a varying energy supply.
- **(iv)** The capital cost of the equipment.
- **(v)** The seawater treatment requirements.
- **(vi)** The land area required or that could be made available for the installation of the equipment, especially for the renewable energy system.

Before the process selection, two basic parameters should be investigated. These concern the evaluation of the available water resources including seawater, brackish water, and freshwater, and the identification and evaluation of the renewable energy resources in the area.

Concerning the water resources, the investigation should be done both in terms of quality and quantity. If brackish water is available in the area, this may be more attractive as its salinity is normally much lower (<10,000 ppm), and thus much lower energy is required to clear the water from salts; for this reason, desalination of the brackish water should be a more viable option. In inland sites, brackish water may be the only option. On a coastal site, seawater is normally available.

As presented in this chapter, renewable energy driven desalination technologies mainly fall into two categories. The first category concerns distillation desalination technologies driven by heat produced by RES, whereas the second concerns membrane desalination technologies driven by electricity or mechanical energy, again produced by RES. Such desalination and RES should be characterized by compactness in size, simplicity of operation, and low maintenance, combined with the need for specialized personnel, easy transportation and installation to site, simple intake and pretreatment system to ensure proper operation and ability of a plant to withstand for long time the difficult conditions often encountered in remote areas. Concerning the combination of desalination and RES, existing experience has shown that there are no significant technical problems that cannot be easily solved [122].

Generally, water production costs from any system include the following [123]:

- Fixed charges, which are dependent on the initial capital cost of the equipment and the depreciation factor; determined from both the expected plant life and the financial parameters and consequently are varying for each country and region;
- Variable charges, which are dependent on the amount of energy consumption and cost, related to the energy source employed, its availability, and location selected; operational costs related to skilled manpower requirement; maintenance cost which depend on both the desalination and renewable energy system type; consumption and cost of chemicals used for pre- and posttreatment of water, especially in RO plants; and the replacement costs which concern mainly the rate at which the membranes in RO plants need replacement and replacement of renewable energy system components; usually, both are site related.

Generally, the amount of TDS (mainly salt) in seawater is practically not important in thermal processes, but it is a crucial factor in RO where the energy demand increases proportionally at a rate of about 1 kWh/m^3 per 10,000 ppm [123]. The percentage of salts in the water produced could be allowed to vary within the allowed value of 500 ppm, fixed by the WHO for drinking water. Normally, this value for the RO process is around 300 ppm, whereas it is much lower for the distillation systems.

Renewable energy sources can provide energy at various forms such as thermal energy produced by solar collectors and geothermal energy, electricity produced by PV, wind energy, and solar thermal power systems, or mechanical energy obtained directly by the wind energy turbines. As was seen in previous sections, all these forms of energy can be used to power desalination plants.

Solar radiation can generally be converted into useful energy either as heat, with various types of solar collectors and solar ponds, or as electricity, with a variety of PV cells and solar thermal power systems. As was seen in the previous section, these are called indirect collection systems and both forms of energy were used to power conventional desalination systems. There is also the other category of direct collection systems, the solar stills, which can utilize solar energy only whenever this is available, and their collection is not very efficient. In the indirect collection systems, since separate systems are used for energy collection and desalination, solar energy is collected by appropriate more efficient solar collectors, in the form of hot water or steam. It should be noted, however, that solar energy, like most other renewable energy sources, is only available for almost half of the 24 h in a day. This implies that the desalination process, powered by such a system, operates for only half the time available unless some storage device of adequate size is used. Alternatively, the storage device, which is usually expensive, can be replaced by a backup boiler or electricity from the grid in order to operate the desalination system during low insolation periods and nighttime. It should be noted, however, that when such a system operates without thermal buffering offered by the storage device, the combined desalination system must be able to follow a variable energy supply, without any problem in its operation.

In all solar energy desalination systems, an optimum performance ratio has to be estimated based on the cost of the solar energy collectors, storage device, and desalination plant [121]. Desalination systems which are sensitive to variable energy supply, such as MSF, are more suitable to be fed with stable energy supplied from a solar pond, which due to its size, it doesn't charge/discharge easily, and thus is more insensitive to variations of the input solar radiation.

As is well known, wind energy is also a highly variable source of energy supply, both with respect to wind speed and frequency. When wind energy is used for electricity generation for desalination, the variation of the wind source can be balanced by the addition of batteries, which act in a similar way to a storage tank in solar thermal systems, that is, the batteries charge when surplus wind energy is available and discharge to the desalination plant when required. In the case of mechanical shaft power production from wind, the desalination plant can operate only when there is wind. Again, as the resource is variable, the desalination plant is usually oversized with respect to water demand and, instead of storing the energy, the freshwater produced when wind is available is stored.

Another parameter to be considered concerning the renewable energy and desalination technologies selection is the type of connection of the two systems. For example, an RO renewable energy powered desalination plant can be designed to operate either coupled to the grid or as a standalone, off-grid, or autonomous system. When the system is grid connected, the desalination plant operates continuously as any other conventional plant and the renewable energy source actually constitutes a fuel substitute. In an isolated system, where no electricity grid is available, the use of autonomous systems is the only option and the system must be capable of following the intermittent nature of the renewable energy source.

Desalination systems are traditionally designed to operate with a constant power input [77]. Usually, a nonsteady power input forces the desalination system to operate in nonoptimal conditions and this may cause serious operational problems [122]. Each desalination system technologies have specific problems when they are connected to a variable power system. For example, the RO system has to cope with the membranes' sensitivity with respect to scaling, fouling, and unpredictable phenomena due to the frequent starting and stopping and partial load operation during periods of variable power supply. In contrast, the VC system has a large thermal inertia and requires considerable energy at the beginning to reach the nominal working point. Therefore, for autonomous systems, a small energy storage system, in the form of batteries or thermal storage, should be employed to offer stable energy supply to the desalination unit. Any candidate option should be examined in detail, especially with respect to the site characteristics, such as accessibility and area requirement, and initial cost requirements [122].

By carrying out a survey of manufacturers' data, the energy required for various desalination processes is obtained and presented in Table 1.4. As can be seen from Table 1.4, the process with the smallest energy requirement is RO with energy recovery. Owing to the high cost of the energy-recovery turbine, however, this is only viable for very large systems. The next lowest

Table 1.4 Energy consumption of desalination systems

Process	Heat input (kJ/kg of product)	Mechanical power input (kWh/m^3 of product)	Prime energy consumption (kJ/kg of product)[a]
MSF	294	2.5–4 (3.7)[b]	338.4
MED	123	2.2	149.4
VC	–	8–16 (16)	192
RO	–	5–13 (10)	120
ER-RO	–	4–6 (5)	60
ED	–	12	144
Solar still	2330	0.3	2333.6

[a]Assumed conversion efficiency of electricity generation of 30%.
[b]Figure used for the prime energy consumption estimation shown in last column.

is the RO without energy recovery and the MED. Similarly, as obtained from a survey of manufacturers' data, a comparison of the desalination equipment cost and the seawater treatment requirement is shown in Table 1.5. The cheapest of all the systems considered is the solar still. As was seen before, this is a direct collection system, which is very easy to construct and operate without any requirement of skilled personnel. The disadvantage of this process is the very low yield, which implies that large areas of ground are required. It is certain that such a process cannot be viable unless cheap, noncultivated land is available near the sea. As can be seen, the MED is the cheapest of the indirect collection systems and also requires the simplest seawater treatment. RO, although requiring a smaller amount of energy, is relatively expensive and requires a complex seawater treatment. This changed recently with the development of new, more efficient, and low cost membranes and the adoption of new filtration systems like nano- and ultrafiltration systems.

Owing to the continuous research and development of RO technology, and the improvements that were made in RO membranes, the energy consumption values of more than 20 kWh/m^3, typical for systems developed in the 1970s, have been reduced today to about 5 kWh/m^3 [123]. Research and development in this area is ongoing worldwide and we may see further reductions in both energy requirements and costs in the coming years. The importance of employing renewable for desalination can be better understood by noting that nearly 3 kg of CO_2 are generated for each m^3 of water produced, at the reduced energy consumption rate of 5 kWh/m^3, which corresponds to the best technology currently used on large scale and could be avoided if the RO is powered by a renewable system.

One alternative for solar-powered desalination is to use an RO system powered with a PV system. This seems to be a good option as RO has higher yields per unit of energy collected. Zarza et al. [124], in a study performed in the early 1990s, compared RO with PV generated electricity with an MED plant coupled to parabolic trough collectors, and concluded that:

(i) The total cost of freshwater produced by an MED plant powered by parabolic trough collectors is lower than that of the RO plant with PV cells because of the high cost of the PV.

(ii) The MED plant operation is highly reliable, so it makes its installation possible in countries with high insolation levels but with no experienced personnel available. It should be noted that any mistake during the operation of an RO plant can ruin its membranes, so skilled personnel must be used to operate them.

Table 1.5 Comparison of desalination plants

ITEM	MSF	MEB	VC	RO	Solar still
Scale of application	Medium-large	Small-medium	Small	Small-large	Small
Seawater treatment	Scale inhibitor Antifoam chemical	Scale inhibitor	Scale inhibitor	Sterilizer Coagulant acid Deoxidizer	–
Equipment price (Euro/m^3)	950–1900	900–1700	1500–2500	900–2500 Membrane replacement every 4–5 years	800–1000

Note: Low figures in equipment price refer to bigger size in the range indicated and vice versa.

The second conclusion is still valid today, but the first is alleviated due to the recent drastic reduction of PV.

It should be noted that as renewable energy is generally expensive to collect and store, an energy-recovery turbine is typically fitted to recover pressure energy from the rejected brine stream, which leads to a considerable increase in the RO plant initial cost. Furthermore, distillation processes are preferred for desalination in polluted waters, because in these the water is boiled, which ensures that the distilled water is cleaned from any microorganisms [121]. Water pollution includes, in addition to the high salinity, manganese, heavy metals, fluoride, bacterial contamination, and pesticide/herbicide residues. If such water is available, thermal processes are preferred instead of the membrane ones. Even the simple direct collection system, that is, the solar still, can show pollutant removal efficiencies of the order of 99% [125].

If both membrane and thermal processes are suitable for a given location, the type of renewable energy available in the area, that is, solar radiation, wind geothermal, etc., and the type energy required electrical/mechanical/thermal by the desalination process, determines the final selection. Finally, the required plant capacity in view of the annual and daily distribution of the freshwater requirement, the final freshwater cost, and any problems related to the combination of the renewable energy and the desalination systems are factors that can influence the selection.

If thermal energy is available from a renewable energy system, it can be used to drive a distillation process such as MSF, MED, or TVC. From the three, the MED plants are more suitable to operate at partial load, less sensitive to scaling, cheaper, and more appropriate for smaller capacity. TVC also has a lower performance than MED and MSF.

If renewable electricity or shaft power is available, RO, ED, or MVC can be selected. As fluctuations of the available energy could ruin the RO system, an energy storage which could smooth the energy supply would be required, but this would increase the initial system costs. The ED is most suitable for brackish water desalination in remote areas due to the fact that it is more robust and its operation and maintenance are simpler than that of the RO systems. In addition, the ED process is able to adapt more easily to changes in the available energy input. On the other hand, although MVC consumes more energy than the RO process, it has fewer problems related to fluctuations in the energy resource than the RO. Additionally, MVC systems are more suitable for remote areas as they are more robust and need less skilled manpower and less quantities of chemicals than RO systems [27].

Furthermore, they need no membrane replacement and generally offer better quality water than the RO because, as already pointed out, in the case of polluted waters, the distillation ensures the absence of microorganisms and other pollutants in the final product.

Solar energy is probably most cheaply collected and used with thermal energy collection systems using a variety of solar thermal collectors. The two systems that could be powered with thermal energy are the MSF and the MED plants and, as was presented in Section 1.4, both have been used with solar energy collectors in various pilot plants worldwide. According to the data presented in Tables 1.4 and 1.5, the MED process is cheaper and requires less specific energy and only a very simple seawater treatment. According to Porteous [126], the additional advantages that the MED process possesses compared to other distillation processes are:

(i) It offers energy economy because the brine is not heated above its boiling point as is the case for the MSF process, which leads to lower irreversibilities as the vapor is used at the temperature at which it is generated.

(ii) The scale formation risks are minimized because the feed is at its lowest concentration at the highest plant temperature.

(iii) The feed flows through the plant from effect to effect and the maximum concentration occurs only in the last effect. For this reason, the worst boiling point elevation is restricted to this effect.

(iv) Owing to the recirculation pump in the MSF or the vapor compressor in the VC systems, these have a higher electrical demand than MED.

(v) MSF is susceptible to equilibrium problems, which are presented as a reduction in the performance ratio. In MED plants, since the vapor generated in one effect is used in the next effect, the performance ratio is not subject to equilibrium problems.

(vi) The MED plant is more simple because the process requires fewer effects for a given performance ratio.

Of the various configurations of possible MED evaporators, the most appropriate for solar energy application is the Multiple Effect Stack (MES) type, shown schematically in Fig. 1.9. This is so because this has a number of advantages, and primarily the stable operation of the unit between virtually zero and 100% output, its ability to adapt to sudden changes, and its ability to follow a varying steam supply without problems. For this system, collectors of proven technology such as good-quality flat-plate collectors, evacuated tube collectors, or low concentration parabolic troughs can be used to produce the input energy to the MES system in the form of low-pressure steam.

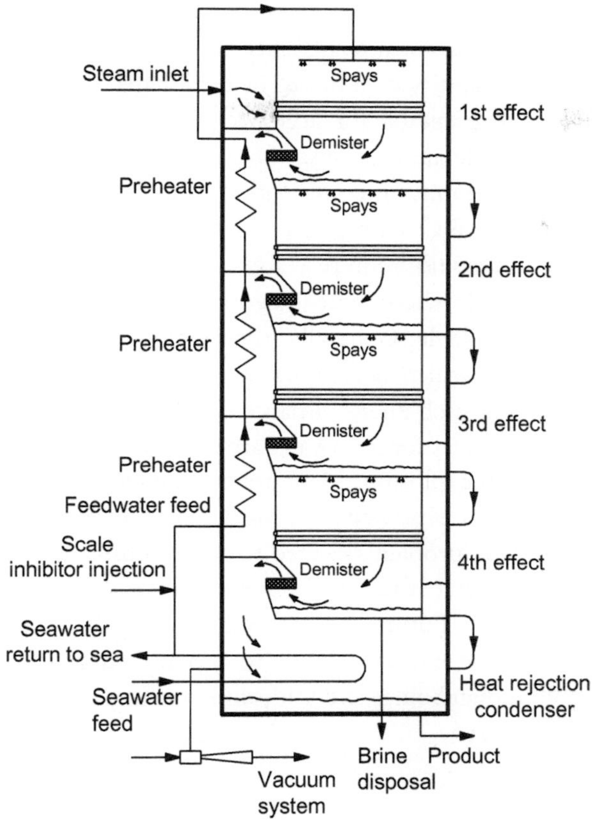

Fig. 1.9 The MES evaporator schematic.

The MED process in general requires temperature for the heating medium between 70°C and 100°C.

1.6 CONCLUSIONS

In this chapter, the characteristics of the various renewable energy desalination systems are presented as well as a review of a number of research or pilot systems erected in various parts of the world. Guidance is also given on the factors to consider for the selection of the appropriate renewable energy desalination technology. The basic ones include feed water salinity, plant size, availability of grid electricity, infrastructure and remoteness of the area, and type of renewable energy resource available. Several possible combinations of desalination and renewable energy technologies have been tried and the lessons learned from these systems are that some are very promising in terms of economic viability. In all cases, the applicability of these systems depends strongly on the local availability of renewable energy resources and the quality and quantity of water to be desalinated. Furthermore, some combinations are more suitable for large size plants, whereas others are suitable only for small-scale applications.

Maybe the most promising combination of technologies is MED with thermal collectors and RO with PV. For large units, wind energy may be more attractive as it does not require a large area to be installed. This is especially suitable for islands where there is a good wind potential but often limited flat ground is available. Generally, the energy cost is one of the most important parameters which determine the product water costs and RES promise the supply of such energy in an environmental-friendly way.

The needs of humanity for freshwater are increasing drastically, especially due to the population explosion. Many renewable technologies that can be used for desalination are nowadays considered as matured technologies. Owing to the climatic problems related to the extensive use of fossil fuels, there is a need to accelerate the development of novel water production systems from renewables. Future water desalination should increasingly be powered by renewable energy sources to provide adequate quantities of freshwater for all human beings and without imposing more environmental problems to the planet.

REFERENCES

[1] Kalogirou SA. Seawater desalination using renewable energy sources. Prog Energy Combust Sci 2005;31(3):242–81.
[2] Delyannis E. Historic background of desalination and renewable energies. Solar Energy 2003;75(5):357–66.
[3] Major JK. Water wind and animal power. In: McNeil J, editor. An encyclopaedia of the history of technology. Bungay, Great Britain: Rutledge, R. Clay Ltd; 1990. p. 229–70.
[4] Delyannis A. Introduction to chemical technology. History of chemistry and technology, Athens; 1960. 410 pp. [chapter 1, in Greek].
[5] Berthelot M. In: Steinheil G, editor. Collection des Anciens Alchemists Greecs. Collections of the ancient Greek alchemists. Paris: Georges Steinheil; 1888.
[6] Delyannis E, Belessiotis V. The history of renewable energies for water desalination. Desalination 2000;128:147–59.
[7] Meteorological. Cambridge, MA: Harvard University Press; 1962. p. 127. Book I [chapter 91].
[8] The works of Aristotle. In: Encyclopaedia britannica. vol. 1. 1956. p. 464. Book II, 354 (25) & 358 (15).
[9] Tiwari GN, Singh HN, Tripathi R. Present status of solar distillation. Solar Energy 2003;75(5):367–73.

[10] Belessiotis V, Kalogirou SA, Delyannis E. Thermal solar desalination. London, UK: Academic Press, Elsevier Science; 2016, ISBN: 978-0-12-809656-7.
[11] Mustacchi C, Cena V. Solar desalination—design, performances, economics. Sogesta; 1981.
[12] Eibling JA, Talbert SG, Lof GOG. Solar stills for community use—digest of technology. Solar Energy 1971;13(2):263–76.
[13] Talbert SG, Eibling JA, Lof GOG. Manual on solar distillation of saline water. R&D progress report no. 546, US Department of the Interior, Battelle Memorial Institute, Columbus Laboratories; 1970. p. 270.
[14] Delyannis A, Delyannis E. In: Solar distillation plant of high capacity. Proceedings of 4th international symposium on fresh water from sea. vol. 4; 1973. p. 487.
[15] Kalogirou SA. In: Solar distillation systems: a review. Proceedings of 1st international conference on energy and the environment, Limassol, Cyprus. vol. 2; 1997. p. 832–8.
[16] Malik MAS, Tiwari GN, Kumar A, Sodha MS. Solar distillation. Oxford, UK: Pergamon Press; 1982.
[17] Tiwari GN. Contemporary physics—solar energy and energy conservation. In: Recent advances in solar distillation. New Delhi, India: Wiley Eastern Ltd.; 1992 [chapter II].
[18] Fath ESH. Solar distillation: a promising alternative for water provision with free energy, simple technology and a clean environment. Desalination 1998;116:45–56.
[19] Mustacchi C, Cena V. Solar water distillation. In: Technology for solar energy utilisation. New York: United Nations; 1978. p. 119–24.
[20] Tleimat MW. Freezing methods. In: Spiegler KS, Laird ADK, editors. Principles of desalination. 2nd ed., Part B. New York: Academic Press; 1980. p. 359–400.
[21] Parekh S, Farid MM, Selman RR, Al-Hallaj S. Solar desalination with humidification-dehumidification technique-a comprehensive technical review. Desalination 2003; 160:167–86.
[22] Lawson KW, Lloyd DR. Review: membrane distillation. J Membr Sci 1997;124:1–25.
[23] Curcio E, Drioli E. Membrane distillation and related operations—a review. Sep Pur Rev 2005;34:35–86.
[24] El-Bourawi MS, Ding Z, Ma R, Khayet M. A framework for better understanding membrane distillation separation process. J Membr Sci 2006;285:4–29.
[25] Dresner L, Johnson J. Hyperfiltration (reverse osmosis). In: Spiegler KS, ADK L, editors. Principles of desalination. 2nd ed., Part B. New York: Academic Press; 1980. p. 401–560.
[26] Shaffer LH, Mintz MS. Electrodialysis. In: Spiegler KS, Laird ADK, editors. Principles of desalination. 2nd ed., Part A. New York: Academic Press; 1980. p. 257–357.
[27] Garcia-Rodriguez L. Renewable energy applications in desalination: state of the art. Solar Energy 2003;75(5):381–93.
[28] El-Dessouky H, Ettouney H. MSF development may reduce desalination costs. Water Wastewater Int 2000;2000:20–1.
[29] Desalination guide using renewable energies, THERMIE programme. Greece: CRES; 1998, ISBN: 960-90557-5-3.
[30] Kalogirou SA. Solar energy engineering: processes and systems. 2nd ed. Oxford, UK: Academic Press, Elsevier Science; 2013, ISBN: 978-0-12-397270-5.
[31] Kalogirou SA. The potential of solar industrial process heat applications. Appl Energy 2003;76(4):337–61.
[32] Kalogirou SA. Solar energy collectors and applications. Prog Energy Combust Sci 2004;30(3):231–95.
[33] Norton B. Solar energy thermal technology. London: Springer-Verlag; 1992.

[34] Matz R, Feist EM. The application of solar energy to the solution of some problems of electrodialysis. Desalination 1967;2(1):116–24.
[35] PVPS. Snapshot of global photovoltaic markets: 2015. Report IEA PVPS T1-29, 2016.
[36] Beurskens J, Garrad A. In: Wind energy. The proceedings of Eurosun'96 conference, Freiburg, Germany. vol. 4; 1996. p. 1373–88.
[37] Ackermann T, Soder L. An overview of wind-energy status 2002. Renew Sustain Energy Rev 2002;6:67–128.
[38] World Wind Energy Association. Available from: http://www.wwindea.org/the-world-sets-new-wind-installations-record-637-gw-new-capacity-in-2015/ [Accessed 14 August 2016].
[39] Popiel C, Wojtkowiak J, Biernacka B. Measurements of temperature distribution in ground. Exp Therm Fluid Sci 2001;25:301–9.
[40] Florides G, Kalogirou SA. In: Ground heat exchangers—a review. Proceedings of 3rd international conference on heat power cycles, Larnaca, Cyprus on CD-ROM; 2004.
[41] Barbier E. Geothermal energy technology and current status: an overview. Renew Sustain Energy Rev 2002;6:3–65.
[42] Barbier E. Nature and technology of geothermal energy. Renew Sustain Energy Rev 1997;1(1–2):1–69.
[43] Baldacci A, Burgassi PD, Dickson MH, Fanelli M. In: Sayigh AAM, editor. Non-electric utilization of geothermal energy in Italy. World Renewable Energy Congress V, Part I, 20–25 September 1998. Florence, Italy: Pergamon; 1998. p. 2795.
[44] Rommel M, Kohl M, Graf W, Wellens C, Brucker F, Lustig K, et al. Corrosion-free collectors with selectively coated plastic absorbers. Desalination 1997;109(2):149–55.
[45] Rommel M. Solar thermally driven desalination systems with corrosion-free collectors. Renew Energy 1998;14(1–2):275–80.
[46] Rajvanshi AK. A scheme for large scale desalination of seawater by solar energy. Solar Energy 1980;24:551–60.
[47] Hermann M, Koschikowski J, Rommel M. In: Corrosion-free solar collectors for thermally driven seawater desalination. Proceedings of the EuroSun 2000 conference on CD-ROM, 19–22 June, 2000, Copenhagen, Denmark; 2000.
[48] Rommel M, Hermann M, Koschikowski J. In: The SODESA project: development of solar collectors with corrosion-free absorbers and first results of the desalination pilot plant. Mediterranean conference on policies and strategies for desalination and renewable energies, 21–23 June, Santorini, Greece; 2000.
[49] Ajona JI. In: ACE-20 Spanish parabolic trough collector. Proceedings of the 6th international symposium on solar thermal concentrating technologies, September 28–October 2, vol. I. Madrid: Ministerio de Industria y Energıa, CIEMAT; 1992. p. 1992.
[50] Miyatake O, Koito Y, Tagawa K, Maruta Y. Transient characteristics and performance of a novel desalination system based on heat storage and spray flashing. Desalination 2001;137:157–66.
[51] Delyannis EE. Status of solar assisted desalination: a review. Desalination 1987;67:3–19.
[52] European Commission. Desalination guide using renewable energies; 1998.
[53] El-Nashar AM. Abu Dhabi solar distillation plant. Desalination 1985;52:217–34.
[54] Hanafi A. Design and performance of solar MSF desalination system. Desalination 1991;82(1–3):165–74.
[55] Abu-Jabal MS, Kamiya I, Narasaki Y. Proving test for a solar-powered desalination system in Gaza-Palestine. Desalination 2001;137:1–6.
[56] Zarza ME. Solar thermal desalination project: first phase and results and second phase description. Madrid: Secretarıa General Tecnicadel, CIEMAT; 1991.
[57] Zarza ME. Solar thermal desalination project, phase II: results and final project report. Madrid: Secretarıa General Tecnicadel, CIEMAT; 1995.

[58] Valverde MV. Planta Desaladora con Energıa Solar de Arinaga (Las Palmas de Gran Canaria). Departamento de Investigacion y Nuevas Fuentes, Centro de Estudios de la Energıa; 1982. April.
[59] Madani AA. Economics of desalination systems. Desalination 1990;78:187–200.
[60] Palma F., Seminar on new technologies for the use of renewable energies in water desalination, Athens, 1991. Commission of the European Communities, DG XVII for Energy, CRES (Centre for Renewable Energy Sources), 1991.
[61] Manjares R, Galvan M. Solar multistage flash evaporation (SMSF) as a solar energy application on desalination processes. Description of one demonstration project. Desalination 1979;31(1–3):545–54.
[62] Bucher W. In: Renewable energy sources and seawater desalination: an assessment of methods and potential. ASRE 89, International conference on applications of solar & renewable energy, Cairo, March 19–22; 1998.
[63] Tleimat MC, Howe ED. Use of energy from salt gradient solar ponds for reclamation of agricultural drainage water in California: analysis and cost prediction. Solar Energy 1989;42(4):339–49.
[64] Hoffman D. The application of solar energy for large scale seawater desalination. Desalination 1992;89(2):115–83.
[65] Caruso G, Naviglio A. A desalination plant using solar heat as s heat supply, not affecting the environment with chemicals. Desalination 1999;122:225–34.
[66] Safi MJ. Performance of a flash desalination unit intended to be coupled to a solar pond. Renew Energy 1998;14(1–4):339–43.
[67] Lu H, Walton JC, Swift AHP. Zero discharge desalination. Int Desalin Water Reuse Q 2000;10(3):35–43.
[68] Szacsvay T, Hofer-Noser P, Posnansky M. Technical and economic aspects of small-scale solar-pond-powered seawater desalination systems. Desalination 1999;122:185–93.
[69] Fisher U. In: Solar pond as an energy source for desalination. Proceedings of ISES solar World Congress 1999 on CD-ROM, Jerusalem, Israel; 1999. p. 1228–32.
[70] Engdahl DD. Technical information record on the salt gradient solar pond system at the Los Banos Demonstration Desalting Facility; 1987.
[71] Safi MJ, Korchani A. Cogeneration applied to water desalination: simulation of different technologies. Desalination 1999;125:223–9.
[72] Rheinlander K, Lippke F. Electricity and potable water from a solar tower power plant. Renew Energy 1998;14(1–4):23–8.
[73] Glueckstern P. Potential uses of solar energy for seawater desalination. Desalination 1995;101(1):11–20.
[74] Rodrıguez G, Rodrıguez M, Perez J, Veza J. In: A systematic approach to desalination powered by solar, wind and geothermal energy sources. Proceedings of the Mediterranean conference on renewable energy sources for water production, 10–12 June. Santorini, Greece: European Commission, EURORED Network, CRES, EDS; 1996. p. 20–5.
[75] Childs WD, Dabiri AE, Al-Hinai HA, Abdullah HA. VARI-RO solar powered desalting technology. Desalination 1999;125:155–66.
[76] Kalogirou SA. Effect of fuel cost on the price of desalination water: a case for renewables. Desalination 2001;138:137–44.
[77] Tzen E, Perrakis K, Baltas P. Design of a stand-alone PV-desalination system for rural areas. Desalination 1998;119:327–34.
[78] Al Suleimani Z, Nair NR. Desalination by solar powered reverse osmosis in a remote area of Sultanate of Oman. Appl Energy 2000;(64):367–80.
[79] Thomson M, Infield D. A photovoltaic-powered seawater reverse-osmosis system without batteries. Desalination 2003;153(1–3):1–8.

[80] ITC (Canary Islands Technological Institute). DESSOL precommercial brochure; 2002.
[81] Hasnain SM, Alajlan SA. Coupling of PV-powered RO brackish water desalination plant with solar stills. Desalination 1998;116(1):57–64.
[82] Gomkale SD. Solar distillation as a means to provide Indian villages with drinking water. Desalination 1988;69(2):171–6.
[83] Maurel A. In: Desalination by reverse osmosis using renewable energies (solar-wind) Cadarche Centre Experiment. Seminar on new technologies for the use of renewable energies in water desalination, Athens, 26–28 September 1991. Commission of the European Communities, DG XVII for Energy, CRES (Centre for Renewable Energy Sources); 1991. p. 17–26. Session II.
[84] AlMadani HMN. Water desalination by solar powered electrodialysis process. Renew Energy 2003;28:1915–24.
[85] Kvajic G. Solar power desalination, PV-ED system. Desalination 1981;39:175.
[86] Adiga MR, Adhikary SK, Narayanan PK, Harkare WP, Gomkale SD, Govindan KP. Performance analysis of photovoltaic electrodialysis desalination plant at Tanote in Thar Desert. Desalination 1987;67:59–66.
[87] Kuroda O, Takahashi S, Kubota S, Kikuchi K, Eguchi Y, Ikenaga Y, et al. An electrodialysis seawater desalination system powered by photovoltaic cells. Desalination 1987;67:33–41.
[88] Ishimaru N. Solar photovoltaic desalination of brackish water in remote areas by electrodialysis. Desalination 1994;98(1–3):485–93.
[89] Doucet G. Energy for tomorrow's world. Renew Energy 2001;2001:19–22.
[90] Garcıa-Rodrıguez L, Romero T, Gomez-Camacho C. Economic analysis of wind-powered desalination. Desalination 2001;137:259–65.
[91] Kiranoudis CT, Voros NG, Maroulis ZB. Wind energy exploitation for reverse osmosis desalination plants. Desalination 1997;109(2):195–209.
[92] Ehmann H, Cendagorta M. Mediterranean conference on renewable energy sources for water production, 10–12 June. Santorini, Greece: European Commission, EURORED Network, CRES, EDS; 1996. p. 84–7.
[93] ITC (Canary Islands Technological Institute). AEROGEDESA precommercial brochure; 2002.
[94] Gonzalez A. In: Proyecto de Desalinizacion por Energıa Eolica en Fuerteventura. Curso: Tratamiento de Aguas Mediante Energıas Renovables, 2 al 4 de junio de 1993. Madrid: CIEMAT; 1993.
[95] Heukelom P, Kursten JPJ, Van Der PS. In: Wind powered desalination with possible combinations of autonomous diesel operation; Economic feasibility and technical optimisation. Seminar on new technologies for the use of renewable energies in water desalination, Athens, 26–28 September 1991. Commission of the European Communities, DG XVII for Energy, CRES (Centre for Renewable Energy Sources); 1991. p. 37–48. Session II.
[96] Habali SM, Saleh IA. Design of stand-alone brackish water desalination wind energy system for Jordan. Solar Energy 1994;52(6):525–32.
[97] Miranda MS, Infield D. A wind-powered seawater reverse-osmosis system without batteries. Desalination 2003;153(1–3):9–16.
[98] Robinson R, Ho G, Mathew K. Development of a reliable low-cost reverse osmosis desalination unit for remote communities. Desalination 1992;86(1):9–26.
[99] Stahl M. In: Small wind powered RO seawater desalination plant design, erection and operation experience. Seminar on new technologies for the use of renewable energies in water desalination, Athens, 26–28 September 1991. Commission of the European Communities, DG XVII for Energy, CRES (Centre for Renewable Energy Sources); 1991. p. 27–36. Session II.

[100] Kostopoulos C., Proceedings of the Mediterranean conference on renewable energy sources for water production, 10–12 June. European Commission, EURORED Network, CRES, EDS, Santorini, Greece, 1996: 20–25.
[101] Vujcic R, Krneta M. Wind-driven seawater desalination plant for agricultural development on the islands of the County of Split and Dalmatia. Renew Energy 2000;19:173–83.
[102] ITC (Canary Islands Technological Institute). Memoria de gestion [in Spanish]; 2001.
[103] Lui CCK, Park JW, Migita R, Qin G. Experiments of a prototype wind-driven reverse osmosis desalination system with feedback control. Desalination 2002;150:277–87.
[104] Karameldin A, Lotfy A, Mekhemar S. The Red Sea area wind-driven mechanical vapour compression desalination system. Desalination 2003;153(1–3):47–53.
[105] Bier C, Coutelle R, Gaiser P, Kowalczyk D, Plantikow U. In: Sea-water desalination by wind-powered mechanical vapour compression plants. Seminar on new technologies for the use of renewable energies in water desalination, Athens, 26–28 September, 1991. Commission of the European Communities, DG XVII for Energy, CRES (Centre for Renewable Energy Sources); 1991. p. 49–64. Session II.
[106] Plantikow U. Wind-powered MVC seawater desalination-operational results. Desalination 1999;122:291–9.
[107] Manolakos D, Papadakis G, Papantonis D, Kyritsis S. A simulation-optimisation programme for designing hybrid energy systems for supplying electricity and fresh water through desalination to remote areas. Case study: Merssini village, Donousa Island, Aegean Sea, Greece. Energy 2001;26:679–704.
[108] Petersen G, Fries S, Mohn J, Muller A. Wind and solar-powered reverse osmosis desalination units—description of two demonstration projects. Desalination 1979;31(1–3):501–9.
[109] Weiner D, Fisher D, Katz B, Moses EJ, Meron G. Operation experience of a solar and wind-powered desalination demonstration plant. Desalination 2001;137:7–13.
[110] Al Malki A, Al Mari M, Al Jabri H. Experimental study of using renewable energy in the rural areas of Oman. Renew Energy 1998;14(1–4):319–24.
[111] Awerbuch L, Lindemuth TE, May SC, Rogers AN. Geothermal energy recovery process. Desalination 1976;19(1–3):325–36.
[112] Boegli WJ, Suemoto SH, Trompeter KM. Geothermal-desalting at the East Mesa test site. Desalination 1977;22(1–3):77–90.
[113] Ophir A. Desalination plant using low grade geothermal heat. Desalination 1982;40 (1–2):125–32.
[114] Karytsas C. In: Low enthalpy geothermal energy driven sea-water desalination plant on Milos Island—a case study. Mediterranean conference on renewable energy sources for water production, 10–12 June. Santorini, Greece: European Commission, EURORED Network, CRES, EDS; 1996. p. 128–31.
[115] Karytsas C. Geothermal energy driven desalination plants. Geotherm Resour Counc Bull 1998;27(4):111–5.
[116] Bourouni K, Chaibi MT, Tadrist T. Water desalination by humidification and dehumidification of air: state of the art. Desalination 2001;137:167–76.
[117] Bourouni K, Martin R, Tadrist L. Analysis of heat transfer and evaporation in geothermal desalination units. Desalination 1999;122:301–13.
[118] Bourouni K, Deronzier JC, Tadrist L. Experimentation and modelling of an innovative geothermal desalination unit. Desalination 1999;125:147–53.
[119] Bouchekima B. Renewable energy for desalination: a solar desalination plant for domestic water needs in arid areas of South Algeria. Desalination 2003;153(1–3):65–9.
[120] Houcine I, Benjemaa F, Chahbani MH, Maalej M. Renewable energy sources for water desalting in Tunisia. Desalination 1999;125(1–3):123–32.

[121] Kalogirou SA. Survey of solar desalination systems and system selection. Energy Int J 1997;22(1):69–81.

[122] Tzen E, Morris R. Renewable energy sources for desalination. Solar Energy 2003;75(5):375–9.

[123] Fiorenza G, Sharma VK, Braccio G. Techno-economic evaluation of a solar powered water desalination plant. Energy Convers Manag 2003;44(4):2217–40.

[124] Zarza E, Ajona JI, Leon J, Genthner K, Gregorzewski A, Alefeld G, et al. In: Solar thermal desalination project at the Plataforma Solar De Almeria. Proceedings of the new technologies for the use of renewable energy sources in water desalination, Athens, Greece; 1991. p. 62–81. Sec III.

[125] Hanson A, Zachritz W, Stevens K, Mimbella L, Polka R, Cisneros L. Distillate water quality of a single-basin solar still: laboratory and field studies. Solar Energy 2004;76(5):635–45.

[126] Porteous A. Saline water distillation processes. Essex: Longman; 1975.

CHAPTER 2

Solar Desalination Potential Around the World

Adrian Pugsley, Aggelos Zacharopoulos, Jayanta D. Mondol, Mervyn Smyth
Ulster University, Newtownabbey, Northern Ireland, United Kingdom

Symbols

F	freshwater stress ratio (quantity abstracted divided by quantity available)
H	global annual horizontal solar insolation (kWh/m^2/year)
N	national freshwater scarcity (m^3/capita/year)
R	rank score quantifying solar desalination applicability ($0 \leq R \leq 1$)
r	rank factors (r_N, r_F, r_S, r_H) used for calculating rank scores (R)
r_F	rank factor quantifying freshwater stress ($0 \leq r_F \leq 1$)
r_H	rank factor quantifying solar insolation levels ($0 \leq r_H \leq 1$)
r_N	rank factor quantifying national water scarcity ($0 \leq r_N \leq 1$)
r_S	rank factor quantifying saline water resource availability ($0 \leq r_S \leq 1$)
S	saline water resource type (S1, S2, S3, S4, and S5 defined in notes to Table 2.7)

Abbreviations

ED	electrodialysis, a type of desalination process
EJ	exajoules, unit of energy (10^{18}J)
HD	humidification-dehumidification, a type of desalination process
MD	membrane distillation, a type of desalination process
MEB	multieffect boiling, a type of desalination process
MENA	Middle East and North Africa region
MSF	multistage flashing, a type of desalination process
MVC	mechanical vapor compression, a type of desalination process
PV	photovoltaic
RO	reverse osmosis, a type of desalination process

2.1 INTRODUCTION

Solar desalination has obvious applicability in arid regions where freshwater scarcity, solar resource abundance, and saline water availability are observed to coincide. This observation seems most commonly referred to in the Middle Eastern and North African (MENA) context [1–13], but also in

Renewable Energy Powered Desalination Handbook
https://doi.org/10.1016/B978-0-12-815244-7.00002-7

relation to other parts of the world including China [14,15], rural regions of India [16,17], and southern Europe [18]. This chapter considers the potential for widespread deployment of solar desalination technologies from a global perspective.

The first part of this chapter reviews the key issues concerning the sustainability and interdependency of our water and energy resources. International data on water scarcity and stress is then presented and the growing need for desalination plants powered by renewable energy is established. A brief review of various types of solar desalination technology is presented and the global availability of saline water and solar energy resources is examined. The main factors affecting solar desalination plant feasibility are briefly reviewed, including technology maturity, economics, and environmental impact. Cartographic data on water scarcity and stress, insolation levels, and saline water resources are then used to identify and rank countries where deployment of solar desalination technologies might be targeted. These results are then reviewed in the context of population densities and local climatic conditions. The analysis could help inform national and international water management policies and assist equipment suppliers in identifying and quantifying emerging markets for solar desalination plant.

2.2 THE WATER-ENERGY NEXUS

Human beings need freshwater, food, and shelter to survive. The production of food stuffs and construction materials requires both freshwater and energy resources. The United Nations World Water Development Report 2014 [19] focused on these interdependencies (commonly referred to as the "Water-Energy Nexus") and highlighted that "*...water use and management and the production of energy can have significant, multifaceted and broad-reaching impacts on each other... drought exacerbates energy crises; energy price volatility contributes to food crises; the expansion of irrigation networks increases water and energy demand; and access to unreasonably inexpensive supplies of energy can lead to the depletion of water resources, further intensifying the impacts of droughts.*" A breakdown of global freshwater use is presented in Fig. 2.1. Normal consumption of drinking water is 2–4 L/day for adults and 0.75 L/day for infants [20]. Personal domestic water consumption for washing and cooking varies significantly in different countries, typically between 50 and 500 L/day [21]. By contrast, agricultural demands for freshwater (primarily for irrigation) are vast. Agricultural water demands are particularly high for

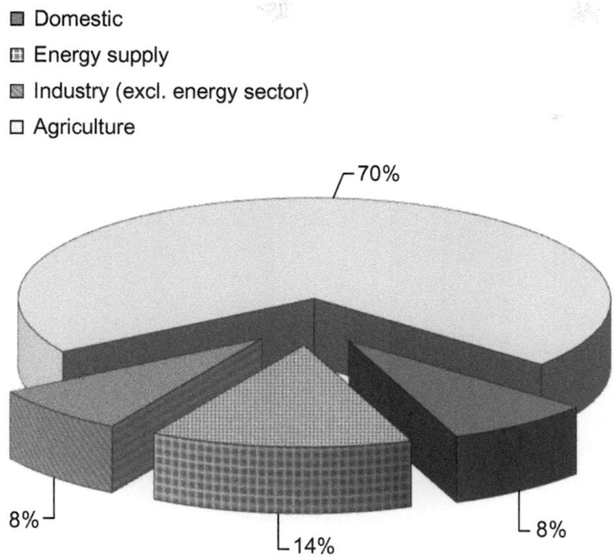

Fig. 2.1 Global use of freshwater. *(Based on data from Coates D, Connor R, Leclerc L, Rast W, Schumann K, Webber M. Water demand: what drives consumption? Managing water under uncertainty and risk—The United Nations World Water Development Report 4. Paris: UNESCO (United Nations Educational, Scientific and Cultural Organization; 2012). p. 44–69. ISBN 978-92-3-104235-5 [chapter 2]. Available at: http://www.unesco.org/new/en/natural-sciences/environment/water/wwap/wwdr/wwdr4-2012/.)*

arable farming in hot climates and for high value products such as grain fed cattle. Global energy supplies including fossil fuel production (extraction and refining processes), biofuels production (irrigation and processing), thermal electricity generation (steam and cooling water), and hydroelectric power (evaporative losses) are also major consumers of water [22]. Paradoxically, water supplies account for a significant share of global energy consumption (see Fig. 2.2). This energy is mainly required for pumping water from bores and through pipelines, for sewerage treatment and desalination.

Annual abstractions of freshwater from the world's lakes, rivers, and ground aquifers amount to ~4000 km^3/year [23]. Considering the differences in national water footprints of the developed and developing countries [24] freshwater demand could conceivably double or perhaps even quadruple by 2050 owing to the population growth and improving living standards. Current global energy consumption is ~350 EJ/year and a 20%–50% increase is expected by 2050 depending on how efficiently

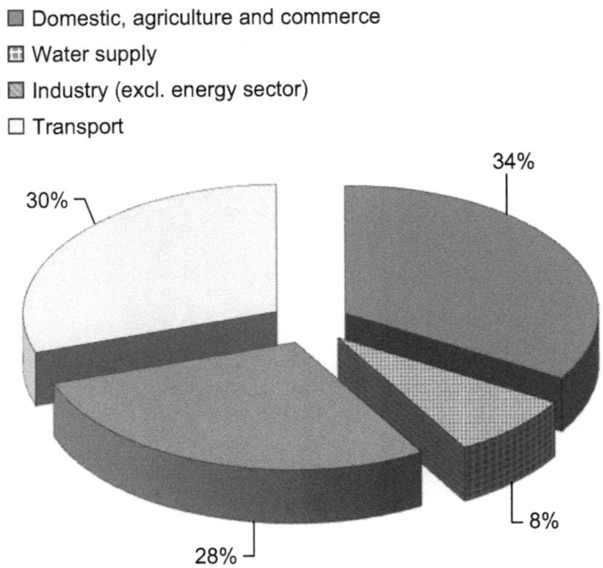

Fig. 2.2 Global use of energy. *(Based on data from Coates D, Connor R, Leclerc L, Rast W, Schumann K, Webber M. Water demand: what drives consumption? Managing water under uncertainty and risk—The United Nations World Water Development Report 4. Paris: UNESCO (United Nations Educational, Scientific and Cultural Organization; 2012). p. 44–69. ISBN 978-92-3-104235-5 [chapter 2]. Available at: http://www.unesco.org/new/en/natural-sciences/environment/water/wwap/wwdr/wwdr4-2012/ and International Energy Agency. Key world energy statistics 2011. Paris: IEA; 2011. Available at: http://www.iea.org/publications/freepublications/publication/name,20938,en.html.)*

we use energy in the future [21,25]. Fig. 2.2 shows that water extraction, treatment, and distribution accounts for 8% of global energy consumption. Average energy intensity of global water supplies can therefore be estimated as $\sim 7\,MJ/m^3$ (calculated by taking 8% of 350 EJ/year and dividing by $4000\,km^3$/year). Energy intensities of various common water supply scenarios are given in Table 2.1. Increasing urbanization, growing populations in water-scarce areas, and climate change will limit the possibilities of reliance on low-energy intensity water supply methods in future. Significant reduction in demand is likely to be achievable by employing more efficient agricultural irrigation techniques and reducing wastage caused by leaks, but increased reliance on wastewater reuse, long pipelines, and desalination, seems inevitable. Water supply system energy intensity will increase correspondingly.

Table 2.1 Energy intensity of selected water supply scenarios

Water supply system	Energy intensity (MJ/m^3)	Notes and references
Rooftop rainwater collection	0	Assumes untreated and gravity fed
Minimally treated nearby river or lake water	0.2	[26] Table 19.2
Groundwater drawn from a depth of 40 m	0.5	[26] Page 471
Activated sludge wastewater treatment	1.2	[26] Page 471
Electrodialysis (ED) desalination (brackish water)	6	[27]
Global average	7	Calculated in preceding text
Inter-basin water transfer	9	[28] Fig. 7.3
Reverse osmosis (RO) desalination (brackish water)	9	[29] Indicator 26
Multieffect boiling (MEB) seawater desalination	10*	[29] Indicator 26
RO desalination (seawater)	15	[29] Indicator 28
Multistage flash (MSF) seawater desalination	18*	[29] Indicator 26
Water transmitted by 1500 km long pipeline	20	[30]

Values marked * are consistent with "Loss of Electrical Power" values reported in Ref. [31] for plants driven by low-grade heat drawn from final stage steam turbines in fossil fueled or nuclear power plants. These values effectively ignore "waste heat" fed to the desalination plant that would otherwise be rejected from the power plant, and are thus somewhat unrepresentative of the real total energy consumption. Typical specific energy consumption reported in Ref. [32] (total input thermal plus electrical) is 223 and 304 MJ/m^3 for MEB and MSF plants respectively.

2.3 WATER SCARCITY AND STRESS

Definitions of the terms "water stress" and "water scarcity" vary in the literature and the terms are often used interchangeably.

If a region is experiencing "water stress" this usually means that freshwater abstractions are occurring at rates higher than natural recharge rates. Consequent reductions in lake and river water levels have obvious visual impacts and can have catastrophic consequences for supported ecosystems, both aquatic and terrestrial. There may also be adverse social impacts associated with fishing and water navigation. Depletion of groundwater aquifers is much less visible but can have equally dire consequences [33] including: reduced flow of natural springs with consequent effects on downstream rivers and lakes; increased groundwater salinity owing to ingress of seawater and subterrain brines into freshwater aquifers; and lowering of the water table which increases depths of wells and bores with consequent increases in energy required for pumping. Water resource depletion is usually temporary in the sense that recovery occurs when abstractions cease. Ecosystem destruction and increased groundwater salinity may however be of long term or of permanent consequences. Certain types of aquifers (such as the Connate aquifers found deep under the Sahara desert) are no longer actively recharged so their depletion would be essentially permanent [34].

The concept of "water scarcity" usually relates to per capita availability of freshwater resources. Scarcity can be caused by a genuine lack of water—"physical scarcity," a lack of water infrastructure—"economic scarcity," or a combination of both. Stress can be defined as a ratio of quantity abstracted divided by quantity of renewable water available. Degrees of water scarcity and stress are defined in Table 2.2.

Table 2.3 lists some of the factors which typically cause water stress and physical scarcity [35]. Fig. 2.3 shows a map quantifying global renewable water resources, color coded to indicate countries where there is physical water scarcity on an overall per capita basis. It should be noted that the national picture hides some important local variations, mainly associated with areas of high population density. A clearer picture of localized water availability problems is the water stress map shown in Fig. 2.4.

Table 2.2 Water scarcity and stress

***Scarcity* of water available for human consumption**		**Environmental water *stress* due to excessive abstraction rates**	
Freshwater availability (N, m^3/capita/year)	**Degree of water scarcity experienced**	**Water stress ratio (F, abstracted/available)**	**Degree of water stress occurring**
$N \geq 2500$	Sufficient supplies	$F \leq 0.3$	Negligible
$1700 \leq N < 2500$	Vulnerable	$0.3 < F \leq 0.5$	Low
$1000 \leq N < 1700$	Strained[a]	$0.5 < F \leq 0.7$	Slightly exploited
$500 \leq N < 1000$	Scarcity	$0.7 < F \leq 1.0$	Moderately exploited
$N < 500$	Absolute scarcity	$F > 1.0$	Heavily or over exploited

[a]UN-Water [35] actually uses the term "stress" here but acknowledges that this is confusing and can be easily muddled with environmental "water stress" associated with excessive abstraction.

Table 2.3 Factors causing water stress and physical scarcity

Demand-based drivers	Climatic and geographical drivers	Anthropological factors
High population density	Arid or semiarid terrain which experiences only minimal rainfall and has no river connection from remote wetter regions	Deforestation and erosion which cause excessive surface water turbidity, siltation and sedimentation of surface water bodies, and reduces aquifer recharge rates
Intensive agriculture or water-demanding industries	Reduced river flows due to glacier disappearance caused by global warming	Reduced river flows owing to riverbed sedimentation or excessive surface or groundwater abstractions upstream
Increased domestic and agricultural demands in hot climates	High prevailing ambient temperatures and winds which cause surface water and soil moisture to evaporate quickly	Aquifer over-exploitation causing saline water ingress which contaminates fresh groundwater, rendering it unusable
Inappropriate pricing or subsidies, for water supplies and/or for energy used for agricultural irrigation pumping	Frequent and extended droughts caused by natural or man-made climate phenomena	Water pollution (e.g., sewerage and industrial wastes) which contaminates surface and groundwater sources, sometimes irreversibly
Wasteful attitudes toward water use	Locations where the surface and/or groundwater has naturally high salinity content	Seasonal physical water scarcity caused by insufficient water storage capacity (reservoirs etc.)

From Balaji R, Bartram J, Coates D, Connor R, Harding J, Hellmuth M, et al. Beyond demand: water's social and environmental benefits. Managing water under uncertainty and risk—The United Nations World Water Development Report 4. Paris: UNESCO (United Nations Educational, Scientific and Cultural Organization); 2012. p. 101–23. ISBN 978-92-3-104235-5 [chapter 4]. Available at: http://www.unesco.org/new/en/natural-sciences/environment/water/wwap/wwdr/wwdr4-2012/.

There are a number of strategies for tackling water supply problems [36] depending on the type of problem and local context, which can be broadly categorized as illustrated in Table 2.4. The importance of demand reduction measures [37,38], loss and leak minimization, effective water

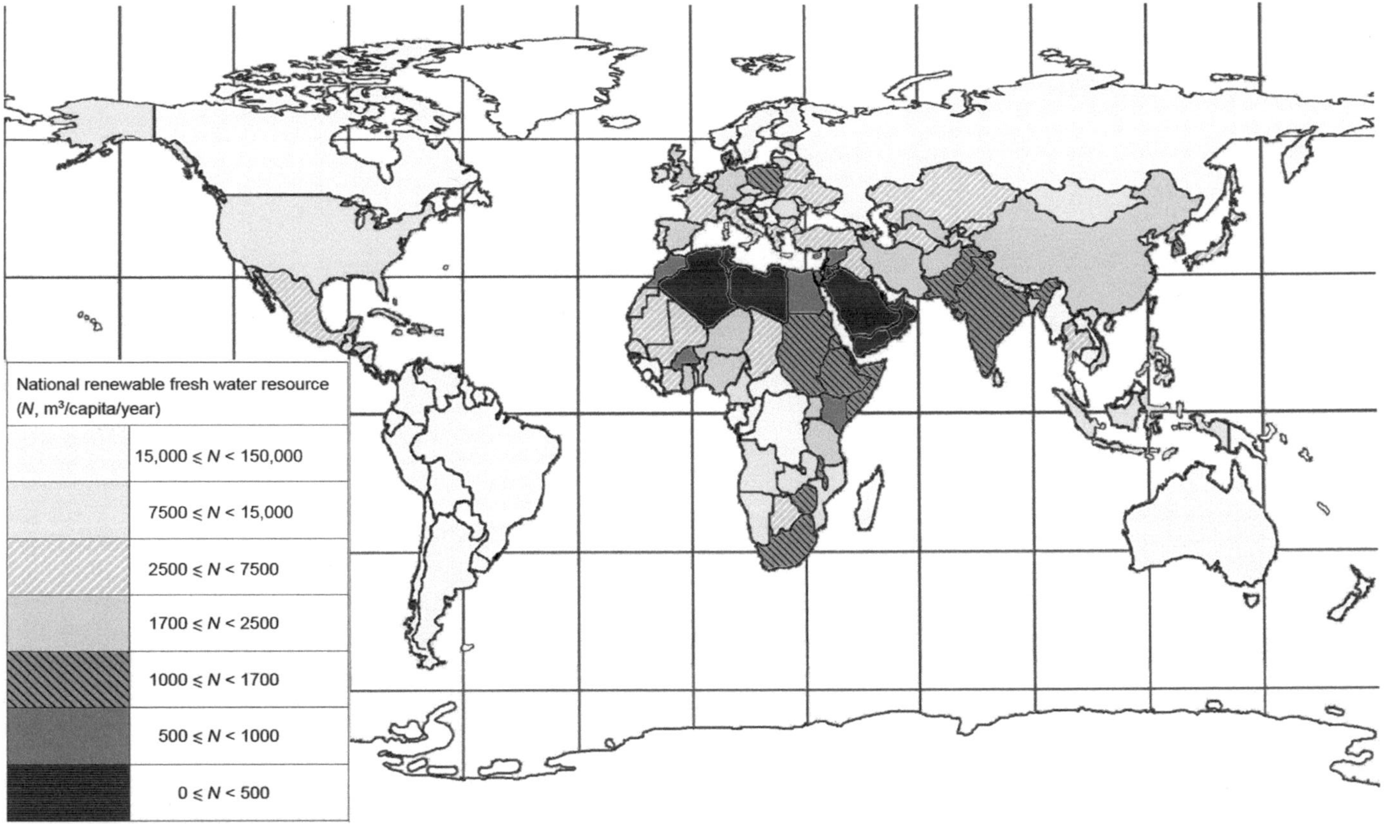

Fig. 2.3 Current annual national renewable water resources and water scarcity. *(Based on cartographic data from Coates D, Connor R, Leclerc L, Rast W, Schumann K, Webber M. Water demand: what drives consumption?, In: Managing water under uncertainty and risk—The United Nations World Water Development Report 4. Paris: UNESCO (United Nations Educational, Scientific and Cultural Organization); 2012. p. 44–69. ISBN: 978-92-3-104235-5. [chapter 2]. Available at: http://www.unesco.org/new/en/natural-sciences/environment/water/wwap/wwdr/wwdr4-2012/.)*

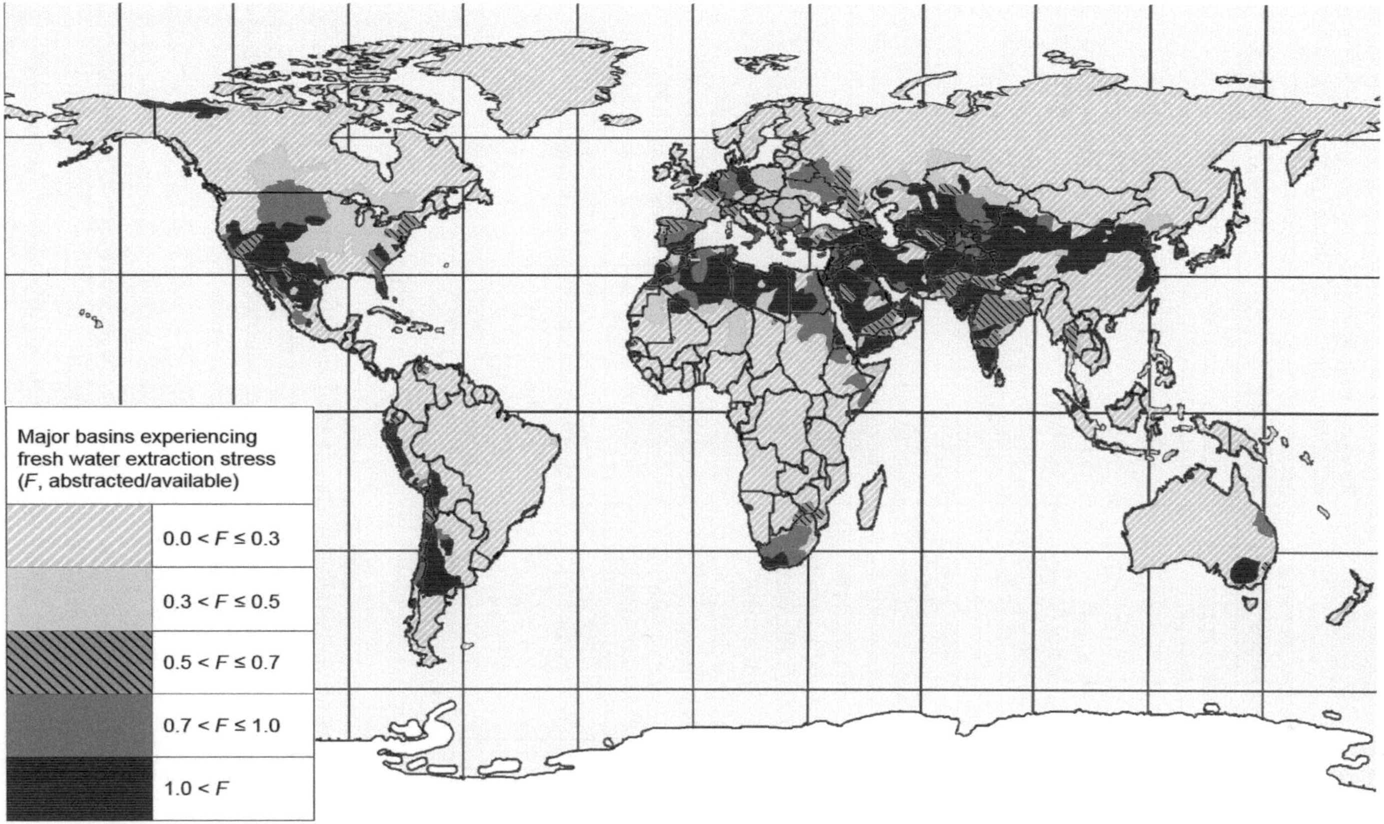

Fig. 2.4 Major basins experiencing water stress. *(Based on cartographic data from Balaji R, Bartram J, Coates D, Connor R, Harding J, Hellmuth M, et al. Beyond demand: water's social and environmental benefits. Managing water under uncertainty and risk—The United Nations World Water Development Report 4. Paris: UNESCO (United Nations Educational, Scientific and Cultural Organization); 2012. p. 101–23. ISBN 978-92-3-104235-5 [chapter 4]. Available at: http://www.unesco.org/new/en/natural-sciences/environment/water/wwap/wwdr/wwdr4-2012/.)*

Table 2.4 Options for tackling water scarcity and stress

	Reduce demand and losses [37,38]	Improve supply reliability and access to supply [35]	Increase the amount of water resource available
Applicability:	Widely applicable to partially alleviate most types of water stress and scarcity problems	Primarily solutions to economic water scarcity problems and situations where seasonal scarcity/stress occurs	Locations experiencing physical scarcity or over-exploitation stress. These solutions are typically energy and capital intensive
Examples:	Water conservation measures and leak reduction	Additional boreholes and wells	Desalination plants [31]
	Water efficient agricultural practices	Reservoirs and surface storage	Long distance pipelines and canals [30,39]
	Reducing evaporative losses from reservoirs	Ground aquifer recharge schemes [36]	Transportation of water by road, rail or sea based tankers
	Wastewater treatment with recycling	Improved water distribution and transmission networks	Fog harvesting [40]

management institutions and systems, as well as water consumer attitudes are emphasized in much of the cited literature and should usually be addressed before building new infrastructure. Many countries which have ample water supplies on a national level suffer severe water resource stresses in densely populated areas. In such cases piping water from remote, less populated areas may be a more economic and energy efficient solution than desalination. For arid regions near the coast, seawater desalination often represents the only reliable source of freshwater. Likewise, desalination is sometimes the sole water supply option for inland arid areas where all nearby aquifers are saline.

2.4 SOLAR DESALINATION

2.4.1 Coupling of Renewable Energy Sources and Desalination Processes

Desalination systems remove or reduce salts from saline water (either seawater or brackish groundwater) using one of the following processes:

- Phase change processes. The most common methods are distillation by multieffect boiling (MEB) and multistage flashing (MSF) which are driven primarily by thermal energy. Membrane distillation (MD) is also thermally driven. Humidification-dehumidification (HD), mechanical vapor compression (MVC), and freezing are phase change processes driven primarily by mechanical energy.
- Pressure-driven membrane processes such as reverse osmosis (RO) and nano-filtration. These processes are primarily driven by mechanical energy, usually in the form of electrically powered pumps.
- Electric charge-driven processes such as electrodialysis (ED) or ion concentration polarization. These processes are driven directly by electricity. With the exception of Ref. [27] there appear to be very few recent examples of ED systems being driven directly by renewable energy sources.

Forward osmosis is a membrane process driven by concentration gradients which can be used to remove freshwater from saline water and transfer it to a draw solution without using additional heat, pressure, or charge. One of the processes listed above must however be used subsequently to remove the freshwater from the draw solution.

Fig. 2.5 presents options for coupling renewable energy and desalination processes. Some of the key research to date has been collated in Refs. [11,41–48]. Wind turbines or solar photovoltaics (PV) driving RO plants are arguably the most mature technology combinations. Two utility scale RO plants near Perth, Western Australia are driven by electricity sourced from nearby wind and PV farms [49]. Practical examples of smaller integrated PV-RO systems are described in Refs. [50–52]. Several authors [53–55] have examined the possibility of driving RO systems using heat engines energized by concentrating solar thermal collectors, although this concept appears much less mature than PV-RO. Several demonstration scale MEB, MSF, and HD seawater desalination plants driven by solar thermal energy have been trialled (see Table 2.5) but do not appear to have been implemented at utility scale.

Solar stills, which operate on phase change principles, have been examined in numerous studies [17,60–62]. Detailed investigations on multistage

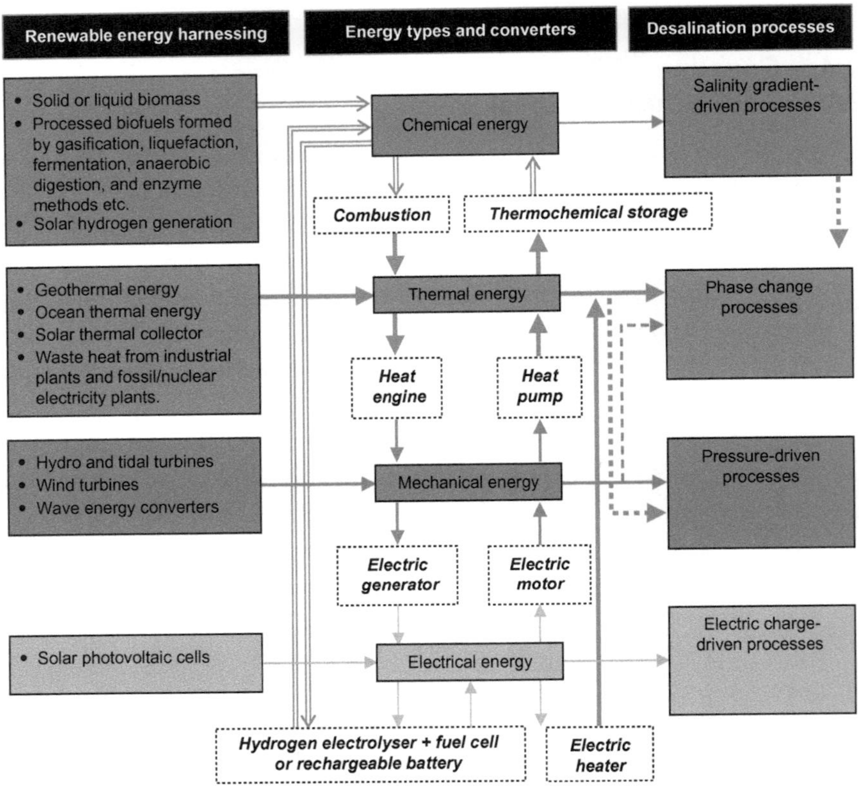

Fig. 2.5 Coupling of renewable energy sources and desalination processes.

Table 2.5 Examples of solar thermal phase change desalination demonstration plants

Location	Production capacity (m^3/day)	Type	References
Gaza, Palestine	0.2	Multieffect boiling (MEB)	[2]
Tunisia	0.5	Humidification-dehumidification (HD)	[7]
Northern China	0.8	MEB	[15]
Dezhou, Shandong, China	1	HD	[14]
El Paso, Texas	19	Multistage flash (MSF)	[11]
Abu Dhabi, UAE	60	MEB	[41,56]
Margarita de Savoya, Italy	60	MSF	[11]
Kuwait	100	MSF	[11]
PSA, Almeria, Spain	20	MEB	[18,57–59]

stacked stills fed by external solar collectors are presented in Refs. [63,64]. Several authors have experimentally investigated solar thermal MD devices [9,10,65–67].

2.4.2 Global Availability of Saline Water Resources

Seas and oceans represent the largest bodies of saline water on earth. Rain and snow melt cause salts in the soils and rocks to dissolve and then eventually be deposited into the seas via river flows. Constant salt deposition (mainly sodium chloride) and water evaporation from the sea surface over many millennia has caused salts to become concentrated. Seawater is essentially an inexhaustible source of feedwater for desalination plants, provided that environmental impacts of intakes are properly addressed. Approximately 60% of metropolitan centers around the world are located on seashores, or in their vicinity [68].

Saline water also occurs in groundwater aquifers and salt lakes, due to one or more different hydrogeological phenomena and are summarized in Table 2.6. In regions suffering from freshwater scarcity remote from coasts, saline groundwater may be the only naturally occurring source of desalination plant feedwater. Fig. 2.6 shows the global distribution of significant occurrences of near-surface saline groundwater aquifers and the locations of major salt lakes. Groundwater toxicity is dependent upon the types and concentrations of salts and other minerals present, which varies depending upon the water source. Toxicity can be caused by naturally occurring fluoride, iodide and silicate compounds, iron, boron, and arsenic, or by industrial and agricultural pollutants such as nitrates, phosphates, sulfates, heavy metals, and hydrocarbons [71]. Saline groundwater typically has lower salt concentration than seawater, which can reduce desalination plant energy consumption and operational costs. Conversely, water found in salt lakes often has much higher salinity than seawater and can thus be more costly to desalinate. Slow natural recharge rates and small volumes may be limiting factors for desalination fed by saline aquifers and salt lakes. Abstractions from ancient "Connate" aquifers and from aquifers formed by geologically historic marine transgressions are clearly not sustainable.

Another source of saline water which can (with suitable pretreatment) be used as feedwater for desalination plants is treated sewerage wastewater [68,72,73]. This is widely available in significant quantities in all densely populated areas. In addition to providing freshwater, desalination of sewerage wastewater can reduce aquatic pollution associated with effluent disposal, which in many cases is dumped into the sea.

Table 2.6 Taxonomy of saline aquifers based on descriptions given in Ref. [69]

Natural saline aquifers of marine origin	Man-made saline aquifers	Natural saline aquifers of terrestrial origin
Connate saline aquifers Formed by seawater being trapped in sedimentary rocks during geological history. These aquifers are no longer actively recharged	**Salinity caused by agricultural irrigation** Irrigation can cause salts to become concentrated in soil due to: (a) water table level increases causing upward migration of saline groundwater and (b) evaporative concentration of salts contained in irrigation water. Saline soils tend to stunt plant growth and reduce yields. More intensive irrigation is then required to flush the salts from the soil. Waste water from the flushing process drains into rivers and percolates into aquifers	**Salt lakes and evaporation-concentration saline aquifers** Salt lakes are formed in inland basins where the outgoing flows (rivers and subsurface streams) are less than the incoming flows. The balance of water (out-going minus incoming) is lost through evaporation, causing a concentration of salts. Saline water produced by evaporation and concentration may percolate into nearby aquifers causing them to become saline
Marine transgression aquifers Caused by seawater flooding low-lying land. Recharge occurs irregularly, perhaps on decade, century, or longer timescales	**Salinity caused by pollutant discharge** Due to waste disposal and land drainage. Minerals contained in industrial wastewater, waste concentrates from desalination plants, animal slurries and human sewerage can percolate into groundwater causing aquifer salinity increases. High mineral content also occurs in water	**Dissolution saline aquifers** Subsurface water flowing through certain types of rock formations (e.g., halites and carbonates) will cause minerals in those formations to be dissolved and carried to nearby aquifers
Sea-spray saline aquifers Freshwater aquifers in coastal areas can		**Geothermal saline aquifers** Highly mineralized water that is produced

Table 2.6 Taxonomy of saline aquifers based on descriptions given in Ref. [69]—cont'd

Natural saline aquifers of marine origin	Man-made saline aquifers	Natural saline aquifers of terrestrial origin
become saline owing to salts deposited on to land due to spray from the sea	drained from fields where excessive fertilizer quantities have been applied, landfill sites and highways (especially when de-icing salt has been applied)	as a side product of igneous activity, either due to rock salts being dissolved by very hot geothermal water, or by geothermal systems which admit seawater
Natural seawater ingression aquifers Naturally occurring in coastal areas due to seawater percolating into coastal freshwater aquifers	**Anthropomorphic seawater ingression aquifers** Excessive freshwater abstractions cause changes in hydrological pressure gradients, allowing seawater to travel inland	**Filtration aquifers** Naturally occurring clays can act as salt filtering membranes. Dissolved salts flowing into the aquifer will not be carried away in the water flowing out, causing a net build-up of salts
Mixed source saline aquifers Where the saline water body is formed by a mixture of the various marine and/or terrestrial saline aquifer types listed above evaporation, dissolution, geothermal, and filtration		

Dark shading indicates unsustainable or unsuitable sources of feedwater for desalination processes.

2.4.3 Global Availability of Solar Energy

The applicability of solar desalination in water scarce and stressed areas depends not only on the availability of saline water, but also on the availability of solar energy. Fig. 2.7 shows a map of typical annual average global insolation levels in the horizontal plane, based on data from Ref. [74]. To achieve comparable yields, solar collectors for desalination systems deployed near polar latitudes (e.g., in the United Kingdom where $H = 900\,\text{kWh/m}^2/\text{year}$) need to be roughly three times those required for the sunniest tropical countries (e.g., in Yemen where $H = 2400\,\text{kWh/m}^2/\text{year}$).

2.4.4 Feasibility of Solar Desalination Technologies

The suitability of solar desalination in any particular location depends on the local freshwater supply and demand context (i.e., scarcity and/or stress) and

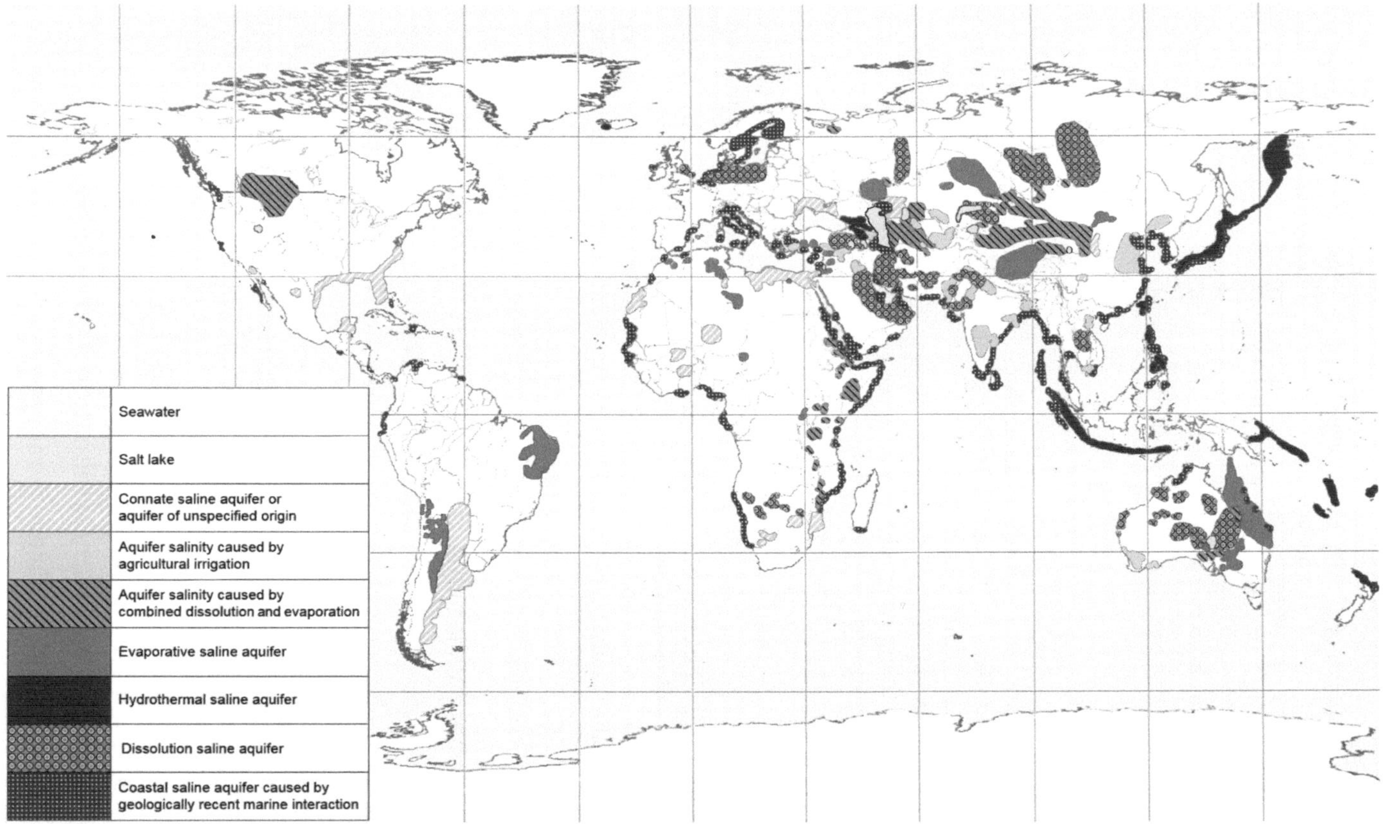

Fig. 2.6 Global distribution of major saline water resources. Based on cartographic data from Ref. [69] showing significant occurrences of saline groundwater aquifers at depths of less than 500 m, and the list of the world's major salt lakes given by Soil & Water Conservation Society of Metro Halifax [70].

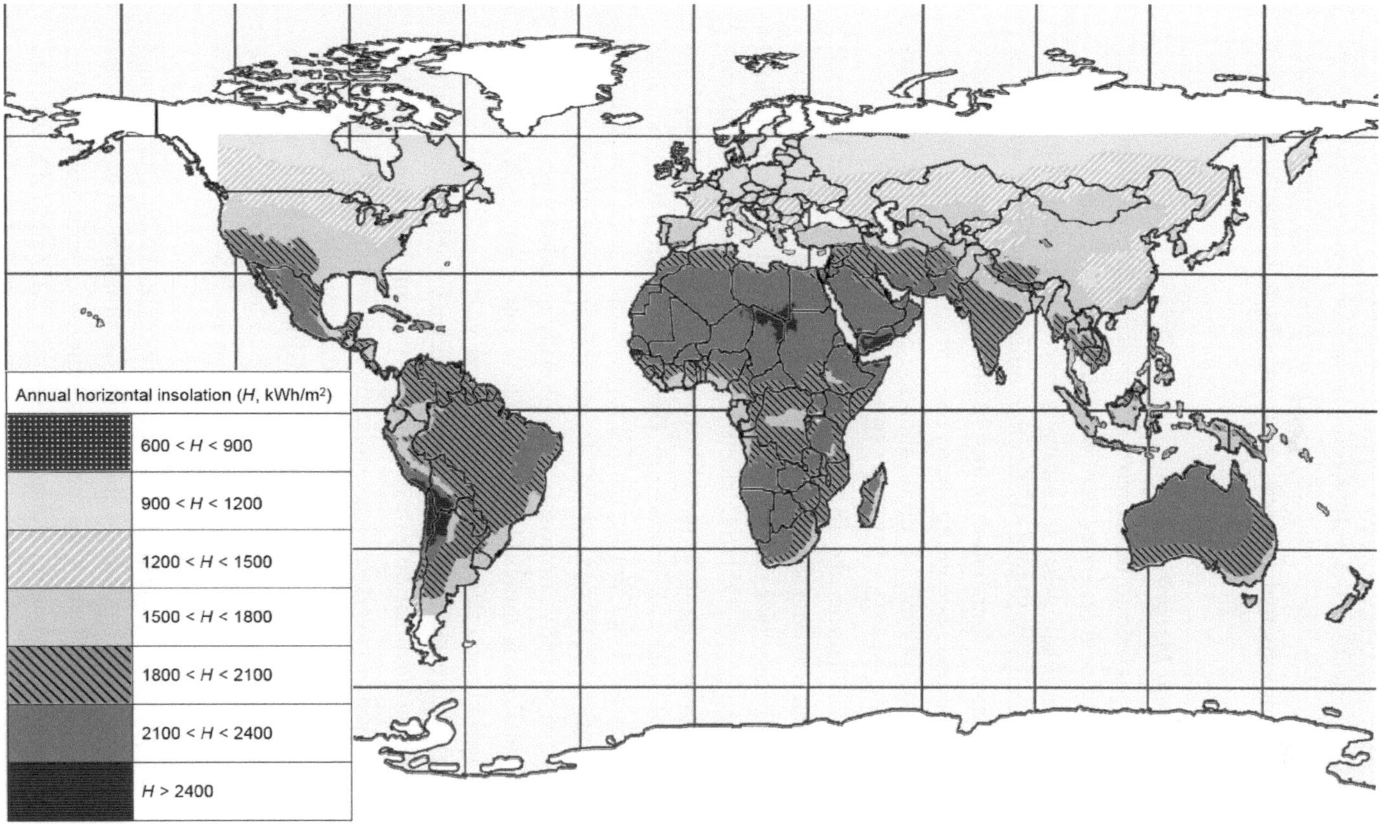

Fig. 2.7 Annual cumulative global horizontal plane solar insolation. No data for polar regions, which are shown in *white*. *(Based on cartographic data from SolarGIS. World map of global horizontal irradiation. Bratislava, Slovakia: GeoModel Solar; 2013. Available at: http://solargis.info/doc/_pics/freemaps/1000px/ghi/SolarGIS-Solar-map-World-map-en.png [Accessed 10 January 2013].)*

the local availability of both the solar energy and saline feedwater resources. From a global perspective, the potential for wide-scale deployment of solar desalination is also dependent on factors such as technology maturity, economics, and environmental impacts, each of which is briefly discussed below.

2.4.4.1 Technology Maturity

Mature technologies tend to be more readily available, more reliable, more efficient, less costly, and less risky than immature technologies.

Large-scale desalination has traditionally been undertaken using phase change methods such as MSF and MEB, often utilizing low-grade waste heat from thermal electricity generation plants. In the last two decades, RO technologies have matured, becoming more cost effective than thermal phase change methods. Most newly built desalination plants in the last few years have utilized RO rather than MSF or MEB [75].

Solar thermal collectors absorb solar energy to generate heat which is typically dissipated from the absorber into a working fluid such as air, water, or oil. PV solar panels utilize semiconductor materials which absorb photons of solar energy and convert them to electrical power. Doped crystalline silicon is by far the most commonly used PV material accounting for about 90% of current global solar cell production [76]. Data from the Renewable Energy Network for 21st Century [76,77] suggests that global solar thermal collector capacity in 2015 was ~445 GW and appears to be increasing at a rate of ~10% per year. By comparison, the globally installed capacity of PV collector installations is much lower at 227 GW but is currently growing much more quickly at a rate of about ~30% per year.

PVs are a mature technology producing electricity which can readily drive any form of desalination process, but are perhaps best suited to RO and ED. PV modules typically have collection efficiencies of about 15% for crystalline silicone and 8% for thin film [78] and typically cost US$1.5/W in Europe during 2013 [79], equivalent to approximately US$225/$m^2$ when solar radiation intensity is 1000 W/m^2. Wholesale costs of PV cells have dropped rapidly in recent years and are now as low as US$0.2/W [80] which equates to approximately US$30/$m^2$.

Various types of solar thermal collector were reviewed by Qtaishat and Banat [81] in terms of their compatibility with desalination processes. Solar ponds and flat plate collectors are mature technologies which appear well suited to emerging desalination technologies such as MD and HD, but tend to operate inefficiently at the temperatures required by MSF/MEB.

Evacuated tube collectors are a mature technology producing heat at temperatures suitable (80–120°C) for conventional MSF and MEB at 50% efficiency and a cost of about US\$100/m^2 [15] which equates to approximately US\$0.2/W when solar radiation intensity is 1000 W/m^2. Parabolic trough concentrators are a moderately mature technology capable of producing heat at temperatures suitable for MSF and MEB, or at higher temperatures suitable for producing mechanical power (via a heat engine) to drive RO systems [53–55,82] but are ineffective under cloudy conditions. Several authors have examined the possibility of utilizing low-grade heat from concentrating solar electricity power plants (based on Heliostat field, Fresnel lens, or Parabolic trough solar concentrator technologies) to drive MSF and MEB desalination systems [83–85].

2.4.4.2 Economics

One of the most thorough and commonly cited works on desalination costs is the study by Karagiannis and Soldatos [86] which considers the effects of plant type, economies of scale, feedwater salinity, and energy supply type, upon specific water costs. According to Ettouney and Wilf [31] large-scale RO, MEB, and MSF plants can all produce water with a similar specific water cost of ~\$0.5/m^3. Smaller systems can be up to four times more expensive and RO costs can be halved if fed by clean low-salinity brackish water.

Fig. 2.8 shows a breakdown of the main components affecting specific water costs for conventional desalination plants, primarily based on data for utility scale plant from Refs. [27,31,32]. Energy usage (typically from fossil fuels) accounts for ~65% of the specific cost of water produced by MEB, MSF, and MVC plant but accounts for much less specific water costs associated with RO and ED plants (~40% and ~25%, respectively). The operation and maintenance costs (other than energy usage) are low for MEB, MSF, and MVC plant (~10% of specific water cost) but are relatively high for ED and RO plant (~35% of specific water cost) because membranes require periodic replacement. Desalination equipment capital and finance costs makes up ~25% of the specific water cost for RO, MEB, MSF, and MVC and almost 40% of the specific water cost for ED. Thorough reviews of energy consumption and specific water costs are given in Refs. [48,75,87].

With respect to solar-driven desalination plants Trieb et al. [88] states that "*Unlike fossil fuel, the 'solar' fuel in the form of a collector field has to be paid upfront and becomes part of the initial debt, with the associated interest and insurance payment leading to a high capital cost.*" Increases in finance costs could be

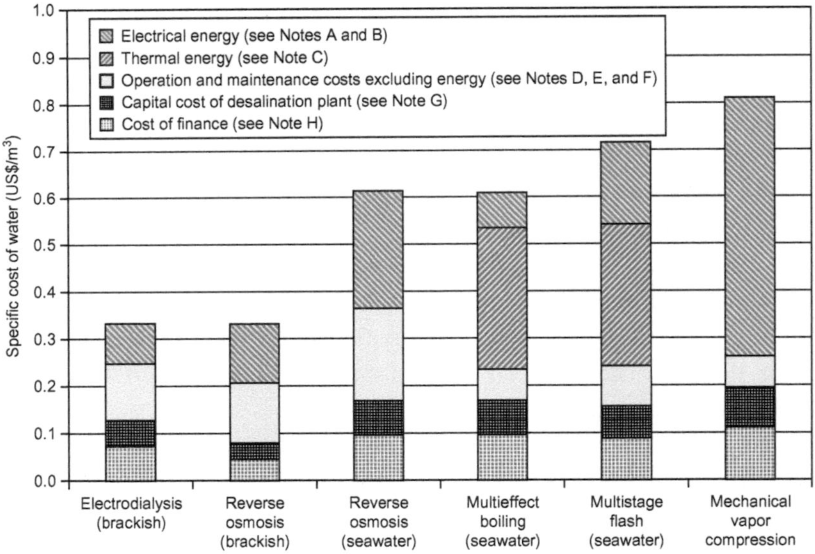

Calculation notes:

A. Electricity consumption values taken from Refs. [27, 29, 31]: 1.7 kWh/m^3 for brackish electrodialysis, 2.5 kWh/m^3 for brackish reverse osmosis, 5 kWh/m^3 for seawater reverse osmosis, 1.5 kWh/m^3 for multieffect boiling, 3.5 kWh/m^3 for multistage flash and 11 kWh/m^3 for mechanical vapor compression.

B. Electricity cost taken from Ref. [31]: US$0.05/kWh.

C. Assumes use of steam from thermal electricity generating plant. Costs based on "Specific cost of power loss" values given in Ref. [31] amounting to 0.3 US$/m^3.

D. Based on data given in Refs. [31,32] the specific cost of operational labor is taken as 0.025 US$/m^3 and the specific cost of treatment chemicals is taken as 0.04 US$/m^3 for all processes except multistage flash. The latter requires antifoaming agents which increases the specific cost of treatment chemicals to 0.06 US$/m^3.

E. According to Ref. [32] membranes for reverse osmosis account for 60% of the initial capital cost and according to Ref. [27] membranes for electrodialysis account for 33% of the initial capital cost. Average membrane life is estimated as about 10 years based on information given in Refs. [27,32,87], resulting in three replacements of the membranes over the 40 year plant life assumed in Ref. [31].

F. Specific cost of spare parts (excluding membranes) taken as 1% of the specific capital cost for all desalination process types based on the average of the values given in Refs. [31,32].

G. Specific capital costs for plant are taken from Refs. [27,31] as 722 US$/m^3/day for brackish electrodialysis, 450 US$/m^3/day for brackish reverse osmosis, 950 US$/m^3/day for seawater reverse osmosis, 951 US$/m^3/day for multieffect boiling, 878 US$/m^3/day for multistage flash and 1100 US$/m^3/day for mechanical vapor compression. Calculations assume a capacity factor of 0.9 as per the value assumed in Ref. [32].

H. Cost of finance taken as 133% of capital cost as per analysis presented in Ref. [31].

Fig. 2.8 Components of specific water cost for utility scale fossil fueled desalination plants.

disproportionate owing to perceived risks. Cleaning of solar collectors would be expected to increase maintenance costs. Several studies [45,89] highlight the economic importance of energy storage for solar desalination plants. Without this, desalination plant capacity factors are limited to 25%–50% due to the intermittent nature of solar energy, which represents poor use of the desalination equipment capital investment. Intermittent operation owing to a lack of energy storage can also result in operational inefficiencies (e.g., MEB and MSF plants) and premature equipment failure (e.g., ED and RO membranes). Connection to the mains electrical grid effectively acts as a form of PV energy storage, or alternatively electrical batteries can be used. Solar thermal energy storage typically takes the form of large insulated hot water tanks.

The capital cost of the solar collector field, and hence also the specific cost of water produced by a solar desalination plant, depends heavily on the required collector area. Solar desalination systems located in climates with low insolation levels will require larger collector areas than those located in climates with high insolation levels. Seasonal variations in irradiance levels and ambient temperatures can be very significant in low insolation climates and solar thermal systems would either cease operation during winter, or suffer from significant seasonal energy storage heat losses. Fig. 2.9 illustrates a broad indication of how specific water costs for different solar desalination methods might be expected to vary depending upon annual solar insolation levels. Calculated costs are again based on data for utility scale desalination plants provided in Refs. [27,31,32] which is supplemented by solar collector cost data from Refs. [15,79]. Land costs have not been accounted for.

Specific water costs in locations with high insolation levels are clearly lower than those in locations with low insolation levels. For example, specific water costs for MVC solar desalination in the United Kingdom ($H = 900$ kWh/year) are almost 2 times that of the same plant deployed in a southern European country such as Spain ($H = 1800$ kWh/year), and 2.2 times that of the same plant deployed in a tropical country such as Yemen ($H = 2400$ kWh/year). Fig. 2.9 also indicates that when deployed in low insolation climates, desalination processes driven by solar thermal energy (MEB and MSF) are significantly more expensive than those driven by PV energy (RO, ED, and MVC). Even when deployed in high insolation climates MEB and MSF plants driven by solar thermal energy are generally less cost effective than RO, ED, and MVC plants driven by PVs. These findings are primarily due to the high-energy intensity of MEB and MSF which results in large solar collector fields at correspondingly high capital cost.

The calculated values presented in Fig. 2.9 can be compared with the ranges of specific water costs for solar desalination plants reported by Al-Hallaj et al. [89] which are 1–5 $\$/m^3$ for MSF, 2–9 $\$/m^3$ for MEB, 3–27 $\$/m^3$ for RO, and 3–16 $\$/m^3$ for ED. The costs of ED and RO reported by Al-Hallaj et al. [89] relate to small-scale demonstration plants and are considerably higher than the costs shown in Fig. 2.9 which relate

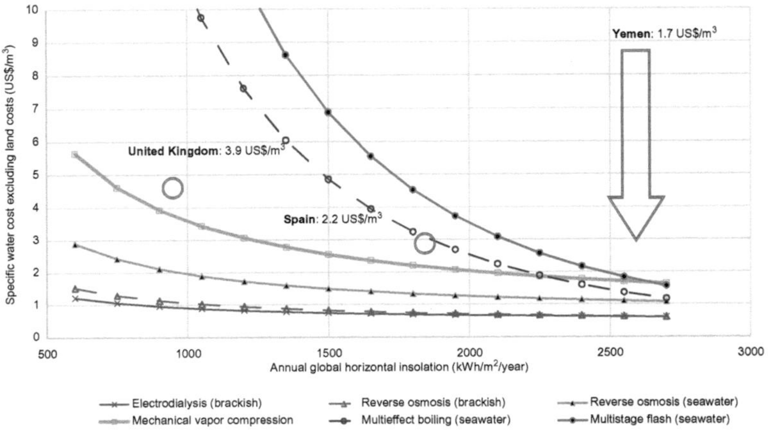

Calculation notes:

A. Electricity consumption values taken from Refs. [27, 29, 31]: 1.7 kWh/m³ for brackish electrodialysis, 2.5 kWh/m³ for brackish reverse osmosis, 5 kWh/m³ for seawater reverse osmosis, 1.5 kWh/m³ for multieffect boiling, 3.5 kWh/m for multistage flash and 11 kWh/m³ for mechanical vapor compression.

B. The total thermal energy consumptions values for multieffect boiling and multistage flash are taken from Ref. [32] as being 60 and 80 kWh/m³ respectively.

C. Solar collector capital costs taken as US$225/m² for photovoltaics with 10% efficiency [78,79] and US$100/m² for evacuated tube collectors with 40% efficiency [15]. The collector field is sized so that its average power output is equal to the power required to continuously operate the desalination plant. Solar collector costs are inflated by 20% to account for costs of energy storage and control devices, as suggested in Ref. [32].

D. Using the same assumption as in Ref. [32] the specific cost of spare parts for the solar collectors and the desalination plant (excluding membranes) is taken as 2% of the total specific capital cost.

E. Specific capital costs for desalination plant are taken from Refs. [27, 31] as 722 US$/m³/day for brackish electrodialysis, 450 US$/m³/day for brackish reverse osmosis, 950 US$/m³/day for seawater reverse osmosis, 951 US$/m³/day for multieffect boiling, 878 US$/m³/day for multistage flash and 1100 US$/m³/day for mechanical vapor compression.

F. According to Ref. [32] membranes for reverse osmosis account for 60% of the initial capital cost and according to Ref. [27] membranes for electrodialysis account for 33% of the initial capital cost. Average membrane life is estimated to be about 10 years based on information given in Refs. [27, 32, 87] resulting in two replacements of the membranes over an assumed 25 year plant lifetime.

G. Based on data given in Refs. [31, 32] the specific cost of operational labour for the desalination equipment is taken as 0.025 US$/m³ and the specific cost of desalination treatment chemicals is taken as 0.04 US$/m³ for all processes except multistage flash. The latter requires anti-foaming agents which increases the specific cost of treatment chemicals to 0.06 US$/m³. Using the same assumption as in Ref. [32] the specific cost of labor for maintaining the solar collector array is assumed to be 0.095US$/m³.

H. Photovoltaic driven desalination systems are assumed to have a capacity factor of 0.9. This value is equivalent to the utilisation ratio assumed in Ref. [32]. Heat losses from seasonal energy storage systems will reduce the annual output from solar thermal desalination systems, particularly when sited in low insolation climates which are typically characterised by low ambient temperatures and reduced irradiance levels during winter. With reference to the solar fraction and capacity factor values reported in Refs. [56,59], capacity factors (CP) for solar thermal desalination systems have been coarsely estimated using the relationship $CP=0.09\exp(0.0009H)$, where H is the annual global horizontal insolation level (kWh/m²/year).

I. Cost of finance taken as 133% of capital cost as per analysis presented in Ref. [31].

Fig. 2.9 Dependence of specific water cost upon insolation for utility scale solar desalination plants.

to utility scale plants. The cost difference stems from recent reductions in PV prices [78,90] and desalination equipment cost reductions associated with utility scale plants [32,47].

2.4.4.3 Environmental Impacts

Environmental impacts associated with desalination plants relate mainly to the following considerations:

- Disturbance and damage to aquatic and marine environments and ecosystems owing to feedwater intakes.

- Thermal and chemical pollution of aquatic and marine environments owing to waste concentrate and cooling water discharges.
- Hydrological disturbances such as saline water intrusion into freshwater aquifers owing to concentrate disposal or excessive feedwater extraction.
- Fossil fuel depletion, greenhouse gas emissions, and air pollution associated with energy consumption and embodied energy.

The first three of these considerations are examined in Refs. [91–95]. Important planning and design aspects such as locations, velocities, salt concentrations, and temperatures of feedwater intakes and discharge outlets are discussed by Lattemann et al. [96].

Life-cycle assessments for RO desalination plants were undertaken by Jijakli et al. [13] and Shahabi et al. [49] and the relative benefits of driving the plants with renewable energy rather than fossil fuels were examined. A key finding of Shahabi et al. [49] was that an RO plant in Western Australia driven by wind and PV would achieve ~90% lower greenhouse gas emissions than a similar plant powered by mains electricity (mainly from coal-fired generators).

2.5 IDENTIFYING LOCATIONS APPROPRIATE FOR SOLAR DESALINATION

2.5.1 Analysis Method

Locations in greatest need of solar desalination plants are those which suffer from insufficient or unsustainable supplies of freshwater, as evidenced by national water scarcity and/or local water extraction stresses. However, the technology is only applicable in locations with good access to saline water sources, and the economic feasibility of plants is heavily influenced by the magnitude of the available solar energy resource. Data in the form of cartographic images which quantify national freshwater scarcity (*N*), freshwater extraction stress (*F*), saline water availability (*S*), and solar insolation levels (*H*) have been obtained and imported into Google Earth software in the form of four separate map overlays shown in Figs. 2.3, 2.4, 2.6, and 2.7, respectively.

The spatial accuracy of the map overlays (e.g., positions of national borders) is typically limited to ±100 km due to image distortions and relatively low resolution of the map images. The analysis presented here is therefore generally limited to nations with contiguous land areas greater than 25,000 km^2 and thus excludes some smaller nations (e.g., Lebanon, Qatar, Slovenia, and Fiji). The map resolution does however allow the biggest of the small island states (e.g., Jamaica and Cyprus) to be analyzed. Eight large countries (United States, Canada, China, Russia, India, Indonesia, Brazil, and Australia) have been split into smaller subnational areas (with boundaries typically congruent with provincial administrative boundaries) in order to better reflect important spatial distributions of water stresses, saline water resources, and solar insolation variations. Some provincial islands are also treated as subnational areas (e.g., Corsica and Sardinia).

A rank scoring system describing the global applicability of solar desalination technologies has been devised by summarizing and correlating these four datasets on a nation by nation basis. The rank scoring system involves determination of an overall rank score (R) for each location which is calculated based on four rank factors (r) which, respectively, quantify national water scarcity (r_N), local water stress (r_F), saline water availability (r_S), and annual solar insolation (r_H) using values between zero and unity.

Rank factors for each location are derived from the cartographic data according to Eqs. (2.1)–(2.3) and the rules set out in Table 2.7 such that

- $r=1$ represents the conditions which give solar desalination ultimately "high applicability." With reference to Figs. 2.3, 2.4, 2.6, and 2.7 and Table 2.3, this applies to locations suffering from "absolute scarcity" of freshwater on a national level, having one or more "heavily overexploited" water stressed basins, having access to both seawater and actively recharged saline groundwater aquifers, and having the highest levels of solar insolation in the world.
- $r \approx 0.75$ represents conditions which are clearly conducive to solar desalination. This applies when the location is "vulnerable" to national water scarcity, has one or more "moderately exploited" water stressed basins, has access to seawater, and has good levels of solar insolation.
- $r \approx 0.5$ represent marginal conditions. This applies when there is no national water scarcity, freshwater resources are "slightly exploited," there is access to saline groundwater but not seawater, and solar insolation levels are moderate.
- $r \approx 0.25$ represents conditions where solar desalination would generally be considered as "not applicable" due to there being no significant evidence of water scarcity or stress, very limited access to sustainable saline feedwater resources, and relatively low solar insolation levels.
- $r=0$ applies in cases where solar desalination would be completely unnecessary owing to an abundance of freshwater, or impossible due to a (hypothetically) complete lack of access to saline feedwater or sunlight.

Table 2.7 Rank scoring system

	Water scarcity	Water stress	Saline water	Solar energy
	National renewable freshwater resource (N, m^3/capita/year)	Freshwater stress ratio (F, extracted/available)	Saline water resources located within borders of country or subnational unit	Annual average global horizontal insolation (H, kWh/m^2/year)
Rank factor	$r_N = \frac{7500}{N+7500}$ (2.1)	$r_F = \frac{F+0.03}{1.33}$ (2.2)	r_S Evaluated based on types of resources present[a]	$r_H = \frac{H}{2700}$ (2.3)
$r=0$	∞	0	If none, then $r_S=0$	0
$r=0.25$	22,500	0.3	If S1/S2/S3 are available but no S4/S5 then $r_S=0.25$[b]	670 (typical global minimum)
$r=0.5$	7500	0.6 "Slightly exploited"	If S4 is available but no S5 then $r_S=0.5$	1350 (e.g., Bordeaux, France)
$r=0.75$	2500 "Vulnerable"	0.9 "Moderately exploited"	If S5 is available but no S2/S4 then $r_S=0.75$	2020 (e.g., Perth, W. Australia)
$r=1$	0 "Absolute scarcity"	1.3 "Heavily over-exploited"	If S5 plus S2/S4 are available then $r_S=1$[c]	2700 (typical global maximum)

[a]Saline water resources are coded: S1 = Connate aquifers, S2 = Salt lakes, S3 = Aquifers where salinity is caused by agricultural irrigation or pollution, S4 = Aquifers where salinity is caused by dissolution, evaporation or hydrothermal processes, S5 = Seawater and coastal saline aquifers caused by recent marine interactions.
[b]Extraction from Connate aquifers, Salt lakes, or aquifers with salinity caused by pollution or agricultural irrigation, are likely to be unsustainable. The rank factors reflect the fact that these resources are broadly unsuitable sources of saline feedwater for desalination plants.
[c]Natural saline lakes and aquifers could potentially be used to dispose of concentrated saline waste water from desalination plants as a method of reducing negative environmental impacts. The rank factors reflect the fact that such resources are therefore advantageous.

$$R = \left(\frac{r_N + r_F}{2}\right) r_S r_H \tag{2.4}$$

Calculation of r_N and r_F is based on mid-range N and F values, respectively, where F is determined by considering the worst affected basin. For example, Fig. 2.3 shows that Germany is "vulnerable" to national water scarcity ($1700 \leq N < 2500$ so mid-range $N = 2100\,\text{m}^3$/capita/year) and Fig. 2.4 shows that the country has at least one "heavily overexploited" stressed water basin ($F > 1$ so mid-range $F = 1.15$). According to Eqs. (2.1)–(2.3) and the rules in Table 2.7 the corresponding rank factor values are therefore $r_N = 0.78$ and $r_F = 0.89$.

r_H values are calculated based on the mid-range H value for the prevailing solar insolation level. For example, insolation of $600 < H \leq 900\,\text{kWh/m}^2$/year occurs in northern United Kingdom whereas $900 < H \leq 1200\,\text{kWh/m}^2$/year occurs in the south so the countrywide value is taken as being $H = 900\,\text{kWh/m}^2$/year, corresponding to $r_H = 0.33$.

Rank factors are combined according to Eq. (2.4), which ensures that the rank score takes on a zero value ($R = 0$) if there is no evidence of water scarcity or stress ($r_N + r_F = 0$), if no saline feedwater is available ($r_S = 0$), or if no solar energy is available ($r_H = 0$). Likewise, Eq. (2.4) causes the rank score to tend toward unity ($R \rightarrow 1$) in cases where there is strong evidence of both water scarcity and stress ($r_N = r_F = 1$), saline feedwater is available from several sources ($r_S = 1$), and solar insolation levels are high ($r_H = 1$).

Substituting the values of $r = 0.75$ and $r = 0.5$ into Eq. (2.4) yields corresponding rank score values of $R = 0.422$ and $R = 0.125$. These values, respectively, serve as obvious thresholds for categorizing whether solar desalination is "highly applicable" ($R > 0.422$) or is "not applicable" ($R \leq 0.125$). Ranks score between these values imply that solar desalination has either "limited potential" ($0.125 < R \leq 0.273$) or "significant potential" ($0.273 < R \leq 0.422$) where the boundary threshold is taken as $R = 0.273$ based on the arithmetic average of the aforementioned thresholds.

2.5.2 Results

The rank score results are presented cartographically in Fig. 2.10.

Solar desalination can generally be considered as "not applicable" in cases where there is no access to a significant sized saline water resource ($r_S = 0$) or where there is minimal freshwater scarcity and abstraction stress ($r_N + r_F < 1$) owing to abundant freshwater resources and/or sparse human populations. Results falling into this category (corresponding to $R \leq 0.125$) are presented in Table 2.8A–C.

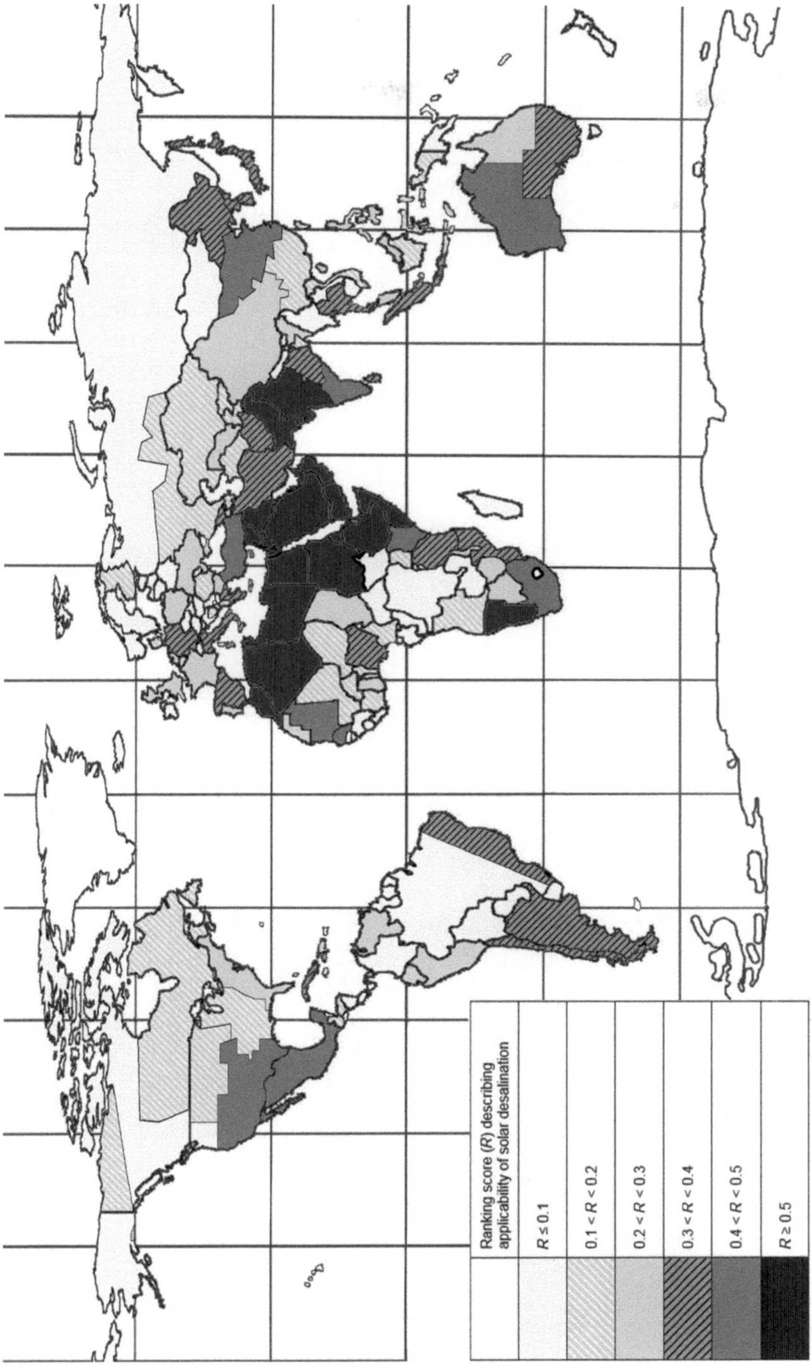

Fig. 2.10 Global applicability of solar desalination based on a rank scoring approach.

Table 2.8 (A)–(C) Solar desalination "not applicable" ($R \leq 0.125$)

Country/region	r_N	r_F	r_S	r_H	R
(A) No significant saline water resources ($r_S = 0$)					
Austria	0.4	0.32	0	0.5	0
Belarus	0.6	0.66	0	0.39	0
Bhutan	0.08	0.89	0	0.72	0
Bolivia	0.08	0.89	0	0.72	0
Burundi	0.08	0.14	0	0.67	0
Central African Rep.	0.08	0.14	0	0.78	0
Czech Republic	0.4	0.89	0	0.39	0
Guinea	0.08	0.14	0	0.78	0
Hungary	0.08	0.32	0	0.5	0
Laos	0.08	0.14	0	0.61	0
Lesotho	0.85	0.32	0	0.72	0
Nepal	0.6	0.14	0	0.83	0
Rwanda	0.08	0.14	0	0.67	0
Serbia	0.08	0.32	0	0.5	0
Slovakia	0.4	0.32	0	0.39	0
South Sudan	0.85	0.14	0	0.78	0
Switzerland	0.6	0.66	0	0.5	0
Zambia	0.4	0.14	0	0.78	0
(B) Minimal water scarcity/stress ($r_N + r_F < 1$) and cloudy ($r_H \leq 0.5$)					
Canada (N, E, and SW)	0.08	0.14	1	0.41	0.04
Canada (Northwest)	0.08	0.89	0.75	0.33	0.12
Estonia	0.4	0.14	0.75	0.39	0.08
Finland	0.08	0.89	0.75	0.28	0.1
Iceland	0.08	0.14	1	0.28	0.03
Ireland	0.4	0.14	0.75	0.28	0.06
Latvia	0.08	0.14	0.75	0.39	0.03
Lithuania	0.4	0.14	0.75	0.39	0.08
Norway	0.08	0.14	0.75	0.28	0.02
Russian Fed. (majority)	0.08	0.14	1	0.39	0.04
Sweden	0.08	0.32	0.75	0.28	0.04
United States (NW and Alaska)	0.4	0.14	0.75	0.41	0.08
Australia (Tasmania)	0.08	0.14	0.75	0.5	0.04
(C) Minimal water scarcity/stress ($r_N + r_F < 1$) and sunny ($r_H > 0.5$)					
Bangladesh	0.4	0.14	0.75	0.61	0.12
Brazil (Inland and North)	0.08	0.14	1	0.72	0.08
Burkina Faso	0.91	0.14	0.25	0.83	0.11
Cambodia	0.08	0.14	0.75	0.72	0.06
Colombia	0.08	0.14	0.75	0.67	0.05
Congo	0.08	0.14	0.75	0.67	0.05
Costa Rica	0.08	0.14	0.75	0.72	0.06
Croatia	0.08	0.32	0.75	0.5	0.08

Table 2.8 (A)–(C) Solar desalination "not applicable" ($R \leq 0.125$)—cont'd

Country/region	r_N	r_F	r_S	r_H	R
Dem. Rep. Congo	0.08	0.14	1	0.67	0.07
Equatorial Guinea	0.08	0.14	0.75	0.61	0.05
French Guiana	0.08	0.14	0.75	0.72	0.06
Gabon	0.08	0.14	0.75	0.61	0.05
Gambia	0.08	0.14	0.75	0.78	0.06
Guatemala	0.4	0.14	0.75	0.61	0.12
Guinea-Bissau	0.08	0.14	0.75	0.78	0.06
Guyana	0.08	0.14	0.75	0.72	0.06
Indonesia (Borneo)	0.4	0.14	0.75	0.61	0.12
Indonesia (Papua and E.)	0.4	0.14	0.75	0.61	0.12
Liberia	0.08	0.14	0.75	0.61	0.05
Madagascar	0.08	0.14	0.75	0.72	0.06
Mongolia	0.4	0.14	0.5	0.56	0.07
Myanmar	0.08	0.14	0.75	0.61	0.05
New Caledonia	0.08	0.14	1	0.56	0.06
New Zealand	0.08	0.14	1	0.5	0.05
Nicaragua	0.08	0.14	0.75	0.61	0.05
Niger	0.78	0.32	0.25	0.83	0.12
Panama	0.08	0.14	0.75	0.72	0.06
Papua New Guinea	0.08	0.14	0.75	0.67	0.05
Paraguay	0.08	0.14	0.25	0.72	0.02
Sierra Leone	0.08	0.14	0.75	0.72	0.06
Solomon Islands	0.08	0.14	1	0.56	0.06
Suriname	0.08	0.14	0.75	0.72	0.06
Uruguay	0.08	0.14	0.75	0.61	0.05
Vietnam	0.4	0.14	0.75	0.61	0.12

With the exceptions of Croatia and Ireland, all countries and subnational units in this category fall within one or more of the following groupings according to global climate zone classifications [97], local population densities (see Fig. 2.11), and coastal access:

- Polar and subpolar regions with sparse human populations (with typically less than 20 people per square kilometer), including large parts of Canada, United States, Russia, Scandinavia, Baltic states, New Zealand, and Tasmania.
- Temperate tropical and equatorial regions, including large parts of Indonesia, Central America, northern South America, inland Brazil, Southeast Asia, western sub-Saharan Africa, Madagascar, and Papua New Guinea.

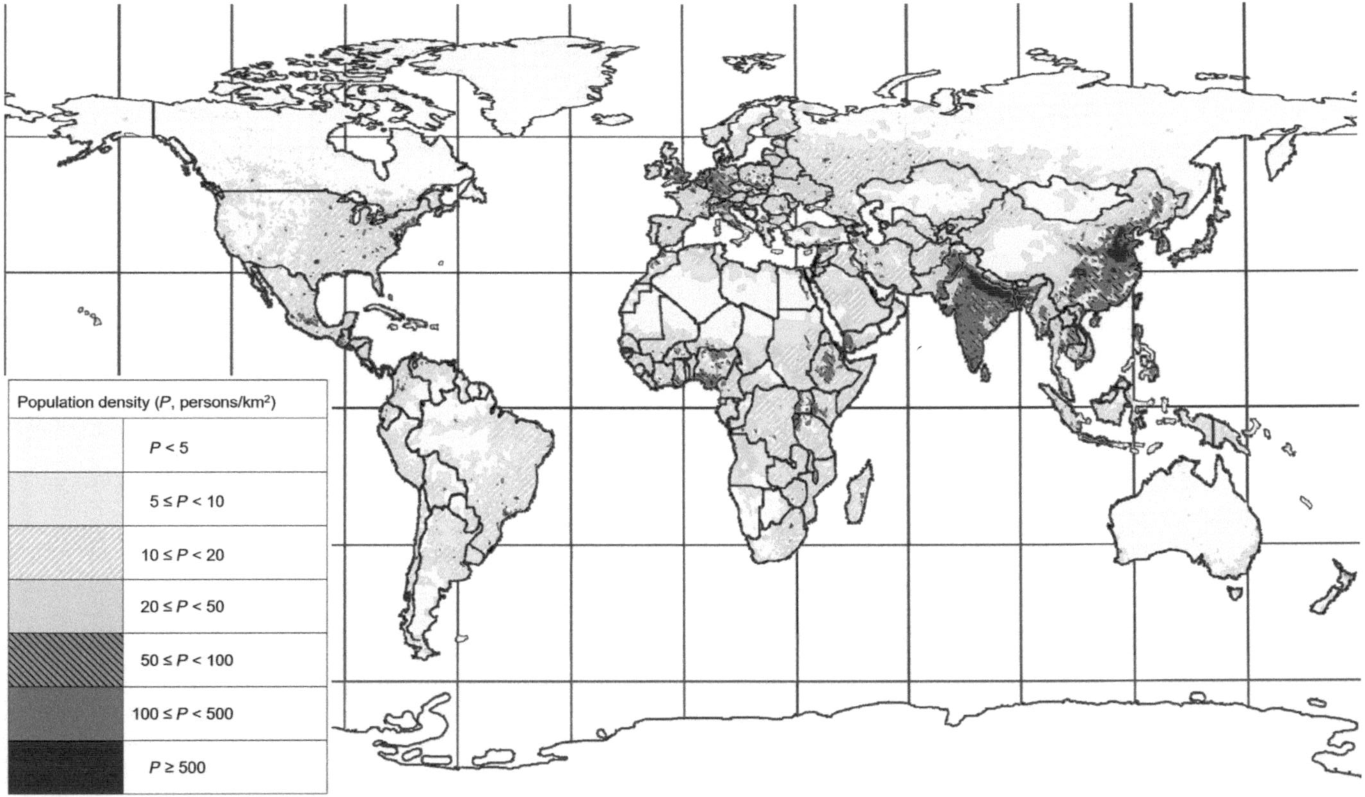

Fig. 2.11 Global population density. *(Based on data from Food and Agriculture Organization of the United Nations (FAO). Global population density estimates 2015 (FGGD). Latest update: GeoLayer. Available at: http://ref.data.fao.org/map?entryId=6d78b970-5883-11db-b1aa-000d939bc5d8 [Accessed 04 June 2015; Last accessed: 06 February 2017].)*

- Landlocked countries which inherently have no access to seawater.

In countries and subnational units achieving rank score results of $0.125 < R \leq 0.273$ solar desalination can reasonably be described as having "limited potential." Table 2.9A–D summarize the cases where

- The scale of solar desalination deployments would be limited by saline feedwater availability because there is no access to seawater ($r_S = 0.5$).
- The solar energy resource is relatively poor ($r_H < 0.5$).

Table 2.9 (A)–(D) Solar desalination has "limited potential" ($0.125 < R \leq 0.273$)

Country/region	r_N	r_F	r_S	r_H	R
(A) No access to seawater (r_S = 0.5)					
Armenia	0.6	0.89	0.5	0.5	0.19
Azerbaijan	0.6	0.89	0.5	0.5	0.19
Botswana	0.6	0.47	0.5	0.83	0.22
Chad	0.6	0.47	0.5	0.83	0.22
China (West and Central)	0.78	0.89	0.5	0.61	0.25
Kazakhstan	0.6	0.89	0.5	0.44	0.17
Kyrgyzstan	0.6	0.89	0.5	0.44	0.17
Malawi	0.85	0.14	0.5	0.72	0.18
Mali	0.6	0.89	0.25	0.83	0.15
Tajikistan	0.6	0.89	0.5	0.44	0.17
Turkmenistan	0.6	0.89	0.5	0.5	0.19
Uganda	0.78	0.14	0.5	0.78	0.18
United States (Inland North)	0.4	0.66	0.5	0.56	0.15
Uzbekistan	0.78	0.89	0.5	0.5	0.21
Zimbabwe	0.85	0.47	0.5	0.78	0.26
(B) National freshwater abundance (r_N < 0.5)					
Angola	0.4	0.14	0.75	0.78	0.16
Bosnia-Herzegovina	0.4	0.32	0.75	0.5	0.14
Cameroon	0.4	0.14	0.75	0.67	0.13
Canada (South and SE)	0.08	0.66	1	0.41	0.15
Ecuador	0.08	0.89	0.75	0.61	0.22
Honduras	0.4	0.14	0.75	0.72	0.14
Malaysia	0.08	0.89	0.75	0.67	0.24
Peru	0.08	0.89	0.75	0.72	0.26
Romania	0.4	0.32	0.75	0.5	0.14
Russian Fed. (S. West)	0.08	0.66	1	0.44	0.17
United States (Central)	0.4	0.32	0.75	0.56	0.15
United States (East Coast)	0.4	0.89	0.75	0.56	0.27
Venezuela	0.08	0.89	0.75	0.72	0.26

Continued

Table 2.9 (A)–(D) Solar desalination has "limited potential" ($0.125 < R \leq 0.273$)—cont'd

Country/region	r_N	r_F	r_S	r_H	R
(C) Minimal abstraction stresses ($r_F < 0.5$)					
Albania	0.6	0.14	0.75	0.5	0.14
Benin	0.6	0.14	0.75	0.67	0.18
China (Southeast)	0.78	0.14	0.75	0.56	0.19
Corsica	0.6	0.14	0.75	0.61	0.17
Denmark	0.85	0.14	0.75	0.39	0.14
Dominican Rep.	0.78	0.14	0.75	0.72	0.25
Ghana	0.78	0.14	0.75	0.67	0.23
India (Assam)	0.85	0.14	0.75	0.56	0.2
Ivory Coast	0.6	0.14	0.75	0.67	0.18
Jamaica	0.6	0.14	0.75	0.72	0.2
Moldova	0.6	0.32	0.75	0.5	0.17
Philippines	0.6	0.14	1	0.56	0.2
Poland	0.85	0.32	1	0.39	0.23
Taiwan	0.78	0.14	0.75	0.5	0.17
Togo	0.78	0.14	0.75	0.67	0.23
Western Sahara	0.6	0.14	0.75	0.83	0.23
(D) Relatively weak solar energy resource ($r_H \leq 0.5$)					
Belgium	0.78	0.89	0.75	0.39	0.24
Bulgaria	0.6	0.66	0.75	0.5	0.24
Netherlands	0.6	0.89	0.75	0.39	0.22
Ukraine	0.6	0.66	0.75	0.44	0.21

- Renewable water resources are sufficiently abundant or suitably managed such that abstraction stress is minimized ($r_F < 0.5$) without relying on desalination.
- Freshwater is abundant on the national level ($r_N < 0.5$) even though local water stresses may be apparent.

Several countries and subnational units in the latter category ($r_N < 0.5$) are densely populated but generally benefit from significant annual rainfall and/or snowfall, including central and eastern United States, southern Canada, southwestern Russia, Malaysia, and Romania. Most other countries in the category have generally low population densities and lie in temperate tropical and equatorial climate zones where significant rainfall occurs. In many cases it may be possible to address local water stress problems by improving infrastructure such as nationwide transmission pipelines rather than relying on large-scale deployment of desalination technologies.

In countries and subnational units achieving rank score results of $0.273 < R \leq 0.422$, solar desalination can reasonably be described as having "significant potential." Table 2.10A–C, respectively, summarize the cases for $r_S < 0.5$, $r_N < 0.5$, and $r_F < 0.5$ in a manner similar to that discussed previously. The applicability of solar desalination in these countries would typically be rather localized. For instance

- Desalination plants in Afghanistan (Table 2.10A) would generally be most appropriate near saline aquifers and large population centers given that the country that has limited access to any water; fresh or saline.
- The greatest potential for deploying large-scale solar desalination plants in large countries with uneven population distributions (such as Australia, Argentina, and Brazil as per Table 2.10B) would be near major coastal population centers.
- While perhaps not currently required, large-scale solar desalination plants may be applicable in the near future for some of the countries and subnational units listed in Table 2.10C (e.g., northeastern India, Nigeria, Thailand, Hainan, Haiti, Kenya, and Tanzania) given that several of these countries have rapidly growing populations and increasing urbanization.

Solar desalination would have local applicability across large parts of Europe and northeastern China as well as Korea, Sri Lanka, and Cuba (Table 2.10D and E). All these countries have notable levels of water stress and/or scarcity and have access to seawater. Most of the countries feature regions of very high population density (>500 persons per square kilometer). France, Germany, Japan, and the United Kingdom have relatively weak solar energy resources ($r_H \leq 0.5$) so in some cases, desalination coupled with other renewable energy resources such as wind or wave energy might be more appropriate than solar.

For countries and subnational units achieving rank scores of $R > 0.422$, solar desalination can be described as "highly applicable." The highest rank scoring countries are listed in Table 2.11B ($R > 0.6$) and mostly located in the Middle East and North Africa (MENA) region where fossil fueled desalination is already widespread, especially in the major oil-producing nations [98] where fuel costs are very low. The other countries in Table 2.11B (Sudan, Somalia, Eritrea, and Pakistan) are all nearby and have similar arid climates and high solar insolation levels. Table 2.11A ($0.422 < R < 0.6$) is likewise primarily made up of countries in the arid belt which covers MENA and many parts of Africa. Notable exceptions (only some of which are arid) are the southwestern

Table 2.10 (A)–(E) Solar desalination has "significant potential" ($0.273 < R \leq 0.422$)

Country/region	r_N	r_F	r_S	r_H	R
(A) No access to seawater ($r_S = 0.5$)					
Afghanistan	0.78	0.89	0.5	0.72	0.3
(B) National freshwater abundance ($r_N < 0.5$)					
Argentina	0.08	0.89	1	0.61	0.3
Australia (NT and WA)	0.08	0.89	1	0.83	0.4
Australia (QLD)	0.08	0.66	1	0.78	0.29
Australia (Southeast)	0.08	0.89	1	0.72	0.35
Brazil (East coast)	0.08	0.89	1	0.72	0.35
Chile	0.08	0.89	1	0.61	0.3
Georgia	0.4	0.89	1	0.5	0.32
Indonesia (SJW)	0.4	0.89	1	0.61	0.39
Mozambique	0.4	0.47	1	0.72	0.32
(C) Minimal abstraction stresses ($r_F < 0.5$)					
China (Hainan)	0.78	0.47	0.75	0.61	0.29
Haiti	0.85	0.32	0.75	0.72	0.32
India (NE excl. Assam)	0.85	0.47	0.75	0.67	0.33
Kenya	0.91	0.14	1	0.78	0.41
Nigeria	0.78	0.14	1	0.72	0.33
Tanzania	0.78	0.14	1	0.78	0.36
Thailand	0.6	0.47	1	0.67	0.36
(D) Relatively weak solar energy resource ($r_H \leq 0.5$)					
France	0.6	0.89	0.75	0.5	0.28
Germany	0.78	0.89	1	0.39	0.32
Japan	0.6	0.89	1	0.5	0.37
United Kingdom	0.78	0.89	1	0.33	0.28
(E) All conditions marginally favorable (all $r > 0.5$)					
China (Northeast)	0.78	0.66	0.75	0.56	0.3
Cuba	0.6	0.66	0.75	0.72	0.34
Cyprus	0.91	0.89	0.75	0.61	0.41
Greece	0.6	0.89	0.75	0.61	0.34
Italy	0.6	0.89	0.75	0.56	0.31
North Korea	0.6	0.89	0.75	0.56	0.31
Portugal	0.6	0.66	0.75	0.61	0.29
Sardinia	0.6	0.89	0.75	0.61	0.34
South Korea	0.85	0.89	0.75	0.56	0.36
Spain	0.78	0.89	0.75	0.61	0.38
Sri Lanka	0.6	0.66	0.75	0.72	0.34

Abbreviations for subnational units: *NT*, Northern Territory; *QLD*, Queensland; *SJW*, Sumatra, Java, and Western islands; *WA*, Western Australia.

Table 2.11 (A) and (B) Solar desalination is "highly applicable" ($R > 0.422$)

Country/region	r_N	r_F	r_S	r_H	R
(A) Rank scores $0.422 \leq R \leq 0.6$					
Mexico	0.6	0.89	0.75	0.78	0.43
Turkey	0.6	0.89	1	0.61	0.45
Djibouti	0.97	0.14	1	0.83	0.46
China (East)	0.78	0.89	1	0.56	0.46
Mauritania	0.6	0.89	0.75	0.83	0.46
Senegal	0.6	0.89	0.75	0.83	0.46
United States (SW)	0.4	0.89	1	0.72	0.46
India (South)	0.85	0.89	0.75	0.72	0.47
South Africa	0.85	0.89	0.75	0.72	0.47
Tunisia	0.97	0.89	0.75	0.72	0.5
Namibia	0.4	0.89	1	0.83	0.54
Iraq	0.6	0.89	1	0.72	0.54
Ethiopia	0.85	0.66	1	0.72	0.54
India (NW)	0.85	0.89	1	0.67	0.58
Iran	0.78	0.89	1	0.72	0.6
(B) Rank scores $R > 0.6$					
Yemen	0.97	0.89	0.75	0.89	0.62
Pakistan	0.85	0.89	1	0.72	0.63
Sudan	0.85	0.66	1	0.83	0.63
Syria	0.91	0.89	1	0.72	0.65
Kuwait	0.97	0.89	1	0.72	0.67
Somalia	0.85	0.89	1	0.78	0.67
Morocco	0.91	0.89	1	0.78	0.7
Eritrea	0.85	0.89	1	0.83	0.72
Israel and Palestine	0.91	0.89	1	0.83	0.75
Egypt	0.91	0.89	1	0.83	0.75
Jordan	0.97	0.89	1	0.83	0.77
Saudi Arabia	0.97	0.89	1	0.83	0.77
Oman	0.97	0.89	1	0.83	0.77
UAE	0.97	0.89	1	0.83	0.77
Algeria	0.97	0.89	1	0.83	0.77
Libya	0.97	0.89	1	0.83	0.77

United States, Mexico, eastern China, southern and northwestern India, and Turkey, where some of the world's largest cities and metropolitan areas (>5 million inhabitants) are located including Shanghai, Beijing, Suzhou, Nanjing, Tianjin, Delhi, Mumbai, Bengaluru, Mexico City, Istanbul, Los Angeles, and Houston. Several of the countries listed in

Table 2.11B also feature very large cities such as Karachi, Lahore, Cairo, and Tehran.

2.5.3 Limitations and Further Work

The results presented above identify the main locations throughout the world where solar desalination technologies are likely to have greatest potential.

One key limitation of the results is the spatial resolution which is primarily constrained by the available data. For countries covering large geographical areas and having apparent national abundance of freshwater resources (high values of N and low values of r_N) the rank scoring results lack fidelity and in some cases, understate the local applicability of local solar desalination. For example, most of mainland Australia is arid but sparsely populated and the analysis identifies solar desalination as having "significant potential" ($0.273 < R \leq 0.422$). A more spatially detailed analysis would probably find it "highly applicable" in all major coastal population centers (and perhaps in the intensively farmed northeast) but of more "limited potential" in the sparsely populated inland areas.

The spatial resolution of the analysis could be improved by using provincial (rather than national) water scarcity data for the largest countries, although such data was not encountered in any of the cited literature (see References list). Likewise, the spatial resolution would be improved if more countries were split into subnational units to improve the fidelity of the r_F, r_S, and r_H parameter analysis.

A further improvement in the analysis could be realized by taking better account of cities and surrounding densely populated municipalities, which

- Generate large quantities of sewerage. Wastewater collection and treatment infrastructure can convert this into significant quantities of saline water which can be used as feedwater for solar desalination plants. This resource could potentially be more useful in many cases than the other inland alternative such as salt lakes (S2) or saline groundwater aquifers (S3 and S4) but has not been accounted for in the calculation of the r_S parameter.
- Generally have high land values and limited space for solar collectors. While solar collectors can be located on building roofs and facades [99], large-scale solar installations are generally much easier and cheaper to realize in sparsely populated areas. This could potentially be accounted for in future analyses by reducing the effective solar energy resource parameters (H and r_H) in densely populated areas.

2.6 SUMMARY

Water plays an essential role in our energy supply systems, yet our water supply system increasingly requires a significant amount of energy to function. Global population growth is stressing our freshwater resources and depleting fossil fuel reserves. Solar desalination systems address these problems by using the sun's energy to remove salts from saline water (either seawater or brackish groundwater) by either phase change, pressure-driven membrane, or electric-charge driven membrane processes.

This chapter has briefly reviewed some of the main considerations affecting the technical and economic feasibility of solar desalination, as well as its environmental impacts. PV-driven RO appears to be the leading technology, but the cost of water produced by a solar desalination plant is currently several times higher than that produced by fossil fueled processes. Costs tend to be lowest for large-scale plants located in areas with high levels of insolation where low-salinity feedwater is available.

A rank scoring system has been devised which quantifies solar desalination applicability based on the objective measures of water scarcity, water stress, the local availability of saline feedwater, and solar insolation levels. Rank scores have been calculated for 154 countries where sufficient data was available. The eight largest countries have been split to form 28 smaller subnational units in order to improve the spatial resolution of the assessment.

Scores of $R > 0.422$ indicate that solar desalination is "highly applicable" for 27 whole-countries and 4 subnational units where freshwater scarcity and stress are problematic, solar energy is abundant, and saline feedwater is readily available. These high scores are primarily apparent in locations with arid climates and very high insolation levels including all MENA countries and most arid nations in Africa. Similar high scores are apparent for Mexico, Turkey, Pakistan, and across large parts of China, India, and the United States where many of the world's largest cities and metropolitan areas are located.

Scores of $0.273 < R \leq 0.422$ indicate "significant potential" for solar desalination in 23 whole-countries and 9 subnational units where analysis suggests that solar desalination would be locally applicable in the most densely populated areas, particularly in (or near) coastal cities. In densely populated countries where solar insolation levels are relatively low (such as United Kingdom, France, Germany, and Japan) it may be more effective to drive desalination plants using wind or wave energy rather than solar.

Scores of $R < 0.125$ indicate that solar desalination is essentially "not applicable" for 57 whole-countries and 8 subnational units. These low scores tend to arise when countries either have abundant freshwater resources and/or relatively low solar insolation levels (e.g., Ireland, New Zealand, and most of Canada, Russia, and Scandinavia), or where freshwater scarcity/stress occurs but there is a lack of suitable saline water resources (e.g., Czech Republic, Nepal, Bolivia, South Sudan). For similar reasons, solar desalination would have "limited potential" in 42 whole-countries and 9 subnational units where rank score results are $0.125 < R \leq 0.273$. Solar desalination of treated sewerage wastewater may be an option in densely populated inland areas (where no seawater is available) of the sunniest countries but other methods of tackling water scarcity and stress, such as fog harvesting and long distance water transport may be more appropriate in sparsely populated areas.

REFERENCES

[1] Chaibi M. An overview of solar desalination for domestic and agriculture water needs in remote arid areas. Desalination 2000;127:119–33.
[2] Abu-Jabal M, Karniyab I, Namsakib Y. Proving test for a solar-powered desalination system in Gaza-Palestine. Desalination 2001;137:1–4.
[3] Sagie D, Feinerman E, Aharoni E. Potential of solar desalination in Israel and in its close vicinity. Desalination 2001;139:21–33.
[4] Weiner D, Fisher D, Moses E, Katz B, Meron G. Operation experience of a solar- and wind-powered desalination demonstration plant. Desalination 2001;137:7–13.
[5] Bouchekima B. A solar desalination plant for domestic water needs in arid areas of South Algeria. Desalination 2002;153:65–9.
[6] Parekh S, Farid M, Selman J, Al-Hallaj S. Solar desalination with a humidification-dehumidification technique—a comprehensive technical review. Desalination 2004;160:167–86.
[7] Houcine I, Banamara M, Guizani A, Maalej M. Pilot plant testing of a new solar desalination process by a multiple-effect-humidification technique. Desalination 2006;196:105–24.
[8] Abdel-Rehim Z, Lasheen A. Experimental and theoretical study of a solar desalination system located in Cairo, Egypt. Desalination 2007;217:52–64.
[9] Banat F, Jwaied N, Rommel M, Koschikowski J, Wieghaus M. Performance evaluation of the "large SMADES" autonomous desalination solar-driven membrane distillation plant in Aqaba, Jordan. Desalination 2007;217:17–28.
[10] Fath H. High performance of a simple design, two effects, solar distillation unit. Energy Convers Manag 1997;38(18):1895–905.
[11] Qiblawey H, Banat F. Solar thermal desalination technologies. Desalination 2008;220:633–44.
[12] Bechki D, Bouguettaia H, Blanco-Galvez J, Babay S, Bouchekima B, Boughali S, et al. Effect of partial intermittent shading on the performance of a simple basin solar still in south Algeria. Desalination 2010;260:65–9.
[13] Jijakli K, Arafat H, Kennedy S, Mande P, Theeyattuparampil V. How green solar desalination really is? Environmental assessment using life-cycle analysis (LCA) approach. Desalination 2012;287:123–31.
[14] Yuan G, Wang Z, Li H, Li X. Experimental study of a solar desalination system based on humidification–dehumidification process. Desalination 2011;277:92–8.

[15] Chen Z, Xie G, Chen Z, Zheng H, Zhuang C. Field test of a solar seawater desalination unit with triple-effect falling film regeneration in northern China. Solar Energy 2012;86:31–9.
[16] Adhikari R, Kumar A. Cost optimization studies on a multi-stage stacked tray solar still. Desalination 1999;125:115–21.
[17] Kaushal A. Solar stills: a review. Renew Sustain Energy Rev 2010;14:446–53.
[18] Alarcón-Padilla D, Blanco-Gálvez J, García-Rodríguez L, Gernjak W, Malato-Rodríguez S. First experimental results of a new hybrid solar/gas multi-effect distillation system: the AQUASOL project. Desalination 2008;220:619–25.
[19] Connor R, Winpenny J. The water–energy nexus, In: Water and energy—The United Nations World Water Development Report. Paris: UNESCO (United Nations Educational, Scientific and Cultural Organization); 2014. p. 12–20. ISBN: 978-92-3-104259-1 [chapter 1]. Available at: http://www.unesco.org/new/en/natural-sciences/environment/water/wwap/wwdr/2014-water-and-energy/.
[20] Micale G, Cipollina A, Rizzuti L. Seawater desalination for freshwater production. In: Cipollina A, Micale G, Rizzuti L, editors. Seawater desalination. London: Springer; 2009, ISBN: 978-3-642-01149-8 [chapter 1].
[21] Coates D, Connor R, Leclerc L, Rast W, Schumann K, Webber M. Water demand: what drives consumption?, In: Managing water under uncertainty and risk—The United Nations World Water Development Report 4. Paris: UNESCO (United Nations Educational, Scientific and Cultural Organization); 2012. p. 44–69. ISBN: 978-92-3-104235-5. [chapter 2]. Available at: http://www.unesco.org/new/en/natural-sciences/environment/water/wwap/wwdr/wwdr4-2012/.
[22] Mielke E, Diaz-Anadon L, Narayanamurti V. Water consumption of energy resource extraction, processing, and conversion (discussion paper #2010-15). Cambridge, USA: Belfer Center for Science and International Affairs, Harvard Kennedy School; 2010. Available at: http://belfercenter.or.
[23] AQUASTAT online database. Food and Agriculture Organisation of the United Nations; 2010. Available at: http://www.fao.org/nr/water/aquastat/dbase/index.stm.
[24] Rajagopalan B, Brown C. State of the resource: quantity. In: Managing water under uncertainty and risk—The United Nations World Water Development Report 4. Paris: UNESCO (United Nations Educational, Scientific and Cultural Organization); 2012. p. 381–95. ISBN: 978-92-3-104235-5. [Chapter 15]. Available at: http://www.unesco.org/new/en/natural-sciences/environment/water/wwap/wwdr/wwdr4-2012/.
[25] International Energy Agency. Key world energy statistics 2011. Paris: IEA; 2011. Available at: http://www.iea.org/publications/freepublications/publication/name,20938,en.html.
[26] Webber M. The global nexus of energy and water. In: Managing water under uncertainty and risk—The United Nations World Water Development Report 4. Paris: UNESCO (United Nations Educational, Scientific and Cultural Organization); 2012. p. 465–77. ISBN: 978-92-3-104235-5. [chapter 19]. Available at: http://www.unesco.org/new/en/natural-sciences/environment/water/wwap/wwdr/wwdr4-2012/.
[27] Ortiz J, Exposito E, Gallud F, Garcia-Garcia V, Montiel V, Aldaz A. Desalination of underground brackish waters using an electrodialysis system powered directly by photovoltaic energy. Sol Energy Mater Sol Cells 2008;92:1677–88.
[28] Tuladhar B, Kitio V, Goodwin R, Dzikus A. Cities. In: Water and energy—The United Nations World Water Development Report. Paris: UNESCO (United Nations Educational, Scientific and Cultural Organization); 2014. p. 12–20. ISBN: 978-92-3-104259-1. [chapter 7]. Available at: http://www.unesco.org/new/en/natural-sciences/environment/water/wwap/wwdr/2014-water-and-energy/.
[29] Koncagül E, Withanachchi S. Part 6—Data and indicators annex. In: Water and energy—The United Nations World Water Development Report. Paris: UNESCO

(United Nations Educational, Scientific and Cultural Organization); 2014. p. 12–20. ISBN: 978-92-3-104259-1. Available at: http://www.unesco.org/new/en/natural-sciences/environment/water/wwap/wwdr/2014-water-and-energy/.
[30] Australian Government. Water for the future—moving water long distances: grand schemes or pipe dreams? The Australian Government Department of Sustainability, Environment, Water, Population and Communities: Canberra; 2010. ISBN: 978-1-921733-06-2. Available at: http://www.environment.gov.au/water/publications/action/moving-water.html.
[31] Ettouney H, Wilf M. Commercial desalination technologies—an overview of the current status of applications of commercial seawater desalination processes. In: Cipollina A, Micale G, Rizzuti L, editors. Seawater desalination. London: Springer; 2009. p. 77–107. ISBN: 978-3-642-01149-8. [chapter 4].
[32] Fiorenza G, Sharma V, Braccio G. Techno-economic evaluation of a solar powered water desalination plant. Energy Convers Manag 2003;44:2217–40.
[33] van der Gun J. Groundwater. In: Managing water under uncertainty and risk—The United Nations World Water Development Report 4. Paris: UNESCO (United Nations Educational, Scientific and Cultural Organization); 2012. p. 756–68. ISBN: 978-92-3-104235-5. [chapter 36]. Available at: http://www.unesco.org/new/en/natural-sciences/environment/water/wwap/wwdr/wwdr4-2012/.
[34] MacDonald A, Bonsor H, O'Dochartaigh B, Taylor R. Quantitative maps of groundwater resources in Africa, Environ Res Lett 2012;7(024009) Bristol: IOP Science. Available at: iopscience.iop.org/1748-9326/7/2/024009/article.
[35] Balaji R, Bartram J, Coates D, Connor R, Harding J, Hellmuth M, et al. Beyond demand: water's social and environmental benefits. In: Managing water under uncertainty and risk—The United Nations World Water Development Report 4. Paris: UNESCO (United Nations Educational, Scientific and Cultural Organization); 2012. p. 101–23. ISBN: 978-92-3-104235-5. [chapter 4]. Available at: http://www.unesco.org/new/en/natural-sciences/environment/water/wwap/wwdr/wwdr4-2012/.
[36] Al-Ismaily H, Probert D. Water-resource facilities and management strategy for Oman. Appl Energy 1998;61:125–46.
[37] Cooley H, Gleick P, Wolff G. California's next million acre-feet: saving water, energy, and money. Oakland, CA: The Pacific Institute; 2010, ISBN: 1-893790-26-6. Available at: http://www.pacinst.org/reports/next_million_acre_feet/next_million_acre_feet.pdf.
[38] Gleick P, Haasz D, Henges-Jeck C, Srinivasan V, Wolff G, Kao-Cushing K, et al. Waste not, want not: the potential for urban water conservation in California. Oakland, CA: The Pacific Institute; 2003, ISBN: 1-893790-09-6. Available at: http://pacinst.org/publication/waste-not-want-not/.
[39] Fort D, Nelson B. Pipe dreams: water supply pipeline projects in the west, New York: Natural Resources Defense Council; 2012. Available at: http://www.nrdc.org/water/management/files/water-pipelines-report.pdf.
[40] Shah A. Water-starved Oman collects monsoon fog. In: GreenSource online magazine, August 2010. New York: Dodge Data and Analytics, GreenSource Magazine; 2010. Available at: http://greensource.construction.com/news/2010/100816oman_monsoon_fog.asp.
[41] Chaibi M, El-Nashar A. Solar thermal processes—a review of solar thermal energy technologies for water desalination. In: Cipollina A, Micale G, Rizzuti L, editors. Seawater desalination. London: Springer; 2009. p. 131–63. ISBN: 978-3-642-01149-8. [chapter 6].
[42] Koschikowski J, Wieghaus M, Rommel M. Membrane distillation for solar desalination. In: Cipollina A, Micale G, Rizzuti L, editors. Seawater desalination. London: Springer; 2009. p. 165–87 ISBN: 978-3-642-01149-8. [chapter 7].
[43] Rheinlander J, Geyer D. Photovoltaic reverse osmosis and electrodialysis—application of solar photovoltaic energy production to RO and ED desalination processes.

In: Cipollina A, Micale G, Rizzuti L, editors. Seawater desalination. London: Springer; 2009. p. 166–211. ISBN: 978-3-642-01149-8. [chapter 8].
[44] Tzen E. Wind and wave energy for reverse osmosis. In: Cipollina A, Micale G, Rizzuti L, editors. Seawater desalination. London: Springer; 2009. p. 213–45 ISBN: 978-3-642-01149-8. [chapter 9].
[45] Papapetrou M, Mohamed E, Manolakos D, Papadakis G, Subiela V, Penate B. Operating RE/desalination units. In: Cipollina A, Micale G, Rizzuti L, editors. Seawater desalination. London: Springer; 2009. p. 247–72. ISBN: 978-3-642-01149-8. [chapter 10].
[46] He T, Yan L. Application of alternative energy integration technology in seawater desalination. Desalination 2009;249:104–8.
[47] Gude V, Nirmalakhandan N, Deng S. Renewable and sustainable approaches for desalination. Renew Sustain Energy Rev 2010;14:2641–54.
[48] Ghaffour N, Lattemann S, Missimer T, Choon-Ng K, Sinha S, Amy G. Renewable energy-driven innovative energy-efficient desalination technologies. Appl Energy 2014;136:1155–65.
[49] Shahabi M, McHugh A, Anda M, Ho G. Environmental life cycle assessment of seawater reverse osmosis desalination plant powered by renewable energy. Renew Energy 2014;67:53–8.
[50] Abdallaha S, Abu-Hilal M, Mohsen M. Performance of a photovoltaic powered reverse osmosis system under local climatic conditions. Desalination 2005;183:95–104.
[51] Peterson E, Gray S. Effectiveness of desalination powered by a tracking solar array to treat saline bore water. Desalination 2012;293:94–103.
[52] Ahmad N, Sheikh A, Gandhidasan P, Elshafie M. Modeling, simulation and performance evaluation of a community scale PVRO water desalination system operated by fixed and tracking PV panels: a case study for Dhahran city. Saudi Arabia Renew Energy 2015;75:433–47.
[53] Childs W, Dabiri A, Al-Hinai H, Abdullah H. VARI-RO solar-powered desalting technology. Desalination 1999;125:155–66.
[54] Delgado-Torres A, Garcia-Rodriguez L. Preliminary assessment of solar Organic Rankine Cycles for driving a desalination system. Desalination 2007;216:252–75.
[55] Peñate B, Garcia-Rodriguez L. Seawater reverse osmosis desalination driven by a solar Organic Rankine Cycle: design and technology assessment for medium capacity range. Desalination 2012;284:86–91.
[56] El-Nashar A, Samad M. The solar desalination plant in Abu Dhabi: 13 Years of performance and operation history. Renew Energy 1998;14(1–4):263–74.
[57] Garcia-Rodriguez L, Gomez-Camacho C. Preliminary design and cost analysis of a solar distillation system. Desalination 1999;126:109–14.
[58] Garcia-Rodriguez L, Gomez-Camacho C. Exergy analysis of the SOL-14 plant (Plataforma Solar de Almeria, Spain). Desalination 2001;137:251–8.
[59] Blanco J, Alarcon D, Zarz E, Malato S, Leon J. In: Advanced solar desalination: a feasible technology to the Mediterranean area. EuroSUN Bologna, Italy, 23–26 June. Freiburg: International Solar Energy Society; 2002. Available at: http://www.psa.es/webeng/aquasol/files/Congresos/JBlanco%20-%20EuroSUN2002.pdf.
[60] Goosen M. Thermodynamic and economic considerations in solar desalination. Desalination 2000;129:63–89.
[61] S. Kalogirou. Solar desalination systems. In: Solar energy engineering, London: Academic Press (Elsevier); 2009, p. 421-468. ISBN: 13: 978-0-12-374501-9 [chapter 8].
[62] Sampathkumar K, Arjunan T, Pitchandi P, Senthilkumar P. Active solar distillation—a detailed review. Renew Sustain Energy Rev 2010;14:1503–26.
[63] Schwarzer K, Vieira da Silva E, Hoffschmidt B, Schwarzer A. A new solar desalination system with heat recovery for decentralised drinking water production. Desalination 2009;248:204–11.

[64] Shatat M, Mahkamov K. Determination of rational design parameters of a multi-stage solar water desalination still using transient mathematical modelling. Renew Energy 2010;35:52–61.
[65] Guillén-Burrieza E, Blanco J, Zaragoza G, Alarcón D, Palenzuela P, Ibarra M, et al. Experimental analysis of an air gap membrane distillation solar desalination pilot system. J Membr Sci 2011;379:386–96.
[66] Banat F, Jwaied N. Exergy analysis of desalination by solar-powered membrane distillation units. Desalination 2008;230:27–40.
[67] Banat F, Jwaied N. Economic evaluation of desalination by small-scale autonomous solar-powered membrane distillation units. Desalination 2008;220:566–73.
[68] Efraty A. CCD series no-22: recent advances in RO, FO and PRO and their hybrid applications for high recovery desalination of treated sewage effluents. Desalination 2016;389:18–38.
[69] van Weert F, van der Gun J, Reckman J. Global overview of saline groundwater occurrence and genesis (report number: GP 2009-1). Utrecht: IGRAC (United Nations International Groundwater Resources Assessment Centre); 2009. Available at: http://www.un-igrac.org/dynamics/modules/SFIL0100/view.php?fil_Id=135.
[70] Soil & Water Conservation Society of Metro Halifax. Saline lakes: the largest, highest & lowest lakes of the world! Halifax, Canada: SWCSMH; 2006. Available at: http://www.lakes.chebucto.org/saline1.htm.
[71] Cotruvo J, Abouzaid H. Overview of the health and environmental impacts of desalination technology. In: Cotruvo J, Voutchkov N, Fawell J, Payment P, Cunliffe D, Lattemann S, editors. Desalination technology: health and environmental impacts. London: Taylor & Francis Group (CRC Press); 2010. p. 1–20 ISBN: 978-1-4398-28908. [chapter 1].
[72] Jasim S, Saththasivam J, Loganathan K, Ogunbiyi O, Sarp S. Reuse of treated sewage effluent (TSE) in Qatar. J Water Process Eng 2016;11:174–82.
[73] Shanmuganathan S, Vigneswaran S, Nguyen T, Loganathan P, Kandasamy J. Use of nanofiltration and reverse osmosis in reclaiming micro-filtered biologically treated sewage effluent for irrigation. Desalination 2015;364:119–25.
[74] SolarGIS. World Map of Global Horizontal Irradiation. Bratislava, Slovakia: GeoModel Solar; 2013. Available at: http://solargis.info/doc/_pics/freemaps/1000px/ghi/SolarGIS-Solar-map-World-map-en.png.
[75] Ghaffour N, Missimer T, Amy G. Technical review and evaluation of the economics of water desalination: current and future challenges for better water supply sustainability. Desalination 2013;309:197–207.
[76] Renewable Energy Network for the 21st century (REN21). Renewables 2016 global status report. ISBN 978-3-9818107-0-7. Paris, France: Renewable Energy Policy Network for the 21st Century (REN21) Secretariat. Available at: http://www.ren21.net/wp-content/uploads/2016/06/GSR_2016_Full_Report.pdf.
[77] Renewable Energy Network for the 21st century (REN21). Renewables 2014—global status report. Paris: REN21 Secretariat; 2014, ISBN: 978-3-9815934-2-6. Available at: http://www.ren21.net/portals/0/documents/resources/gsr/2014/gsr2014_full%20report_low%20res.pdf.
[78] Deutsche Gesellshaft fur Sonnenenergie (DGS). Planning and installing photovoltaic systems—a guide for installers, architects and engineers. 2nd ed. London: Earthscan; 2008, ISBN: 978-1-84407-442-6.
[79] Kost C, Mayer J, Thomsen J, Hartmann N, Senkpiel C, Philipps S, et al. Levelized cost of electricity renewable energy technologies (study edition November 2013). Freiburg, Germany: Fraunhofer Institute for Solar Energy Systems; 2013.
[80] ENF Solar Ltd. Solar cell directory. Available at: https://www.enfsolar.com/pv/cell.
[81] Qtaishat M, Banat F. Desalination by solar powered membrane distillation systems. Desalination 2013;308:186–97.

[82] Shalaby S. Reverse osmosis desalination powered by photovoltaic and solar Rankine cycle power systems: a review. Renew Sustain Energy Rev 2017;73:789–97.
[83] Hamed O, Kosaka H, Bamardouf K, Al-Shail K, Al-Ghamdi A. Concentrating solar power for seawater thermal desalination. Desalination 2016;396:70–8.
[84] Palenzuela P, Alarcón-Padilla D, Zaragoza G. Large-scale solar desalination by combination with CSP: techno-economic analysis of different options for the Mediterranean Sea and the Arabian Gulf. Desalination 2015;366:130–8.
[85] Demir M, Dincer I. Development of an integrated hybrid solar thermal power system with thermoelectric generator for desalination and power production. Desalination 2017;404:59–71.
[86] Karagiannis I, Soldatos P. Water desalination cost literature: review and assessment. Desalination 2008;223:448–56.
[87] Ali M, Fath H, Armstrong P. A comprehensive techno-economical review of indirect solar desalination. Renew Sustain Energy Rev 2011;15:4187–99.
[88] Trieb F, Muller-Steinhagen H, Kern J. Financing concentrating solar power in the Middle East and North Africa—subsidy or investment? Energy Policy 2011;39: 307–17.
[89] Al-Hallaj S, Parekh S, Farid M, Selman J. Solar desalination with humidification–dehumidification cycle: review of economics. Desalination 2006;195:169–86.
[90] Renewable Energy Network for the 21st century (REN21). Renewables 2012—global status report. Paris: REN21 Secretariat; 2012. Available at: http://www.ren21.net/Portals/0/documents/Resources/GSR2011_FINAL.pdf.
[91] Hoepner T. A procedure for environmental impact assessments (EIA) for seawater desalination plants. Desalination 1999;124:1–12.
[92] Cooley H, Gleick P, Wolff G. Desalination, with a grain of salt—a California perspective, Oakland, CA: The Pacific Institute; 2006, ISBN: 1-893790-13-4. Available at: http://pacinst.org/publication/desalination-with-a-grain-of-salt-a-california-perspective-2.
[93] Liu T, Sheu H, Tseng C. Environmental impact assessment of seawater desalination plant under the framework of integrated coastal management. Desalination 2013;326:10–8.
[94] Dawoud M, Mulla M. Environmental impacts of seawater desalination: Arabian Gulf case study. Int J Environ Sustain 2012;1(3):22–37.
[95] Lattemann S. Protecting the marine environment. In: Cipollina A, Micale G, Rizzuti L, editors. Seawater desalination. London: Springer; 2009. p. 273–99 ISBN: 978-3-642-01149-8 [chapter 11].
[96] Lattemann S, Mancy K, Damitz B, Khordagui H, Leslie G. Environmental impact assessment of desalination projects. In: Cotruvo J, Voutchkov N, Fawell J, Payment P, Cunliffe D, Lattemann S, editors. Desalination technology: health and environmental impacts. London: Taylor & Francis Group (CRC Press); 2010. p. 153–78 ISBN: 978-1-4398-28908. [chapter 6].
[97] Met Office. Weather and climate guide. Exeter, UK: Met Office; 2012. Available at: http://www.metoffice.gov.uk/binaries/content/assets/mohippo/pdf/4/d/weather_and_climate_guide.pdf.
[98] Koncagül E. Desalination in Gulf Cooperation Council countries. In: Water and energy—The United Nations World Water Development Report. Paris: UNESCO (United Nations Educational, Scientific and Cultural Organization); 2014. p. 12–20. ISBN: 978-92-3-104259-1. [chapter 20]. Available at: http://www.unesco.org/new/en/natural-sciences/environment/water/wwap/wwdr/2014-water-and-energy/.
[99] Kalogirou S, editor. Building integrated solar thermal systems, design and applications handbook; 2017, ISBN: 978-9963-697-22-9. EU COST Action TU1205 (BISTS). March.

FURTHER READING

[1] Voutchkov N, Sommariva C, Pankratz T, Tonner J. Desalination process technology. In: Cotruvo J, Voutchkov N, Fawell J, Payment P, Cunliffe D, Lattemann S, editors. Desalination technology: health and environmental impacts. London: Taylor & Francis Group (CRC Press); 2010. p. 21–90. ISBN: 978-1-4398-28908. [chapter 2].

[2] Thomson M, Infield D. Laboratory demonstration of a photovoltaic-powered seawater reverse-osmosis system without batteries. Desalination 2005;183:105–11.

[3] Food and Agriculture Organization of the United Nations (FAO). Global population density estimates 2015 (FGGD). Latest update: GeoLayer. Available at: http://ref.data.fao.org/map?entryId=6d78b970-5883-11db-b1aa-000d939bc5d8.

CHAPTER 3

Wind-Powered Desalination—Principles, Configurations, Design, and Implementation

Eftihia Tzen
Centre for Renewable Energy Sources & Saving, CRES, Pikermi, Greece

3.1 INTRODUCTION

The idea of using renewable energy sources (RES) to drive desalination processes is fundamentally attractive, as a considerable number of in-depth studies and real-life applications demonstrate [1]. An effective integration of these technologies will allow countries to cope the problems of water shortage with domestic energy sources and with the use of environmentally friendly technologies, contributing to greenhouse gas reductions. Additionally, RES desalination is uniquely suited to provide water and electricity in remote areas where water and electricity infrastructures are lacking [2]. Most of these systems are well established and reliable, with a significant number of applications worldwide (see Fig. 3.1). Numerous combinations such as photovoltaic units (PV) with reverse osmosis (RO) plants, wind turbines (WTs) with RO, electrodialysis or electrodialysis reversal (ED or EDR), and solar thermal distillation plants have been in operation, most of them within research programs [3]. In addition, most of these combinations are of small scale, purposing to cover the water needs of rural off-grid areas' where seawater or brackish water sources are available.

The selection of the most suitable desalination technology for a specific site as well as the technological combination of RES desalination is an important factor for the success of a project. In 1998, the first guide on RES desalination was published by CRES in Greece within the framework of the THERMIE Project, SUP-0940-96-HE [4]. This document was the first guide on the coupling of the two technologies, RES and desalination. After 17 years, the number of applications has been significantly increased and an important knowledge and experience has been achieved. As it is shown in Fig. 3.1 the most common and favor technologies combination for small-scale off-grid applications is PV with RO for seawater or brackish

Renewable Energy Powered Desalination Handbook
https://doi.org/10.1016/B978-0-12-815244-7.00003-9

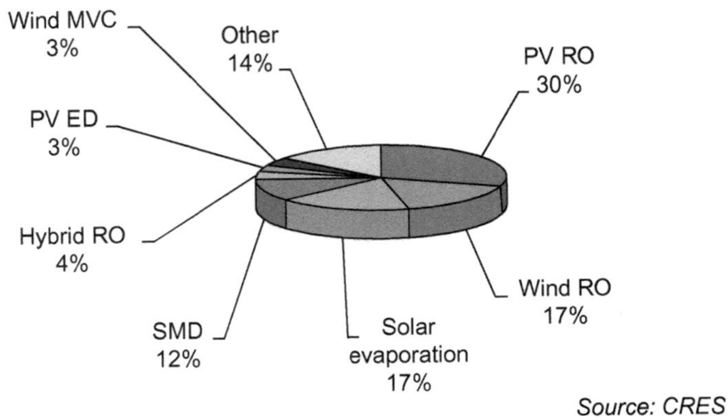

Fig. 3.1 Share of the stand-alone RES DES installations for seawater and brackish water desalination (as a percentage of the total installations) [3].

water desalination. Most of these combinations have been implemented during 1990s and beginning of 2000s, in a period where photovoltaics had high capital cost, leading to a relative high unit water cost from the PV desalination units. The second more applicable combinations seem to be wind RO and solar thermal evaporation systems.

RES potential, feedwater type, the required quantity and quality of produced water, electricity grid and water network existence, land and staff availability, and budget obtainability are some of the parameters that should be taken into consideration before any decision is made and during the design of a project.

In recent years, due to the tremendous increase in desalination installations around the world and due to their huge energy requirements, there is a tendency to cover the consumed energy with the employment of grid-connected RES units, mostly WTs. Moreover, the cost of desalination and renewable energy systems are steadily decreasing, while fossil fuel prices are rising and its supplies are depleting. These issues render the proper RES desalination combination a favorable and promising alternative. The cost reduction of RE and desalination systems contribute to the vision that now is the right time for their expansion. For both technologies the market is ready; societies have accepted these technologies; and politicians have agreed to a better and cleaner environment. In the coming years, a significant progress on RES desalination is expected mostly on the high consumers, the large-scale and grid-connected desalination applications.

Three desalination processes require only electricity for their operation (instead of thermal processes, such as multistage-flash distillation (MSF) and multieffect distillation (MED) that require thermal and electrical energy) and can be easily coupled with wind turbine(s). These are RO for the desalination of brackish and seawater and ED for the desalination of brackish water and mechanical vapor compression (MVC) for the desalination of seawater. Wind energy to drive RO units is the second most used combination, followed by PV/RO systems. Several applications have been implemented using wind energy to drive RO units and a few to drive MVC and ED units [1]. The technologies combination and their application are presented analytically in the following paragraphs.

3.2 WIND ENERGY

WTs have the tremendous advantage that it can be harnessed at a very wide range of scales; installations can vary from few kilowatt (kW)-scale wind generators up to megawatt (MW)-scale wind farms. WTs can be used as stand-alone (autonomous) applications, or they can be connected to a utility power grid. They can also be combined with a PV system or a diesel generator and energy storage systems, thus forming "hybrid" systems, which are typically used in remote locations, where connection to a utility grid is not available.

In recent years, megawatt WTs have formed the mainstream of the international wind-power market. As the technology is mature it appears that the most important potential development is the further increase in WT capacity. Although focus has been given to large utility-connected wind farms, small WTs have also their position in the world market. There are several applications of in-grid and off-grid areas for small-scale WTs, such as the autonomous wind energy systems for water desalination, water pumping, as well as local power supply systems in conjunction with diesel generators. In addition, the installation of individual small WTs continues in most countries at homes, farms, and small industrial users. For example in the United States, the distributed capacity totals 934 MW, representing more than 75,000 turbines, of which 28 MW (more than 1700 turbines) were installed in 2015. In Italy, small size WTs of less than 200 kW nominal power received easier permitting and better market conditions and at the end of 2015 a total of 50 MW (2000 turbines) were in operation [5].

3.2.1 Wind Energy Turbines: State-of-the-Art

Today the development of the so-called "modern WTs," such as the variable-speed WTs with full power-processing capability, enables their use in a variety of conditions, for example, into a weak or a stiff grid. Modern WTs are typically of horizontal type. The principal elements of a horizontal axis turbine are: the rotor, the multiplier-gear, the electrical generator, the tower, and the control system. The rotor consists of the blades, the hub, and the spin axis. The surface swept by the blades forms the collection area of the turbine. WTs are also classified as fixed or variable pitch, depending on whether the blades embedded in the hub are permanently fixed or not. The inside of the hub houses the hydraulic devices, which change the pitch of the blades, if applicable.

In a typical WT, the kinetic energy of the wind is converted to rotational motion by the rotor. The rotor turns a shaft, which transfers the motion into the nacelle (the large housing at the top of a WT tower). Inside the nacelle of a typical WT, the slowly rotating shaft enters a gearbox that greatly increases the rotational shaft speed. The output (high-speed) shaft is connected to a generator that converts the rotational movement into medium-voltage electricity. The electricity flows down heavy electrical cables inside the tower to a transformer, which increases the voltage of the electric power to the distribution voltage. The distribution-voltage power flows through underground lines to a collection point where the power may be combined with that of other turbines [6,7].

The amount of energy which the wind transfers to the rotor through the blades depends on the density of the air, the rotor area, and the wind speed. The wind speed is extremely important for the amount of energy a WT can convert to electricity. The maximum attainable power from a WT is estimated by the following formula:

$$P_{WT} = 1/2 \cdot \rho_{air} \cdot C_p \cdot \pi r^2 \cdot V^3$$

where

P_{WT}: wind turbine output, kW;

r: rotor radius, m;

V: wind speed, (m/s);

C_p: power coefficient (or wind turbine efficiency);

ρ_{air}: air density at the site, kg/m^3.

As shown by the formula, wind speed has the greatest influence in determining the potential wind power. Moreover, the International electrotechnical commission (IEC) sets international standards for the wind speeds each wind class must withstand (see Table 3.3).

Modern WTs are available on the market in a wide range of nominal power and are classified according to their class [1]. Table 3.1 provides

Table 3.1 Technical data of different models of medium and large wind turbines (WTs) [8]

Technical data	Gamesa G58	VESTAS V90	ENERCON E70	NORDEX SE N117	GE ENERGY GE 2.5–100
Rated power, kW	850	2000	2300	2400	2500
Hub height, m	44/55/65/74	60/65/67/78/80/100	57/64/85/98/113	91/120/141	75/85/98
Tower construction	Conical tubular steel tower	Conical tubular steel tower	No data available	91/20M:tubular steel 141:hybrid tower	Tubular steel tower
Cut in wind speed, m/s	3	4	No data available	3	3
Cut out wind speed, m/s	23	25	28–34	20	25
Rated wind speed, m/s	12	16	N/S	12,5	12
Rotor diameter, m	52	80	71	116,8	100
Number of blades	3	3	3	3	3
Swept area, m^2	2642	5027	3959	10,715	7854
Blade tip speed, m/s	44.3–93.5	45.2–80	No data available	72	73.6
Rotor speed, rpm, U/min	19.44–30.8	10.8–19.1	6–21.5	7.5–13.2	4.7–14.1
Rotor speed control	Variable	Yes	Variable	Variable via MC-system	Active blade pitch
Overspeed control	Blade pitch control	Blade pitch control	Blade pitch control	Blade pitch control	Active blade pitch

Continued

Table 3.1 Technical data of different models of medium and large wind turbines (WTs) [8]—cont'd

Technical data	Gamesa G58	VESTAS V90	ENERCON E70	NORDEX SE N117	GE ENERGY GE 2.5–100
Generator construction	Asynchronous, doubly fed	Asynchronous	Directly driven	Doubly fed ansychronous generator	Permanent magnet
Generator cooling system	Forced convection	Liquid cooling comb	No data available	Liquid-air cooling	Air–air cooling
Grid connection	Directly connected with soft start	2MW	ENERCON converter	Via IGBT converter	Full converter
Generator voltage, V	690	690	No data available	660V ± 10%	690
Gearbox steps	3	3	Gearless	N/S	3
Main brake system	Blade feathering	Individual blade pitch	Individual blade pitch control	Blade pitch control	Aerodynamic brake
2nd Brake system	Disc brake	Disk brake	Pitch control	Disc brake	Pitch, hydr. HSS (high speed shaft) brake pitch
Distance control	Gamesa Windnet			Remote control surveillance system	PC based interface to remote control

the technical data on the medium and large capacity WTs on the market. Table 3.2 provides the technical data of some small WTs available on the market. Table 3.3 provides the WT classes,[1] as provided by VESTAS, one of the largest WT manufacturers.

Table 3.2 Technical data of small wind turbines (WTs) [8]

Technical data	Southwest Windpower Whipser 500	Fortis wind energy Fortis Alize	Jonica Impianti Jimp30
Rated power, kW	3,2	10	20/25/30
Hub height, m	6–20	12–20	18/24/30
Tower construction	Tubular, lattice must	Guyed or free standing mast	Tubular steel tube
Cut in wind speed, m/s	3,1	3	3
Cut out wind speed, m/s	None	None	26
Rated wind speed, m/s	12.5	12	11.5
Rotor diameter, m	4.5	6.4	10.4/11.6
Number of blades	2	3	3
Swept area, m^2	15.9	32	91.6/105.6
Blade tip speed, m/s	N/S	100	N/S
Rotor speed, rpm, U/min	100–600	25–300	162
Rotor speed control	Electronical	Furling	Blade pitch control
Overspeed control	Controller EZI	Dump load	Passive pitch and stall control
Generator construction	Permanent magnet	N/S	Permanent magnet

Continued

[1] IEC classification of turbines: specification of classes of wind turbines (class IA, class IB, etc.) in order to determine which turbine is suitable for the wind conditions of a particular site. Wind classes are mainly defined by the average annual wind speed (measured at the turbine's hub height), the speed of extreme gusts that could occur in a 50-year period, and the degree of turbulence at the site.

Table 3.2 Technical data of small wind turbines (WTs) [8]—cont'd

Technical data	Southwest Windpower Whipser 500	Fortis wind energy Fortis Alize	Jonica Impianti Jimp30
Generator type	SWWP	Direct drive, permanent magnet	Multipole, synchronous, axial-flux
Grid connection	Sliprings	Inverter	Yes
Generator voltage, V	24/48/120	Depending on application	400 max
Gearbox type	Gearless	Gearless	Gearless
Gearbox steps	–	–	–
Main brake system	Tilt rotor	Short circuit switch	Blade pitch control
2nd Brake system	Electronical	Overvoltage control	Electrical brake
Distance control	N/S	N/S	Modem Gsm

Table 3.3 Wind turbine (WT) classes—VESTAS WTs

Turbine class	IEC I High wind	IEC II Medium wind	IEC III Low wind
Annual average wind speed ($m s^{-1}$)	10	8.5	7.5
Extreme 50-year gust ($m s^{-1}$)	70	59.5	52.5
Turbulence classes	A 18% B 16%	A 18% B 16%	A 1.8% B 16%
Turbine/IEC Wind Class	**IEC I High wind**	**IEC II Medium wind**	**IEC III Low wind**
V90-1.8–2.0 MW	x	x	xx
V100-1.8–2 MW	x		
V100-2 MW		x	xx
V112-3.45 MW			×

x: standard IEC conditions, xx: site dependent.
(Vestas, 2016)

Large WTs (of more than 2 MW) can be used in onshore and offshore applications (Fig. 3.2). Mini (in the range of few Watts to 10 kW) and small WTs (in the range of 10–100 kW) are also efficient and can be used for stand-alone or mini-grid applications. The term of medium WTs mostly referred to a range of more than 100 kW and less than 1000 kW nominal power.

Fig. 3.2 (A, B) Foundation for large wind turbines (WTs) in a wind farm in Greece. *(Source: CRES)*

At present, the typical generator capacity is 2–3.5 MW for onshore applications, while in the 1990s the typical generator capacity was in the range of 500–750 kW (see Fig. 3.6). Moreover, the tendency of the wind manufacturers to develop larger wind turbines is constantly continues. Recently wind turbine manufacturer ENERCON GMBH announced the installation of prototype onshore WTs of 4.2 MW nominal power in Germany. According to the manufacturer, E-126 EP4/4.2 MW WTs was developed for sites with moderate wind speeds (Wind Class IIA), while E-14 EP4/4.2 MW are low-speed converters of Wind Class III. According to ENERCON, E-141 can generate more than 13 million kWh per annum at locations where the average wind speed is 6.5 m/s [9]. Moreover, WTs with nominal power of more than 7000 kW for onshore applications are now available. An example is the E-126, 7580 kW gearless WT by ENERCON GMBH. E-126 WT has a swept area of 12.668 m^2, a rotor diameter of 127 m, and a hub height of 135 m.

Fig. 3.3 presents blades and nacelles of 3 and 3.3 MW wind turbines (WTs) before their installation at Agios Georgios Island in Greece. The 73.2-MW wind farm at Agios Georgios consists of nine, V90–3.0MW turbines and 14 V112–3.3 MW machines.

A significant advantage of the technology is that large modern WTs can produce huge amounts of energy with the less possible use of land. An estimation of the energy produced from medium and large WTs is presented in Tables 3.4 and 3.5.

On the basis of the data of Table 3.4 and assuming an average annual energy consumption of a house to be at around 4500 kWh, it is clear that a WT of 2 MW nominal power can cover the energy requirements of 2365 houses.

(A) (B)

Fig. 3.3 (A, B) Wind turbine (WT) blades and nacelles of 3 and 3.3 MW WTs. *(Source: CRES)*

Table 3.4 Estimated energy production from wind turbines (WTs) of different power capacities at 10 m/s average wind speed at 10 m anemometer height

WT nominal power	Estimated annual energy production, kWh	Estimated annual revenues in EURO (US$)[a]
850	4,871,308	438,417 (490,838)
1800	9,800,559	882,050 (987,516)
2000	10,644,125	957,971 (1,072,515)
2300	13,378,471	1,204,062 (1,348,031)
2600	14,328,683	1,289,581 (1,443,776)
3000	16,466,749	1,482,007 (1,659,210)
Weibull shape factor, *k*	2	
Average wind speed at 10 m height	10	
Feed-in tariff (€/kWh)	0.09 (0.10 US$)	

[a] 1 EURO = 1.12 US$.

The above figures will be changed in case of lower wind potential, for example, for average wind speed of 7 m/s at 10-m anemometer height. Based on the calculations the energy production will be as follows (see Table 3.5):

As shown the wind potential of a site is the most vital parameter on the planning of a wind farm (Fig. 3.4). Wind maps provide the information of the average wind speed, which is often used as a general indication of the wind-power potential at a particular site.

Table 3.5 Estimated energy production from wind turbines (WTs) of different power capacities at 7 m/s average wind speed at 10 m anemometer height

WT nominal power	Estimated annual energy production, kWh	Estimated annual revenues in EURO (US$)
850	3,951,149	355,603 (398,275)
1800	9,733,681	876,031 (981,154)
2000	9,866,135	887,952 (994,506)
2300	10,812,180	973,096 (1,089,867)
2600	12,324,758	1,109,228 (1,242,335)
3000	12,936,419	1,164,277 (1,397,132)
Weibull shape factor, *k*	2	
Average wind speed at 10 m height	7	
Feed-in tariff (€/kWh)	0.09	

(A) (B)

Fig. 3.4 (A, B) A 73.2 MW wind farm in Agios Georgios Island, in Greece. *(Source: CRES)*

In Europe, wind potential varies significantly across the continent, as totally different climates can be found from the British Isles and Ireland to the Mediterranean climate of Greece or the Spanish Levant. The wind conditions in Europe are determined by three general elements: water and land distribution, the occurrence of great mountain barriers, and the

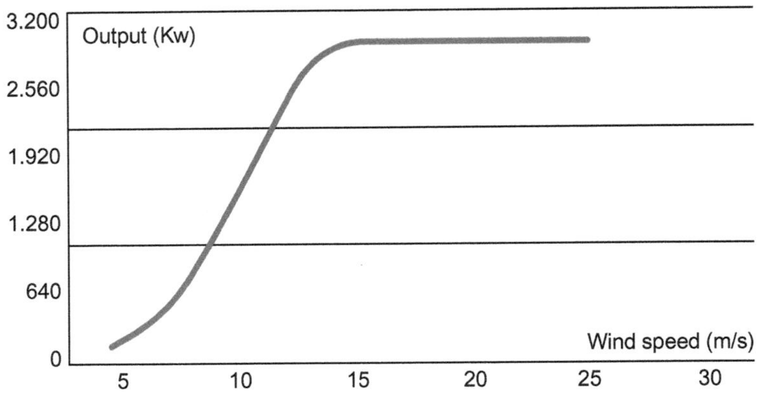

Fig. 3.5 The power curve of the V90 3 MW wind turbine (WT). *(Available at https://www.vestas.com)*

considerable difference in temperature between the polar air of the North and the subtropical air from the South. Furthermore, the effect caused by the presence of great differences in temperature over short distances, coinciding with specific mountainous features, can substantially modify the possibilities of exploiting wind power within a particular area. Although the typical mean annual wind speed required for a viable development is around 7 m/s, turbines can be found on sites characterized by mean speeds of about 5 m/s [7].

The actual WT power output is determined by its power curve. The power curve of a WT is a graph that indicates the electrical power output for the turbine at different wind speeds. Fig. 3.5 presents the power curve of a 3 MW WT.

Moreover, a WT is characterized by the cut-in wind speed (i.e., the speed at which it starts to generate power), the rated wind speed (i.e., the speed at which it starts to generate power at the rated level), and the cut-out wind speed (i.e., the speed at which the unit is shut down for safety reasons). The cut-in speed is around 3–4 m/s. The cut-out speed typically lies between 25 and 30 m/s (see Tables 3.1, 3.2).

In addition, the energy efficiency of a WT is defined by the capacity factor (%) and the availability (%).

The annual capacity factor is the amount of energy a generating plant (e.g., a wind farm) produces over the year divided by the amount of energy that would have been produced if the plant had been running at full capacity during that same time interval. For WTs, the capacity factor is dependent on

- the quality of the wind resource,
- the technical reliability, and
- the size of the generator in comparison with the length of the rotor blades. Long blades improve the capacity factors especially at low wind sites.

Most wind-power plants operate at a capacity factor of 25%–40%. Offshore WTs generally have higher capacity factors due to excellent winds [5].

Finally, the availability (%) of a wind power plant is the percentage of time during the year that the WT is able to operate.

3.2.2 WTs Progress

The use of WTs to produce electricity is already playing an important role in the European countries and the rest of the countries. Globally, a record of more than 63 GW of wind power was added in 2015, reaching a total of about 433 GW installed capacity (Fig. 3.6). Wind power was the leading source of new power generating capacity in Europe and the United States as well as in Canada in 2015, and the second largest in China. Wind-generated electricity encountered close to 4% of the world's electricity demand in 2015.

Wind power is playing a major role in meeting the electricity demand in an increasing number of countries, including Denmark (meeting more than 40% of demand); Portugal, Spain, and Ireland (close to 20%); Germany (close to 15%); and Sweden and the United Kingdom (close to 10%) [5]. According to the European Wind Energy Association Annual Statistics in 2015, a total of 141.6 GW was installed in the European Union, a record growth of 9.7% [10]. Fig. 3.7 presents the cumulative wind power installations in EU.

(A) (B)

Fig. 3.6 (A, B) View of a medium wind turbine (WT) of 500 kW (E-40) at CRES wind farm in Keratea Attikis, Greece.

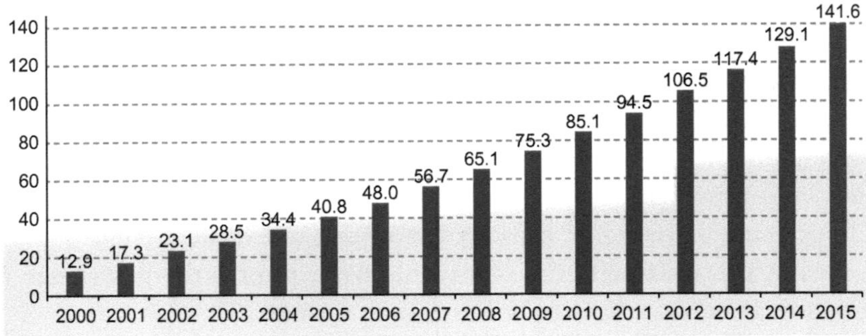

Fig. 3.7 Cumulative wind power installations in GW per year in EU [10].

3.2.3 WT Economics

The economics of a wind project depend on the technical, resource, and cost parameters. The latter two vary from country to country. The total installed project cost mentioned in Table 3.6 includes the costs for turbines, roads, electrical equipment, installation, development, and grid connection.

In general, the cost of the equipment represents the highest portion of the total investment cost. The costs for electrical and civil engineering infrastructure, transport, management and administration, land, etc. can vary between 25% and 35% of the cost of the installed turbines. The annual operating and maintenance cost accounts for around 1.5% of the initial investment. In addition to the above-mentioned cost, a percentage of around 1% per annum of the initial investment should be considered for insurance.

The levelized cost of wind electricity depends mainly on the wind potential of the installation site and the size of the WT used. The levelized cost of electricity (LCOE), in euro cents per kWh of wind-generated power, is calculated as a function of the wind speed (number of full load hours) and the discount rate. The LCOE varies from approximately 0.08 €/kWh to 0.10 €/kWh at sites with low average wind speeds to about 0.05 €/kWh

Table 3.6 Estimated average wind turbine (WT) cost and total project costs for 2015 [5]

Country	Turbine cost (€/kW)	Total installed project cost (€/kW)
China	639	1327
Ireland	900	1650
Spain	935	1320
Belgium	1251	1686
Portugal	1308	1635
Japan	1660	2280

to 0.065 €/kWh at windy coastal sites, and to approximately 0.075 €/kWh at a wind site with average wind speeds.

Wind energy has developed to such an extent that for many sites and many countries it is becoming the most economic new electricity supply option. In Quebec, the latest contracts have an average price of 0.045 €/kWh for the generated energy. In China, under the current technology, without considering the cost of long distance transmission, the cost of wind power is higher than that of coal-fired power by 0.027 €/kWh. If resources and environmental benefits are taken into consideration, the cost of wind power was nearly equal to that of coal-fired power generation. The targets for wind power are aiming at cost reductions, to reach the coal power plant cost of energy level in 2020 [5].

Despite the decrease in the cost of medium and large WTs, small WTs cost is still high. As an example, small wind generators in the range of 1.5–15 kW cost from around 1500 to 60,000 €, meaning from around 1000–4000 €/kW. WTs with nominal power from 15 to 100 kW cost from 60,000 to 220,000 €.

3.2.4 WT Current Market

The prospects for the global wind industry are promising. Regarding the WT industry, the average rated capacity of new WTs installed in 2015 continued the tendency for larger installations. Several countries have also reported increased productivity: the turbine designs with larger rotors for same installed capacity mean more generation per installed power. The trend of taller turbines reaching better wind resource is still continuing. European wind manufacturers have a significant market share during 2015, while non-European WTs are gaining increased shares in the global market.

Small-sized turbines, of less than 100 kW, are also being manufactured in Italy, Spain, the United Kingdom, United States, and China.

3.3 CONVENTIONAL DESALINATION SYSTEMS

The major commercially available desalination technologies are split into two main categories: thermal (or distillation) and membrane processes. The most widely used thermal technologies for seawater desalination are: the MSF, the MED or ME, and the vapor compression [thermal vapor compression (TVC) and MVC] process.

Membrane technologies for seawater and brackish water desalination mainly involve RO and ED, EDR technologies. ED is confined to the desalination of brackish water, while RO can be used for both, brackish and seawater desalination.

A critical issue of a desalination system is the energy required to drive the desalination plant, the energy availability, and the energy cost. In general, the energy requirements of thermal processes are considered as higher than those of RO for plants with the same water capacity and for seawater desalination.

An example of the energy requirements of several desalination technologies is presented in Table 3.7 [11]. The presented values include plant auxiliaries (such as seawater intake, potabilization, and chlorination systems, etc.).

The energy consumption of RO technology is related to the state of the art of the membranes that are used for the desalination system. New membranes are continuously developed with lower transmembrane pressures (TMP) and therefore lower specific power consumption. In addition, the seawater RO technology has been able to benefit from important improvements of the energy recovery devices (ERD). At present the ERD are highly efficient techniques, are market available at a large range of brine water flows at a significant cost. With the use of ERD, SWRO technology in large plants is capable of achieving low energy consumptions, in the range of 2.5–2.0 kWh/m^3. ERD for brackish water RO plants are also available in the market.

Based on the RO technologies improvements a tremendous increase of RO plant installations have been achieved worldwide during the last decade, even in the Gulf region, where great advances in thermal processes had been taking place for many years. An important number of large seawater projects, mostly in the Gulf region have been accomplished, as 2016 marked the third yearly increase in seawater capacity, from 1 million m^3/day additional capacity in 2014 to 1.6 million m^3/day so far in 2016, exceeding 2015s total by 200,000 m^3/day. Overall, the global cumulative installed and online capacity equated to 73% membrane-based desalination plants and 27% thermal. In contracted capacity for 2015 to June 2016, 93% was membrane and 7% thermal (73% MSF and 27% MED) [12].

3.4 WIND DESALINATION: STATE-OF-THE-ART

Most existing desalination systems use fossil fuels, or electrical power derived from fossil fuels, as the main energy source, creating rising environmental concerns. Over the last few decades, wind-power technology has come

Table 3.7 Desalination technologies energy consumption

	Specific electric power kWh/m^3	Specific heat consumption kJ/kg	Steam extraction pressure Bar abs	Thermal energy Thermal kWh/m^3	Equivalent power loss Electric kWh/m^3	Total energy requirements kWh/m^3
SWRO (MED Sea)	3.5	0	n.a.	0	0	3.5
SWRO (Gulf)	4.5	0	n.a.	0	0	4.5
MSF	4–5	287	2.5–2.2	78	10–20	14–25
MED–TVC	1.0–1.5	287	2.5–2.2	78	10–20	11–21.5
MED	1.0–1.5	250	0.35–0.5	69	3	4–4.5

of age and in some cases is even competitive with conventional energy sources and is therefore one of the most promising RES [13].

A significant number of units coupling WTs with desalination technologies have been designed, installed, and tested around the world (see Table 3.8]. The most common choice is the stand-alone WT systems to drive RO units for seawater or brackish water desalination. Other combinations are wind turbine(s) to drive MVC and ED, EDR plants. Moreover, WTs can be also used in conjunction with other conventional or renewable power sources (e.g., diesel, PV), known as hybrid systems (see Fig. 3.8) [16].

The history of wind desalination combination most likely started in the decade of 1900s. As early as 1982, a small system was set at Ile du Planier, France; a 4 kW WT coupled to a 0.5 m^3/h RO desalination unit. The system was designed to operate via either a direct coupling or batteries. Another case where wind energy and RO combined is that of Island of Drenec in France, in 1990. The WT was of 10 kW nominal power and was used to drive a seawater RO unit. Another application was carried out at a test facility in Lastours, France, where a 5 kW WT provides energy to a number of batteries (1500 Ah, 24 V) and via an inverter to an RO unit of 1.8 kW installed power [17].

The first applications were followed by a significant number of wind desalination projects which were implemented within EU and non-EU programs and by several research institutes and universities around the world. A great work concerning with feasibilities studies, demonstration and pilot units development and testing in laboratories and real conditions, logistic tools, etc., have been implemented within several EU, non-EU, and national projects such as APAS, ADU RES, IRESMED, INTERSUDMED, PRODES, AERODESA, SDAWES, AEROGEDESA, MORENA, OPRODES, DeReDes, etc. The National Renewable Energy Laboratory, NREL, in USA, the Canary islands Institute ofTechnology, ITC, and the University ofLas Palmas in Spain; the Centre for RES, CRES, and the Agricultural University of Athens, AUA, in Greece; the Fraunhöfer Institute in Germany; the Middle East Desalination Research Center, MEDRC, in Oman; the King Abdullah University of Science and Technology, KAUST, in Saudi Arabia; and many other organizations around the world have investigated and/or in funding the combination of the two technologies mostly in off-grid applications.

The most applicable wind desalination combinations are reviewed in the following paragraphs.

One of the drawbacks of utilizing wind power and in general RES in desalination stand-alone (off-grid) applications is the variable nature of

Table 3.8 Wind desalination installations around the world

Plant name	Location	Water capacity m^3/day	Desalination technology	RE source type	Nominal power kW	Year commissioned
Wind RO plant	Canada	10	RO	Wind turbine	-	1974
PV-Wind, RO, ED	Borj-Cedra, Tunisia	6 (RO)	RO, ED	PV, WTs	0.38 + 3.5 PV	1981
Wind RO plant	Coast of the North sea, Germany	4	RO	WT	WT:11	1982/83
Wind RO plant	Ile du Planier, Pacific Islands	12	RO	WT, 2 coupling systems (direct and with battery)	WT:4	1982/83
Wind RO plant	Island of Suderoog, Germany	6–9	3 RO-modules	Wind turbine	WT: 6	1983
Hybrid RO desalination plant	Western Australia	0.2	RO	WT, PV, Diesel genset	PV:32	1988
Wind MVC Plant	Borkum Island - North Sea, Germany	48	MVC	WT	WT: 45	1993
Wind SWRO plant	Patras Science Park, Patra, Greece	5–25	RO	WT	WT: 30	1994/5
Wind–MVC	Rugen Island, Germany	120–300	MVC	WT (Tacke TW 300)	WT: 300	1995

Continued

Table 3.8 Wind desalination installations around the world—cont'd

Plant name	Location	Water capacity m^3/day	Desalination technology	RE source type	Nominal power kW	Year commissioned
Wind powered-diesel system (VALOREN project)	Punta Jandia, Fuerteventura, Canary Islands, Spain	56	RO	WT V27, Diesel, Flywheels	WT: 225, Diesel:120	1995
Wind distillation sea water plant	Layang layang Island, Malaysia	16.8	Evaporator/ Distiller	WT, Diesel genset	WT: 150, Diesel:150	1995
Wind RO desalination plant, Tenerife (JOULE program)	Tenerife, Spain	60–110	RO	WT	WT: 30 kW	1997/8
Hybrid BWRO plant [14]	Maagan Michael, Israel	9.6	RO	WT, PV, Diesel genset	WT: 0.6, PV: 3.5, Diesel: 3	1997
Wind RO plant (JOULE program)	Syros island, Greece	60–900	RO	Gearless WT	WT:500	1997/8
SWRO-wind plant (APAS Project)	Therasia island, Greece	4.8	RO	WT	WT:15	1997
Wind MVC plant (SDAWES project, JOULE)	Gran Canaria, Spain	8 × 25, 190, 50	RO, EDR, MVC	WT	WT:2 × 230	1999–2002

Hybrid SWRO (PAVET project)	CRES Wind Farm, Greece	3.12	RO	Hybrid with energy storage	3.9kWp PV, WT:0.9	2001/2002, the RO unit redesigned in 2004
Wind RO	Loughborough, UK	12	RO	WT	WT:2.5	2003
Wind RO (AEROGEDESA project)	Pozo Izquierdo	19.2	RO	WT	WT:15	2003/2004
Windmill RO plant [15]	University of Delft, The Netherlands	4.8–9.6	RO	Windmill	No energy storage	2007/2008
Wind RO, Grid connected (OPC project)[a]	Milos ilsand, Greece	3600	RO	WT	WT: 850	2008
Wind MVC (OPC project)	Symi island, Greece	280–360	MVC	WT NORDWIND	WT: 750	2009

[a]OPC: Operational program for competitiveness, Greek Ministry of Development.

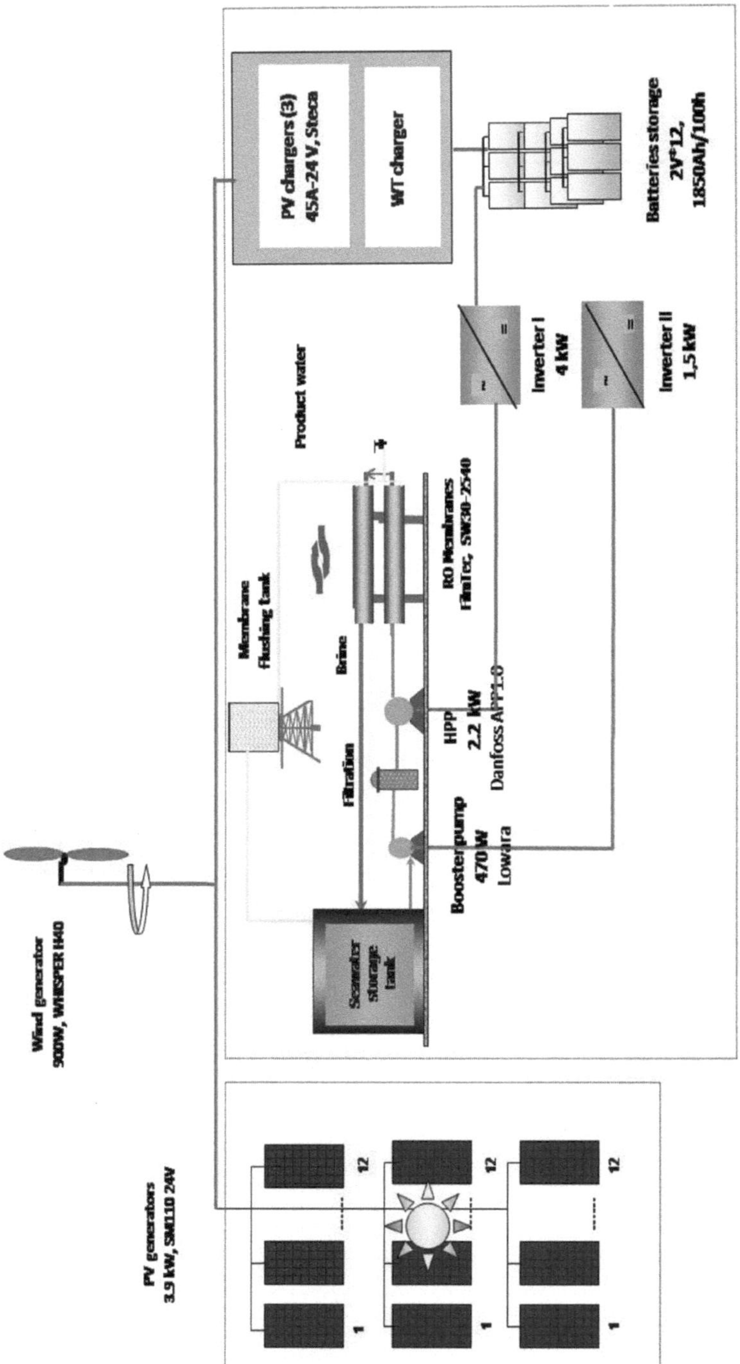

Fig. 3.8 Schematic of CRES pilot hybrid RO unit [16].

the resource. Variable power input forces the desalination plant to operate in nonoptimal conditions, such as unstable feed flow rate and pressure, and this may cause operational problems. Especially for RO systems the reduction of the membrane lifetime could be a consequence of the nonstable operation. Several small autonomous units with operation under variable conditions have been tested in a laboratory environment by research centers and universities [18–20]. To achieve reliable conclusions regarding the operation of an RO unit under variable conditions a sufficient period of almost 1 year of daily operation is recommended. In real applications, parameters such as the availability of potable water at the required quality and quantity to the consumers, the reliability of the system and the necessity of proper operation and low maintenance cost should be considered during the design of an autonomous RES desalination unit.

To smooth power variations battery storage systems are commonly used. A battery system is typically used for medium-term storage, whereas flywheels and supercapacitors can be used for short-term storage [21]. In addition, batteries are not only used to protect the system from fluctuations and intermittent operation, but also can be utilized to run the desalination unit in periods where wind energy is not sufficient. The battery systems that have been mostly used in the existing wind RO applications are lead–acid batteries.

Lead–acid batteries are ideal for long-term energy storage due to their high energy density, efficiency (85–90%), and their low self-discharge rate of 2% of rated capacity per month. Lead–acid batteries have relatively low cost, of about 120 $/kWh. The main disadvantage associated with lead–acid batteries is their lifespan, characterized by a low cycle lifetime (500–1800 cycles) and reduced life caused by deep discharging or operation at high temperatures [21]. The depth of discharge (DOD) is the amount of the battery's capacity that has been consumed and it is expressed as a percentage of the battery's full energy capacity. The deeper is a battery's discharge, the shorter will be the expected lifetime. This is true of several cell-based batteries due to cell degradation, including lead–acid and lithium-ion ones. For example, if a battery discharges 10% of its full energy capacity, 90% of the full capacity is unused. This corresponds to 10% DOD. This battery will be able to complete more charging cycles (defined above) than a battery cycled at deep discharge. Deep cycle is often defined as 80% or more DOD [22]. Moreover, recent data show that a lead–acid battery would have to be 2.5 times larger in capacity than a lithium-ion battery to achieve a comparable cycle life.

Lithium-ion battery technology was introduced in the 1970s, and took 20 years for it to gain popularity [23]. Lithium-ion batteries live longer,

Table 3.9 Lifetime cost comparison of VRLA lead–acid batteries and Lithium-ion batteries

	VRLA (moderate climate: 25°C	**VRLA (hot climate: 33°C)**	**Lithium-ion**
System size	100 kWh	100 kWh	62.5 kWh
Battery cost	$12,000 ($120/kWh)	$12,000 ($120/kWh)	$37,500 ($600/kWh)
Cycle life	1000 at 50% DOD	500 at 50% DOD	1900 at 80% DOD
Installation cost	$20/kWh	$20/kWh	$3.6/kWh
Transportation cost	$28/kWh	$28/kWh	$5/kWh
Lifetime cost	$0.34/kWh throughput	$0.67/kWh throughput	$0.40/kWh throughput

recycle better, require much less maintenance, and their cycle lifetime is not affected by environmental conditions such as temperature. A lithium-ion system costs around 600 $/kWh (and more in some countries), with high expectations of dramatic cost reduction in the coming years. The lifetime cost, which is measured in $/kWh or €/kWh, is the critical number to understand the system economics. Table 3.9 presents a comparison of the lifetime cost of lead–acid VRLA and lithium–ion batteries [24].

Furthermore, charge controllers are used to protect the batteries from overcharging, and inverters are used to convert the DC current from the battery output to AC for the load. A diesel genset can also be used as backup to charge the battery bank or to drive the SWRO unit directly.

The supercapacitors are energy storage devices that operate by building up positive and negative charge within an electrolytic solution, similar to the chemical reaction utilized in batteries. They are also referred to as ultracapacitors or electrochemical double-layer capacitors and are one of the most promising devices for buffering short-term fluctuations from RE sources [21]. The supercapacitors are able to provide sufficient energy during periods of no wind (intermittency) and enhance the power quality delivered to the membrane by absorbing turbulent wind (fluctuations).

Fig. 3.9 presents the configuration of an isolated microgrid model including WTs (doubly fed induction generator DFIG), seawater desalination devices, diesel genset, and hybrid energy storage (battery and supercapacitor), as studied by the State Key Laboratory of Alternate Electrical Power

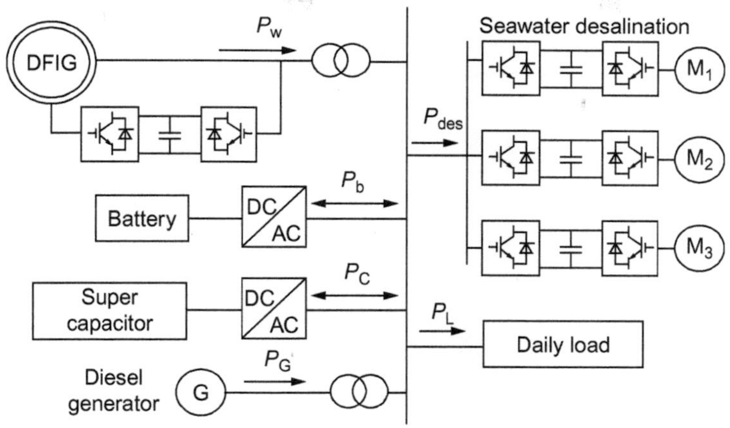

Fig. 3.9 Configuration of a microgrid system with wind turbines (WTs) and seawater desalination [25].

in China [25]. Fig. 3.10 presents the configuration of a pilot wind RO plant for brackish water desalination developed by NREL in USA.

In addition, the desalinated water can be used as temporary energy storage, thus providing a means for the "regulation" of one of the most important inherent characteristics of RES, that is, their intermittence. When the system is grid connected, the desalination plant can operate continuously as a conventional plant and the renewable energy source merely acts as a fuel substitute. In such cases, the desalination load could be also used to moderate the amount of energy injected into the grid. This can be a very useful concept for hybrid electricity systems, with high use of intermittent renewable electricity sources.

According to the available data from the existing installations the levelized cost of water from small wind RO systems ranges between 1.5 and 7 €/m^3, with lower values relevant to larger installations. The already installed hybrid systems, like the one developed by CRES in 2001 for seawater desalination (SWRO), and the one for brackish water desalination (BWRO) developed by the Planning, Development, and Technology Division of Israel Electric Corporation at Maagan Michel in 1997, presented higher unit water costs (>7€/m^3), mainly because of the high capital cost of PVs at that time and the small size of the RO plants (see Table 3.8). At that time, small-scale ERD were not available in the market or were available at some countries markets at very high cost, and the consumed energy of the small RO units was extremely high. The cost of small WTs was also high in the order of 3000 €/kW.

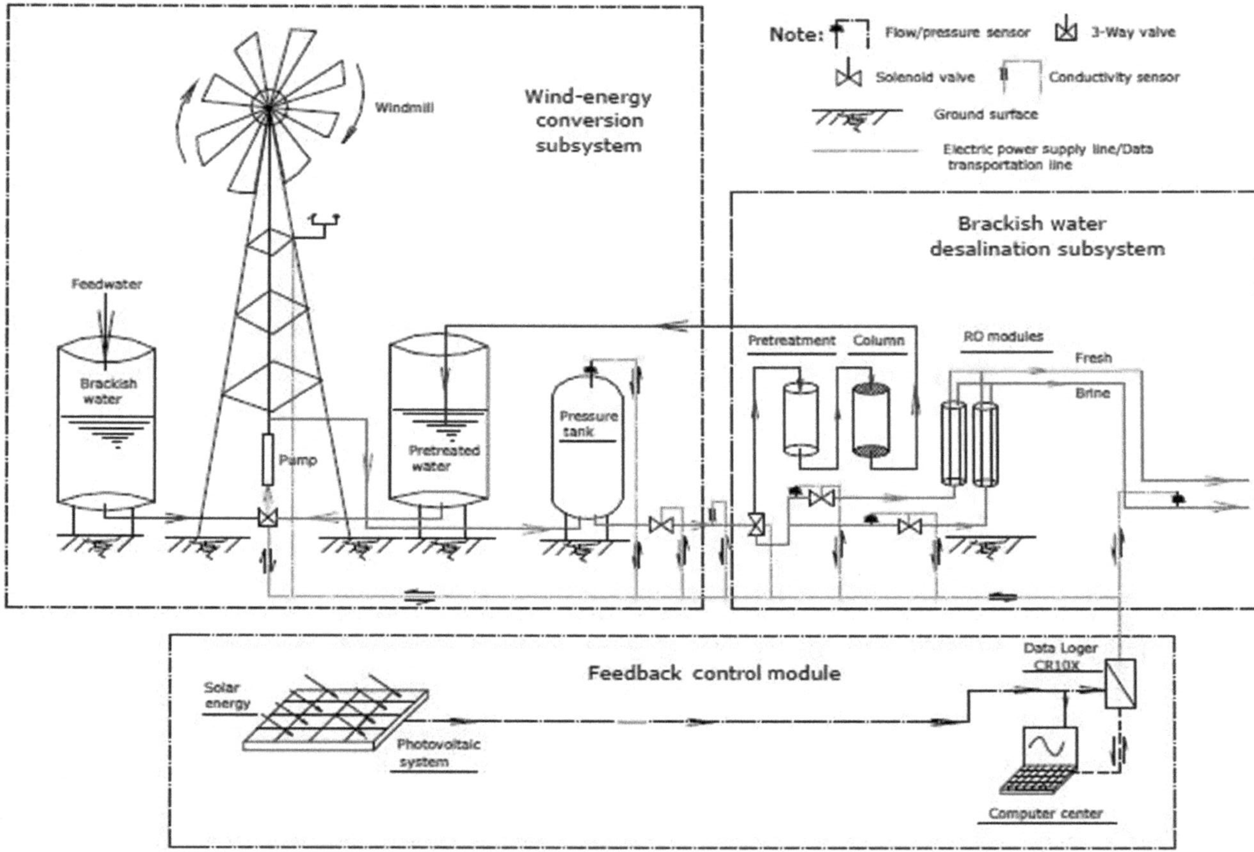

Fig. 3.10 Schematic of NREL wind RO pilot plant, 2009 [26].

However, due to the lack of detailed information on the calculation of the levelized cost of water from the several cases, it is not possible to provide specific values. Moreover each case, especially for off-grid installations, is site specific and mainly depends on the site characteristics and the design of the system.

3.4.1 Wind MVC

MVC technology can be driven by RE technologies providing electrical power. So far, the only combination of MVC with RES concerns its matching with wind turbine(s) [1,7,27–30]. The WT supplies electricity to the compressor, to the auxiliary heater, and to all the peripheral equipment, allowing the system to operate stand-alone (Fig. 3.11). Few applications have been implemented using wind energy to drive a MVC unit. A pilot plant was installed in 1991 at Borkum Island in Germany, where a WT with

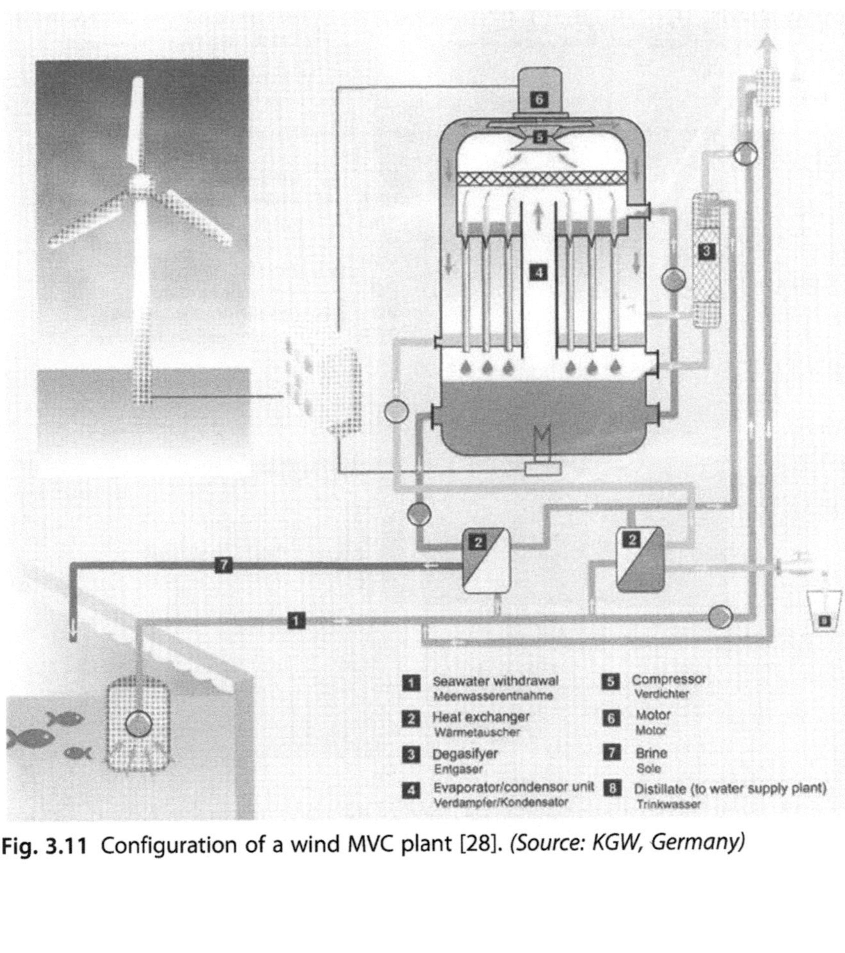

Fig. 3.11 Configuration of a wind MVC plant [28]. *(Source: KGW, Germany)*

a nominal power of 45 kW was coupled to a 48 m^3/day MVC unit. A 36-kW compressor was required. The experience was followed in 1995 by another plant at the Island of Rügen. A WT with a nominal power of 300 kW was coupled to a 300 m^3/day MVC unit. Additionally, in 1999, a 50 m^3/day wind MVC installed within SDAWES project at Gran Canaria in Spain (Fig. 3.12), while in 2008 a Wind MVC plant has been built on Symi Island in Greece to cover the water needs of the area (Figs. 3.13 and 3.14). In general, the water capacities of wind MVC units that have been installed and tested are in the range of 50 to around 500 m^3/day. The specific energy consumption of MVC units is high ranging from 9 to 20 kWh/m^3. According to

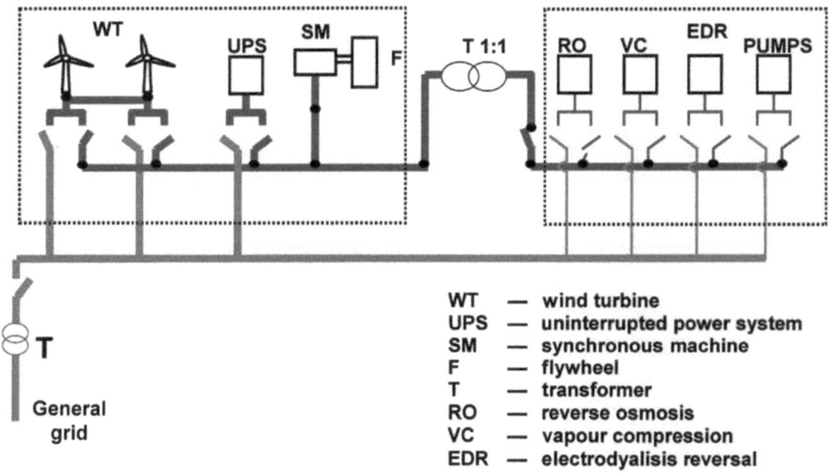

Fig. 3.12 SDAWES system configuration [27].

(A) (B)

Fig. 3.13 (A, B) Transportation and installation of the MVW unit in Symi Island. *(Source: WME-GmbH)*

(A) (B)

Fig. 3.14 (A, B) Installation of the wind turbine (WT) at Symi Island, Greece.

the literature, the most recent achievement (2010–2011) regarding the coupling of MVC and wind energy concerns with the research and development of an MVC system powered exclusively by wind and solar energy for the treatment of brackish groundwater. The MVC system utilizes wind power to drive the system mechanically (operation of the main load-compressor) and solar power to preheat the feedwater. The overall goal of the project was to develop and construct a wind-driven MVC system to demonstrate that the technology can be used for the treatment of brackish groundwater to provide drinking water at remote locations. The research was implemented by Montana Tech of the University of Montana in The United States. The project was awarded by United States Environmental Protection Agency, EPA [31].

3.4.2 Wind ED

Most of the RES ED/EDR applications have been considered until today concern with the coupling of ED/EDR with PVs. Regarding wind ED/EDR matching, the most known and long-term testing application is the one implemented within the project "Seawater Desalination with an Autonomous Wind Energy System—SDAWES project," (Joule III Program), (Fig. 3.15). Commissioning and testing period was carried out from 1999 to 2002. The main objective of the project was to identify the most suitable desalination system for connection to a stand-alone wind farm, and to assess the influence of the variations of the wind energy on the operation of desalination plants [32]. The EDR plant was tested in on-grid and off-grid operation. The two operation modes were examined and compared. A control system has been developed for automatic management and optimization of plant operation, for both modes. The control system was capable of setting the optimum operating condition, based on the available power and the

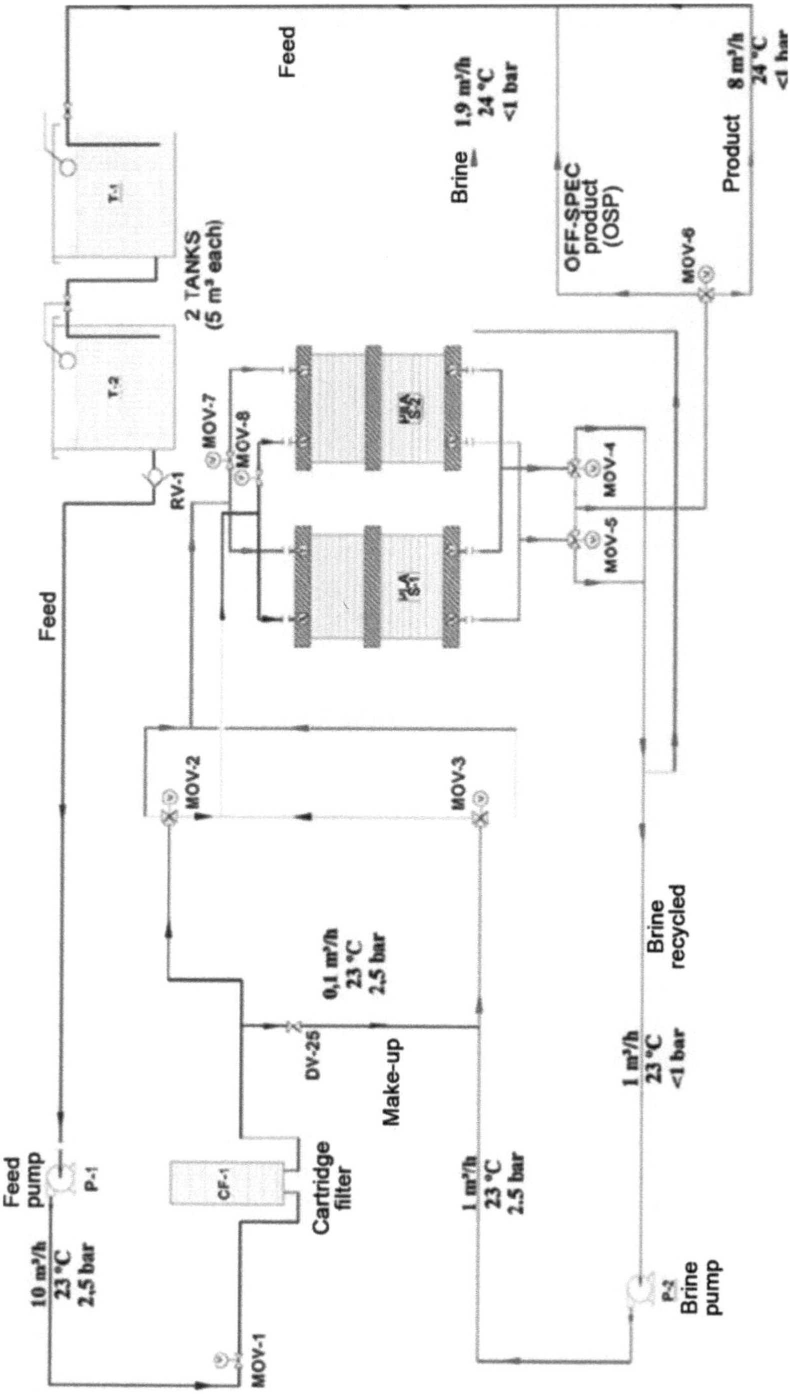

Fig. 3.15 Configuration of the EDR unit, SDAWES project.

criteria predefined by the plant operator [33]. The EDR brackish water desalination unit under testing involved two stacks and its hourly capacity ranged from 3 to around 8 m^3. The electrical conductivity of the feedwater ranged from 2500 to 7500 μS/cm, the product water conductivity ranged between 200 and 500 μS/cm, and the power requirements ranged between 4 and 19 kW. The average specific energy consumption was about 3.0 kWh/m^3. The EDR unit was operated at a conversion ratio of 35%–75%. The unit was a modified design of a standard module (Aquamite V-II, Ionics), in order to account for the variable input from the WT.

Each of the two EDR stacks had four hydraulic stages and two electrical stages, consisting of 340 cell pairs. More analytically, the design of the unit was modified in order to include DC drivers for each electrical stage (E1, E2), in each one of the membrane stacks, variable frequency drivers (VFD) for both the feed pump and brine recycling pump, and a revised control system are provided. Specifically, the electrical stacks are provided with drivers capable for 38 kW (0–170 V_{DC}) for stage E1 and 19 kW (0–170 V_{DC}) for stage E2. The feed pump (3.5 kW max) and brine recycling pump (2.5 kW max) operated via VFD, which modified the motor speed in order to control the feed pressure and the stack inlet differential pressure, respectively. The WTs used to drive the desalination unit have a nominal power output of 230 kW each.

According to the researchers, the desalination unit presented good flexibility, adapting smoothly to variations in wind power, even when sudden drops or rises occurred. The control system, slightly modified from a standard design, can cope with such sudden variations. Good agreement between performance simulated with software and the actual operating performance was observed. However, it is mentioned that the presence of harmonics in the electric system due to DC drivers and due to the VFD became harmful for the control and electric system, and care must be taken through appropriate mitigating measures [34].

3.4.3 Wind RO

The coupling of WTs with RO units is the most common combination and an important number of installations exist. RO technology for seawater and brackish water desalination is market available at a wide range of water capacities, is modular, easy in operation, and is the less energy intensive desalination technology with the use of ERDs. Most of the wind RO applications are related with small-scale off-grid systems for seawater desalination which are using energy storage systems to ensure proper operation of the RO unit. Fig. 3.16 presents the configuration of a typical RO unit which can be fitted in container(s) even at large water capacities (Fig. 3.17).

A typical stand-alone wind RO system involves the following equipment:

- Wind generator(s),
- Charge controller, used for the protection of batteries from overcharging,
- Energy storage, usually a battery bank is used for power stability and to provide some days of autonomy to the RO unit,
- Inverter(s), used to convert the DC current from the battery output to AC for the load (RO),
- An RO unit.

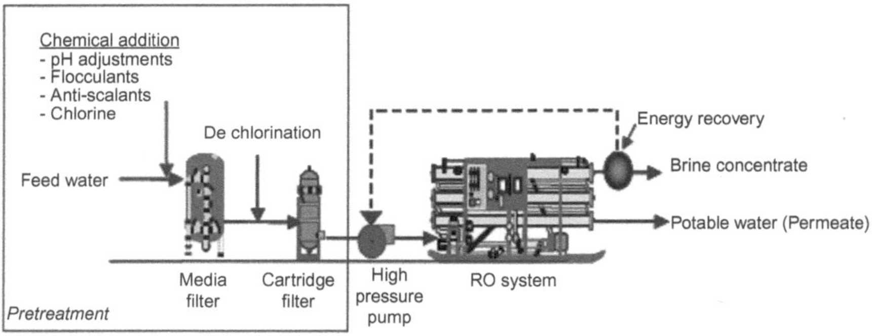

Fig. 3.16 Typical configuration of an RO plant [41].

Fig. 3.17 Containerized SWRO plant in Tinos Island, Greece. *(Source: CRES)*

A diesel backup system can also be used to charge the battery bank or to drive the RO unit directly.

The water capacity of the already installed wind RO plants is in the range of some liters to around 1000 m^3/day. Several plants were installed and tested during the 1900s mostly within the EU projects, as presented in Table 3.8. Since then, there is a significant market progress of the RE and desalination technologies but less support on the research and demonstration of the more mature technologies and their matching. Regarding stand-alone off-grid cases there is a tremendous need to decrease the cost of the water produced in order to make the technologies matching more competitive.

As it has been mentioned the cost from the existing small-scale wind RO plants is in the range of 1.5–7€/m^3, which is still high. ERDs are used to recover the energy from the pressurized reject brine, thus improving the overall efficiency of the system. Their use increases the initial cost of the system but effectively reduces the energy requirements. Another factor that affects the unit water cost is the high capital cost of the small WTs in the range of 15–100 kW nominal power.

Traditionally, off-grid applications involve the use of battery banks which also increase the capital and running costs of the system. However, their use ensures, at a high percent, the quality and quantity of the produced water and the proper operation of the system.

To avoid the use of batteries and reduce the water cost from RES desalination units, ITC has tested five different off-grid RE-powered RO units, either wind powered or PV powered, since 1998. For the case of wind RO the operation under variable energy availability is managed as follows:

(a) When wind power increases: pitch control in wind generators.

(b) When wind power decreases: reduction of the power consumed in desalination units and pumps (partial load operation and/or stop of units).

(c) Transitory periods are possibly due to the use of energy storage systems: flywheel for short periods (seconds) and batteries for long periods (hours).

A more recent project entitled SODAMME was implemented at Pozo Izquierdo in Spain by ITC and the University of Las Palmas. The project started in 2009 and ended in December 2015. In SODAMME, an SWRO unit of 0.1–0.8 m^3/h variable product water capacity operates with the use of a 15 kW WT without a battery system and under variable capacity control. The small-scale prototype desalination plant was designed to continuously adapt its energy consumption to the variable power supplied by the WT, with mass energy storage in batteries and with the use of a supercapacitor bank as a dynamic regulation system [35]. According to the researchers, one of the main conclusions from the implemented project is the viability of adapting the consumption of the prototype of the SWRO desalination plant to widely varying WT-generated power. Despite using various time interval lengths in which it was assumed that the WT output power remained constant, a perfect fit was not obtained between the theoretical WT-generated power and the power consumed by the SWRO desalination plant, nor was it possible to maintain a constant permeate recovery rate at each instant.

Moreover, ITC has worked on the design and comparison of two innovative large-scale wind SWRO plants for off-grid operation with fixed and gradual water capacity of the desalination unit [36]. Three commercial WTs of 100, 225, and each of 300 kW nominal power, are selected for the comparative analysis. The SWRO plant with the gradient water capacity consists of three RO trains (or racks–Rack 1, 2, 3) with water capacities of 200, 400, and 400 m^3/day, respectively (see Fig. 3.18).

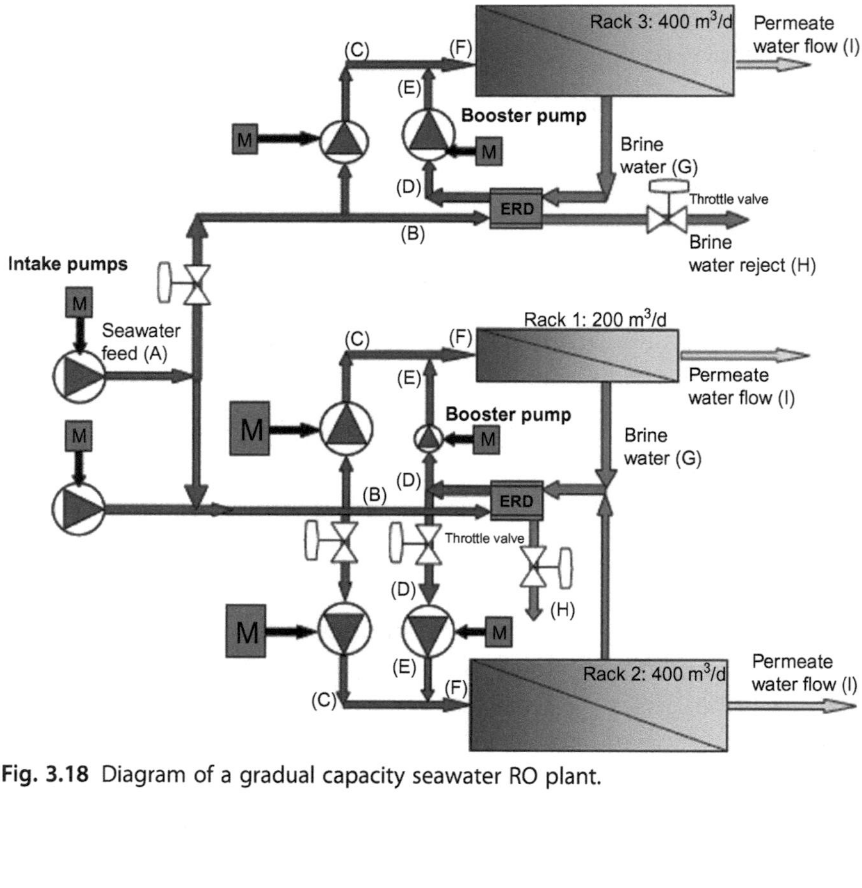

Fig. 3.18 Diagram of a gradual capacity seawater RO plant.

According to the scientists in the first case (RO with fixed water capacity), the plant can operate within a more restrictive range of power consumption when the wind resource is enough to supply all the energy required by the RO plant. In the second case, the plant is more versatile because it is able to operate the RO trains independently or jointly, depending on the amount of energy available.

Another recent application concerns with the coupling of a 50 kW WT to an RO unit for brackish water desalination in the City of Seminole in Texas. The demonstration project has been accomplished by Researchers from the Texas Tech University (TTU), the National Wind Institute (NWI), and the Water Resources Center (WRC) with the aim to explore the potential for connecting locally owned renewable wind energy systems with the electrical needs for pumping and treating the available brackish groundwater (Figs. 3.19 and 3.20).

The project objectives included the installation of an around 550 m deep well in the Dockum Aquifer, a BWRO system of 9.3 m^3/h permeate water capacity, and a 50 kW WT, as well as other related infrastructure to collect and report useful data from the demonstration project. The operation began in April 2013 and ended in August 2014. The results provide operational experience and data for planning of larger-scale systems.

The RO unit operated in high recovery of about 73%. The feedwater source was brackish water of 7980 mg/L, while the permeate water has a low salinity of 520 mg/L.

The EW50 WT generated 37,054 kWh in 4276 h of operation. The well and RO system energy demand during their operation was about 78,000 kWh. The cost of energy per volume of permeate produced at an

(A) (B)

Fig. 3.19 (A, B) Wind RO units at the demonstration site, City of Seminole, Texas [2].

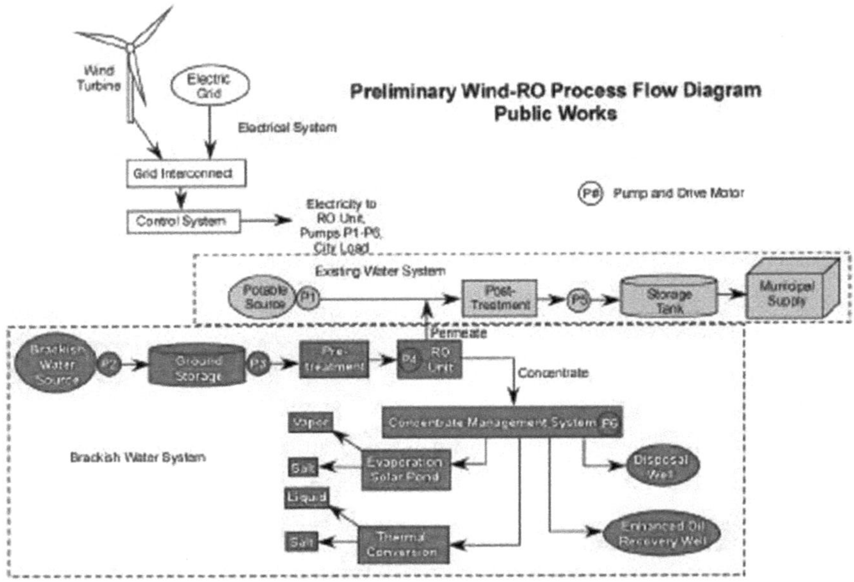

Fig. 3.20 Integrated wind-RO layout for an inland municipality [26].

average of 9.31 m^3/h (41 gpm) was about \$0.33/1000 gal[2] at the average \$0.062/kWh charged locally by Xcel Energy.

The amount of energy generated by the WT was 47% of the demand of the well and RO system. According to the researchers, the exact times of operation of the turbine and well and RO system did not always coincide. Nevertheless, the amount of energy generated was significant and encouraging as a renewable energy source alongside the energy from the grid [2].

The cost of the 50 kW WT was \$250,000, in which is included the turbine and tower, site engineering, foundation, WT assembly and start up, and a five-year warranty. The turbine's nacelle and blades are supported by a 30.5 m (100-ft[3]) galvanized steel lattice tower at a centerline hub height of 31 m (102 ft). Fig. 3.21 presents the power curve of the 15 kW WT. Fig. 3.22 shows the installation of the nacelle at the top of the lattice tower. More technical specifications for the WT are presented in Table 3.10.

Moreover, in 2014 two pilot projects regarding wind SWRO desalination were implemented by the University of Stavanger in Norway [37]. Pilot

[2] 1000 US gallons = 3.7854 m^3, 1 US gpm = 0.2271 m^3/h.
[3] 1 foot = 0.3048 m.

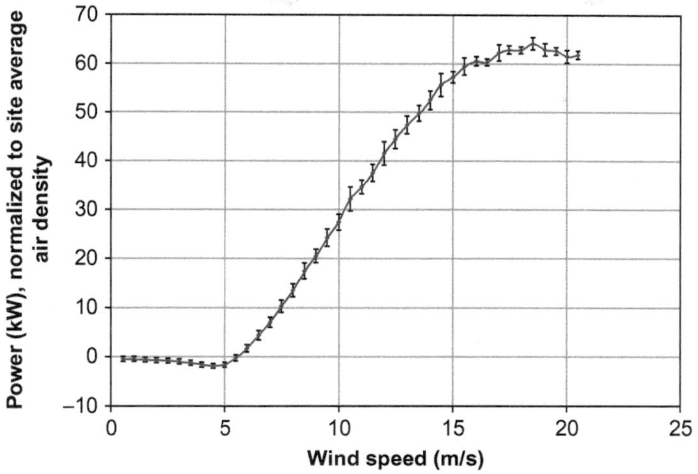

Fig. 3.21 EW50 power curve at Golden site average air density, 1,050 kg/m^3.

Fig. 3.22 Installation of the 50 kW wind turbine (WT) in the City of Seminole in Texas [2].

project 1 had a WT connected to a battery bank, while project 2 had a WT connected directly to a crank mechanism, a compressor, and a water pump. According to the authors, both the projects showed that a combination of RO desalination and wind power is technologically feasible when the RO

Table 3.10 Technical data of the EW50 WT [2]

Technical data	
Rated electrical power	50 kW
Cut-in speed	4.0 m/s
Cut-out speed	25 m/s
Design speed	59 m/s
Average annual output at sea level:	Class 2 wind speed: 115,000 kWh Class 3 wind speed: 149,000 kWh Class 4 wind speed: 177,000 kWh
Type of Hub	Fixed pitch
Rotor diameter	15 m (49.2 ft)
Swept area	579.7 m^2 (1902 ft^2)
Number of blades	3
Material of blades	Epoxy/glass fiber
Rotor speed	65 rpm
Location relative to tower	Downwind
Rotor tip speed	114 mph
Generator type	3 Phase/4 pole asynchronous
Frequency	60 Hz
Voltage	3 Phase @ 50/60 Hz, 415–600 V

system is operated with fluctuating and intermittent loads following the energy supply characteristic of the WT.

Furthermore, there are some examples of "grid-connected," wind-powered systems that are in operation." In these cases, the RE technology has to provide the energy consumed by the desalination unit(s) via a power node connected to the grid, without the need of power backup sources. In the case of excess of produced energy from the RES, this excess is provided to the grid with the benefit of feed-in-tariff issue (depending on the procedures and legislations of the country). These cases mostly related to the use of a WT or a wind farm to cover the power requirements of RO plants.

For example, in Australia many desalination plants cover their power requirements by the use of wind farms connected to the grid. In 2006, a large-scale SWRO plant with a daily water capacity of 143,000 m^3 was constructed in Kwinana, south of Perth, in Western Australia. The Perth plant was the first large RO plant in the world to be fed by renewable energies. The specific energy consumption of the RO plant is of 3.5 KWh/m^3 thanks to the use of ERD. The plant buys its power from electricity generated by the Emu Downs Wind Farm, located 200 km north of Perth. This wind farm has a capacity of production of 80 MW, more than three times the maximum consumption of the desalination plant. The 48 WTs contribute more than 272 GWh per year into the grid, fully offsetting the Perth SWRO plant's estimated electrical requirement of 180 GWh per year. The Emu Downs Wind Farm consists of 48 Vestas (V82) WTs each with 1.65 MW nominal or generating capacity.

Another example is the Kurnell RO Desalination Plant, which has a daily water capacity of 250,000 m^3 and is powered using 100% renewable energy from the 140 MW Capital Wind Farm. Australia is one of the countries of the world which is increasingly dependent on seawater water desalination to supply freshwater to its 22 million people. Today, Australia has the capacity to produce as much as 35% of its water through desalination technology; which means high energy requirements for their operation [38]. For this reason, the country joined several of these plants to WTs, creating the largest wind-powered desalination systems in the world. Yet with coal providing 85% of Australia's energy demand (Australia is one of the largest exporter of coal in the world), adding energy-intensive desalination plants to the grid is increasing Australia's carbon footprint.

Another example of an on-grid wind RO project exists in a Greek Island in Cyclades group. In Milos Island the water needs are mainly covered by the use of a 3600 m^3/day SWRO plant driven by an 850 kW (VESTAS-V52) WT connected to the grid. The system has a successful operation and no water has been transferred via tankers after its installation in 2008 (Fig. 3.23). The RO plant is located at the old betonies' mine at "Vouno Triovassalou" on the Island of Milos. The project consists of the following units:

(A) (B)

Fig. 3.23 Milos Island SWRO wind plant [39].

- 3600 m^3/day (6 × 600 m^3/day) RO unit for seawater desalination.
- Potable water storage tanks of 3000 m^3 capacity.
- Transfer pipeline for potable water from the RO unit to the storage tanks.
- An 850 kW WT (the WT is installed in an already existing wind park, of 3 WT); the average yearly electricity generation of the WT is estimated to be 1,800,000 kWh.
- Supervisory Control and Data Acquisition Systems (SCADA) to balance the operation of the units according to the electrical load of the Island of Milos and the operating status of the thermal plant (Public Power Corporation, PPC). With the use of ERD and of high-efficiency pumps, the specific energy consumption of the RO unit is around 2.5 kWh/m^3 [39].

It is important to mention that until 2008, the island covered its water needs with the transportation of water with tankers, like several other Greek islands. The water transportation cost up to 2009 for Cyclades and Dodecanese Aegean islands ranged from 5 to 8 €/m^3. During 2010, the water transported cost in several Aegean islands, increased to around 12.5 €/m^3. The project is designed to cover the continual increasing needs for potable water on the island, supplying high-quality water to comply with the highest standards on a 24-h basis, at low cost. At present the owner of the wind desalination plant offers the potable water produced by the wind RO plant to the Municipality of Milos at a price of 2 €/m^3.

Even if this plant is small compared with those in Australia, its operation is of high importance. Milos Island is a self-power producer (non-connected to the mainland electricity network) with the use of diesel engines. The existence of any form of renewable energy to cover the extra loads of the island is vital from an environmental and economical perspective.

Furthermore, according to a study performed by CRES for the Municipality of a Greek Cyclades Island, based on parameters such as the wind potential of the island and that all RO units installed in the islands contain ERDs, the conclusion was that a 5 MW wind farm (capacity factor of up to 50%) is able to cover the energy requirements of 8.500 m^3/day SWRO units (24 h/day operation). The total daily power required for the operation of SWRO units is 50,000 kWh. The yearly cost to cover all the energy requirements (feed pumps, high-pressure pumps, and potable and transportation water pumps) of the RO units is estimated at 1,500,000 €. The cost for the procurement and installation of a WT of around 2–2.5 MW is estimated at around 1350 €/kW.

3.5 DESIGN AND IMPLEMENTATION OF WIND DESALINATION PROJECTS

Among the several possible combinations of desalination and wind energy technology, some seem to be more promising in terms of economic and technological feasibility. However, their applicability strongly depends on site characteristics, site remoteness, the type of application (off-grid or on-grid), and on the water requirements regarding the required/contracted quality and quantity of the produced water.

Local circumstances always play a significant role in determining the most appropriate process for an area. The "best desalination" system should be more than economically reasonable in the study stage. It should work when it is installed and continue to work and deliver suitable amounts of freshwater at the expected quantity, quality, and cost for the life of a project.

The decision between off-grid or on-grid application is obvious. Off-grid projects involve remote and rural areas without any access to the electricity grid and with water scarcity problems. Owing to the dispersed population in several rural areas around the world relatively small autonomous wind desalination systems can be used to cover the water needs and the basic power supply. The main desirable features for such systems are the low cost, low maintenance requirements, simple operation, as well as high reliability and sustainability.

On-grid applications mostly involve larger units where the energy requirements of the desalination units can be covered by a WT or a wind farm. This technique is applied in areas where the electricity grid is available. The advantage of these wind desalination applications is that the desalination unit can still have a continuous and dependable operation. Any energy excess from the WT or wind farm can be sold back to the electricity grid. Another important factor is the monitoring control and full automation of the system. In both, on-grid and off-grid systems, a special energy management system should be designed. The wind potential and siting issues are major technical parameters for the design and installation of a WT or a wind farm. The capacity of the desalination system should be defined according to the energy availability, the energy and water storage capability, and the freshwater or potable water requirements. The final decision, with the exception of any technical issues, will be determined according to the available budget [40].

For off-grid applications, the selection of very innovative wind desalination couplings is not recommended to solve the water needs in remote and rural areas. The simplicity and the full automation of the system is an important parameter affecting the daily operation and the life of the system.

In general, some of the parameters that should be considered during the design and implementation of a wind desalination project are presented below:

- Wind Assessment

  A very important step is assessing the wind resource. For that purpose, meteorological masts are used to measure wind speed and other climatic conditions for at least one year. This data is then used to estimate the amount of energy the wind farm will produce.
- Water Resource Assessment

  The determination of the water resources in order to decide on the type of the desalination unit is the second important step for the design of a project. Usually, in coastal areas the operation of seawater desalination units is preferred since seawater is available in large quantities, has a relatively stable concentration, and the brine disposal is an easier issue in comparison with the brine disposal from an inland brackish-water plant. The composition of brackish waters is of extremely wide variation, and a water analysis is a must for a good process design.
- Land Acquisition

  Early in the process, developers usually approach landowners to negotiate "option" agreements to use their land. In case of a public project, in most cases there are public areas available and their use has to be examined.
- Municipal Consultations

  For private projects, the consultations with municipalities help in making a project successful. It is extremely important to contact the municipality before undertaking any of the mentioned steps.
- Social Acceptance

  For both private and public projects, the information and involvement of the local people is an essential issue for the development and operation of the plant. Social acceptance is becoming an issue in nearly every country that has wind and desalination projects development.
- Creation of Employment

  Both the wind and the desalination sector create opportunities and job positions. Local engineers and/or technicians could be trained for the proper operation and in sometimes the partly maintenance of the system. The availability of personnel from the local region or the area around should be investigated and considered in the running costs of the project.

- Wind Farm and Desalination Unit Design

  Before the design of the project, several information of the site has to be collected such as the availability of several infrastructures (roads, water storage tanks, electricity grid, water network, etc.).

  In case where the system will be connected to the grid the steps are similar with those as required for the design and implementation of conventional systems. The WT or the wind farm should be capable to cover the energy requirements of the desalination system by 100% or less according to the agreement. Wind data is combined with topographical information to design the wind farm. Engineers use this data to model wind flow, turbine performance, sound levels, and other parameters to optimize the location of the WTs. They also design the access roads, turbine foundations, and the local electric network, as well as the connection to the electricity grid. For the desalination system, the water intake system, the brine disposal system, and the selection of the optimum pretreatment procedure are vital issues for the successful operation of the system and must be carefully designed. The type of the water intake system (open intake or beach wells) and the distance of the water intake and brine disposal point from the desalination plant should be taken in consideration according to environmental and economic issues (e.g., large amounts of energy consumption from the feed SW pumps have been mentioned in plants due to the long distances between the water intake point and the plant). Whenever it is possible, beach wells as intake systems are preferred since they aid pre-filtration of the feedwater.

  In case of off-grid systems, the design of the wind desalination system is more complicated and several parameters should be considered. For example, an important parameter is the quantity of the required produced water. Usually desalination systems have a daily operation of about 24 h and their nominal water capacity based on a 24-h operation. For instance, a 20 m^3/day RO unit provides 20 m^3/day during its 24-h operation. Suppose the hours of energy availability from the off-grid autonomous system is 8 h per day, the RO unit will be capable to produce around 6.6 m^3. The designer has to examine several alternatives such as an oversized RO unit, a larger power supply system, bigger storage system, or the use of a backup system, depending also on the budget availability. A larger capacity RO unit with the use of ERD device could be more energy efficient than one of small capacity. For example, the specific energy consumption of the 20 m^3/day SWRO unit with the use of ERD is estimated at about 4.8 kWh/m^3. For a larger SWRO of 100 m^3/day water capacity, the specific energy consumption of the unit with the use of ERD is estimated at around 3.6 kWh/m^3.

In general, the selection of the WT and of the energy storage system should be calculated according to the energy requirements and the required hours of operation of the desalination unit. Before any decision is made the several alternatives regarding the use and the type of energy storage devices, PV, and/or diesel backup, etc., should be examined according to the purpose of the project and the available budget.

- Environmental Study

  Environmental assessments are conducted to identify any impacts on landscape, plants and wildlife, soil, and water, land use, etc. The benefits of the use of wind energy to drive the desalination plant should be considered and published.

- Permitting Process

  As with any other major project, developers must seek municipal, provincial, and federal permits before the project can go ahead. They also meet the local communities to present the project, solicit their feedback, and seek community support. The permitting process varies greatly from country to country, state to state, and in some cases, even from town to town. It is intended to ensure that land is used appropriately, and that operation of the installed WT or wind farm and of the desalination unit will be safe and environmentally benign.

- Economic and Financial Analysis

  The developers must demonstrate the economic viability of their project to raise the funds to build the wind desalination plant. On one hand, they work to estimate the cost of turbines and of the desalination plant and their installation, as well as roads, electrical system, operation and maintenance, etc. On the other hand, they estimate the income they will get from the energy and water production over the lifetime of the project. Then they make sure everything balances.

- Manufacturing

  The WT parts are manufactured and preassembled into the main components at the factory, and then shipped to the wind farm site where the final assembly will take place. Regarding the desalination units, in most cases, they are fitted in containers ready for installation at the site.

- Site Preparation

  Once permits have been obtained, the site needs to be prepared for the installation and operation of the WT and of the desalination plant.

- Construction

  Once all components have been received, the assembly can take place. Regarding the erection of a WT a crane is used to erect the tower and install the nacelle and the rotor with its hub and blades. On the ground, the electrical collection network is installed and connected to the grid through the substation. Once the turbine is installed and connected to the grid, it is ready for operation. Regarding the desalination unit, in most cases a crane is used for the installation of the unit. The assembly depends on the desalination process, but in general after the installation of the desalination plant the connection with the intake and brine disposal systems as well as the electrical connections is installed.
- Commissioning

  Finally, the WT or the wind farm and the desalination system are tested before becoming fully operational. Commissioning consists of appropriate tests to ensure correct turbine and desalination operation and maintenance and operation training for the plant owner or operator.
- Operation and Maintenance

  A variety of skill sets are required to ensure the effective operation of a wind desalination plant. The activities that have to be performed on a regular basis throughout the project's life include monitoring and analyzing performance, conducting environmental surveys, and performing preventive maintenance and repairs on the turbines and desalination plant and other components of the facility.

3.6 R&D RECOMMENDATIONS

The lessons learnt from the installed wind/RO units have hopefully been passed on and incorporated into plants currently being built and tested. The main scope of the research work that has been implemented until now is to provide reliable and sustainable wind desalination systems and to promote this knowledge to the industry and to the market. Hopefully, such environmentally friendly systems should be soon potentially available at done on the new generation products, such as ERDs, new-generation small WTs, energy storage systems (lithium-ion batteries, supercapacitors, etc.) and, on the development of more advanced control and monitoring systems in order to develop autonomous systems able to supply freshwater to remote areas in a reliable, sustainable, and affordable way. Demonstration/pilot activities of improved wind desalination competitive costs. However, to achieve that more research has to be combinations in real environment, with the collaboration of research organizations and the market (major companies, SMEs, etc.) and with the contribution of all stakeholders is necessary.

3.7 CONCLUSIONS

Renewable energy technologies are rapidly emerging with potential economic and environmental viability for desalination. There is a need to accelerate the development of novel water production systems from renewable energies. Wind energy and such other RES could provide a viable source of energy to power seawater and brackish water desalination plants. It has been recognized that part of the solution to the world's water shortage lies in water production via desalination techniques, in an environmentally sustainable way. Any decision on the matching of the two technologies, the selection of the WT and the desalination technology, the use of battery storage systems, etc., has to be designed carefully according to the site characteristics and existing experience.

Although, there is experience on the technologies matching, there is much room for the improvement of technologies combination, from a technical and economical point of view. Further reduction of the cost of small ERDs and of small WTs, as well as the testing of other types of batteries (than the commonly used lead–acid batteries) at pilot/demo wind-powered desalination plants, are some prospects for the future for wind desalination.

Finally, there is a need for much stronger effort on R&D and real installations in the Mediterranean area, Middle East, and Africa; close collaboration between the R&D institutes and the industry as well as provision of reliable compact and mobile RES desalination devices by the market at a reasonable cost. Finally, the acceleration of the dissemination of the existing knowledge, education, and training is also of vital importance.

REFERENCES

[1] Tzen E. Wind Energy powered technologies for freshwater production: Fundamentals and Case Studies, Sustainable Energy Developments 2. In: Renewable Energy Applications for Freshwater Production. CRC Press, Taylor & Francis Group; 2012. ISBN 9 780415 620895.

[2] Rainwater K, Song L, Lehman T, Schroeder J. An integrated wind-water desalination demonstration plant for an inland municipality. Final Report, Texas Tech University; 2015.

[3] Cipollina A, Tzen E, Subiela V, Papapetrou M, Koschikowskie J, Schwantese R, et al. Renewable Energy Desalination: Performance analysis and operating data of existing RES-desalination plants. Desalin Water Treatment 2014;55(11):3120–40.
[4] Desalination Guide Using Renewable Energies, Center for Renewable Energy Sources, CRES, 1998; ISBN 960-90557-5-3.
[5] International Energy Agency, 2016;IEA Wind TCP, ISBN 978-0-9905075-3-6.
[6] American Wind Energy Association. Wind Power Today. AWEA, 2007.
[7] Tzen E. Renewable Energy Sources for Seawater Desalination—Present Status and Future Prospects. NY: NOVA Science Publishers Inc.; 2008. ISBN 978-1-60456-567-6.
[8] Wind Turbines. Special Report, Sonderdruck-45852, 2012. Hannover, Germany: SunMedia Verlags GmbH; 2012.
[9] Windblatt-the ENERCON magazine, March 2016, enercon.de/new-products.
[10] The European Wind Energy Association, EWEA Wind in power 2015 European statistics February 2016.
[11] Sommariva C. Desalination and Advanced Water Treatment; 2010, ISBN 086689069-6.
[12] WaterWorld. http://www.waterworld.com/articles/wwi/2016/11/desalination-and-water-reuse-markets; 2016.
[13] Forstmeier M, Mannerheim F, D'Amatob F, Shah M, Liub Y, Baldea M, et al. Feasibility study on wind-powered desalination. Desalination 2007;203:463–70.
[14] Weiner D, Fisher D, Moses EJ, Katz B, Meron G. Operation experience of a solar- and wind-powered desalination demonstration plant. Desalination 2001;137:7–13.
[15] Heijmana SGJ, Rabinovitcha E, Bosa F, Olthofa N, van Dijka JC. Sustainable seawater desalination: Stand-alone small scale windmill and reverse osmosis system. Desalination 2009;248:114–7.
[16] Tzen E, Theofiloyianakos D, Kologgios Z. Autonomous reverse osmosis units driven by RES, experiences and lessons learned. Desalination 2008;221:29–36.
[17] Tzen E. Wind and Wave Energy for Reverse Osmosis. In: Cipollina A, Micale G, Rizzuti L, editors. Chapter 9 of the book entitled Seawater Desalination; 2009, ISBN 978-3642-01149-8.
[18] Miranda M, Infield D. A wind-powered seawater reverse-osmosis system without batteries. Desalination 2002;153:9–16.
[19] Infield DG. A wind-powered seawater reverse-osmosis system without batteries. Desalination 2002;153(1-3):9–16.
[20] Lai W, Ma Q, Hui L, Weng S, Fan J, Fang H. Effects of wind intermittence and fluctuation on reverse osmosis desalination process and solution strategies. Desalination 2016;395(3):17–27.
[21] Schaefer GLA, Richards BS. Renewable energy-powered membrane technology: Supercapacitors for buffering resource fluctuations in a wind-powered membrane system for brackish water desalination Park. Renewable Energy 2013;50:126–35.
[22] Battery storage for Renewables: Market Status and Technology Outlook, 2015; International Renewable Energy Agency (IRENA), www.irena.org.
[23] Gretz A, et al. Lead Acid vs. Lithium-Ion Batteries: Which is better for storing solar?, Home Battery.
[24] Albright G, Edie J, Al-Hallaj S. A comparison of lead acid to lithium-ion in stationary storage applications. Chicago, USA: AllCell Technologies LLC; 2012.
[25] Qing L., Shitong F., 2015; Energy Management of Isolated Micro-grid With Wind Turbines and Seawater Desalination Devices Asia-Pacific Energy Equipment Engineering Research Conference (AP3ER) 2015.
[26] Swift A, Rainwater K, Chapman J, Noll D, Jackson A, Ewing B, et al. Wind Power and Water Desalination Technology Integration; 2009. DWPR Report No. 146, Bureau of Reclamation, USA.

[27] Subiela VJ, Juan A, de la Fuente PG, Peñate B. Canary islands Institute of Technology (ITC) Experiences in desalination with renewable energy sources (1996–2008). Desalin Water Treatment 2009;7:220–35.
[28] Plantikow U. Production of potable water by a wind driven seawater desalination plant — Experience with the operation World Wind Energy Conference (WWEC), Cape Town, South Africa, 2003.
[29] Karameldin A, Lotfy A, Mekhemar S. The Red Sea area wind-driven mechanical vapor compression desalination system. Desalination 2003;153(1–3):47–53.
[30] Witte T., Siegfriedsen S., Magdy El-Allawy, WindDeSalter® Technology Direct use of wind energy for seawater desalination by vapor compression or reverse osmosis, Desalination, Volume 156, Issues 1–3, 1 2003; p. 275-279.
[31] Final Report: Use of Mechanical Vapor Compression (MVC) System for Development of Water Resources in Rural Areas of Developing Countries, EPA Grant Number: SU834712 EPA, USA, https://cfpub.epa.gov/ncer_abstracts/index.cfm/fuseaction/display.highlight/abstract/9211/report/F.
[32] Veza J, Peñate B, Castellano F. Electrodialysis desalination designed for wind energy (on-grid tests). Desalination 2001;141:53–61.
[33] Carta JA, Gonzalez J, Subiela V. The SDAWES project: an ambitious R&D prototype for wind powered desalination. Desalination 2004;161:33–48.
[34] Veza J, Peñate B, Castellano F. Electrodialysis desalination designed for off grid wind energy. Desalination 2004;160(3):211–21.
[35] Carta JA, González J, Cabrera P, Subiela VJ. Preliminary experimental analysis of a small scale prototype SWRO desalination plant, designed for continuous adjustment of its energy consumption tothe widely varying power generated by a stand-alone wind turbine. Applied Energy 2015;137:222–39.
[36] Peñate B., Castellano F., Bello Al., Lourdes García-Rodríguez, 2011; Assessment of a stand-alone gradual capacity reverse osmosis desalination plant to adapt to wind power availability: a case study, Energy, https://doi.org/10.1016/j.energy.2011.04.005
[37] Bilstad T, Protasova E, Simonova A, Stornes S, Yuneizi I. Wind-powered RO desalination. Desalin Water Treatment 2015;55(11).
[38] Win4Water. Wind-Powered Desalination: The Australian Story; 2016. USA.
[39] Bouzas I., Desalination with the use of wind energy in Milos island, International Technological Applications (ITA) S.A., PRODES Workshop, Renewable energy Technologies with Desalination: Technologies Progress-Legislation-Funding Schemes CRES, Athens, 2010.
[40] Tzen E, Rossis K., González J., Cabrera P., Subiela V., Penate B., Penate Suarez B., Subiela Ortín V. 2016, Chapter 5 of the book Renewable Energy Technologies for Water Desalination, Mahmoudi H., Ghaffour N., Goosen M., Bundschuh J., (under press).
[41] Schreck S. Integrated wind energy/desalination system. National Renewable Energy Laboratory; 2006. NREL/SR-500-39485.

FURTHER READING

[1] Desalination and water reuse markets continue to recover – WaterWorld, http://www.waterworld.com/articles/wwi/2016/11/desalination-and-water-reuse-markets, 2016.
[2] Lindemann JH. Wind and solar powered seawater desalination, Applied solutions for the Mediterranean, the Middle East and the Gulf Countries. Desalination 2004;168:73–80.2.
[3] Shatat M, Saffa BR. International Journal of Low-Carbon Technologies Advance Access published April 6, http://ijlct.oxfordjournals.org/; 2012.

[4] Garcia-Rodriguez L, et al. Economic analysis of wind-powered desalination. Desalination 2001;137:259 – 265.
[5] Integrated Wind Energy/Desalination System, October 11, 2004–July 29, 2005, GE Global Research Niskayuna, New York, Subcontract Report NREL/SR-500-39485.
[6] Witte T, Siegfriedsen S, El-Allawy M. WindDeSalter Technology, direct use of wind energy for seawater desalination by vapor compression or reverse osmosis. Desalination 2003;156:275–9.
[7] Liu CCK, Park J-W, Migita R, Qin G. Experiments of a prototype wind-driven reverseosmosis desalination system with feedback control. Desalination 2002;150:277–87.

CHAPTER 4

Geothermal Source for Water Desalination—Challenges and Opportunities

Veera Gnaneswar Gude
Mississippi State University, Mississippi State, MS, United States

4.1 INTRODUCTION

Providing clean water for human consumption has become a major challenge at local, regional, national, and global levels [1]. This is mainly due to increasing demand prompted by population growth and urbanization [2]. The projected global demand for clean water supply for the future will significantly accelerate not only depletion of fossil fuel reserves but also concomitant environmental damage and emission of greenhouse gases [3]. This situation provides the basis for the utilization of renewable energy for water production to solve the energy–water–environment trilemma [4–6]. This chapter provides a general overview of the geothermal energy uses around the world and describes a specific case for geothermal source-based desalination and its potential for energy- and pollution-free desalination. This resource has been underutilized for various reasons. The benefits of geothermal desalination over other renewable energy-driven desalination and the present status of worldwide geothermal desalination, selection of desalination technologies, and challenges in implementation of geothermal source powered desalination are discussed in detail.

4.2 GEOTHERMAL ENERGY UTILIZATION AROUND THE WORLD

Geothermal source can be utilized for various direct uses including bathing and swimming, cooling and snow melting, agricultural drying, aquaculture pond heating, greenhouse heating, space heating, industrial process heat and use, geothermal heat pumps, and power generation. Worldwide geothermal utilization trend for various applications between 1995 and 2015 is shown in Fig. 4.1 (data taken from Ref. [7]). It can be noted that geothermal heat

Renewable Energy Powered Desalination Handbook
https://doi.org/10.1016/B978-0-12-815244-7.00004-0

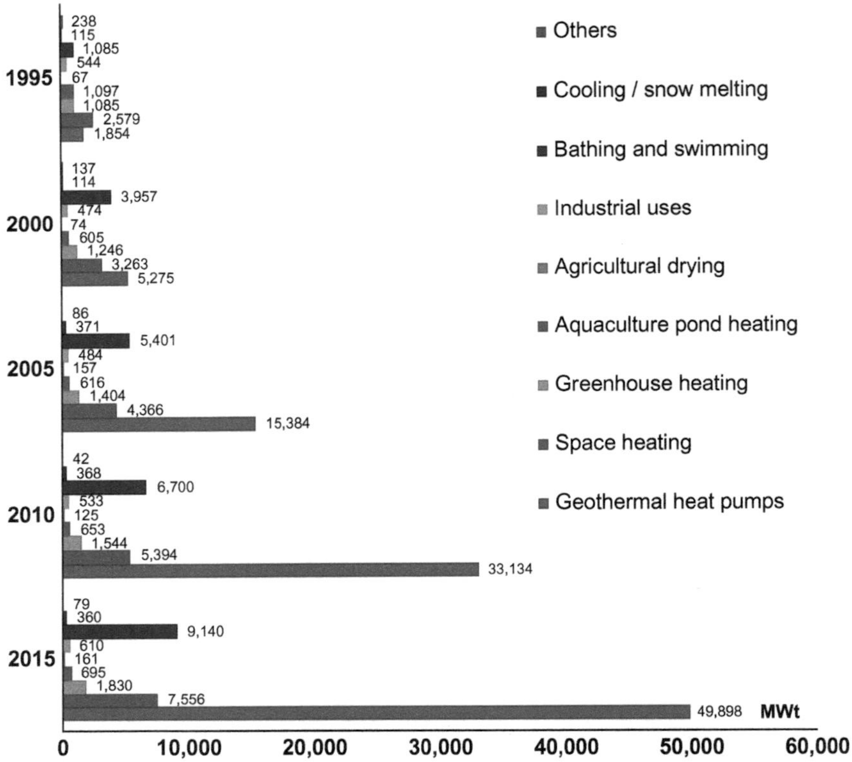

Fig. 4.1 Geothermal energy source utilization around the world.

pump application has been increasing consistently over the years. Other significant uses include bathing and swimming, and space heating. The countries with the largest installed capacity are the United States, China, Sweden, Norway, and Germany, accounting for about 63% of the installed capacity and the five countries with the largest annual energy use were: China, United States, Sweden, Turkey, and Japan, accounting for 55% of the world use. On the basis of geothermal utilization per land area, the top countries can be listed as Netherlands, Switzerland, Iceland, Norway, and Sweden (TJ/area). On the basis of the population, the top countries include Iceland, Norway, Sweden, Denmark, and Switzerland (TJ/population). The largest increases in geothermal energy use (TJ/yr) over the past 5 years are in the United Kingdom, Netherlands, Korea (Republic), Norway, and Iceland; and the largest increases in installed capacity (MWt) are in the United Kingdom, Korea (Republic), Ireland, Spain, and Netherlands, mostly due to the increased use of geothermal heat pumps [8]. Acknowledging the wide applications of geothermal sources, this review article will focus on the feasibility of geothermal energy-based desalination for freshwater production. The following sections discuss the rationale, critical features, feasibility studies, and global potential for geothermal desalination in the near future.

4.3 WHY GEOTHERMAL DESALINATION?

There are several benefits associated with the use geothermal sources for various domestic and industrial applications. Geothermal energy is a proven and well-established commercial technology for electricity production, district heating and cooling, and industrial process applications. Geothermal energy desalination provides the following advantages [4,9–12]:

(i) Geothermal energy sources have a high capacity factor which provides a stable and reliable heat supply ensuring stability of thermal desalination as well as reverse osmosis (RO). Capacity factor is defined as the resource availability both in terms of quantity and quality over a period of time of application.

(ii) Geothermal production technology (extraction of hot water from underground aquifers) is mature. It is unaffected by the seasonal changes and weather fluctuations.

(iii) Typical geothermal source temperatures are in the range of 70–90°C in most parts of the world, which are ideal for low-temperature multieffect desalination (MED) desalination. High-grade sources above 100°C can be used for power generation and other process heat application.

(iv) Geothermal desalination is cost effective, and simultaneous power and water production is possible.

(v) Geothermal desalination is environmentally friendly because it is the only renewable energy used in the process with no emissions of air pollutants and greenhouse gases related to fossil fuels.

(vi) Geothermal desalination saves imported fossil fuels which can be used for other purposes improving local energy security and environmental sustainability.

(vii) Geothermal sources have relatively lower surface area or land requirements per unit (MW) of all renewable energy sources (e.g., $20\,MW_{th} = 10\,m \times 10\,m$ well size) and energy demand can be matched from smallest to the largest energy-consuming utilities.

4.3.1 Capacity Factor

Unlike other renewable sources, geothermal sources can be relied upon for their availability both in quantity and quality. Geothermal energy can be considered inexhaustible if it can be operated in a closed-loop configuration. Geothermal sources vary in temperatures providing flexibility for various process applications. These applications may include integrated configurations to provide additional benefits while supporting desalination process. Additional benefits could be air-conditioning, district heating, process heating and cooling applications. Another important factor of the geothermal sources is the capacity factor which is related to the availability of an energy source in both quantity and quality. It refers to the ratio of actual energy generated by a system to the available energy source. Capacity factor is an important variable to be considered for scalability and operation of a technology. When compared with other renewable energy sources such as solar, wind, and biomass sources, the capacity factor for geothermal source is very high. The capacity factors for various renewable energy sources are shown in Fig. 4.2 [13,14].

The capacity factor for a resource can be influenced by resource availability, processing equipment or transmission variability, and other external factors such as power market fluctuations [12,13]. In solar energy applications, the amount of energy that can be extracted is limited by the harvesting capacity of the process equipment and materials in use. Similarly, the wind energy capture potential will be diminished by its intensity and fluctuations and the mechanical efficiency of the turbine unit and electrical energy

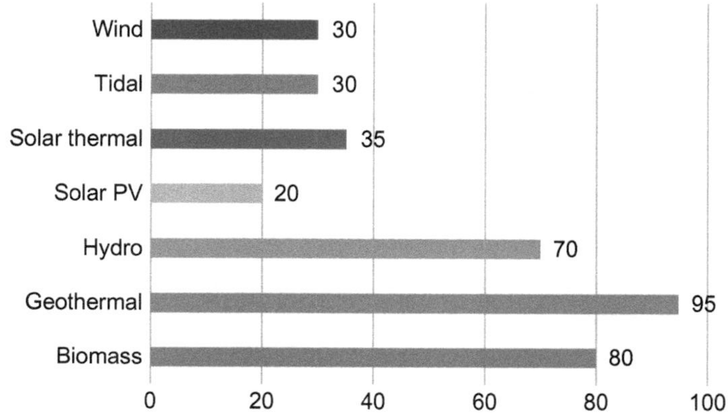

Fig. 4.2 Capacity factor (in percent) ranges reported for various renewable energy sources.

converter. Biomass availability is seasonal and depends on the climate and other resource availability such as water and land. Geothermal sources are not an exception but suffer to a less extent from these constraints. For geothermal sources, the capacity factor may be influenced by the constrained resources, a clogged well, change in subsurface conditions, and loss in water availability.

4.3.2 Comparable Costs

Geothermal sources compare well with other renewable and nonrenewable electricity production technologies [13]. As shown in Fig. 4.3, the electricity production costs (for OECD countries in 2011 electricity prices) for geothermal source-driven power plants are within comparable range for wind, biomass, and hydroelectric power plants [15]. This range is also comparable with the costs for conventional energy sources such as natural gas, coal, and oil sources. This indicates that water production driven by either thermal energy or electricity derived from geothermal sources could be economical.

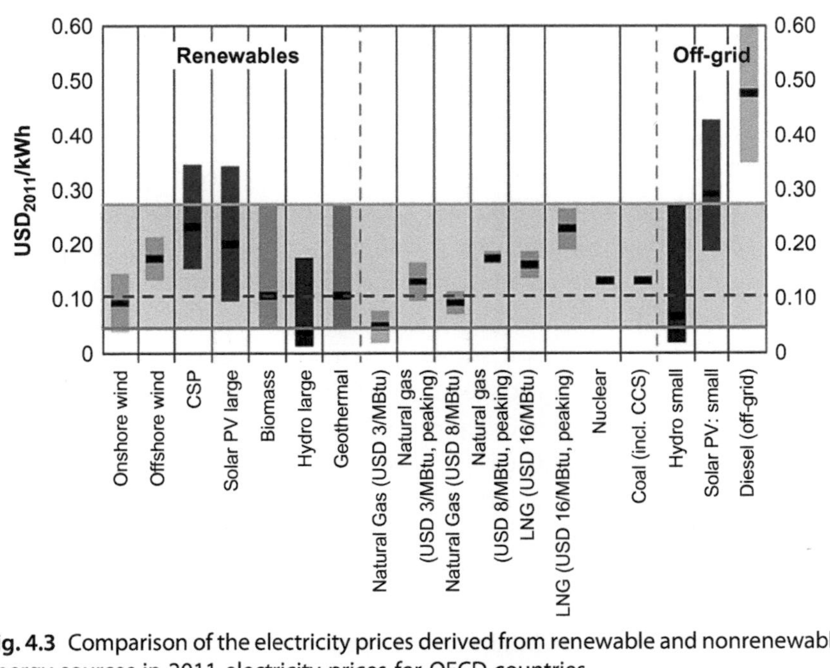

Fig. 4.3 Comparison of the electricity prices derived from renewable and nonrenewable energy sources in 2011 electricity prices for OECD countries.

4.3.3 Efficient Resource Utilization

Various renewable energy resources are available which suit the energy needs of different desalination processes. Currently, the resources which are well explored and exploited for desalination applications include solar energy (harvested by solar collectors or photovoltaic modules), wind energy, geothermal energy, and wave energy [2]. Sustainable use of these resources depends on the conversion technology employed and the end-user process configuration. Sustainability, in this context, can be interpreted as how the available energy resource is being utilized. Conserving, recycling, and increasing the efficiency of the conversion technologies are some of the approaches which result in a sustainable use of an energy resource. For instance, daily solar energy available on the surface of the earth is roughly 15,000 times more than the daily energy consumption of the world population [16], which means that only much less than 0.001% of the daily solar energy captured in energy devices could solve the energy problems of the world. However, the maximum energy conversion rate of the current photovoltaic modules in the market has not exceeded 15% to date. In this context, geothermal sources present as an efficient and excellent energy source which can be utilized for both energy and heat generation with minimum energy losses.

4.3.4 Energy Savings in Geothermal Applications

Energy savings otherwise not possible with other renewable sources are shown in Fig. 4.4 [17,18]. Geothermal sources can be used for various process heat applications, cooling and desalination applications. Potential energy–cost savings using geothermal sources for process heat are shown in Fig. 4.4A. The cases are presented for oil prices $90/bbl and $40/bbl and compared with cost savings by geothermal sources at 100°C, 115°C, and 130°C. This comparison shows potential savings of 50%, 66%, and 75% for geothermal sources at 100°C, 115°C, and 130°C, respectively. Also, possible electricity savings in different cooling applications are compared for electricity-driven decentralized (packaged or splits) and water cooled electric chillers with geothermal-driven absorption and adsorption chiller configurations at 100°C, 115°C, and 130°C (Fig. 4.4B). This comparison indicated potential electricity savings of 45%, 50%, and 62% for geothermal sources at 100°C, 115°C, and 130°C, respectively. Similarly for desalination applications, MED with geothermal sources at 100°C, 115°C, and 130°C as heat sources will result in savings of 46%, 85%, 88%, and 89% for hybrid

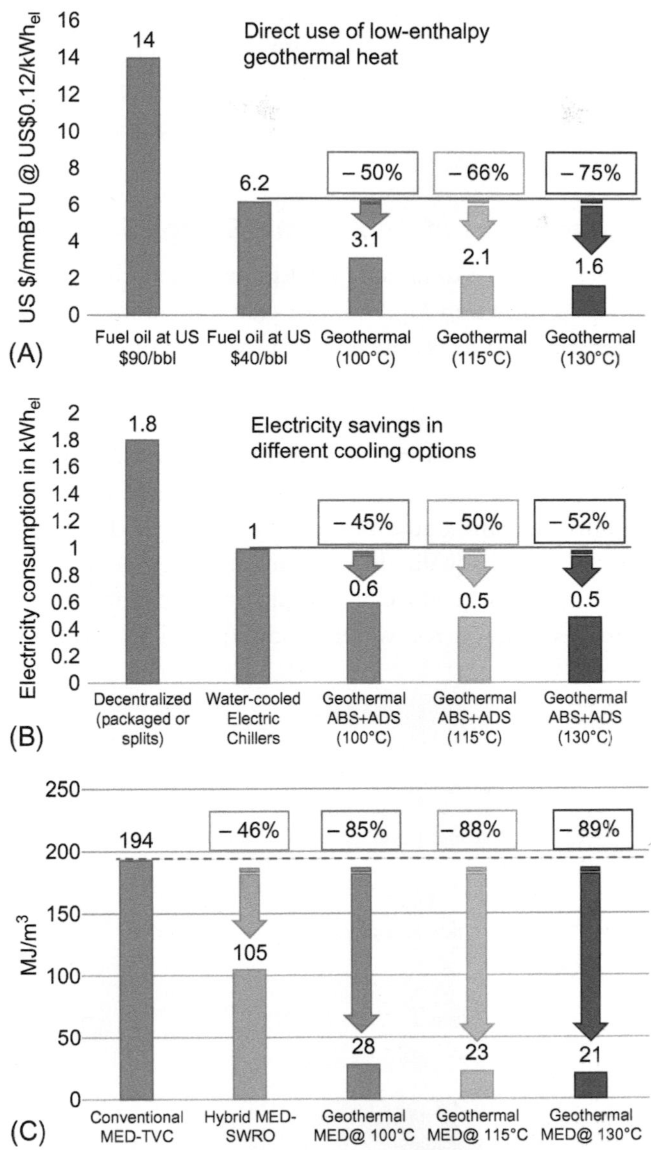

Fig. 4.4 Potential oil (A), electricity (B), and energy (C) savings for geothermal source-driven process heat, cooling, and desalination applications. Thermal energy requirements for an MED process can be reduced significantly corresponding to the source temperature.

MED with seawater reverse osmosis (SWRO) and for geothermal sources at 100°C, 115°C, and 130°C, respectively (Fig. 4.4C). MED unit is compared due to possibility of low-enthalpy geothermal sources in most parts of the world and for its lower thermal energy requirements compared with multistage flash (MSF) process.

4.3.5 Integrated Uses of Geothermal Energy Sources

Desalination plants can be designed to integrate other beneficial uses of geothermal sources as shown in Fig. 4.5. For example, higher quality geothermal energy can be used to produce power, followed by desalination using both thermal and membrane processes, then for applications in food processing, refrigeration plants and district heating or cooling systems, heating of building spaces, greenhouses and soil heating, industrial process heat and agricultural drying, and fish farming. When the source temperature is too low, a heat pump scheme could be utilized to extract the enthalpy which can be converted to high-quality heat. If the source temperature is very high, then it can be used to produce electricity in a cogeneration scheme to accommodate desalination for water production. This would reduce the

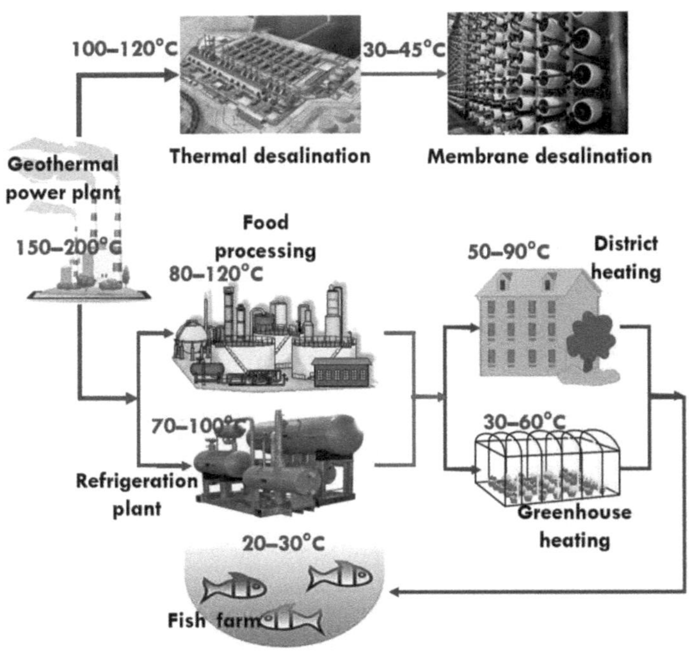

Fig. 4.5 Integrated configurations for geothermal energy sources—poly generation for multiple benefits

overall costs, environmental impacts, and the land footprint. As shown in Fig. 4.5, agricultural and aquacultural uses require the lowest temperatures, with values from 25°C to 90°C. Space heating requires temperatures in the range of 50–100°C, with 40°C useful in some marginal cases and ground-source heat pumps extending the range down to 5°C. Cooling and industrial processing normally require temperatures over 100°C [20].

4.4 GLOBAL GEOTHERMAL DESALINATION—CURRENT STATUS

Fig. 4.6 shows the hottest geothermal source locations around the world [19]. It can be noted that these locations match with the water scarce and

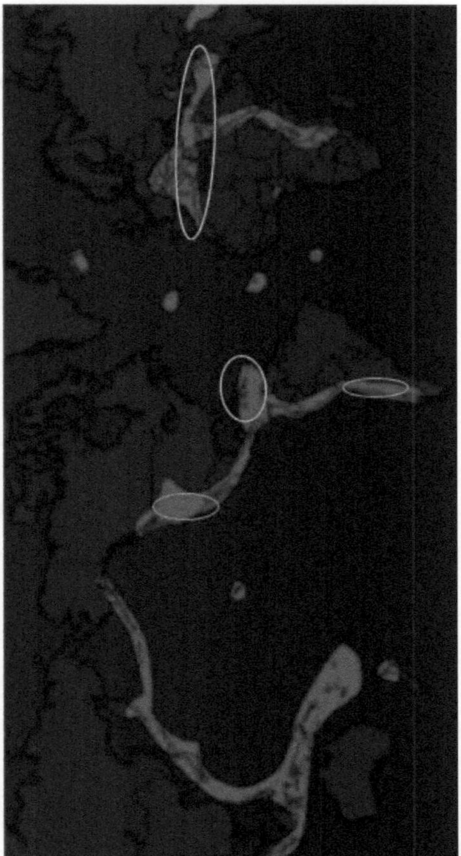

Fig. 4.6 Hottest geothermal energy sources around the world which are suitable for desalination, cogeneration, and poly-generation schemes [19].

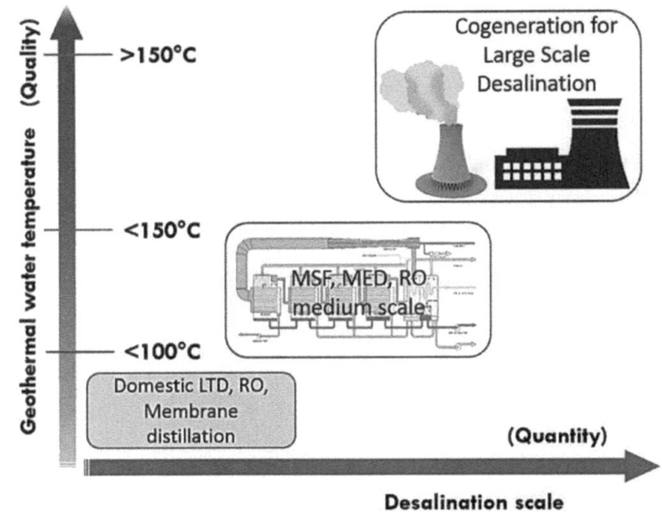

Fig. 4.7 Geothermal energy applications in desalination and cogeneration (water and power production).

desert regions around the world in most cases. This map also resembles the solar map for the worldwide resources where water sources are limited. This clearly indicates that water-scarce regions can definitely benefit from locally available geothermal sources. Hottest geothermal spots are found in northwestern and southwestern parts of the United States, Mexico, Central America and Caribbean Islands, Middle Eastern countries, and North African regions. Depending on the source temperature, various desalination technologies can be integrated with the geothermal sources. Geothermal sources with temperatures between 40°C and 70°C can be used for low-temperature desalination applications which include simple evaporation basins (such as solar stills), membrane distillation or low-temperature multi-effect distillation units (LTMED), and humidification–dehumidification processes. Higher grade heat sources (>70°C) can be used for either MED or MSF desalination processes. Much higher source temperatures between 120 and 200°C are suitable for cogeneration schemes (Fig. 4.7). The choice of desalination technology also depends on the desired capacity and other factors which are discussed in the following sections.

4.4.1 Geothermal Water Composition

The geothermal water composition is characterized by the macroelements of the reservoir rock and the subsurface environment to which it is exposed most of the time. The most frequently observed elements with high concentrations are Na^+, K^+, Ca^{2+}, Mg^{2+}, HCO_3^-, CO_3^{2-}, SO_4^{2-}, and CO_2. Other micropollutants are heavy metals such as mercury, copper, lead, silver, iron, zinc, arsenic, manganese, chromium, beryllium, selenium, vanadium, cadmium, nickel, strontium, uranium, cobalt, gallium, and antimony. Some other elements of boron and silica could be present in geothermal waters as well [21]. Boron and silica causes serious problems if present in high concentrations.

Dissolved gas and solid forms of the components have different equilibria, which can be affected by different factors (changes in temperature or pressure) and shifted toward scaling. Geothermal fluids usually contain high dissolved salts and other ion concentrations, generally reported as total dissolved solids (TDS). Sodium chloride is the most predominant salt component present in dissolved form adding salinity to the geothermal waters. A wide range of TDS concentrations were reported for various geothermal wells across the United States as shown in Fig. 4.8 [12]. TDS for

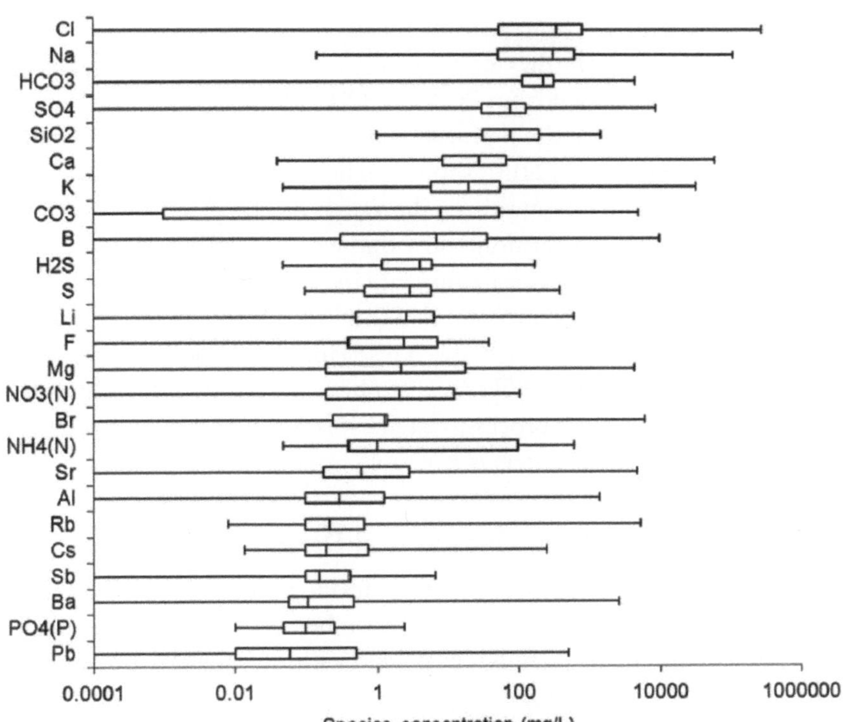

Fig. 4.8 Major constituents of geothermal waters in the United States [21].

high-temperature geothermal waters (>150°C) was reported between 2500 and 81,000 mg/L while for medium-temperature geothermal waters (90–150°C), the TDS range was between 1100 and 8200 mg/L. Most of the TDS concentrations were reported between 500 mg/L and 5000 mg/L; very high concentrations between 260,000 and 280,000 mg/L have been reported for a geothermal plant (Hudson Ranch I and II) in California. Fig. 4.8 was adapted from a recent review article by Finster et al. [21] on geothermal fluids.

4.4.2 Geothermal Water for Membrane Desalination

Feedwater temperature influences the production rates in membrane desalination. Low-temperature feedwater has higher viscosity and higher resistance to pass through the membrane while high-temperature feedwater produces high flux (higher production) due to lower viscosity. Membrane processes require higher mechanical energy to pump the cold feedwater in winter seasons to meet the daily production rates or higher quantities of feedwater needs to be processed. Geothermal waters with high salinity (TDS) can serve as feedwater from which freshwater can be produced. An increase in permeate flux of 60% was reported when the feedwater temperature was increased from 20°C to 40°C [5].

The relation between the permeate flux rate and the feedwater temperature is shown in Fig. 4.9 [22]. It should be noted that the temperature tolerance of the RO membranes is in the range of 20–35°C. Fig. 4.9 inset shows the fraction of the permeate flow that would be affected by the feedwater temperature. The permeate flux rate increases by 34% when the feed temperature is increased from 25°C (1000 m^3/d) to 35°C (1344 m^3/d) theoretically [5]. In other words, roughly a 6.1% increase in permeate flow rate for every 2°C temperature difference can be achieved by utilizing process waste-heat sources. When process waste heat is not available, utilizing solar collectors or geothermal waters is a feasible option.

4.4.3 Recent Studies on Geothermal Desalination

The application of geothermal water in desalination is a relatively unexplored technical concept [23]. Limited number of studies evaluating the potential of geothermal water as a heat source for desalination are available. High-temperature geothermal energy sources are suitable for electricity production while low-temperature sources can be used for desalination. Hot brines may be fed directly to power or large distillation plants. Here, a brief

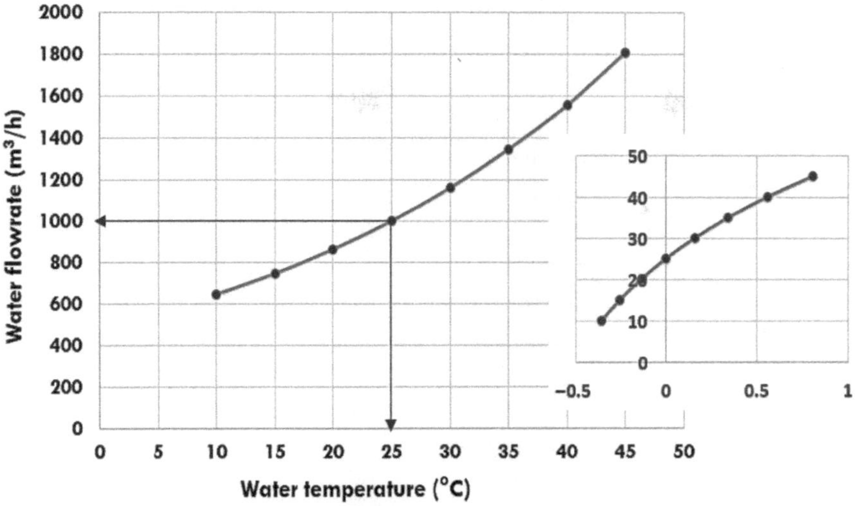

Fig. 4.9 Relationship between the permeate flow and the feedwater temperature in an SWRO process [5].

review on the geothermal energy-driven thermal, membrane, and hybrid (membrane desalination, humidification, and dehumidification) desalination technologies is presented.

The first study of geothermal desalination was proposed and analyzed by Awerbuch et al. in 1976 to produce power and water from geothermal brines in a novel process [24]. In this process, a separator, steam turbine, and a MSF unit were included. The separator was used to make sure that the steam flashed from the hot brine extracted from the geothermal production well was circulated in the steam turbine while the non-evaporated hot brine was used as the feedwater to the MSF unit to produce freshwater.

An earlier study at East Mesa test site evaluated the feasibility of geothermal energy-driven vertical tube evaporator (VTE), MSF, and a high-temperature electrodialysis (HTED) desalination processes [25]. Smooth and enhanced heat transfer surfaces (tubes) were used in the distillation units while previously tested Teflon-backed membranes, dacron-, and polypropylene-backed membranes in thin cell configurations were investigated in HTED process. Data was collected to analyze the heat transfer coefficients, tube fouling and scaling effects, feed, brine and product compositions, cell-pair resistance, current efficiency, membrane fouling, and scale effects. To control the effects of calcium carbonate, silica, and barium sulfate scaling, a pretreatment consisting of poly-phosphonate addition in distillation units and acidification was done for electrodialysis. MSF and VTE units were tested at high and low temperatures at 270°F and 190°F while HTED operation included two- and three-stage operations at 140° F and 160°F, respectively.

The advantage with geothermal sources is that energy output is generally invariant with less intermittence problems making them ideal for thermal desalination processes. Another possible advantage with geothermal waters is that the feedwater itself can be replaced by the geothermal waters; in other words, the geothermal water can serve both as feed and heat transfer medium for desalination. Karystsas has described a case study of a low-enthalpy geothermal energy-driven seawater desalination plant on the Milos Island in Greece [26]. The proposed design consists of coupling MED units to a geothermal groundwater source with temperatures ranging from 75°C to 90°C. The study showed that the exploitation of the low-enthalpy geothermal energy would help save the equivalent of 5000 TOE/year for a proposed plant capacity of 600–800 m^3/day of freshwater. Even in the case of limited geothermal energy, thermal desalination processes such as MED, thermal vapor compression (TVC), single-stage flash (SF) distillation, and MSF can benefit greatly when coupled to geothermal sources by economizing considerable amounts of energy needed for preheating.

Bouchekima [27] analyzed the performance of solar still in which the feedwater is brackish geothermal water in South Algeria. Most geothermal sources in South Algeria have low enthalpy with maximum temperatures of 60–70°C. A solar distillation (capillary film solar distiller) system was developed and its performance was studied. Theoretical analyses of the heat and mass transfer mechanisms inside this solar distiller were compared with experimental results from the distillation unit. With heat recovery in the capillary film distiller was able to produce up to 20 L/m^2/d while a conventional solar still could produce 5–6 L of freshwater per square meter per day of collector surface.

Bourouni [28] demonstrated an aero–evapo–condensation process which was found to be promising for cooling as well as for desalting geothermal water. Heat and mass transfer in the process was evaluated through modeling and experimental studies with geothermal temperatures in the range of 60–90°C. Two experimental pilot studies were used. The latter, installed, respectively, in France and the south of Tunisia, was supplied by fuel and geothermal energy. In another study, a geothermal spring with a water temperature of about 70°C was used to evaluate the heat transfer of air–water–vapor mixtures in the aero-evapo-condensation process [29].

The influence of different thermal and hydrodynamic parameters on the unit performances was also investigated. The experimental results correlated well with the theoretical predictions. The geothermal desalination costs for this process were compared with other desalination options. The costs for this process were $\$1.15/m^3$ when powered by geothermal energy sources; however, when heat is supplied by a fuel source, the cost was $\$4.80/m^3$. The costs for other options were estimated as follows: MSF with back-pressure steam-turbine—$\$1.57/m^3$; MSF with gas turbine and waste-heat boiler—$\$1.44/m^3$; MSF/TVC with gas turbine and waste-heat boiler—$\$1.31/m^3$; and RO single stage with energy recovery—$\$1.39/m^3$. Mohamed and El Minshawy have conducted theoretical and experimental studies on desalting water using humidification–dehumidification processes similar to Bourouni et al. in Egypt [30].

A new brackish water greenhouse desalination process powered by geothermal energy source was proposed by Mahmoudi et al. [31]. This process is suitable for cold and dry weather common in arid regions, such as Algeria. This process was also similar to the humidification and dehumidification of air. In this process, a ground heat exchanger was used to increase the temperature of the geothermal waters acceptable for the proposed desalination process which included evaporative and condensing surfaces while heating the greenhouse. They have reported that geothermal energy is the most promising application for inland brackish water desalination.

An integrated configuration including a multieffect boiling unit and a MSF unit was evaluated in a feasibility study utilizing geothermal sources in Baja California, Mexico [32]. The geothermal source was available as a heat source at 80°C. In this process, an optimum desalinated water (freshwater) to geothermal source (heat source water) ratio was found to be 1:14. In another study, coastal geothermal desalination plant including MSF–MED configuration was studied. The prototype was developed and tested at a geothermal source temperature of 80°C. This study reported a much smaller ratio of 1:5.9 for the freshwater to geothermal source. About $20\,m^3/d$ of freshwater was produced from a geothermal heat flow of $118\,m^3/d$ [33].

Sephton Water Technology developed a pilot project for simulating process conditions for a demonstration project for showcasing its viability under commercial conditions [34,35]. They performed several tests that simulate process conditions in several effects across the range of a 15-effect VTE plant. The VTE distillation process applied to an MED plant design with low-cost noncommercial geothermal steam as an energy source may prove to be cost competitive compared with other desalination technologies. A 15-effect plant could produce up to 14 pounds of distilled water for each pound of geothermal steam used. VTE technology could also have the advantage of discharging much lower volumes of concentrated brine than other methods. A recovery rate in excess of 80% from a seawater source is typical in this process. They used heat (noncommercial low-pressure steam; temperature: 100°C) from a geothermal power plant to reduce salinity of Salton Sea using MED in which a VTE distillation process was applied for geothermal desalination [34,35]. The freshwater production of the pilot unit (MED/VTE 2-effects) and the demonstration unit (MED/VTW 15-effects) were 18.9 and 79.5 m^3/day, respectively, and the respective consumptions were 454 and 3402 kg/h [34,35]. Table 4.1 provides a summary of the process conditions for various geothermal installations around the world. This table shows the geothermal water temperatures and the flow rates required in different desalination processes including MED, MSF, VTE, humidification–dehumidification, and membrane distillation. The geothermal water flow requirements and the costs depend on the desalination process configurations.

Kamal [37] had shown that enhancement of feedwater temperature for seawater RO plants located in southern California induced a substantial reduction in the cost of potable water. The membrane productivity increase is about 2%–3% per 1°C increase of the feeding temperature. Most of the membranes commercialized for RO desalination processes can tolerate temperatures up to 40°C. However, few membrane suppliers offer new membranes for high-temperature applications. For example, Backpulseable membranes (polypropylene tubular membranes) can tolerate temperatures up to 60°C. It is important to note that an increase in the temperature of the feeding water of the RO desalination plant of Gabes to 40°C (temperature tolerance of most commercialized membranes) will increase its productivity by about 20%–30%.

A fluidized bed crystallizer and an air-gap membrane distillation (AGMD) unit were investigated for the suitability of geothermal energy sources by Bouguecha and Dhabi [36]. The membrane surface area in the AGMD was $64 \pm 0.04\,cm^2$ per cell in a three-cell AGMD unit. The membranes were supported by polypropylene grids in 12 mm thick Plexiglas frames. Cooling plates were included to provide condensation of the water vapor. Salt water concentrations were tested between 3 and 35 g/L concentrations and the permeate flux decreased with increase in the salt concentration. This was attributed to vapor reduction due to salt interference effect,

Table 4.1 A summary of geothermal desalination installations around the world

Location	Process description	Process conditions	References
Baja California, Mexico	Combination of multieffect distillation (MED) and multistage flash (MSF), called "multiflash with heaters" (MFWH).	An initial temperature of 150°C, 4 m^3 of seawater were required to produce 1 m^3 of desalinated water. At an initial temperature of 80°C, 14 m^3 of geothermal water was required.	[33]
Kimolos, Greece	A two-stage MED with distillation under vacuum in vertical tubes.	Geothermal water flow rate of 60 m^3/h at a wellhead temperature of 61–62°C from a 188-m deep borehole. Total production of freshwater is 80 m^3/day Produced water cost is estimated of the order of 1.6 €/m^3 (including only annual operation costs)	[26]
Tunisia	Humidification and dehumidification process (HD)	At 75–90°C geothermal temperatures, 75% energy savings were reported for a geothermal powered desalination process. Produced water cost is estimated as $1.2/$m^3$	[29]
Tunisia	MD coupled with multiple-effect distiller	The study found that the best operating parameters are 85°C for a feed brine temperature at the evaporator inlet and a circulation flow of about 170 kg/h. Under these conditions, a GOR value of 3.7 and a water production of 16 kg/h may be reached. The integration of one membrane module distiller as a second step at the MED outlet permits an increase of distilled water production by about 7.5% and improvement in the energetic efficiency by practically 10%	[36]
Salton sea/ Imperial valley (USA)	MED/VTE 2 effects MED/VTE 15 effects	T=100 (steam) at 454 kg/h; Freshwater production rate of 18.9 m^3/d T=100 (steam) at 3402 kg/h; Freshwater production rate of 79.5 m^3/d	[34,35]

increased temperature polarization, and concentration polarization. Further, the effect of process parameters such as temperature and feed flow rate were evaluated. An increase in the permeate flow with feed velocity seems to reach asymptotically a maximum value of 7.5 $kg\,m^{-2}h^{-1}$.

In a recent study, geothermal potential of the Greek island of Nisyros located in the southeastern part of the Aegean Sea for desalination of seawater and power production, in separate and coupling schemes was evaluated [38]. A multieffect distillation was preferred in this study due to its lower specific energy requirements and lower scaling potentials due to low-temperature operation. Energy analysis of the desalination plant was also conducted to understand thermodynamic performance of the desalination plant. Further, a life cycle analysis was performed on the MED desalination-power production coupling scheme. The environmental impacts were reported to be negligible when compared with a fossil fuel powered cogeneration plant. Moreover, in this process, waste heat from the power plant was utilized in the desalination process eliminating associated environmental emissions.

Boron concentration seems to be an issue with geothermal waters limiting their applications and other beneficial uses. Since boron at elevated levels can cause reproductive and developmental toxicity in animals as well as affecting crops, additional boron removal processes must be added to desalination plants. Significant amount of work was performed by researchers in Poland and Turkey on various geothermal waters, even containing radioactive compounds such radium, radon, and tellurium. In Poland, due to increasing water stress levels, geothermal waters are increasingly evaluated for their beneficial uses [39]. Several hybrid technologies integrating membranes [microfiltration, ultrafiltration (UF), and RO] and ion-exchange resins and electrode ionization processes were evaluated to decrease the boron concentrations to the acceptable levels (usually below 1.0 mg/L).

Desalination of spent geothermal waters for beneficial uses was pursued in Poland. Three different geothermal waters were treated using the two-stage membrane process. The sites were Podhale basin (GT-1), Polish Lolands (GT-2), and Western Carpathian Mountains (GT-3). These geothermal waters had a wide range of concentrations of minerals including high iron, strontium, boron, and silica [40,41]. A process schematic of the membrane process is shown in Fig. 4.10. The geothermal water characteristics are given in Table 4.2 [40,41].

A hybrid membrane desalination technology consisting of UF–RO membranes was evaluated for boron removal [42]. The purpose of this study

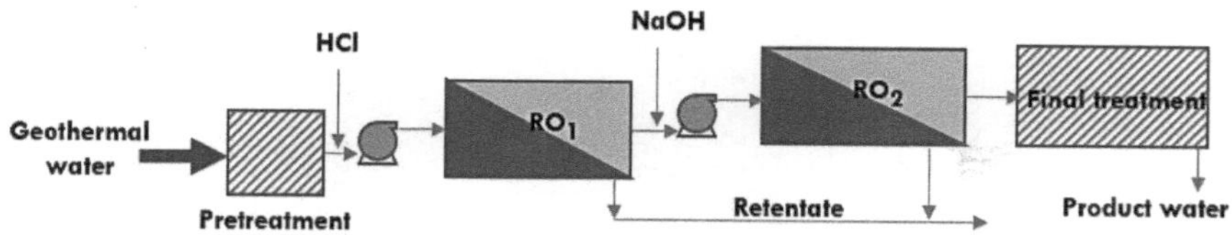

Fig. 4.10 Schematic of a two-stage RO process for recovering spent geothermal waters for drinking purposes in Poland.

was to reduce the concentrations of salinity and micropollutants such as arsenic, iron, and fluoride to make them suitable for drinking water or surface discharge. This system consisted of a pretreatment facility which has a mechanical filter, iron removal stage, and an UF module. Spiral wound DOW FILMTEC BW30HR-440i polyamide thin-film composite RO membranes were used. Considering the presence of high boron concentrations, a two-stage RO system with water reaction adjustment before the RO-2 stage was designed. Geothermal water was split into two streams in the membrane process, a permeate stream (treated water) and a retentate stream that contained separated salts and dissolved mineral contents.

Different boron concentrations (2.5, 8.98, and 96.73 mg B/L) of spent geothermal waters were tested [42]. The removal efficiencies were reported as 56%, 48%, and 12%, respectively, for 2.5, 8.98, and 96.73 mg B/L concentrations. High pH conditions (around 10 and 11) resulted in up to 97% removal. Other removal efficiencies were 96%–97% with respect to conductivity and 94% with respect to SiO_2, 92% for fluoride and not less than 84% for arsenic. Scaling studies on the membranes after 8 months of operation through morphological evaluation of sediments in the electron scanning microscope image and through the analysis of the chemical composition of the sediments crystallized on the membrane revealed silicates (SiO_2) and aluminosilicates (Al_2SiO_5), barite sulfate ($BaSO_4$) and copper sulfate ($CuSO_4$) crystalline structures and calcium phosphate, and celestite ($SrSO_4$) and strontianite ($SrCO_3$) sediments. A high rejection rate of radionuclides was also obtained, ranging from 70.7% to 99.6% [43,44].

A bench scale ion-exchange–microfiltration integrated process was evaluated for boron removal from geothermal waters containing high concentrations (8.9 mg/L) [45]. A small flow of 100–200 mL/h of geothermal water was processed through the unit to determine the effect of process parameter and their optimization. A boron exchange resin Diaion CRB02 was used in this process. In this study, 4.2 g/L of resin with 45–125 mm particle size range was found to be optimum for a flow rate of 0.5 mL/min to achieve

Table 4.2 Geothermal desalination feed and product water compositions studied in Poland and Turkey

Composition	GT-1 Feed	GT-2 Feed	GT-3 Feed	GT-1 Product	GT-2 Product	GT-3 Product	Balcova, Turkey Feed	Izmir-Turkey (1) Feed	Izmir-Turkey (2) Feed
Na^+ (mg/L)	466.8	2297	9492	40.88	151.8	575.1	353.5	280–310	324–363
K^+ (mg/L)	45.2	27.2	83.1	0.83	1.76	8.19	26.55	26–31	27–45
Ca^{2+} (mg/L)	196	146.8	71.24	<10	0.24	7.11	28.17	14–22	20–22
Mg^{2+} (mg/L)	42.7	26.2	36.6	<0.1	0.1	1.905	2.14	1.4–3.9	2.3–5.8
Cl^- (mg/L)	536	3574	12,815	7.6	11.2	1294	208.96	189–190	160–169
SO_4^{2-} (mg/L)	938.2	193.7	3	6.4	<3.0	<3.0	176.56	162–164	153–174
HCO_3^- (mg/L)								561–586	617–666
F^- (mg/L)	2.6	0.696	1.050	0.137	0.078	0.104	7.72	–	–
As (mg/L)	0.026	0.0005	0.094	0,001	<0,005	0,006			
NO_3^- (mg/L)									
pH								8.9–9.0	7.2–7.4
EC (μS/cm)	3550	10,960	35,500	417	1887	4500		1746–1836	1679–1772
TDS (mg/L)	2561.8	6556	2447	181.5	291.6	2588		861–916	840–887
Salinity (‰)								0.88–0.93	0.7
Turbidity (NTU)								0.43–1.54	11.5–25.5
B (mg/L)	8.98	2.53	96.73	0.24	0.159	32.98	8.5–9.5	10.3–11.4	8.8–11.4

Uranium 238U, mBq/L	3.7 ± 0,3	2.5 ± 0.3	10.6 ± 1.0	≤0.50	≤0.50	≤0.50			
Uranium 234U, mBq/L	5.9 ± 0,3	2.2 ± 0.3	9.4 ± 0.9	≤0.50	≤0.50	≤0.50			
Radium 226Ra, mBq/L	514 ± 60	272 ± 10	615 ± 14	18.0 ± 2	≤2.0	≤2.0			
Radium 228Ra, mBq/L	157 ± 55	242 ± 38	607 ± 45	≤47.0 ± 5	≤10.0	≤10.0			
Radon 222Rn, Bq/L	5.4 ± 0.4	5.0 ± 0.6	0.7 ± 0.3						
Tritium, TU/L	0.0 ± 0.4	0.7 ± 0.4	0.1 ± 0.4	0.0 ± 0.3	0.0 ± 0.4	0.3 ± 0.4			
Gross α, mBq/L	620 ± 150	300 ± 60	910 ± 180	≤50	≤50	≤50			
Gross β, mBq/L	1150 ± 340	370 ± 110	1250 ± 380	≤100	≤100	≤100			

boron removals that make them suitable for beneficial uses. For low boron concentration, low particle sizes seemed to be optimum. In another study [46], a hybrid process coupling ion exchange with UF was developed with ion-exchange resin Dowex XUS-43594.00 with an average particle diameter of 20 μm. The UF membrane module was (ZW-1, GE) used to filter the resin suspensions. A RO process coupled with ion-exchange and UF unit was tested for boron removal in Turkey using brackish geothermal water [47]. Boron selective chelating ion-exchange resin Dowex (XUS 43594.00) with a particle size of 20 μm was used for the removal of boron. A submerged hollow fiber-type UF membrane module (ZW-1) was used for filtration. Effect of such process variables as resin concentration in the suspension, flow rates of fresh and saturated ion-exchange resins, and flow rate of permeate on the efficiency of the hybrid process for boron removal from RO permeate of geothermal water was studied. The geothermal water at around 60°C was transferred to the feed tanks and kept for 1–2 days until the geothermal water cooled down to the room temperature (around 25°C). Then, it was filtered through the sand filter and cartridge filters. The anti-scalant to prevent precipitation of silicate was dosed to the filtered geothermal water. The pretreated geothermal water was fed to RO membranes via a high pressure pump. The characteristics of geothermal water used for RO tests are given in Table 4.2. An UF module (GE-Zenon ZW-1 UF module with an effective membrane surface area of 0.047 m^2 and nominal pore diameter of 0.04 μm) was used at the same flow rate of feed solution (RO permeate of geothermal water). The RO permeate samples containing 4.75–5.15 mg B/L were collected from the mini pilot-scale BWRO plant.

Electrodeionization was also used to remove boron and silica from the permeate of a RO (BW-30-2540, Dow FilmTech) process [48]. The concentration of boron remaining in the permeate was 5.9 mg/L which was reduced to 0.4 mg/L and silica was decreased from 0.3 mg Si/L to 0.1 mg Si/L with a layered bed configuration of EDI system when a 40 V of voltage was applied to the EDI system [49]. Further, this group also investigated the potential of a novel chelating resin poly(N-(4-vinylbenzyl)-N-methyl-D-glucamine) (P(VbNMDG)) [50]. The sorption performance of this resin was compared with boron selective commercial resin Diaion CRB02 containing N-methyl-D-glucamine (NMDG) groups for boron removal from geothermal water. The P(VbNMDG) resin gave a higher sorption capacity and faster kinetics than that of Diaion CRB02 for boron removal from geothermal water. Experimental data obtained from boron selective ion-exchange resins Diaion CRB 02 and Dowex (XUS 43594.00) in fixed-bed column studies were analyzed using Yoon–Nelson and Thomas models. It was shown that both models were successful in predicting the process efficiencies due to a good fit with experimental data.

4.4.4 Selection of Desalination Process

The following factors need consideration when selecting a desalination process suitable for geothermal applications [5,51]: (a) plant size; (b) availability of geothermal energy sources and other renewable energy sources; (c) desalination processes; (d) feedwater characteristics; (e) product water quality requirements; (f) brine disposal options; and (g) energy intensity of the desalination process.

4.4.4.1 Plant Size

The plant size depends on the water demand, feedwater concentrations, sources, and water storage capacity. If feedwater quality is high, then a bypass flow can be introduced in the plant design which may decrease the size of the desalination plant. A survey of available feedwater sources should be conducted to evaluate the desalination potential for a specific community. It is desirable to identify suitable quality and quantity of feedwater sources suitable for desalination. On the basis of the historical trends on water utilization for a specific community, the water demand per capita can be estimated which can be further translated into daily plant capacity. Regional statistics can be used in cases the water demand data is not available.

4.4.4.2 Geothermal Energy Quality and Quantity and Other Renewable Energy Sources

A survey of available geothermal water sources is important in determining the desalination plant type, capacity, and the economics. The chemical and physical characteristics of geothermal waters are critical in the design of a heat extraction and utilization scheme [21]. High-temperature geothermal waters are suitable for cogeneration schemes whereas low-temperature geothermal waters are suitable for low-temperature desalination processes described in case studies earlier. Other renewable energy sources such as solar, wind and wave, and biomass sources should also be identified in a case where integrated and mixed energy sources are desirable for process flexibility. For the solar and wind energy sources, estimates of possible thermal and electrical generation capacities and storage capacities should be available.

4.4.4.3 Desalination Technology

The selection of appropriate desalination process (whether a membrane or a thermal process) depends on the geothermal energy source characteristics such as the temperature and the flow capacities [2]. The selection process is also influenced by the feedwater quality (e.g., salinity), technical readiness of the community, plant size, infrastructure and maintenance skills for the desalination plant and the type and potential of other renewable energy sources and the availability of grid electricity and the remoteness of the community as well as the local climate conditions. Low-temperature geothermal sources are suitable for small-scale desalination processes such as low-temperature MED, membrane distillation, humidification–dehumidification, and high-temperature RO processes. High-temperature geothermal waters also allow for cogeneration schemes integrating desalination with power production.

4.4.4.4 Feedwater

The feedwater characteristics should be evaluated to determine a suitable desalination process. If low-concentration TDS feedwater is available, then membrane processes can be favored. This also depends on the geothermal energy source temperature. Nanofiltration and RO processes can be employed for low-temperature geothermal energy and low TDS ground water combinations. Electrodialysis process may also fit this purpose better for low TDS feedwater. For high TDS concentrations, MSF or MED processes are preferred when high-quality geothermal waters are available. SWRO process can be considered for low geothermal water temperatures.

4.4.4.5 Product Water

The desired product quality should be determined based on the end use. Industrial processes may require high-quality distillates which can be produced by thermal desalination plants. Potable water from low TDS waters can be produced from membrane processes. Blending options can be considered where possible as long as the requirements for the product water can be met with some reliable process configuration. Other agricultural and fish farm and animal feed operations may not require high-quality potable waters. Desalination processes may prove to be cost prohibitive for the agricultural applications.

4.4.4.6 Brine Disposal

Brine disposal and management issues cannot be ignored for any desalination plant. Inland desalination plants would have to consider different disposal and management options from the coastal desalination plants. It was shown that inland desalination has less environmental impacts when compared with seawater desalination. Disposal into oceans is a commonly practiced method for coastal desalination plants while evaporation ponds or sewer disposal is a better option for inland desalination plants. Zero liquid discharge configurations can be used to enhance the water recovery thereby minimizing the waste brine disposal issues albeit at a higher cost.

4.4.4.7 Technoeconomic Requirements

Energy source and the desalination energy requirements play a critical role in determining the desalination process. Considering the aforementioned information on the match of desalination process energy requirements and the type and quality of the geothermal sources, a process that has lowest energy footprint with reliable and robust performance should be chosen for a freshwater supply option. Technoeconomic analyzes should also consider the feedwater characteristics such as concentration and temperature to be connected with the process performance in terms of performance ratio, heat losses, and other economic considerations and process-related issues.

4.5 CHALLENGES AND CONSIDERATIONS FOR GEOTHERMAL DESALINATION

An ideal geothermal source should be available at reasonable depths, easily accessible, and at a quality suitable for intended application and close to the point of application [52]. The key requirements for identifying a good geothermal resource can be summarized as follows [11]:

- A high temperature for good power plant efficiency (for a cogeneration scheme).
- A large quantity of stored heat for resource longevity.
- A low rate of liquid production per unit of energy reinjection well sites available at a lower elevation than production for disposal under gravity.
- Produced fluids with a near-neutral pH for low corrosion rates in wells and plants.
- Adequate permeability to ensure adequate outputs from individual wells.
- A low tendency for scaling in pipelines and wells.
- Low elevation and easy terrain for access roads.

- A low risk of volcanic activity and hydrothermal eruptions.
- Proximity to electrical load or transmission lines.

The basic steps involved in geothermal source utilization for desalination applications are as follows: identification of geothermal reservoir beneath the surface, estimation of geothermal reservoir volume and its type (temperature, flow, and pressure, chemical characteristics of the fluid, saturated steam, or hot water), and estimation of future availability of geothermal reservoir. Some other important factors need to be considered for successful implementation of geothermal desalination plants. The major issues are related to land use, geological hazards, waste heat, atmospheric releases, water footprint, noise, and socioeconomic issue which are discussed in the following sections [53]. Table 4.3 lists the different types of environmental, socioeconomic impacts associated with geothermal power plants and possible desalination schemes and their probability of occurrence, severity, and the duration of impacts [54]. Some events and their impacts have low probability of occurrence with short- and long-term impacts. A number of environmental regulations may apply for the environmental impacts caused by geothermal power and water facilities. These regulations and

Table 4.3 Environmental impacts of geothermal source-based power production and desalination

Impact	Probability of occurring	Severity of consequence	Duration of impact
Air quality emissions	Low	Medium	Short term
Surface water discharge	Medium	Low to medium	Short term to long term
Underground contamination	Low	Medium	Long term
Land subsidence	Low	Low to medium	Long term
High noise levels	High	Medium to high	Short term
Well blowouts	Low	Low to medium	Short term
Conflicts with cultural and archeological features	Low to medium	Medium to high	Short term to long term
Socioeconomic problems	Low	Low	Short term
Chemical or thermal contamination	Medium	Medium to high	Short term to long term
Solid waste disposal	Medium	Medium to high	Short term

the requirements may vary from region to region. A summary of environmental regulations applicable for geothermal operations is given in Table 4.4 [55]. Regulations are covered under "Clean Water Act," "Clean Air Act," "Solid Waste Management," noise control, "Occupational Safety and Health Act," and "Endangered Species Act" enforced by the United States Environmental Protection Agency (USEPA). These regulations correspond to the environmental impacts presented in Table 4.3.

4.5.1 Land Use

Geothermal energy systems require significant land areas for energy extraction as well as the geothermal water treatment and management. About 1.2–2.7 tm^2 of land area is required per each MW energy production. Higher land footprints are reported for those facilities managing the geothermal brines. High capacity flash-steam plants require low land footprints. Plants with reinjection of geothermal waters also require low land footprints.

4.5.2 Geological Hazards

Geothermal power plants involve excessive extraction and circulation of geothermal waters and excavation of land deep into soils is a must. This might cause seismic instability at some locations although not a large impact. Continued extraction of geothermal waters may also cause land subsidization. In some extreme conditions, although uncommon, hydrothermal explosions or a sudden release of pressure (steam) caused by earth quakes may take place.

4.5.3 Waste-Heat Releases

Geothermal wells for either direct use or power production release significant amounts of high-temperature waters. This is mainly due to the large quantities of geothermal flows and inefficient electricity conversion. In the case of power plants with cooling towers, the waste-heat releases are even higher than other conventional power plants. In most cases, the waste heat is released in the form of vapors into the atmosphere, into any artificial ponds or other receiving water bodies.

4.5.4 Atmospheric Emissions

Geothermal wells release significant amounts of noncondensable gases such as H_2S, CO_2, and CH_4, mainly associated with flash steam and dry steam power plants. Other minor pollutants are Hg, ammonia, radon, and boron.

Table 4.4 Environmental regulations to be considered for geothermal source-based power production and desalination in the United States

Laws and regulations	Air	Surface water	Geothermal fluids	Solid waste	Liquid waste	Noise	Subsidence/ seismicity	Cultural resources	Biological resources
Federal Water Pollution Control Act (NPDES)		×			×				
Safe Drinking Water Act (Underground Injection Control Regulations)			×		×				
Clean Air Act	×								
Resource Conservation and Recovery Act (RCRA)				×	×				
Toxic Substance Control Act				×	×				
National Environmental Policy Act (NEPA)	×	×	×	×	×	×	×	×	×
Noise Control Act						×			
Geothermal Resource Operational Order #4	×	×	×	×	×	×	×	×	×
Occupational Safety & Health Act (OSHA)	×		×	×	×	×			
Endangered Species Act									×

A wide range of 4–740 g/kWh of CO_2 are released from a variety of power plants [56]. Some studies report as low as 0–40 g/kWh with a common average range of 80–100 g/kWh. Concerning sulfur dioxide emissions, between 0.159 and 1 g/kWh have been reported [55,57]. Greenhouse gas, CH_4, is also another common pollutant which is released in very small concentrations. Other gases of interest are nitrogen oxides. Other pollutants such as arsenic, boron, ammonia, and mercury are of major concern in some locations.

4.5.5 Water Footprint

Water is required from the initial stage of drilling a geothermal well to its operation. The water requirements for drilling have been reported between 5 and 30 m^3 which depends on the geology, technology, and the depth [58]. Power generation with air or hybrid cooling plants consume water between 0 and 1.5 m^3/MWh while water cooling towers consume up to 17 m^3/MWh [56]. In addition, the brines released from these wells are substantial often reinjected into the wells but spills often occur which introduce the pollutants belonging to heavy metals and other toxic substances. Geothermal brines in addition have the disadvantage of high salt concentration that creates in general operational problems, hard scale formation, and concentrated brine disposal, if not near the sea [59]. The scale formation could be mainly due to the hydrogen sulfide and ammonia in geothermal heat sources [34,35].

4.5.6 Noise and Social Impacts

Well excavation, drilling, and operations all involve noise issues, some aesthetic issues can be anticipated in some locations. The noise levels from each well and the combined impacts depend on the number of wells and drilling technology in use. Social impacts would be creation of job opportunities for local communities. This may be a limited benefit since skilled labor will be required for geothermal operations. Geothermal energy and water production can benefit remote communities which lack access to other renewable energy resources and connection to the grid. In some cases where geothermal wells are operated for tourism and other transient community uses, the local economies can be improved.

4.6 TECHNOECONOMICS OF GEOTHERMAL DESALINATION

Desalination costs depend on various factors which include the desalination process configuration, energy source, financial package and the local capital, and operation and maintenance costs. Table 4.5 presents the desalination cost for thermal and membrane desalination processes at different capacities. Desalination costs decrease with increasing capacity in general. The costs will be significantly affected by the energy source and the source water characteristics. For example, brackish water desalination can be cost effective while seawater desalination is both energy intensive and cost intensive as illustrated in Table 4.5. A recent evaluation of geothermal powered desalination costs suggests that geothermal RO is more cost effective than geothermal MED, despite the low efficiency of the geothermal electricity power plant that provides the electricity to the RO [61]. It was found that geothermal RO has a levelized water cost of ($2.06/$m^3$) that is 17% lower than geothermal MED ($2.48/$m^3$).

Table 4.5 Desalination costs for various desalination processes for various capacities

	Capacity of desalination plant (m^3/day)	**Desalination cost (US $/$m^3$)**
Desalination process		
Multieffect distillation (MED)	Less than 100	2.5–10
	12,000–55,000	0.95–1.95
	Greater than 91,000	0.52–1.01
Multistage flash (MSF)	23,000–528,000	0.52–1.75
Vapor compression	1000–1200	2.01–2.66
Type of feedwater		
Reverse osmosis (RO)—Brackish water	Less than 20	5.63–12.9
	20–1200	0.78–1.33
	40,000–46,000	0.26–0.54
Reverse osmosis (RO)—Seawater	Less than 100	1.5–18.75
	250–1000	1.25–3.93
	15,000–60,000	0.48–1.62
	100,000–320,000	0.45–0.66
Brackish water [60]	Conventional	0.21–1.06 (€/m^3)
	Photovoltaic	4.50–10.32 (€/m^3)
	Geothermal	2.00 (€/m^3)

Renewable energy-driven desalination processes are in general more expensive than the conventional energy-driven desalination processes. This is due to the low cost of fossil fuel-derived energy sources. Among the renewable energy sources, the capital costs for geothermal electricity is very competitive, although it is a recovered over a longer period of time (see Fig. 4.11A) [60,62–64]. The capital costs are comparable to coal-based electricity production. As discussed in this article, geothermal sources are more suitable for locations where other renewable energy sources have more challenges and in remote locations. These communities are usually small in size requiring low desalination plant capacities. For small communities, the energy demands would be low. A comparison of diesel generated electricity with geothermal electricity is shown in Fig. 4.11B. It shows that the electricity costs decrease with the production capacity but the geothermal electricity costs can be comparable at lower end of capital costs and better financial packages. This shows the potential for cost-competitive desalination powered by geothermal sources. More detailed discussions on the feasibility and potential developments for geothermal desalination around the world are discussed by the author elsewhere [65].

4.7 SUMMARY

Renewable energy sources, though freely available, still have a very high value and need to be utilized efficiently or utilized in a "sustainable manner." Among the renewable energy sources, geothermal sources allow for efficient resource utilization due to their high capacity factor, especially in desalination applications. Since geothermal sources vary in quality and quantity, suitable and energy-efficient desalination or cogeneration schemes should be integrated for efficient resource utilization. High-temperature (high-enthalpy) geothermal sources can be used in cogeneration schemes for simultaneous power and water production using multistage flash distillation, multieffect distillation technologies. Low-temperature desalination processes can be coupled with low-temperature (low-enthalpy) geothermal sources. Low-temperature desalination processes have lower specific energy requirements and a higher thermodynamic efficiency. Apart from the above, other advantages include lower corrosion rates, low-cost construction materials with a longer plant life, lower scaling, lower heat losses, and shorter start-up periods.

Geothermal source utilization around the world is expected to increase significantly in the future, mainly in direct heat utilization and electricity

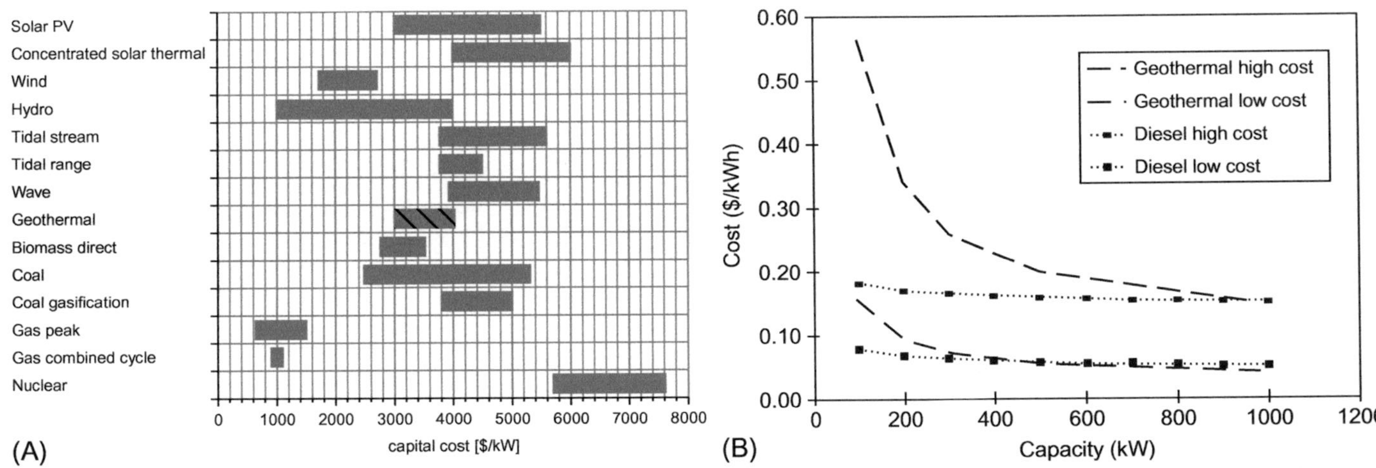

Fig. 4.11 (A) Geothermal energy capital cost comparison with other renewable energy sources; (B) geothermal electricity cost comparison with diesel generators for small capacities [62–64].

production and industrial application uses. In conjunction with this development, geothermal source utilization in desalination applications should be assigned priority in water stressed areas to combat the negative environmental impacts by the nonrenewable energy powered desalination. However, a suitable regulatory framework and specific incentives and subsidies should be provided to encourage the development of geothermal-driven desalination industry. Not all geothermal sites are suitable for desalination application which needs to be carefully evaluated case by case. Uncertainties related to resource characteristics and its availability (reservoir volume and other flow, pressure and temperature characteristics) should be determined in advance to match the life span of the desalination plants. Owing to the inefficiencies in the power generation scheme, cogeneration schemes may not suitable in all locations. For this scenario, a suitable combination of a desalination process and a geothermal source should be identified with a preliminary assessment and evaluation procedure. The economics and environmental impacts of the desalination plants should be assessed with site-specific information to eliminate future failures.

REFERENCES

[1] Gude VG, Nirmalakhandan N. Sustainable desalination using solar energy. Energy Convers Manage 2010;51(11):2245–51.
[2] Gude VG. Energy storage for desalination processes powered by renewable energy and waste heat sources. Appl Energy 2015;137:877–98.
[3] Gude VG, Nirmalakhandan N, Deng S. Desalination using solar energy: Towards sustainability. Energy 2011;36(1):78–85.
[4] Gude VG, Nirmalakhandan N, Deng S. Renewable and sustainable approaches for desalination. Renew Sustain Energy Rev 2010;14(9):2641–54.
[5] Gude VG. Energy consumption and recovery in reverse osmosis. Desalin Water Treatment 2011;36(1–3):239–60.
[6] Khawaji AD, Kutubkhanah IK, Wie JM. Advances in seawater desalination technologies. Desalination 2008;221(1):47–69.
[7] Lund JW, Boyd TL. In: Direct Utilization of Geothermal Energy 2015 Worldwide Review. Proceedings World Geothermal Congress 2015 Melbourne, Australia, 19-25 April 2015; 2015.
[8] World Energy Council 2013. World Energy Resources 2013 Survey. https://www.worldenergy.org/wp-content/uploads/2013/09/Complete_WER_2013_Survey.pdf.
[9] Goosen M, Mahmoudi H, Ghaffour N. Water desalination using geothermal energy. Energies 2010;3(8):1423–42.
[10] Barbier E. Geothermal energy technology and current status: an overview. Renew Sustain Energy Rev 2002;6(1):3–65.
[11] Li K, Bian H, Liu C, Zhang D, Yang Y. Comparison of geothermal with solar and wind power generation systems. Renew Sustain Energy Rev 2015;42:1464–74.

[12] Goldstein B, Hiriart G, Gutierrez-Negrin L, Tester J, Bertani R, Bromley C, et al. Great Expectations for Geothermal to 2100–Messages for Now. GRC Trans 2011;35:1175–83.
[13] Evans A, Strezov V, Evans TJ. Assessment of sustainability indicators for renewable energy technologies. Renew Sustain Energy Rev 2009;13(5):1082–8.
[14] Chamorro CR, Mondéjar ME, Ramos R, Segovia JJ, Martín MC, Villamañán MA. World geothermal power production status: Energy, environmental and economic study of high enthalpy technologies. Energy 2012;42(1):10–8.
[15] IRENA (2014). Rethinking Energy: Towards a new power system, September 2014. IRENA, Abu Dhabi. http://www.irena.org/rethinking/Rethinking_FullReport_web.pdf.
[16] Szacsvay T, Hofer-Noser P, Posnansky M. Technical and economic aspects of small-scale solar-pond-powered seawater desalination systems. Desalination 1999;122(2):185–93.
[17] Geothermal energy: An option for Oman? http://www.iqpc.com/media/8328/11981.pdf.
[18] SAAD, Mohammad Amin (2012), International Conference on Desalination and sustainability Morocco 2012 & RG-TES calculations.
[19] Sanner, B. Geothermal energy opportunities for desert regions. Global Conference on Renewable Energy Approaches for DEsert Regions [GCREADER] Le Royal Hotel Amman, Jordan 18-22 September 2006.
[20] Lund JW. Direct utilization of geothermal energy. Energies 2010;3(8):1443–71.
[21] Finster M, Clark C, Schroeder J, Martino L. Geothermal produced fluids: Characteristics, treatment technologies, and management options. Renew Sustain Energy Rev 2015;50:952–66.
[22] Cardona E, Piacentino A, Marchese F. Energy saving in two-stage reverse osmosis systems coupled with ultrafiltration processes. Desalination 2005;184(1):125–37.
[23] Davies PA, Orfi J. Self-powered desalination of geothermal saline groundwater: technical feasibility. Water 2014;6(11):3409–32.
[24] Awerbuch L, Lindemuth TE, May SC, Rogers AN. Geothermal energy recovery process. Desalination 1976;19(1):325–36.
[25] Boegli WJ, Suemoto SH, Trompeter KM. Geothermal desalting at the East Mesa test site. Desalination 1977;22(1):77–90.
[26] Karytsas. Mediterranean conference on renewable energy sources for water production, European commission, EURORED Network. Santorini, Greece: CRES; 1996. p. 128–131.
[27] Bouchekima B. Solar desalination plant for small size use in remote arid areas of South Algeria for the production of drinking water. Desalination 2003;156(1):353–4.
[28] Bourouni K, Martin R, Tadrist L. Analysis of heat transfer and evaporation in geothermal desalination units. Desalination 1999;122(2):301–13.
[29] Bourouni K, Deronzier JC, Tadrist L. Experimentation and modelling of an innovative geothermal desalination unit. Desalination 1999;125(1):147–53.
[30] Mohamed AMI, El-Minshawy NAS. Humidification–dehumidification desalination system driven by geothermal energy. Desalination 2009;249(2):602–8.
[31] Mahmoudi H, Spahis N, Goosen MF, Ghaffour N, Drouiche N, Ouagued A. Application of geothermal energy for heating and fresh water production in a brackish water greenhouse desalination unit: A case study from Algeria. Renew Sustain Energy Rev 2010;14(1):512–7.
[32] Rodrıguez G, Rodrıguez M, Perez J, Veza J. In: A systematic approach to desalination powered by solar, wind and geothermal energy sources. Proceedings of Mediterranean Conference on Renewable Energy Sources for Water Production, European Commission, EURORED Network, CRES, EDS, 10–12 June 1996, Santorini; 1996. p. 20–5.

[33] H. Gutiérrez, S. Espíndola, Using low enthalpy geothermal resources to desalinate seawater and electrucity production on desert areas in Mexico, GHC Bull 19-24, (Aug. 2010) (http://geoheat.oit.edu/bulletin/bull29-2/art5.pdf).
[34] Sephton Water Technology, VTE Geothermal Desalination Pilot/Demonstration Project, http://sephtonwatertech.com/DocumentsPDF/VTE_GeothermalDesalination ProjectSummary_2010_04_15.pdf2010. Accessed online 08/16/2015.
[35] Sephton Water Technology, VTE Geothermal Desalination Pilot/Demonstration Project. Project Summary, http://sephtonwatertech.com/DocumentsPDF/VTE_Geothermal_Desalination_Project_Summary_2012_02_05.pdf2012. Accessed online 08/16/2015.
[36] Bouguecha S, Dhahbi M. Fluidised bed crystalliser and air gap membrane distillation as a solution to geothermal water desalination. Desalination 2003;152(1):237–44.
[37] Kamal I. Cogeneration desalination with reverse osmosis: A means of augmenting water supplies in Southern California. Desalination 1992;88(1):355–69.
[38] Koroneos C, Roumbas G. Geothermal waters heat integration for the desalination of sea water. Desalin Water Treatment 2012;37(1–3):69–76.
[39] Bujakowski W, Tomaszewska B, Bodzek M. Geothermal water treatment—preliminary experiences from Poland with a global overview of membrane and hybrid desalination technologies. In: Bundschuh J, Hoinkis J, editors. Renewable Energy Applications for Freshwater Production (Sustainable Energy Developments). London, United Kingdom: Taylor & Francis Ltd. CRC Press; 2012. p. 181–206.
[40] Tomaszewska B. The use of ultrafiltration and reverse osmosis in the desalination of low mineralized geothermal waters. Arch Environ Protect 2011;37(3):63–77.
[41] Tomaszewska B, Bujakowski W, Szczepański A, Chowaniec J. Geothermal water desalination as the possible solution for the decentralization of the fresh water production. Melbourne, Australia: Proceedings World Geothermal Congress; 2015.
[42] Tomaszewska B, Bodzek M. Desalination of geothermal waters using a hybrid UF-RO process. Part I: Boron removal in pilot-scale tests. Desalination 2013;319:99–106.
[43] Tomaszewska B, Bodzek M. Desalination of geothermal waters using a hybrid UF-RO process. Part II: Membrane scaling after pilot-scale tests. Desalination 2013;319:107–14.
[44] Tomaszewska B, Bodzek M. The removal of radionuclides during desalination of geothermal waters containing boron using the BWRO system. Desalination 2013;309:284–90.
[45] Kabay N, Yilmaz-Ipek I, Soroko I, Makowski M, Kirmizisakal O, Yag S, et al. Removal of boron from Balcova geothermal water by ion exchange–microfiltration hybrid process. Desalination 2009;241(1):167–73.
[46] Kabay N, Köseoğlu P, Yapıcı D, Yüksel Ü, Yüksel M. Coupling ion exchange with ultrafiltration for boron removal from geothermal water-investigation of process parameters and recycle tests. Desalination 2013;316:17–22.
[47] Kabay N, Köseoğlu P, Yavuz E, Yüksel Ü, Yüksel M. An innovative integrated system for boron removal from geothermal water using RO process and ion exchange-ultrafiltration hybrid method. Desalination 2013;316(1–7):151–7.
[48] İpek İY, Kabay N, Yüksel M. Modeling of fixed bed column studies for removal of boron from geothermal water by selective chelating ion exchange resins. Desalination 2013;310:151–7.
[49] Arar Ö, Yüksel Ü, Kabay N, Yüksel M. Application of electrodeionization (EDI) for removal of boron and silica from reverse osmosis (RO) permeate of geothermal water. Desalination 2013;310:25–33.
[50] Santander P, Rivas BL, Urbano BF, İpek İY, Özkula G, Arda M, et al. Removal of boron from geothermal water by a novel boron selective resin. Desalination 2013;310:102–8.

[51] Andrianne J, Alardin F. Thermal and membrane processe economics: Optimized selection for seawater desalination. Desalination 2003;153(1):305–11.
[52] Zheng B, Xu J, Ni T, Li M. Geothermal energy utilization trends from a technological paradigm perspective. Renew Energy 2015;77:430–41.
[53] Bayer P, Rybach L, Blum P, Brauchler R. Review on life cycle environmental effects of geothermal power generation. Renew Sustain Energy Rev 2013;26:446–63.
[54] Lunis B, Breckenridge R, McClenahan Hietter L. Environmental considerations. Chapter 19, In: Lund J, et al., editors. Geothermal direct-use engineering and design guidebook. 3rd ed. Geo-Heat Center, Oregon Institute of Technology; 1998. p. 443.
[55] Bloomfield KK, Moore JN, Neilson RN. Geothermal energy reduces greenhouse gases. Geotherm Resour Council Bull 2003;32(2):77–9.
[56] EPAUS. Public review draft—National Water Program Strategy: Response to Climate Change. Office ofWater2008.
[57] Kagel A, Bates D, Gawell K. A guide to geothermal energy and the environment (p. 75). Washington, DC: Geothermal Energy Association; 2005.
[58] Clark CE, Harto CB, Sullivan JL, Wang MQ. Water use in the development and operation of geothermal power plants. Argonne National Laboratory; 2011.
[59] Belessiotis V, Delyannis E. Water shortage and renewable energies (RE) desalination—possible technological applications. Desalination 2001;139(1):133–8.
[60] Karagiannis IC, Soldatos PG. Water desalination cost literature: review and assessment. Desalination 2008;223(1–3):448–56.
[61] Loutatidou S, Arafat HA. Techno-economic analysis of MED and RO desalination powered by low-enthalpy geothermal energy. Desalination 2015;365:277–92.
[62] Vimmerstedt LJ. Opportunities for Small Geothermal Projects: Rural Power for Latin America, the Caribbean, and the Philippines; 1998. National Renewable Energy Laboratory.
[63] Entingh DJ, Easwaran E, McLarty L. Small geothermal electric systems for remote power. GRC Bull 1994;331–8.
[64] Abergas R. Strategic Power Utilities Group. Philippines: National Power Corporation; 1998.
[65] Gude VG. Geothermal source potential for water desalination—current status and future perspective. Renew Sustain Energy Rev 2016;57:1038–65.

CHAPTER 5

Desalination of Seawater Using Geothermal Energy for Food and Water Security: Arab and Sub-Saharan Countries

Dornadula Chandrasekharam*, Arif Lashin†, Nassir Al Arifi†, Abdulaziz M. Al-Bassam†
*India Institute of Technology Hyderabad, Hyderabad, India
†King Saud University, Riyadh, Saudi Arabia

5.1 INTRODUCTION

With exponential increase in population and decreasing supply of freshwater for drinking and irrigation, the future concern of all the countries is food security. As of today, nearly 884 million people have no access to drinking water [1] and many countries are either importing large quantities of food or supporting agricultural produce outside the country there by promoting food security outside the country (Fig. 5.1). This is especially so in Gulf countries [Gulf Cooperation Council (GCC) countries: Saudi Arabia, Kuwait, Qatar, Oman, Bahrain, UAE], Middle East and North Africa (MENA), and Sub-Sahara countries (e.g., Eritrea, Ethiopia, Djibouti, Kenya, Egypt).

Over 87% of the landmass occupied by the Arab region (including northern Africa) is desert. These regions experience extreme climate events like severe droughts, high evaporation rates, and very poor rainfall leading to decrease in annual freshwater supply (Fig. 5.2). Thus out of 20 most water stressed countries, 14 fall within this region. This problem has to be addressed with all the seriousness if the Arab region intends to achieve the millennium development goals (MDG). This situation may result in economic collapse and regional conflicts. Any country that is able to produce large quantities of food will be the most powerful country in the future. Besides irrigation and drinking, 15% of available water is being used for energy production [2]. These countries are devoid of renewable water resources due to scarce rainfall. However, the Gulf countries are able to

Renewable Energy Powered Desalination Handbook
https://doi.org/10.1016/B978-0-12-815244-7.00005-2

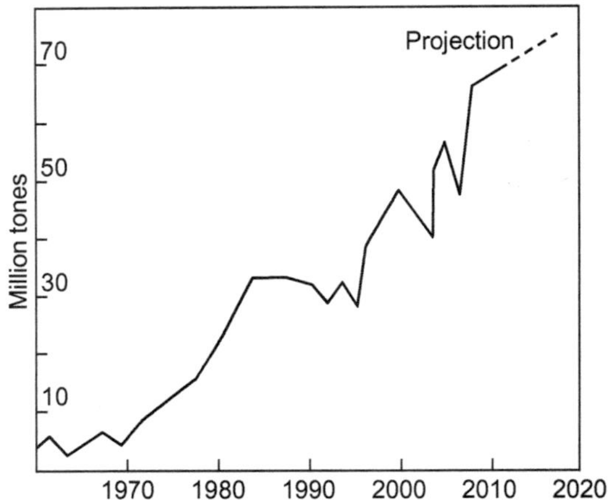

Fig. 5.1 Food imports by Arab and Middle East and North Africa (MENA) countries. *(Adapted from UNDP, 2013. Water governance in the Arab region: Managing scarcity and securing the future. 182 p.)*

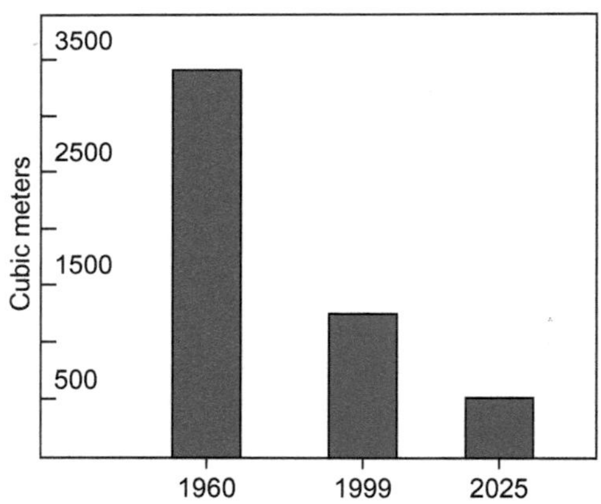

Fig. 5.2 Annual per capita water supply in the Arab countries.

exploit nonrenewable fossil groundwater for irrigation because these aquifers are trans-boundary aquifers (e.g., The Cambrian Saq Ram sandstone aquifer shared between Saudi Arabia and Jordan. The Ummer-Raduma Eocene limestone aquifer shared between Saudi Arabia, Bahrain, Qatar, Kuwait, and Oman [3–6]). Over exploitation of these aquifers resulted in decline in productivity of the aquifer over the years. Besides trans-boundary aquifers, surface waters are also trans-boundary and are shared. For example, Jordan River is shared by Jordan, Israel, Lebanon, Syria, and Palestine; Shabelle River is shared by Ethiopia and Somalia. These waters are loci of regional political conflicts and threat to regional stability. To circumvent this grave situation, oil rich countries around the sea, such as, Saudi Arabia, Oman, Bahrain, Qatar, and Kuwait have large desalination plants and meet freshwater demand for domestic and agricultural purpose using conventional energy sources. Currently, this process is not cost effective and results in the emission of large quantities of CO_2 there by adversely affecting the global climate. Thus, although desalination is the only solution to secure water and food security, the energy source used for this process is adversely affecting the climate and defeating the very purpose of solving freshwater problem. This barrier can be overcome if low-carbon emitting energy source is utilized for desalination process. In the present paper, we discuss the current water and food security status and geothermal energy resources available in the Gulf and MENA countries to generate low-cost freshwater for domestic and agricultural purpose through desalination and to reduce CO_2 emissions that are detrimental to these countries in particular and the world in general.

5.2 WATER RESOURCES OF GULF COUNTRIES

5.2.1 Saudi Arabia

Saudi Arabia is an oil-based economy where the country earns 90% of its revenue from foreign export earnings. The population has grown from 6.9 million in 1972 to over 30 million in 43 years and is expected to cross 40 million by 2025 [7], exerting tremendous pressure on agricultural sector (Fig. 5.3). To reduce this stress, Saudi Arabia adopted advanced methods in irrigation methods and in reducing production quantities.

Currently the water extracted from 2000 m deep Paleozoic-Mesozoic-Cenozoic sandstone aquifers consumes energy more than that consumed by domestic sector [8].

The annual agricultural water demand is 20,800 MCM/yr (million cubic meters) that is 10 times more than the domestic demand [9]. Agriculture-related energy use will increase exponentially with growth in population and food demand, resulting in emission of large quantities of emissions CO_2 that may further deteriorate the climate and weather pattern of Saudi Arabia that is already in adverse state. The major aquifers that are supporting agricultural water demand are the Saq Ram sandstone aquifer with a capacity of 258 400 MCM, the Wajid sandstone aquifer with a capacity of 237,500 MCM, and the Tawila aquifer with a capacity of 109,800 MCM. Since they are all trans-boundary aquifers, they are not in a position to support agricultural demand of Saudi Arabia [9]. To fulfil this demand, Saudi Arabia uses energy intensive desalination process to generate freshwater, consuming

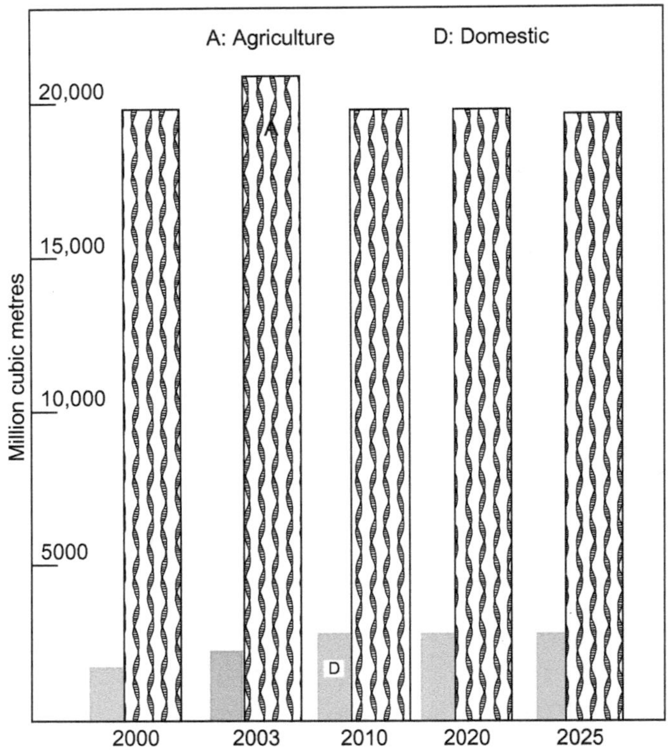

Fig. 5.3 Water demand by domestic and agricultural sectors since 2000 by Saudi Arabia. *(Adopted from Abderrehman WA. Groundwater resources management in Saudi Arabia. Proceed Water Conservation Workshop, Khober, Saudi Arabia, 2006.)*

about 5695 kWh/ton (5.7 MWh/ton) of energy for cultivating 5.56 tons/ha of wheat [8], emitting 4656 kg of CO_2 [10]. Since the techniques used for desalination are multistage flash (MSF) and multistage effect distillation (MSED), the energy consumed is considerable compared with that used for lifting water from borewells 1000–2000 m deep. On the domestic front, Saudi Arabia is consuming 134×10^6 kWh of fossil fuel generated electricity to generate 275 L/day per capita of freshwater through desalination process [11,12]. With declining water table levels in these aquifers, Saudi Arabia has no option but to increase freshwater output through desalination to meet the domestic and agricultural demand.

The energy intensive desalination technique that Saudi Arabia is adopting is very expensive compared with the reverse osmosis (RO) method. The energy consumed is about 12×10^9 kWh to generate 1 m^3 of desalinated water [13]. This emits an amount of about 13 Mt of CO_2 [5]. The cost to generate 20,000 m^3/day of freshwater through vacuum membrane technique is about US$ 0.53/m^3 while the cost to generate similar volume using conventional energy source is US$ 1.22/m^3 [14]. The CO_2-related temperature effect on the local weather that is currently experienced in Saudi Arabia [15] has direct impact on the agricultural production, especially with respect to wheat production. Fossil fuels-based agricultural economy has to be replaced with renewable energy-based economy to ensure the country's food and water security. Declining water levels in deeper aquifers, high cost involved in the production of freshwater through desalination process, and escalating financial burden due to subsidies compelled Saudi Arabia to invest in other countries (Manila for cultivating rice, Argentina for farm products) for agricultural produce and increase food imports. Wheat cultivation in Saudi Arabia is being phased out since 2016 and wheat imports have increased to >3.25 million tons in MY 2013–2014 from 1.92 million tons MY 2012–2013 [9]. Effectively the country is into virtual water trade (VWT) importing large quantities of higher water intensive agricultural and farm products from other countries. There are good and bad issues with such global VWT because countries that are economically poor, for example, African countries are losing their precious land to economically rich countries and in certain cases, enter into VWT pact at the cost of deforestation, for example, Brazil is cultivating soya in large scale to meet China's demand [16]. Such situation arises due to serious water stressed conditions in 33 countries around the world and 14 of them fall within the MENA and Gulf regions (Fig. 5.4).

5.2.2 Kuwait

In Kuwait, cogeneration power-desalination plants account for more than half of total energy consumption; the energy required to meet desalination plant demand is expected to equal the country's current fuel oil production by 2035 [18].

Kuwait consumes more than $45{,}2000 \times 10^6$ kWh of electricity since 2008 and this will increase in future due to population growth (~4% annually) and high per-capita consumption of electricity [19,20] has already overtaken the population growth and stands at 6.8% annually. In 2014 the CO_2 emission was 84×10^6 tons [21] which has increased by a factor of 2 in a decade. Most of the emissions come from the combustion of oil and gas. The per-capita CO_2 emission is 23 tons. Together with energy, water demand has increased, especially during the postwar period. The groundwater and

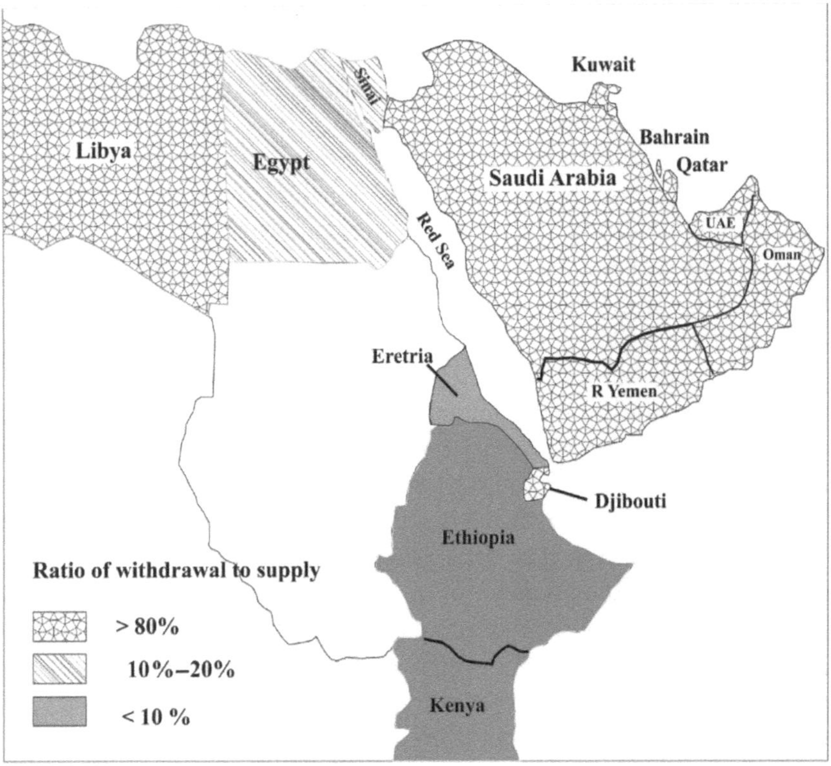

Fig. 5.4 Water stressed countries focused in this chapter in 2040 [17].

surface water resources are insufficient to meet the growing demand of the population that is growing at the rate of 2.5% [22]. With very low rainfall (~110–130 mm/yr, [22–24]) and hence very low recharge ~160 CMC/yr, [24]).

Compared with MENA and GCC countries, Kuwait's internal renewable water resources are minimum and the country does not have any trans-boundary aquifers such as Saudi Arabia, Jordan, and Yemen which share common aquifers to extract groundwater. In fact, this groundwater is non-renewable and is formation waters stored during the deposition of old geological formations (see section on Saudi Arabia). Absence of trans-boundary aquifer makes Kuwait one of the most water stressed states in this region with water withdrawal to supply ratio greater than 80% (Fig. 5.4). However, within Kuwait, the Kuwait group and the Dammam formation, separated by an impermeable chert bed are the main aquifers supplying high total dissolved salts (TDS) groundwater, which is saline (some time contaminated

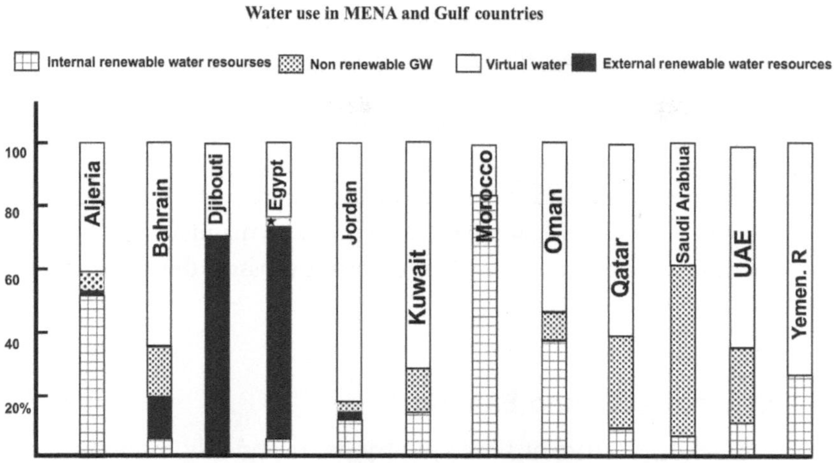

Fig. 5.5 Water use in Middle East and North Africa (MENA) and Gulf countries [22].

by hydrocarbons) and unfit for human consumption [25,26]. To overcome this quality issue, Kuwait's attempt to draw water from Iran and increase its external water resources content (Fig. 5.5) did not materialize due to socio-economic and political reasons [22,24] and hence depends heavily on virtual water resources (Fig. 5.5). Owing to its geographical location, only 0.6% of 17,818 km^2 land is arable with about 80%–90% of sand [27]. The agricultural land in this area is insufficient to meet food demand of growing population and hence the country has no option but to enter into VWT to import large quantities of food to be food secure (Fig. 5.5). Thus, agriculture accounts for 0.24% of the gross domestic product which is the lowest in a country rich in oil and gas reserves. The only option for the country to increase the impact of agriculture on gross domestic product (GDP) is to be "food independent" so that the imports will decrease and shape the country's food and water supply. This can be achieved by obtaining freshwater through desalination, that the country has already adopted in a small way. Kuwait started its desalination plants way back in 1960 with three MSF units generating 0.082 million m^3/day of freshwater each at low and high temperatures. Currently more than 44 MSF plant are in operation generating over 2.4 million m^3/day of freshwater [23,28] consuming $210{,}000 \times 10^9$ BTU of energy, equivalent to 62×10^6 MWh. The amount of CO_2 emitted by these plants at this rate of energy consumption, assuming an emission of 817 kg of CO_2/MWh for oil and 193 kg of CO_2/MWh for gas consumed [10], is of the order of 11 to 50×10^6 tons. This amount will increase as the demand for water and food increases in future, unless a more cost effective and low-carbon emitting energy source, such as solar pv or geothermal source, replaces oil or gas energy source for desalination process. Owing to Kuwait's geographic location, the country has no access to geothermal energy sources. However, the country has abundant solar radiation but this source is not cost effective for desalination as a large part of the freshwater generated will be used for cleaning the solar panels to maintain efficiency and the maximum efficiency at which the solar pv-based desalination units operate is below 20% [29].

5.2.3 Oman, Qatar, and Bahrain

Oman, although is a part of Gulf country, is strategically located and has several advantages over other countries in terms of water resources and food. It has considerable internal water resources (Fig. 5.5) compared with other Gulf countries. Owing to its geographical location, it receives both southwest and northeast monsoon rain. The country receives about 100 mm of rain annually during these seasons and over the mountain region it is 350 mm and is not very frequent. The volume of water the country receives from monsoon is of the order of 19,250 Mm^3 and more than 80% of it evaporates, thus leaving only about 1400 Mm^3 that goes as runoff and part of it infiltrates into the ground. The surface runoff and Wadi flow are estimated to be of the order of 1050 Mm^3. During a period of intense rainfall, the groundwater recharge from the surface runoff is considerable [30]. Check dam are constructed to increase infiltration. Owing to such dams only about 119 Mm^3 of the surface runoff drains into the sea. There are more than 59 such dams in Oman. Maximum amount of groundwater and water from Aflaj are used for agricultural purposes. The amount of water withdrawn from these two sources for agricultural purposes is about 1131 Mm^3/yr while only 169 Mm^3/yr of water from these sources is used for domestic purpose [30]. The remaining domestic demand is supported by water from desalination plants. Nearly 52% of groundwater is utilized for date cultivation and only 7% is used for vegetable cultivation. Apparently, Oman has to rely heavily on food imports (Fig. 5.5) and the import of virtual water in 1 year is about 1700 m^3/capita. According to FAO [31], Oman needs 3000 kcal/capita/day of food by 2030. Although a part of this is supplied by meat, even for growing animal/birds for meat, freshwater is required. Estimates indicate that Oman may need 1300 m^3/capita/yr of water to support its kcal requirements [32,33]. Being water stressed country, by 2050, the per-capita freshwater requirement may not reach the above figure. Besides water, land available for agricultural activity is meagre. For example, the total land area of Oman is about 320×10^3 km^2. About 7% of it is arable and only 0.2% of it is fit for agriculture. Thus, even if water is manageable through desalination for irrigation, there is no land available to make the country food secure [33]. The first desalination plant, MSF plant, in Oman was installed in 1976 with a capacity of about 23,000 m^3 per day. Subsequently, six plants with similar capacity started functioning. Recently, RO plant with a capacity of 120,000 m^3/day and in future 281,000 m^3/day will soon be in place to supply drinking water to the growing population of Oman (Membrane Technology 2016). It is projected that the production of desalinated water will cross 3 Mm^3/day mark by 2025 consuming electricity of the order of 12,150 GWh [34] releasing 2 million tons of CO_2 equivalent to one-tenth of per-capita emissions of CO_2 of the country [21,34]. Oman lacks geothermal systems which can support power generation but it can use geothermal heat pumps for space cooling and space heating thereby reducing the consumption of oil/gas and CO_2 emissions. The saved energy can be used for desalination thereby reducing the dependence on the volume of virtual water (Fig. 5.5) and eventually achieve self-sufficiency in food production. In addition, the country can adopt hydroponic method of cultivation and greenhouse technology over the sea to meet water demand. This will to some extent solve the land problem for agriculture and increase food production and self-sufficiency [33]. Further CO_2 mitigation can be accomplished using RO technology for desalination. This will be further discussed elsewhere in this chapter. Qatar, with a surface area of 11,651 km^2, is one of the highly water stressed countries in the Gulf (Fig. 5.5) with meagre annual rainfall of 80 mm [35]. The population of Qatar is growing at 10% annually and currently it is 2 million. The country's GDP stands at 7%. The groundwater reserves estimated from the northern aquifer is about 2500 Mm^3 and this water is mainly used for irrigation. It is about 33% of water that is being utilized by the country and rest 57% is supplied by five desalination plants generating about 987,000 m^3/day. Two more desalination plants are being constructed with production capacity of 490,000 m^3/day. MSF distillation units generate about 81% of freshwater while only 19% is generated by multieffect distillation (MED). In addition, two RO plants with generation capacity of 1880 m^3/day is being planned to increase freshwater supply by 60% although the demand is 87% [35]. Nearly 36% of the available water is utilized by the agricultural sector. A major part of it goes for growing fodder for the animals and flood irrigation, although flood irrigation wastes a large amount of water. Groundwater balance reported by MDPS [35] indicate that the total water that is being infiltrated into the aquifer is about 63 Mm^3/yr from the precipitation and 2.2 Mm^3/yr flows from the aquifers in Saudi Arabia. Out of this volume about 18 Mm^3 of water flows into the sea. Hence, in effect the total water being recharged into the aquifer is about 47 Mm^3/yr. The total water extracted from the aquifer annually for agriculture is about 230 Mm^3/yr. Besides groundwater, 55 Mm^3/yr of treated sewage effluent (TSE) is also used for irrigation. The current policy of the government is to store 7 days of water to tide over any emergency situations. Qatar has a vision to produce 40% of its food within the country without depending on VWT by 2025. Achieving self-sufficiency in agricultural production may be expensive relative to VWT since it has to generate large volume of desalinated water. However, depending on VWT over a long period time is detrimental to the country as this may lead to food risk issue due to geopolitical conflicts [36,37]. Increasing the desalinated water using MSF will emit large CO_2 compared with SWRO. The reported CO_2 emission from MSF units is 3.6 million tons/yr while SWRO units emit 0.9 M tons/yr. The CO_2 emission per cubic meter of water generated from MSF units is 8.2 kg while for SWRO units it is 2 kg/m^3 [38]. The total CO_2 emissions by Qatar from fuel combustion is 78 million tons (oil 17 million tons; gas 61 million tons, electricity generated is 5×10^{11} GWh, per-capita CO_2 emissions is 36 tons) and share of emissions from residential sector is 8 million tons [21]. Owing to the geographic location, Qatar uses about 60% of its energy for space cooling and heating [39]. This amounts to about 3×10^{11} GWh of electricity. Assuming minimum CO_2 emission quantity of 198 kg/MWh (from oil combustion), cooling and heating of buildings will emit 594×10^9 tons of CO_2. Since Qatar's energy demand grows with increase in population, to meet the water and food demand, the country has to increase the number of desalination plants to generate freshwater. Part of this energy can be drawn from energy used for space cooling and heating by using geothermal heat pumps for space cooling and heating. There is no in-depth data on the heat flow and heat conduction data for Qatar. Since Oman and Qatar fall within similar geographic domain, there should be suitable sites to tap this source of energy to establish GHPs. A large part of CO_2 emitted from the energy source used for air conditioning of the building can be saved using geothermal energy.

Bahrain consists of a group of 40 islands with a total surface area of 762 km^2. Bahrain's population has increased by 10 fold since 1940 and stands at 1.34 million today [40]. The Dammam and Rus Umm Eradhuma are the two main aquifers supplying 57% of water to domestic and agricultural sector. The industrial and agricultural sector use desalinated water (37%) and treated sewage water (9.7%). The Dammam aquifer's production is declining while the Rus Umm Eradhma is able to supply 70% of the water to domestic and agricultural sectors [40]. The per-capita consumption is about 318 L/day. Besides normal irrigation, Bahrain adopted hydroponic cultivation to decrease food imports and move toward food self-sufficiency. The agricultural products include dates, alfalfa feed stock, and variety of fruits. Cereals are main food imported to the country. The contribution of agriculture to the GDP is about 3%. To decrease food imports, the government is extending heavy subsidies on agricultural machineries, irrigation equipment, pesticides, and fertilizers.

Owing to issues related to land availability, salinity of the soils, and man power, Bahrain imports 84% of food from Saudi Arabia and Australia [40]. This puts the country on high risk in terms of price and supply due to uncertain political equations between these countries. Desalination using low-carbon energy sources such as geothermal and hydrophonic cultivation may reduce this risk in future. Currently 36 gas turbines generate 4938 MWe to meet the current demand contributing 23×10^3 tons of CO_2 annually. Industrial CO_2 contribution is about 12×10^3 tons and waste contributes is 2×10^3 tons [41]. Bahrain gives heavy subsidies to its population and in 2012 the country spent US$ 1.7 billion on energy subsidy which amounts to 5.6% of the country's GDP. The average monthly power consumption by one house hold is 3000 kWh (number of houses: 173,069). The solar pv contributes 12% of this consumption [41] offsetting about 413×10^6 of CO_2 emissions (assuming 0.553 kg of CO_2/kWh from natural gas thermal power plants, EIA [42]).

5.3 WATER RESOURCES OF MENA AND SUB-SAHARAN COUNTRIES

The countries included in this chapter are Kenya, Ethiopia, Eretria, Egypt, Libya, Djibouti, and Yemen Republic. These countries have good geothermal energy resources and some of them heavily depend on desalination process to support domestic and agricultural activities.

5.3.1 Kenya

Kenya, with GDP of 994 US$/yr, compared with other countries listed above, is gifted with abundant water and food resources and has vast geothermal resources that make the country self-sufficient in energy and water. Kenya's total population is 45 million. About 62% of the population have water supply (82% urban and 55% rural). The total area is about 580,370 km^2 which includes 11,230 km^2 of water bodies. Overall 270,000 km^2 of land is cultivable but only 60,000 km^2 is cultivated and the remaining land is preserved for nationals parks. Kenya falls within the Great Rift Valley with diverse climate provinces and receives annual rainfall of 200–1800 mm [43,44]. From the current available water for irrigation (Groundwater 1162 MCM, and 1602 MCM of surface water) the projected available water (groundwater and surface water) for irrigation by 2030 will be of the order of 8063 MCM.

Currently, Kenya draws its power from oil and coal. The total CO_2 emission is 12.4 million tons: 1.3 million tons from coal and 11.1 million tons from oil [21,45]. The per-capita emissions are about 280 kg. By 2030 Kenya will be in a position to offset major part of its primary energy supply by geothermal energy and may be in a position to support all the rift valley countries (Ethiopia, Eritrea, and Djibouti) to meet their growing demand of water, and mitigate CO_2 emissions. The estimated total potential in the rift geothermal province is about 10,000 MWe and currently the installed production capacity stands at 573 MWe. By 2030 the total generation is expected to reach 5000 MWe [46]. The Oserian green house facility in Kenya is using both geothermal power and geothermal heating for green house cultivation with 50% reduction in conventional electricity cost and 70% saving in conventional fuel heating [47].

For example, in the year 2010 the total agricultural produces from green house cultivation using geothermal energy was 7.6 million tons while in 2012 it was 12.6 million tons [47]. It is apparent that Kenya is the only country among the sub-Saharan countries poised to achieve 100% self-sufficiency in power and food by 2030 and further support the other neighboring countries.

5.3.2 Ethiopia

Ethiopia with a population of 95 million has good surface and groundwater resources. The Awash River basin is the nerve center of Ethiopia supporting domestic and agricultural electricity demand [48]. The river originates from Lake Victoria catchment and drains the entire rift valley before discharging into the Mediterranean Sea through Egypt. Besides water, the Awash River is a major source of power to Ethiopia, supplying nearly 86%

of electricity [49]. Besides hydroelectric power, the country has wind and geothermal energy resources and the total power generation potential of the country is about 60,000 MWe. However, at present the Awash hydroelectric power plant has an installed capacity of 2300 MWe (Power Africa in Ethiopia 2016). Besides surface water, nine major aquifers, with an estimated potential of 12–30 billion cm^3 support domestic and agricultural activity. The Awash River basin is currently providing 2285 million m^3 of water for agricultural sector [50]. Ethiopia is planning to enhance the installed capacity of hydroelectric project by 24,000 MWe by 2030 to meet the future demand of 14,200 MWe by 2030. The surplus power will be sold to the neighboring countries. The country has sufficient geothermal energy resources which have not been exploited till now. Ethiopia is a land locked country and has no provision to generate freshwater through desalination. However, it has sufficient surface and groundwater which will meet the domestic and agricultural demand for the future and further by 2030, it will have surplus electricity to support its neighbors [51,52].

Thus, although agriculture in Ethiopia depends on the vagaries of monsoon [51], with the development of hydroelectric projects and geothermal, the country will be in a comfortable situation with respect to food security and may be in a position to offset biomass energy that supplies nearly 334 TWh of electricity to rural population [49]. The major food and water security risks the country may face in the future is from VWT. Several countries are investing in Ethiopia to grow food to provide food security to their respective country. For example, India, China, Saudi Arabia, Turkey are investing heavily in Gambella region to grow soya bean, rice, sugarcane, cotton, and cereals [53]. This may erode part of Ethiopia's water resources if sufficient water resources management practices are not implemented.

5.3.3 Eritrea

Eritrea lies along the western margin of the Red Sea, spreading over a length of 1000 km, with an area of about 118,000 km^2. Physiographically the land consists of savannah, steppes, and desert. The highlands reach an elevation of 2000 m and support cultivation since ancient times. Although the cultivable area is about 16,000 km^2 only 5030 km^2 is cultivated currently to support a population of 6.1 million. Agriculture is the main source of income and about 76% of the population is active in agriculture. Only half the population have access to safe drinking water. Arid lowland constitutes 33% of the land area and semidesert area constitutes 39% of the land area. Eritrea receives torrential rains mostly from the Indian monsoon winds extending from July to September. The rainfall varies from 50 to 1000 mm with an average of 380 mm [54]. More than 80% of the population depend on agriculture for their income and due to vagaries of monsoon followed by droughts, the country depends on food imports, especially wheat which amounts to 46% of the total goods imported. Eritrea never practiced organized farming and being rice eaters, most of the rice fields are located along the hill slopes as run-off irrigated rice fields [55]. Rainfall is also affecting the domestic animals that constitute a part of the agricultural activity. The country, however, has abundant geothermal energy resources surrounding Aldi and Dubbi provinces which can be exploited. The potential of these energy fields is discussed later in this chapter [56].

Eritrea made commendable progress in socioeconomic structure with its economy growing at about 11%, after independence in 1991 [57] and continues to grow. Its current GDP growth is about 14%. Only 32% of population has access to electricity. The annual per-capita income is only US$ 403. Since agriculture is rain fed, frequent droughts and change in weather pattern is pushing the country below the poverty line and hence it is not able to secure food and power supply. The main source of electricity is supported by oil-based power plants which generates 338×10^6 kWh putting the per-capita electricity consumption at 49 kWh which is far below its neighboring countries [58]. The rural population still rely on biomass as energy source. The oil-based power plants are the major source of CO_2, with the emission levels touching 514 kt (kilo tons). Eritrea is endowed with rich natural resources which can lift the country's economy. Both hydropower and geothermal energy, if managed properly, can reduce imports of 5000 bpd of oil and refined products and make the country food and water secure under the recently launched MDG [56]. Geothermal energy can support desalination process and enable the country to grow more food.

5.3.4 Egypt

Nearly 97% of Egypt's water requirement is met by the Nile. Egypt, with a population of 86.9 million (WB [57], Pacini and Harper [59]; projected to grow to 106 million by 2030, ICARDA [60]), utilizes 57 billion m^3 of discharge from the Aswan dam and 12 billion m^3 of groundwater (shallow aquifer) thus consuming maximum quantity of Nile water compared with other riparian countries through which the river flow. Its per-capita water consumption is about 636 m^3/yr. Egypt's GDP is 20% greater than the combined GDP of all the countries falling within the Nile River basin [57,59]. In addition, Egypt is planning to tap the Mesozoic Nubian Sandstone aquifer, a trans-boundary aquifer shared by Egypt and Libya which is estimated to contain 200 trillion m^3 of nonrenewable freshwater. The total area of Egypt is around 1 million km^2 and the cultivable land is 3% within the Nile River basin. Irrigation in this cultivable land consumes about 85% of the Nile water supply (57 billion m^3 is the water released from the Aswan dam). In addition, a part the land along the northern coast supports traditional rain-fed cultivation and pasturing animals [60]. Owing to expanding population and insufficient water for irrigation and land, 40% of Egypt's population is under poverty and a large percentage of the poor lie in rural area. To meet the growing demand for water Egypt has commissioned several desalination plants, sourcing seawater from the Red Sea and saline groundwater from the coastal aquifers and irrigation return water. RO desalination is expected to grow in solving the freshwater problems since the RO technology is very cost effective at 0.53 euros/m^3 of freshwater [61]. Pessimism surrounds the growth of desalination plants in solving freshwater issue in Egypt. Under the current population growth trend and food security, Egypt needs to install 1200 plants as large as the world's largest desalination facility at a whooping cost of US$ 1.7 trillion that is three times more than Egyptian GDP [61].

Assuming the current cost of desalinated water at 50 US$ cents/m^3 is maintained over the next 100 years, it is argued that desalination may not be the right solution to meet water and food demand of Egypt [61]. However, the cost of desalinated water in Egypt is quite competitive and in some cases lower than the cost prevalent in the world. For example, generation of m^3 of freshwater from seawater using RO costs about 1.6 US$ in Egypt while the world cost varies from 0.6 to 2.1 US$. In the case of MSF, the cost is more or less similar to the world cost [62]. Hence in the future desalination may be the best solution for Egypt's water problems and food imports. For example, in 2000 Egypt saved 5.8 billion m^3 of water by importing 5.2 tons of maize [63]. However, what is not considered by the critics here is the free and abundant geothermal energy [hydrothermal and enhanced geothermal systems (EGS)] that can completely offset oil and gas energy source, its potential in Egypt, discussed elsewhere in this chapter, in generating freshwater from Red Sea water.

5.3.5 Libya

Libya is the northern most country in MENA, with an area of 1.76 million km^2 and population of about 6.3 million. Overall 80% of the population live along the Mediterranean coast. Libya's major water life line is the trans-boundary Mesozoic Nubian Sandstone aquifer (area: 26 million km^2; non-renewable groundwater) that traverses Egypt (see section above) Sudan and Chad. Only 34% of the water is shared by Libya ([64]; 4 268 million m^3). In addition, Libya shares Northwest Sahara aquifer System (area 1.2 million km^2) with Algeria and Tunisia and Murzuq-Djado basin (450,000 km^2) with Algeria and Niger. The Mediterranean coast receives 100–300 mm of rain annually and supports food cultivation. The remaining area receives scanty rainfall not exceeding about 20 mm annually [64]. Hence, the country has to depend on the nonrenewable groundwater sharing with other neighboring countries. Libya has been under importing 80% of its food requirement and 90% of the cereals (wheat, maize). Libya installed desalination plants since 1960s and as date more than 400 desalination plants are in operation with an installed capacity of 226 Mm^3/yr but until 2012, 70 Mm^3/yr desalinated water was generated using thermal and membrane desalination process. In addition, 79 wastewater treatment plants were in operation producing 40 Mm^3 water for the irrigation sector. With abundant availability of oil, agricultural sector was neglected. Inefficient water management practices and geographic conditions brought the country to a "water stressed" status. It is difficult for the country to regain the agricultural status that it enjoyed before the country's GDP is controlled by oil. With the growing population and static state of water resources, the country has to depend heavily on VWT and desalinated water resulting in high CO_2 emissions and increased food security situation [65]. Although renewable energy sources such as wind and solar pv may help the country to tide over the situation (Mohamed et al., 2013), due to the geographic location of the country solar pv may not be able to render any help. Solar thermal may be a good option to provide hot water in winters. However, ground source heat pumps may help in reducing the dependence on oil for space cooling and heating and the energy saved may help in increasing the production of freshwater through desalination and reducing CO_2 emissions [66].

5.3.6 Djibouti

Djibouti, with a population of about 712,000, is the smallest country in the sub-Saharan region with a land area of 23,200 km^2. The cultivable land is only 6000 ha but only 400 ha in Dikhil in the southern part of the country, bordering Somalia, is being cultivated. Agriculture accounts for only 3% of GDP. The climate in Djibouti is arid with an annual rainfall of 147 mm. There are no rivers and hence the surface water resources are meagre. Occasional flash floods drain the coastal and Lake Abhe regions. The renewable water resources are limited to about 300 million m^3/yr. Although part of the Awash River drains into Djibouti, the water flows into Lake Abhe, which is a saltwater lake. The groundwater recharge takes place whenever the regions experiences flash floods and the groundwater is stored in basalt aquifers that extends between Ethiopia and Djibouti. The groundwater is not suitable for irrigation due to high boron content. However, the top alluvial aquifer, with a yield of 21 million m^3/yr supports small-scale irrigation. More than 80% of water is used for agricultural purpose, considering the population, it is an insignificant quantity [67]. Hence, Djibouti heavily depends on VWT (Fig. 5.5). Thus, the country's food, energy, and water security lies outside the country. However, the country has significant geothermal energy resources which are not being exploited. This energy if utilized will make the country energy and food independent and can offset 100% imported fossil fuels [10]. The estimated water demand is 29 million m^3. Djibouti, like other neighboring countries, entered into VWT and relies on the food imports to meet local demand. Considering the amount of geothermal energy resources the country has, it is not in a helpless state. It is a matter of government policy and donors like World Bank and African Development Bank to emphasize the exploitation of the geothermal energy resources. Since the country is small with small population, this energy will drive the country from its present poor state to a self-sufficient country with respect to food and water. The geographic location of the country (located within Afar triangle) is itself a great asset to the country with respect to energy.

5.3.7 Yemen Republic

Yemen Republic, bordering the Gulf of Aden and the Red Sea and with a population of 24 million, has geographic area of 528,000 km^2 with cultivable area of 36,200 km^2. It has 2200 km long coastline. Yemen is the most populous country in the Arabian Peninsula and nearly equals Saudi Arabia. Only 119,000 km^2 area is cultivated with cereals and qat. Rainfall varies from 1000 to 300 mm/yr. Although the agricultural activity is very aggressive, the cereal produced supports only 21% of the domestic demand. It imports sugar, vegetables, fruits, milk products and exports coffee, fish products, and banana. The import expenditure is more than export value [68]. Rainfall is high along the western mountain ranges while it gradually decreases toward the eastern plains that are covered by desert. In the regions with poor rainfall, farmers depend on scanty rains and poor groundwater resources and lived in poverty. About 7 million people fall under this category. Owing to the lack of organized abstraction of groundwater process, the withdrawal often exceeded recharge thereby resulting in fall in water table. Nearly 70% of the available groundwater is used for irrigation. On an average, the water table declined at the rate of 6 m/yr. Nearly 55,000 borewells fitted with high-power diesel/electric pumps (diesel is subsidized, IMF [69]) pumping indiscriminately at 50 L/s (about 1.5 km^3 of water) resulted in lowering of the water table. There is no legislation which controls groundwater abstraction, hence the groundwater levels in several regions have fallen below the specific storage levels due to indiscriminate pumping and rendered the aquifers unsuitable for future recharge [54]. For example, in Sana'a, the state capital, water wells have reached 2 km in basalts with poor yield [70]. The next option for Sana'a is to transport desalinated water from the coast at a whopping cost of US$ 6.60/m^3. In the case of coastal aquifers such as the Tihama Plains, which consist of quaternary alluvial deposits, the recharge and discharge rate are very high due to high hydraulic conductivity of the aquifers. For example, with an average annual rainfall of 138 mm/yr, the recharge and discharge rates are 100–150 m^3/yr [71]. Owing to the indiscriminate abstraction, the quality of coastal aquifer is deteriorating due to saltwater intrusion.

Yemen has already installed desalination plant with a capacity of 28 MCM/yr in 2002 to supply freshwater to urban population (e.g., Taiz). This is, like other Gulf countries, a viable option to circumvent issues related to declining groundwater resources [68]. According to USAID report [72], Yemen in collaboration with Saudi Arabia is installing US$ 31 million desalination facility along the Red Sea coast. This plant appears to be the largest in the world and expected to ease the water crisis in Yemen Republic. Yemen has no option but to strengthen its VWT or support internal irrigation sector augmenting it through desalination process. Desalination is very expensive for Yemen, considering its GDP, but has no option if it wants to provide food security to its growing population.

5.4 DESALINATION: PROCESS, EMISSIONS, AND COST

The preceding sections have clearly demonstrated that all the GCC and MENA countries are heavily water stressed and have no option either to get into VWT, that all the courtiers have already entered into or adopt desalination methods to meet the growing demand for food and water and to maintain country's GDP above poverty line. Although these countries have good coastlines (except Ethiopia) and free access to seawater, the process involved in generating freshwater through desalination to support domestic and agricultural sectors comes with a cost. Fig. 5.6 shows the number of desalination plants in operation in GCC countries. Among the MENA and sub-Saharan countries, Libya (400 plants) and Egypt have already desalination plants in operation and Eritrea and Yemen have plans to install desalination plants to tide over the water crisis.

Perhaps, Egypt was the first country which started desalination process in 1912 generating 75 m^3/day. This was followed by Kuwait which installed first multiflash desalination plant in 1957 on a commercial scale and subsequently Kuwait started generating desalinated water on large scale at Shuwaikh (4550 m^3/day). With the development of RO technology, currently desalination plants are able to generate >100,000 m^3/day freshwater [73]. Although desalination plants can satisfy country's population and agricultural demand, the process is energy intensive (except RO) and expensive

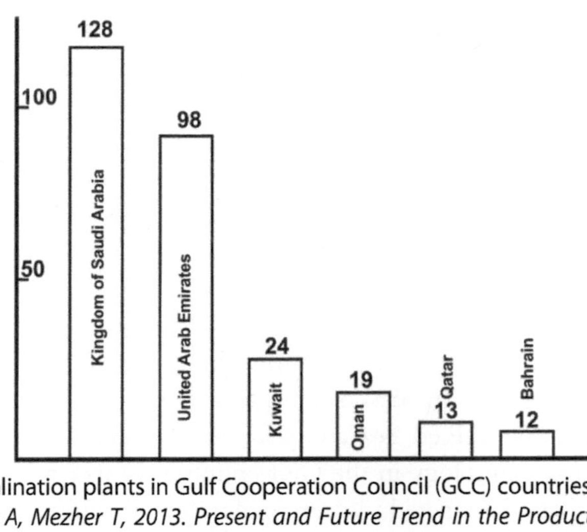

Fig. 5.6 Desalination plants in Gulf Cooperation Council (GCC) countries. *(Adapted from Fath H, Sadik A, Mezher T, 2013. Present and Future Trend in the Production and Energy Consumption of Desalinated Water in GCC Countries. Int J Thermal Environ Eng 5, 155–165.)*

resulting in large CO_2 emissions and consuming high quantity of oil and gas. For example, production of 1000 m^3/day of water through desalination consumes 10,000 tons/yr, damaging the environment with large CO_2 emissions and brine discharge [74,75]. Currently there are about 12,500 desalination plants in operation around the world with production capacity of 26 million m^3/day [73]. However, several countries, especially those that are covered in the present chapter, have reasonably rich geothermal resources (discussed in the later sections) that can offset the high cost and provide sustainable socioeconomic growth. The cost of desalinated water can be brought down substantially by replacing fossil fuels with geothermal. Perhaps this could be the main focus for these countries to become energy and food independent. In fact, since 1960s countries across the world have desalination plants with installed capacity of 0.12 million m^3/day consuming more than 0.02 million metric tons of oil/day and emitting 0.36 metric tons of CO_2/day. By 2008 the generation has exceeded 52 million m^3/day, consuming 1.42 million tons of oil/day and emitting 156 metric tons of CO_2/day [76]. The present global freshwater generation is about 98 million m^3/day [77]. In 2015, desalination plants in Abu Dhabi alone has emitted 50 m^3 metric tons of CO_2 [78], which is equivalent to the global emissions in 2000 [76].

Multieffect evaporation/distillation (MED), MSF distillation, thermal vapor compression (TVC), mechanical vapor compression (MVC), RO, and electrodialysis (ED) are the six common desalination technologies available today and are widely used depending on specific conditions. The last two methods are known as membrane separation technology. There other technologies that combine one or two of the above processes to form hybrid technology such as a combination of membrane distillation (RO) with MES of MSF. The energy requirement and GHG emissions for different desalination technologies vary considerably with MED being relatively conservative technology with regard to CO_2 emissions and electricity consumption (Table 5.1)

Considering temperature variation and salinity levels of the Red Sea, MSF and MED desalination technologies are most suited compared with RO and in fact more than 65% of the plants are of MSF type. The high salinity content in the Red Sea and the Persian Gulf is not suitable for RO desalination technology in the Gulf countries (Table 5.2).

The cost of desalinated water generated through conventional energy source is still economical compared with solar thermal and PV technology-based desalination plants. While the desalinated water

Table 5.1 Energy requirement and CO_2 emissions for different desalination technologies

	MSF	MED	MVC	RO
Average capacity (m^3/d)	25,000	10,000	3000	6000
Electrical energy (kWh/m^3)	4	2	7	5
Cost of plant ($\$m^3/d$)	1300	1200	1000	1000
Production cost ($\$m^3$)	1.1	0.8	0.7	0.7
kg CO_2m^3 (oil)[a]	4.1	2.1	9.8	7.3

[a]$kgCO_2/MWh = 817$.
Adapted from Fiorenza G, Sharma VK, Braccio G. Techno-economic evaluation of a solar powered water desalination plant. Energy Convers Manage 2003;44:2217–40; Gude VG, Nirmalakhandan N, Deng S. Renewable and sustainable approaches for desalination. Renew Sustain Energy Rev 2010;14:2641–54; Chandrasekharam D, Bundschuh J. Low enthalpy geothermal resources for power generation. CRC Press, 2008. 149p.

Table 5.2 Installed capacity of desalination plants compared with world share (%) and technology share (%) in Gulf Cooperation Council (GCC) countries

Country	Capacity m^3/d	World share%	MSF%	MSD%	MVC%	RO%	ED
Saudi Arabia	>5 000,000	20	64	0.3	1	33	1.7
UAE	>3000,000	11	87	8		5	
Kuwait	>1600,000	6	97			3	
Libya	>700,000	2	66	11		16	7
Qatar	>500,000	2	94	4		2	
Bahrain	>473,000	2	63	10		27	
Oman	>378,000	1	87	1	4	8	

Blank denotes data not available.
Adapted from Fiorenza G, Sharma VK, Braccio G. Techno-economic evaluation of a solar powered water desalination plant. Energy Convers Manage 2003;44:2217–40.

costs <US\$1/m^3, the solar technology using the same amount of water costs >US\$ 1/m^3 with a plant with a capacity of 5000 m^3/day. Even with more than 50% subsidy there is not much cost difference between conventional and solar energy (thermal and pv) sourced desalination plants [73]. This high-water cost is inherent since the electricity generated through solar pv varies between 14 and 18 US cents/kWh and the cost is expected to come down to 12 US cents/kWh by 2020 [79]. However, the unit cost of electricity generated through geothermal energy is less by several orders relative to solar and wind (Fig. 5.7). In future, once EGS matures, the unit cost will further fall and may compete with electricity generated from fossil fuels. Besides the cost, in all the GCC, MENA, and sub-Saharan countries, as described in the later sections of this chapter, geothermal energy sources are available along the Red Sea [56] coast and this source is yet to be

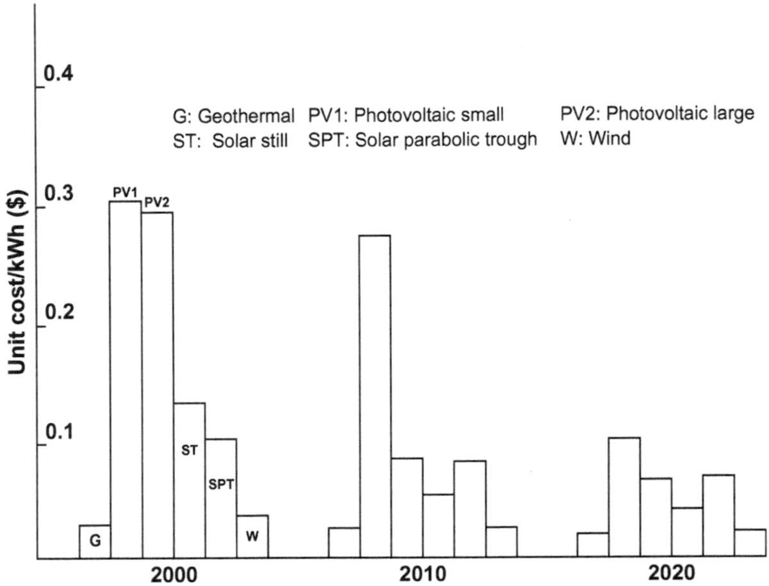

Fig. 5.7 Unit cost of electricity generated through renewables.

developed in many countries falling in the above group. This solves several issues like grid accessibility to the desalination plants, suitable sites for installing solar power establishments, and other issues described and highlighted by Fiorenza et al. [73].

5.5 GEOTHERMAL ENERGY RESOURCES

Geothermal provinces with surface manifestations in the countries around the Red Sea coast are shown in Fig. 5.8. Surface manifestations are represented by the presence of thermal springs, fumaroles, and gas emissions (known as hydrothermal manifestations). Besides the hydrothermal manifestations, Saudi Arabia, Egypt, and Libya have large EGS systems with very large potential and are the future energy sources.

5.5.1 Saudi Arabia

Among the GCC countries, Saudi Arabia and Republic of Yemen have very high geothermal resources which can be utilized for power generation. The western margin of Saudi Arabia is a loci of high-temperature geothermal systems represented by more than 50 thermal springs with issuing temperatures varying from 31 to 79 °C and a flow rate of 5–20 L/m

Fig. 5.8 Geothermal provinces in Gulf Cooperation Council (GCC), Middle East and North Africa (MENA), and sub-Saharan countries.

[3,4,6,11,12,56,80–82]. These thermal springs are associated with volcanic (active) fields (known as "Harrats") and with high heat-generating granites (Fig. 5.9). These "Harrats" (Fig. 5.9) occupy an area of about 90,000 km^2 [56,83]. A few of these Harrats are still active [84,85], due to the spreading of the Red Sea rift and plume activity below the Arabian Shield [56,86], and emitting fumaroles associated with hot water springs (Fig. 5.9). These Harrats cover a large part of ancient river channels giving rise to geothermal reservoirs below the Harrats. These areas record high geothermal gradient (>90 °C/km, [56,83]) and high heat flow. The heat flow value recorded over the western Arabian shield geothermal provinces is $> 80\ mW/m^2$ [56,87,88] while the normal heat flow value over the shields around the

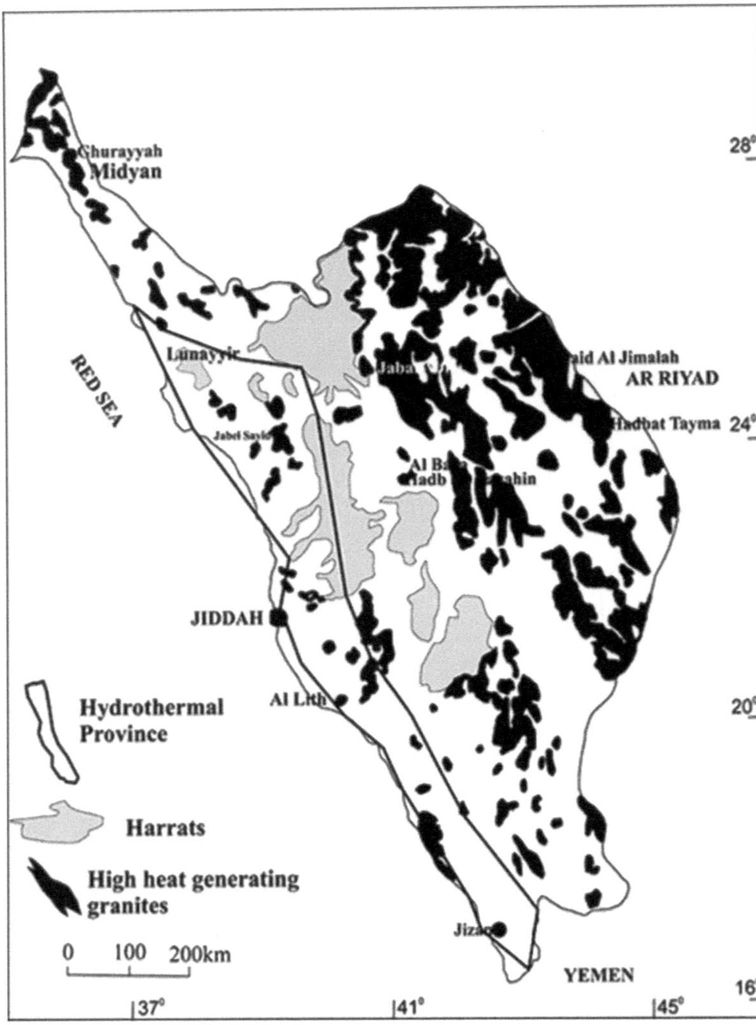

Fig. 5.9 Western Arabian shield geothermal provinces showing the location of the thermal springs associated with Harrats and high heat generating granites. *(Adapted from Chandrasekharam D, Lashin A, Al Arifi N, Al-Bassam M, 2016. Red Sea Geothermal Provinces. CRC Press, 221p.)*

world is ~ 50 mW/m^2 [89]. A detailed description of the province is given by Chandrasekharam et al. [6,56].

The evolution of the entire geothermal systems on either side of the Red Sea is related to the opening of the Red Sea and subsequent spreading of the Red Sea axis due to continuous magmatic activity along the axis represented on land by the Harrats in the Arabian shield (Fig. 5.10) and live volcanoes

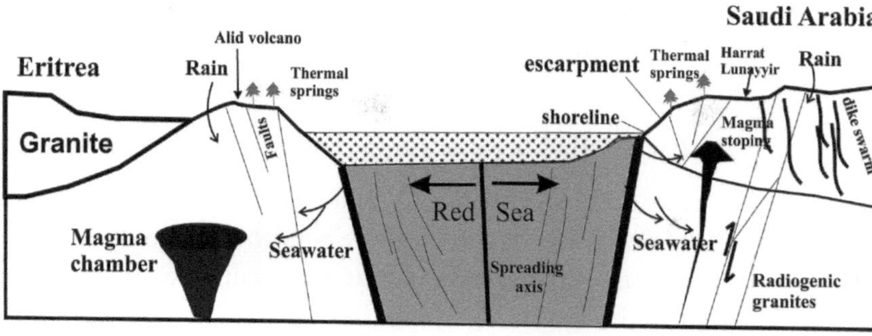

Fig. 5.10 Evolution of the geothermal systems due to active spreading of the Red Sea axis. *(Adapted from Chandrasekharam D, Lashin A, Al Arifi N, Al Bassam AA, Varun C, 2015. Evolution of geothermal systems around the Red Sea. Environ Earth Sci 73, 4215–4236; Chandrasekharam D, Lashin A, Al Arifi N, Al-Bassam M, 2016. Red Sea Geothermal Provinces. CRC Press, 221p.)*

such as the Alid volcano in Eritrea. The Red Sea tectonic and magmatic activity is also responsible for the activation of the East African Rift system which extends from Djibouti to Kenya (Fig. 5.8) and for the evolution of geothermal systems in these countries. On the basis of the isotope and geochemical signature of the thermal springs, it is estimated that the Jizan hydrothermal system (Fig. 5.9) can generate 134×10^6 kWh of electricity. Even if half of this amount is generated, this energy can provide three times the amount of desalinated water being consumed per day by the population (235 L/day per-capita of desalinated water using 17 million kWh of electricity; [4,90]). Similarly, the Al-Lith hydrothermal site, located NW of Jizan along the cost, represents a very matured and highly potential geothermal system where thermal fluids are circulating with in a granite reservoir, with temperature $> 200°C$, for > 75 years, and has the potential to generate 120×10^6 TWh of electricity [11,12,81].

Besides the hydrothermal systems represented by surface manifestations, the western Arabian shield is endowed with large volume of high heat-generating granites (Fig. 5.9). The surface area occupied by these granites is about 161,467 km^2 while the Harrats occupy 90,000 km^2 [11,83,91]. These granites are very special, in the sense, that they contain high concentration of radioactive minerals which contain uranium (4–363 ppm), thorium (8–625 ppm), and potassium (1–4%) and the heat generation by these granites is very high (2–134 $\mu W/m^3$) and consequently the heat flow values are very high (64–1382 mW/m^2) [5,11,92,93]. The radioactive heat generation of the granites (RHP in $\mu W/m^3$) is calculated following the procedure suggested by Raybach [89] and Cermak et al. [94] taking into account the heat generation constant (amount of heat released per gram of U, Th, and K in per unit time) and the concentrations of U, Th, and potassium (C_U, C_{Th}, and C_K):

$$RHP = \rho(9.52C_U + 2.56C_{Th} + 3.48C_K) \times 10^{-5}$$

where ρ is the density of rock in kg/m^3, C_U and C_{Th} are the concentration of U and Th in mg/kg, respectively, and C_K is the concentration of K in weight percentage in the granites. The surface heat flow values were calculated using the equation of Lachenbrough [95]:

$$Q = Q_0 + D \times A$$

where Q is the heat flow at the surface, Q_0 the initial value of heat flow unrelated to the specify decay of radioactive element at certain time, D the thickness of rock over which the distribution of radioactive element is more or less homogeneous, and A is the radioactive heat production. Since the Moho depth along the western Arabian coast varies from 18 to 25 km [96] the background heat flow (Q_0) value along the shelf is taken as 250 mW/m^2 and along the coastal region (between the shore line and the escarpment) the value is taken as 100 mW/m^2 (Fig. 5.1) [97].

Such granites are present in other land masses on either side of the Red Sea (i.e., in Egypt, Eritrea, Republic of Yemen, Fig. 5.1). The heat flow values, calculated based on the uranium, thorium, and potassium content in granites of the Arabian shield is similar to the heat flow values measured from deep borewells drilled into granites along the western margin of the Red Sea in Egypt and over the geothermal sites along the eastern margin of the Gulf of Suez (see Fig. 5.7) (175 mW/m^2; [98,99]). Overall 1 km^3 of such granite can generate 79 × 10^6 kWh of electricity for a period of 30 years [100]. Adopting the procedure followed by the Cooper Basin EGS project of Australia and assuming 2% heat recovery from such granites [100], the Midyan granites (Fig. 5.2) with heat-generating capacity of 134 μW/m^3 can generate about 160 × 10^{12} kWh of electricity.

5.5.2 Yemen Republic

The opening of the Red Sea triggered intense tectonic and volcanic activity on the neighboring landmasses such as Saudi Arabia, Yemen Republic, Eritrea, Djibouti, Ethiopia, and Kenya. Volcanism was active over Yemen between 15 and 31 Ma (Million years). The volcanic activity continues to be active as evident by the eruption of Dhamar volcano in 1937 (Fig. 5.11) and evolution of volcanic cones near Al-Lisi and Isbil villages (Fig. 5.11).

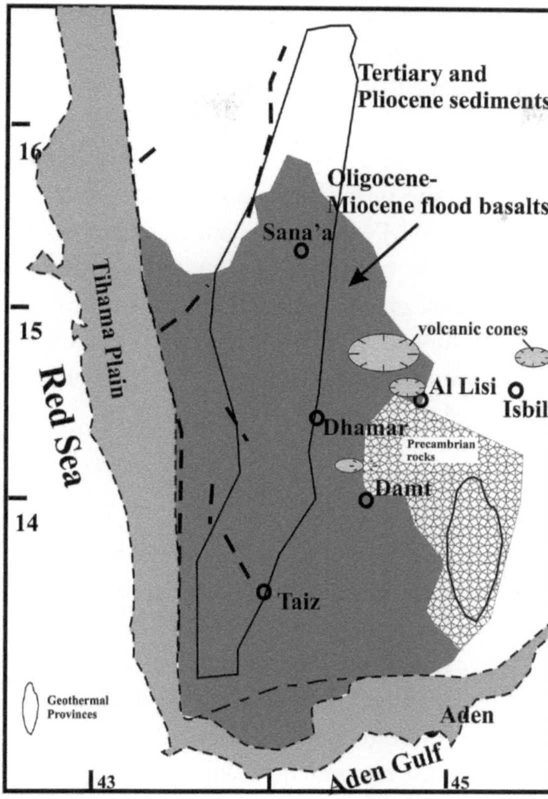

Fig. 5.11 Geothermal provinces in Yemen Republic. *(Adapted from Chandrasekharam D, Lashin A, Al Arifi N, Al-Bassam M, 2016. Red Sea Geothermal Provinces. CRC Press, 221p.)*

Such volcanism and associated tectonic activities related to the rifting of the Red Sea axis gave rise to several geothermal provinces along the western margin of Yemen and these provinces are aligned parallel to the geothermal provinces of Saudi Arabia (Fig. 5.8; [6,56]).

The geothermal provinces of Yemen are characterized by the presence of more than 100 thermal springs with issuing temperatures varying from 42 to 96°C. Majority of them are aligned parallel to the coast and a few of them occur within the granites, east of Damt (Fig. 5.11). At places the thermal springs are associated with fumaroles and boiling pools [6,56,101]. The geothermal sites are characterized by high geothermal gradient (77°C/km) and high heat flow (154 mW/m^2). The presence of epithermal alteration zones around the thermal sites and anomalous He^3/He^4 ratios (3.2; [101]) indicates the presence of magma chambers within the crust at shallower levels [102].

The geothermal provinces in Yemen are an extension of the regional geothermal belt extending along the western margin of the Arabian shield (Fig. 5.8) and the presence of active young silicic volcanoes near Damt, Dhamar, and Al-Lisi (Fig. 5.11) makes these provinces highly potential for generating electricity. In 2009 techno-economic feasibility study has been completed over a site near Damt and Dhamar with financial support from the World Bank. A 1-MWe power plant was proposed in this site. This project is delayed due to the political situation in this country [56]. Since the heat flow and geothermal gradient, geochemical, and isotope signatures of the thermal waters in the geothermal province of Yemen are similar to that occurring in Jizan, these sites (Damt, Dhamar) have the potential to generate $>134 \times 10^6$ kWh of electricity [56].

5.5.3 Djibouti

Djibouti is located strategically at the junction of the two active ocean ridges, the Aden and the Red Sea ridges and falls within the Afar depression. The East African Rift System initiates from the Afar depression and extends down to Kenya enclosing several geothermal provinces (Fig. 5.8). High temperature, shallow geothermal systems are well developed in Djibouti and are represented by three distinct geothermal provinces: The Lake Asal on the east, the Hanle at the center, and Lake Abhe on the east, at the foot of Damh Ale Volcano (Fig. 5.12).

The most promising sites are Lake Asal and Lake Abhe provinces which represent high-temperature geothermal systems in Djibouti.

Lake Asal geothermal province is represented a group of thermal springs with issuing temperatures varying from 30 to 99°C. The Lake is situated 150 m below the sea level and seawater is the main fluids circulating in the deep geothermal reservoir. Bottom hole temperature measured from a 1500 m deep exploratory borewell is 260°C [103,104].

The Lake Abhe geothermal province is located toward western side of Djibouti (Fig. 5.12) at the border between Djibouti and Ethiopia, east of Damah Ale volcano, and south of Erta Ale volcano in Ethiopia. This province is characterized by high-temperature thermal springs with issuing temperatures varying from 88 to 98 °C with fluid discharge of 18 L/s [6,56,105,106]. Several gas vents occur together with the thermal springs degassing large volumes of CO_2. The geothermal system is located within the volcanic flows that are rich in calcium bearing minerals. Owing to water-rock interaction between the calcium bearing minerals and CO_2,

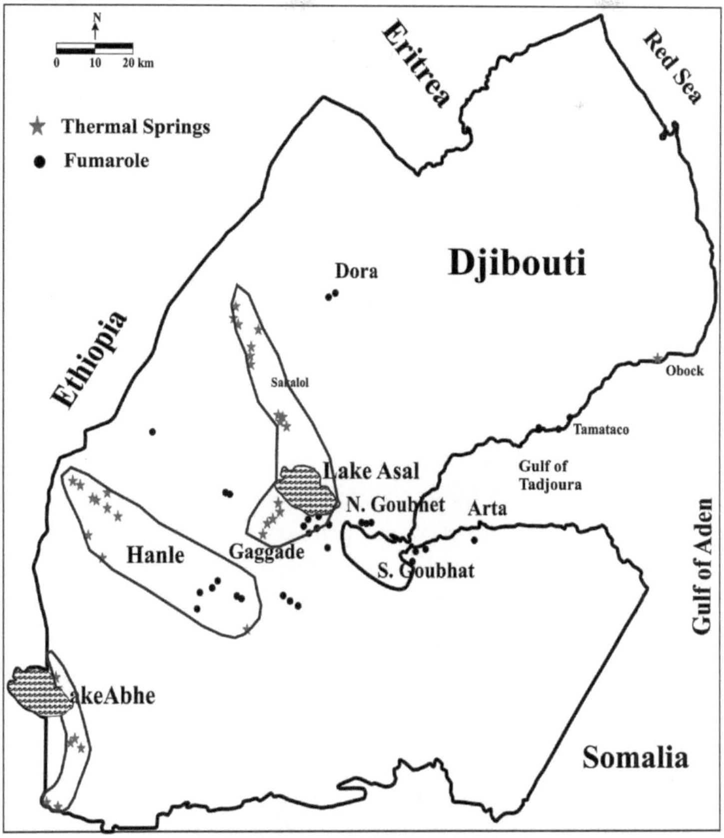

Fig. 5.12 Geothermal provinces in Djibouti. *(Adapted from Chandrasekharam D, Lashin A, Al Arifi N, Al-Bassam M, 2016. Red Sea Geothermal Provinces. CRC Press, 221p.)*

calcium carbonate is precipitated on the surface in the form of mounds, which often attain a height of 15 m, placed linearly along the fractures channelizing the thermal fluids. Earlier several exploratory borewells were drilled to a depth of about 1500 m and the results of this exploratory phase of the geothermal development plan were very encouraging to establish geothermal power plant. It is proposed to build a 100 MWe (less than about 891×10^6 kWh) plant at Lake Asal by 2020 and several 1 MWe (9×10^6) plant around Lake Abhe and Hanle provinces [107]. This amount of electricity is surplus to Djibouti considering its population (900,000) and need in domestic and agricultural sectors. The current electricity usage by Djibouti is 730×10^6 kWh [56]. Geothermal development of Lake Asal alone will make the country surplus in electricity.

5.5.4 Eritrea

The geothermal systems in Eritrea evolved due to the presence of active volcanoes: Musa Ali, Alid, Asale, and Dubbi (Fig. 5.13), located within the Afar depression [6,56,86].

The acidic volcanic activity induced several faults and shears and tensional cracks that controlled the fluid circulation. Active volcanism associated with seawater circulation through the faults gave rise to high-temperature geothermal systems (reservoir temperatures > 200°C) along the coast. A large number of fumaroles are present all around the base of the volcanoes [5,108]. The three-dimensional modelling of magneto-telluric (MT) data over the Alid indicate the presence of magma body at a depth of 5 km and hence the presence of high-temperature (225°C; [109,110]) geothermal system in this region. UNEP is planning to extend financial support to Eritrea to develop its geothermal fields (Meseret Teklemariam, UNEP Personal communication). At present Eritrea's primary energy consumption is 330 $\times$ 10^6 kWh [56]. With the development of that above described geothermal sites, the country can completely offset fossil fuels for power generation and secure freshwater for domestic and agricultural sectors. The temperatures of the geothermal reservoirs and the geochemistry of the thermal fluids indicate

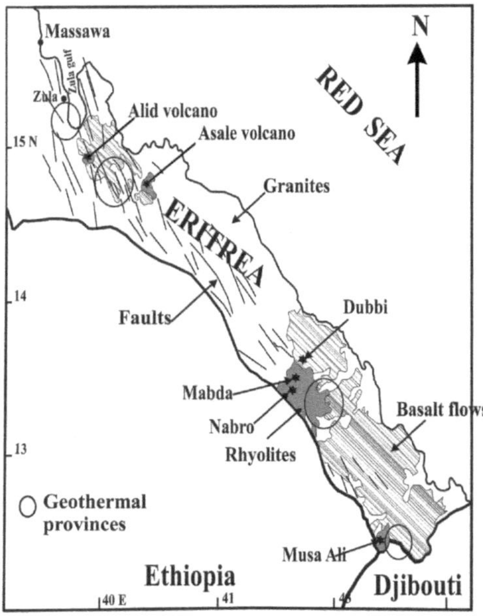

Fig. 5.13 Geothermal provinces of Eritrea. *(Adapted from Chandrasekharam D, Lashin A, Al Arifi N, Al-Bassam M, 2016. Red Sea Geothermal Provinces. CRC Press, 221p.)*

that the Alid geothermal site alone can generate 100 MWe (891×10^6 kWh), similar to Lake Asal geothermal system in Djibouti. This amount is double the current usage by the country.

5.5.5 Egypt

Geothermal provinces in Egypt are distributed on the banks of the Gulf of Suez, along the western margin of the Red Sea coast and at Kharga, Dakhla, Bahariya, and Siwa in the western part of the Nile River (Fig. 5.14). The issuing temperatures of the thermal waters are high along the Suez Gulf and Red Sea coast(51–70°C) while those occurring on the western side of the Nile are low (37–40°C; [111]). The heat flow values within the geothermal provinces around Gulf of Suez and along the Red Sea coast are

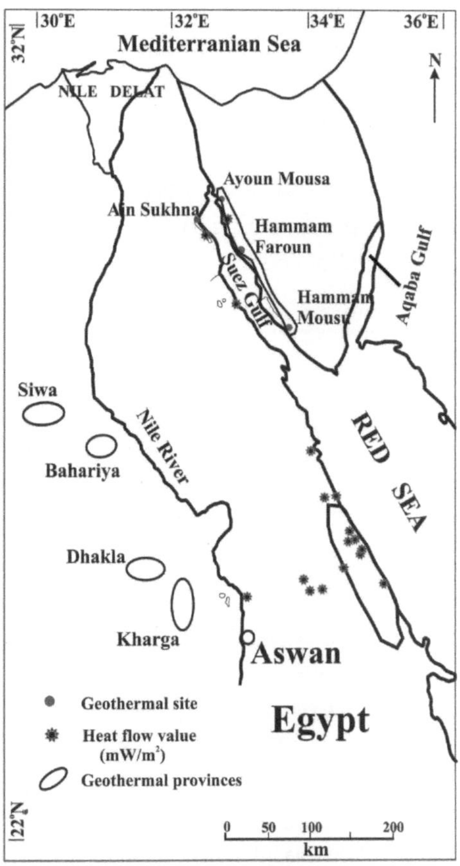

Fig. 5.14 Geothermal provinces in Egypt. High heat flow values (100–1136 mW/m^2) of sites are indicated by stars.

higher than the average value (>95 mW/m^2) and the bottom hole temperatures measured in certain oil wells around the Suez Gulf range between 120 and 260°C at 2 km depth [99,112,113].

The hydrothermal sites located around the Suez Gulf and the Red Sea coast can generate 28 MWe (221 × 10^6 kWh; [113,114]). In addition to the hydrothermal sources, in Egypt, like Saudi Arabia, substantial EGS resources in the radiogenic granites are exposed along the Red Sea coast east of Aswan. The heat flow values reported for these granites vary from 45 to 1136 mW/m^2 [56,115]. Thus, these granites, like the western Arabian shield granites (e.g., Midyan granites), are potential EGS source in Egypt. For example, the El Faliq granites located east of Aswan has a surface outcrop of 95 km^2 and can generate about 7 × 0^9 kWh [115] electricity. Both hydrothermal and EGS sources can be utilized for desalination process and make the country food and water independent.

5.5.6 Ethiopia

The Ethiopian geothermal province falls within the East African Rift, which is an extension of the Red Sea and Aden rifts (Fig. 5.8). The most promising geothermal sites within this rift are Dallol, Tendaho, Aluto Langano, and Corbett (Fig. 5.15). The Dallol and Tendaho sites are part of the Lake Abhe

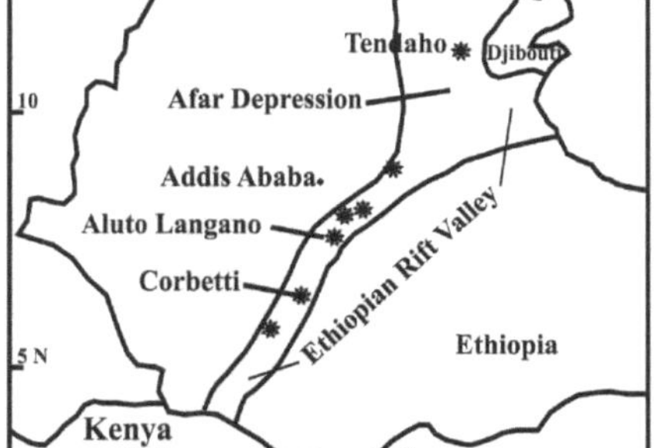

Fig. 5.15 Geothermal sites in Ethiopia. *(Adapted from Chandrasekharam D, Lashin A, Al Arifi N, Al Bassam AA, Varun C, 2015. Evolution of geothermal systems around the Red Sea. Environ Earth Sci 73, 4215–4236.)*

geothermal province of Djibouti as they are closely related to the Damah Ale and Erta Ale volcanoes in Ethiopia.

Geophysical exploration around Tendaho Site indicates reservoir temperature of 200°C. Geophysical investigations indicate the presence of a thermal reservoir with temperature of 200°C [116] with heat source lying at a depth of 5 km within the Stratoid basalt flows [117]. Owing to large hydroelectric projects, Ethiopia's geothermal projects are being developed at slow pace. The Tendaho and Aluto Langano geothermal sites have been assessed and have an installed capacity of 100 MWe. These two sites will be developed for power generation in the next decade [118].

However, the East African Rift valley geothermal provinces in Kenya are well developed. These provinces may become the energy provider to the entire East African countries in the next decade.

5.5.7 Kenya

The Kenyan geothermal provinces are located within the East African Rift valley that extends from Djibouti, through Ethiopia, to Kenya. Kenya's geothermal industry is growing fastly and at present >50% of Kenya's electricity is supplied by geothermal power plants. In the year 2015 the Olkaria power plant generated > 380 MWe in December alone.

The Kenyan rift valley is a loci of several high-temperature geothermal systems supported by an active spreading ridge, with its floor directly overlying hot shallow mantle (Fig. 5.16). Kenya's geothermal power projects have been under development since 1950 and the current estimated potential is about 10,000 MWe (79×10^9 kWh). At present, the installed capacity is 573 MWe (4×10^9 kWh), and being located in Olkaria, it is the largest field in the rift valley. Menengai field is being developed to generate 105 MWe (828×10^6 kWh) of electricity. The geothermal production at Olkaria will cross 1500 MWe (12×10^9 kWh) in another couple of years [46].

According to the Geothermal Development Corporation of Kenya (GDC) funds have been allotted to develop other sites such as Silali in the north and Suswa in the south (Fig. 5.16) By 2020 the installed capacity will be 2765 MWe (27,800 GWh). When fully developed Kenya will be in a position to support domestic and agricultural sectors in the country as well as the growing population demand of the entire MENA and a few sub-Saharan countries such as Ethiopia and Egypt for the next 30 year, which is an average life cycle of geothermal power plants.

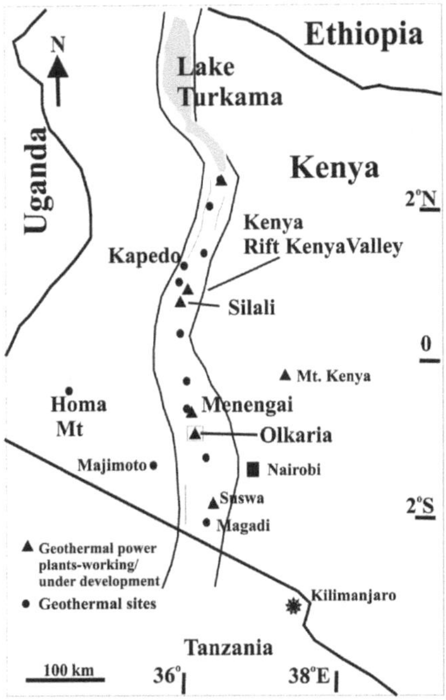

Fig. 5.16 Geothermal provinces, Kenya. *(Adapted from Chandrasekharam D, Bundschuh J, 2008. Low enthalpy geothermal resources for power generation. CRC press, 149p.)*

The existence of surface geothermal manifestations has not been reported in other GCC countries such as Kuwait, Bahrain, Oman, UAE, and Qatar. However, the geothermal gradients in these countries are capable of supporting geothermal direct applications such as space heating and cooling and refrigeration. By using ground source heat pumps, these countries can save oil and gas resources that are being used for such purposes and the saved fuels can be used for desalination purpose. Such direct geothermal application will reduce CO_2 emissions by the countries listed above.

5.6 DISCUSSIONS

The water-related issues in GCC countries are not new and its geographic location makes them water starved countries. Countries that share surface and groundwater face greater issues in resolving water-related issues. In certain cases, the respective governments have to change or adopt new water policy/water laws to suit other countries; a solution that is difficult to implement [22]. Counties that are water stressed adopt viable solution without impinging into the others' sovereignty; adopt VWT policy, and are investing heavily on food imports (see Fig. 5.5) and implemented desalination plants to meet growing demand for freshwater. Although desalination appears to be a permanent solution for the GCC, MENA, and sub-Saharan countries, the problems that centers around desalination process is the cost of freshwater generated and the CO_2 emissions associated with the process. Although a viable and cost-effective solution is available; this solution did not attract the attention of the water managers. All the issues related to trans-boundary aquifer issues, VWT, agricultural growth and GDP, and socioeconomic conditions can be resolved if energy sources low in CO_2 emissions and capable of supplying baseload electricity is used for desalination process. At present, large amount of electricity is being used for desalination process by all the GCC countries to meet the demand for freshwater. Among the GCC countries, by 2020 the demand for freshwater in Saudi Arabia alone will be more than 1.6 billion m^3 (Fig. 5.17). The amount of electricity required to generate this volume of desalinated water is 6400×10^6 kWh (MSF), 3200×10^6 (MED), $11{,}200 \times 10^6$ (MVC), and 9000×10^6 kWh (RO) for the four commonly operating desalination plants [77,119]. The electricity cogenerated through desalination process will be around $29{,}000 \times 10^6$ kWh for 1.6 billion m^3 of desalinated water. Although there is an advantage in desalination process, the results come with a heavy price tag. The desalination process (e.g., MSF and MED) emits significant

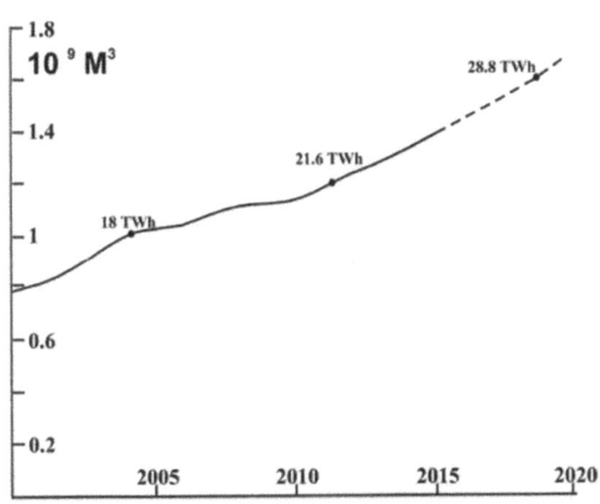

Fig. 5.17 Current and projected freshwater generation from desalination of seawater by Saudi Arabia. *(Adapted from Chandrasekharam D, Bundschuh J, 2008. Low enthalpy geothermal resources for power generation. CRC press, 149p.)*

quantities of CO_2 (about 3.8 kg/m^3 of water generated; [119]) hence the process is a deterrent to the environment.

The nuclear power option for desalination systems was discussed by several researchers [120–122]. Using DEEP software, economics of different oil and gas supported desalination systems have been compared with modelled nuclear power supported desalination systems [122]. According the DEEP model, nuclear energy supported RO desalination systems appears to be very cost effective compared with other desalinations systems. Owing to the issues related to safety, waste disposal, and volume of systems needed to operate desalination plants, countries may not replace fossil fuel-based desalination plants with nuclear powered desalination plants.

However, geothermal energy-based desalination power plants were not in focus and did not get its deserved attention although this energy source is low carbon emitter and is available in abundance in all the countries around the Red Sea. Since the energy source is free, unlike the fossil fuel-based desalination plants, such plants will be cost effective. Geothermal-based desalination plants were operating prior to 2004 and the best example is one that is operating in Milos Island, in the Aegan Sea, SE of Greece which is generating 80 m^3 of freshwater from the seawater at a cost of 1.5 euro/m^3 (1.61 US \$ at the current rate of exchange; ~ 16 US cents/L). The only source of energy in these islands is geothermal [123]. These islands being volcanic in origin have high-temperature geothermal systems with reservoir temperatures exceeding 230°C. However, shallow wells are able to produce 100–125 m^3/h of hot water with temperature varying from 50 to 99°C. The desalination plant supported by this energy is producing 80 m^3/h (MED-TVC system) of freshwater for the inhabitants and the waste heat generated from this plant is being utilized to generate additional 179 kWe of electricity using ORC unit. All the production wells are located near the sea. This case study demonstrates that all the GCC, MENA, and sub-Saharan countries can utilize geothermal energy around the Red Sea to generate freshwater through desalination and become energy and food independent. The geothermal energy resources of GCC, MENA, and sub-Saharan countries are summarized in Table 5.3 and the volume of freshwater that can be generated using various desalination techniques are given in Table 5.4.

Electricity generated from geothermal power plants is cost competitive compared with electricity generated from fossil fuels, nuclear, solar pv, and wind (Fig. 5.7, Table 5.5) and the capacity factor of geothermal power plants are higher compared with other energy source-based power plants. The

Table 5.3 Geothermal energy status of Gulf Cooperation Council (GCC), Middle East and North Africa (MENA), and sub-Saharan countries

	Installed $\times 10^6$ kWh	Development stage $\times 10^6$ kWh	EGS estimate $\times 10^9$ kWh
Saudi Arabia		254	160,000[a]
Y Republic		134	
Djibouti		900	
Eritrea		891	
Egypt		921	7[b]
Ethiopia		788	
Kenya	1228	27,800	

[a]generation capacity for Midyan granite only.
[b]generation capacity for El Faliq granite only.
Adapted from Omenda P, Simiyu S. Country update report for Kenya 2010–2014. Proceed. World Geothermal Congress 2015, Melbourne, Australia April 19–25, 2015 (CD); Chandrasekharam D, Bundschuh J. Low enthalpy geothermal resources for power generation. CRC Press, 2008. 149p; Chandrasekharam D, Lashin A, Al Arifi N, Al-Bassam M. Red Sea Geothermal Provinces. CRC Press, 2016. 221p.

Table 5.4 Volume of freshwater that can be generated using different desalination methods using geothermal energy (assuming 1 kWh/m^3, see Table 5.1)

	Multistage flash (MSF) million m^3	Multieffect distillation (MED) million m^3	Mechanical vapor compression (MVC) million m^3	Reverse osmosis (RO) million m^3
Saudi Arabia	64	117	33	52
Y Republic	34	67	20	27
Djibouti	225	450	113	180
Eritrea	223	445	129	199
Egypt	230	460	119	185
Ethiopia	197	394	113	158
Kenya	307	614	175	206
Kenya (27,800[a])	6950	13,900	3956	5560

[a]when Kenya's installed capacity becomes $27,800 \times 10^6$ kWh, see Table 5.3.

advantage of geothermal source unlike coal, oil, gas, and hydro is that only geothermal power plants can supply baseload electricity (Table 5.5) and are online > 90% of the time and the land required for geothermal power plants is much smaller (1–2 acre/MWe plant) than that for solar pv (12 acre for MWe) and wind-based power plants (65 acre/MWe) [11]. The geothermal

Table 5.5 Levelized cost of electricity generated from various energy sources

	1	2	3	4	5	6[a]
Conventional coal	85	66	4	29	1.2	10
Nuclear	90	83	12	12	1	11
Combined cycle	87	16	2	48	1	7
Geothermal	93	76	12	0	1	9
Wind	34	70	13	0		9
Solar pv	25	130	10	0	4	14
Solar thermal	20	214	41	0	6	3
Hydro	52	78	4	6	2	9

1, Capacity factor; 2, levelized capital cost; 3, Fixed O & M; 4, Variable O & M; 5, Transmission investment; 6, Levelized cost US$cents/kWh; [a]2011 US$ value.
Adapted from Chandarasekharam D, Lashin, A, Al Arifi N. CO2 mitigation strategy through geothermal energy. Saudi Arabia Renew Sustain Energy Rev 2014;38:154–163; Zarrouka SJ, Moon H. Efficiency of geothermal power plants: A worldwide review. Geothermics 2014;51:142–153.

Table 5.6 CO_2 emissions from different energy sources (kg CO_2/MWh)

Coal	Oil	N.Gas	Solar PV	Hydro	Wind	Nuclear	Biomass	Geothermal
953	817	553	71	25	31	14	86	67

Adapted from Chandrasekharam D, Bundschuh J. Low enthalpy geothermal resources for power generation. CRC Press, 2008. 149p; Shrestha E, Ahman S, Johnson W, Shrastha P, Batista R. Carbon footprint of water conveyance versus desalination as alternatives to expand water supply. Desalination 2011;280:33–43; IEA, 2016. CO2 emissions from fuel combustion, Statistics, International Energy Agency, Statistics, 166p.

power plants emit less CO_2 compared with plants operated by conventional energy sources (Table 5.6).

Thus, considering the levelized cost of electricity, CO_2 emissions, and efficiency of the renewable energy sources, geothermal stands apart compared with other renewables. The geothermal power production has reached 13.3 GW of operating capacity as of January 2016, spread across 24 countries and expected to reach about 18.4 GW by 2021. The geothermal power plant capacity factor [ratio of the actual output of the power plant and its potential output, it operated at full capacity (design capacity)] varies between 80% and 93%. This is the highest among any renewable energy resource (Table 5.5; [124]). Although the unit cost of desalinated water generated through conventional energy source is far less compared with other energy sourced desalinated water (Table 5.7), taking into account the penalty that the countries have to pay in terms of large CO_2 emissions (Table 5.6), and assuming that the countries have to pay carbon tax in future, desalination plants supported by geothermal energy stands out most

Table 5.7 Energy source and cost of desalinated water (US$/m^3)

Water type	Energy source	Cost/m^3 (1000 L)
Brackish	Fossil fuels	0.22–1.13
	Solar pv	4.8–10.99
	Geothermal	2.13
Seawater	Fossil fuels	0.37–2.87
	Wind	1.06–5.32
	Solar pv	3.34–9.59
	Solar collectors	3.73–8.5
	Geothermal	1.61

Adapted from Karagiannis IC, Soldatos PG. Water desalination cost literature: review and assessment. Desalination 2008;223:448–456. Karytsas C, Mendrinos D, Radoglou G. The current geothermal exploration and development of the geothermal field of Milos Island in Greece. GeoHeat Centre Quart Bull 2004;25(2):17–21.

economical (especially when EGS technology matures and provides least expensive energy) in terms of unit cost of freshwater generated. The cost savings comes from the energy source.

The advantage of using geothermal energy is net savings in fuel/energy costs. Since geothermal power plants are online >90% and operate for more than 30–40 years, the initial drilling costs gets absorbed during the initial few years there by reducing the operation cost. This reduces the capital cost of desalination plants that are now linked to the fuel costs (in the case of fossil fuel-based desalination plants) (Table 5.8). The capital cost of fossil fuel-based desalination plants further varies depending on the fluctuation in the oil and gas price that is linked to the international markets. On the other hand, geothermal energy-based desalination plants can maintain cost structure for a longer period as long as the geothermal wells are in production/operation.

In the case of geothermal energy-based desalination plants, the carbon saving (or the tax from carbon emissions) further aids in reducing the capital cost as the CO_2 emissions from geothermal power plants are nearly 10 times lower relative to the fossil fuels (Table 5.6).

Table 5.8 Capital cost of desalination plants using fossil fuels [125]

Plant type	Capacity (m^3/d)	Unit capital cost US$/m^3/d
MED	37,850	1860
MSF)	37,850	1598
MVC	24,000	1131
RO	37,850	1313

Table 5.9 Renewable water availability in Gulf Cooperation Council (GCC), Middle East and North Africa (MENA) and sub-Saharan countries

	Total (km^3/yr) Current	Per-capita (m^3/yr) Current	Per-capita (m^3/yr) 2015
Renewable water availability and future demand			
Saudi Arabia	2	85	63
Kuwait	0	8	0
Yemen	4	120	89
Oman	1	311	196
Libya	1	163	107
Egypt	69	636	789
Kenya	30	837	739
Ethiopia	110	1680	1020
Eritrea	9	1470	1302

Adapted from Miller JE. Review of water resources and desalination technologies, Sandia National Laboratories, Sand 2003-0800, 2003. 54p.

Thus, both electricity generated from geothermal energy and direct heat from geothermal energy are becoming a viable and environmentally friendly option to generate freshwater from seawater or brackish water to meet the growing demand for freshwater in agricultural and domestic sectors. Table 5.9 illustrates the status of availability of current renewable water resources and future demand in the GCC, MENA, and sub-Saharan countries

By the year 2025, these countries will be under highly water stressed condition and will be in danger of losing their sovereignty due large food and water imports from neighboring countries and pushing their food and water security out of their country's boundaries. However, these countries, as discussed above, have large geothermal energy reserves (Table 5.3) which need to be exploited. In geothermal energy development, Kenya is a forerunner and soon will be in a position to support future energy demand of a large part of MENA countries. Other countries have to follow this example to become water and food independent. Geothermal energy is being recognized as an economical and environmentally sound choice for generating power and for direct applications, and to meet the future electricity demand and to guarantee energy security and energy independency in developing as well as developed countries [10,126–133].

5.7 CONCLUSIONS

The water and food security are the primary concerns of all the countries in the world. In the case of GCC, MENA, and sub-Saharan countries, due to their geographic location, arid climate, and poor rainfall these countries are labelled as water stressed countries. These countries have excellent solutions at their hands which the respective governments are not able to execute. Comparing the energy resources availability, future water requirement, and the present water availability (Tables 5.3, 5.4, and 5.9) these countries are not in a helpless condition as far as food and water securities are concerned. A good example is Kenya which surged ahead with the development of geothermal energy and is able to go to the top of the socioeconomic development ladder (Table 5.4). Saudi Arabia, with high heat flow and active magmatic activity and large EGS resources did not pay attention to these resources until recently. With its excellent knowledge in oil and gas exploration and drilling technology, it did not pay due attention to geothermal resources. As demonstrated above it is the geothermal energy resources that can bail these countries from future water and food crisis. In comparison to solar pv, as shown above, geothermal is cost effective in supporting (US$ 1.6/1000 L) freshwater supply to all the GCC countries. In addition to free energy source, these countries can save CO_2 and earn additional revenue through carbon trade. No doubt the GCC countries today are enjoying the subsidy on oil as the resources are available in plenty but the penalty they pay in terms of environmental pollution and related climate issues are huge. VWT is good to some extent but with exponential growth in population in these countries, water rich countries will eventually exit from VWT due to local demand leaving the receiver countries in distress. Shifting the food security outside the boundaries of any country is disastrous to that country. The respective governments need to rethink and reframe their trade, food, and water security policies. EGS is going to be future reality with the development of plasma drilling technology. Every country will become energy independent in future once this technology matures. It took nearly 160 years for the oil industry to come to the current stage and the industrial revolution acted as a catalyst for its fast development. The demand for energy for growing industries around the world necessitated the technological development in oil industry. Now water and food crisis will necessitate the countries to invest more on geothermal energy resources. The demand for water, food, and affordable energy to generate the required water for the domestic and agricultural purposes is the prime future focus of the countries. Oil is finite while heat from earth is infinite. Countries have to prioritize their developmental plans to generate freshwater for the growing millions by developing geothermal energy resources available in their regions. The extraction of heat from the earth and generation of electricity is a proven technology today and future technologies will fine-tune the technology to further reduce the cost of electricity generation and make the countries energy and food independent. The cost of desalinated water will further fall below the current cost of US$ 1.61/1000 L.

ACKNOWLEDGMENTS

The authors extend their sincere appreciation to the Deanship of Scientific Research at King Saud University for the funding of this research group No. (ISPP-0099). The corresponding author thanks the Director of the Indian Institute of Technology, Hyderabad for providing the facilities for this work.

REFERENCES

[1] Papapetrou M, Wieghaus M, Biercamp C. (etds) Roadmap for the development of desalination powered by renewable energy. Stuttgart, German: Fraunhofer Verlag Pub.; 2010. 75p.
[2] IEA. Water for energy, Chapter 17 in World Energy Outlook. IEA pub 2012. 672 p.
[3] Chandrasekharam D, Lashin A, Al Arifi N, Chandrasekhar V, Al B. A. 2015a. Clean Development Mechanism through Geothermal, Saudi Arabia. Proceed. CD: World Geothermal Congress; 2015.
[4] Chandrasekharam D, Lashin A, Al Arifi N, Al Bassam A, Ranjith PG, Varun C, et al. Geothermal energy resources of Jizan, SW Saudi Arabia. J African Earth Sci 2015;109:55–67.
[5] Chandrasekharam D, Aref L, Al Arifi N, Varun C, Al Bassam A. High heat generating granites of Western Saudi Arabian shield. Proceed. Melbourne, Australia (CD): World Geothermal Congress; 2015. p. 2015.
[6] Chandrasekharam D, Lashin A, Al Arifi N, Al Bassam AA, Varun C. Evolution of geothermal systems around the Red Sea. Environ Earth Sci 2015;73:4215–36.
[7] Abderrehman, W.A. 2006. Groundwater resources management in Saudi Arabia. Proceed. Water Conservation Workshop, Khober, Saudi Arabia, 2006
[8] Grindle K, Siddiqi A, Anadon LD. Food security amidst water scarcity: Insights on sustainable food production from Saudi Arabia. Sustainable Production and Consump 2015; https://doi.org/10.1016/j.spc.2015.06.002.
[9] Chandrasekharam D, Lashin A, Al Arifi N, Al Bassam A, Varun C. Desalination of seawater using geothermal energy to meet future fresh water demand of Saudi Arabia. Water Resour Manage 2016; https://doi.org/10.1007/s11269-016-1419-2.
[10] Chandrasekharam D, Bundschuh J. Low enthalpy geothermal resources for power generation. CRC press; 2008. 149 p.
[11] Chandarasekharam D, Lashin A, Al Arifi N. CO_2 mitigation strategy through geothermal energy, Saudi Arabia. Renew Sustain Energy Rev 2014;38:154–63.
[12] Chandrasekharam D, Lashin A, Al Arifi N. The potential contribution of geothermal energy to electricity supply in Saudi Arabia. Int J Sustain Energy 2014; https://doi.org/10.1080/14786451.2014.950966.
[13] Ghaffour N, Bundschuh J, Mahmoudi H, Goosen MFA. Renewable energy-driven desalination technologies: A comprehensive review on challenges and potential applications of integrated systems. Desalination 2015;365:94–114.
[14] Sarbatly R, Chiam CK. Evaluation of geothermal energy in desalination by vacuum membrane distillation. Appl Energy 2013;112:737–46.

[15] Almazroui M, Islam MN, Athar H, Jones PD, Rahman MA. Recent climate change in the Arabian Peninsula: annual rainfall and temperature analysis of Saudi Arabia. Inter J Climatol 2012;32:953–66.
[16] Dalin C, Konar M, Hanasaki N, Rinaldo A, Rodriguez-Iturbe I. Evolution of the global virtual water trade network. Proc Natl Acad Sci (PNAS) 2012;199:5989–94.
[17] Maddocks, A., Young, R.S. and Reig, P. 2015. Ranking the world's most water-stressed countries in 2040. Water Resour Inst (accessed on 20 Feb 2017).
[18] UNDP. Water governance in the Arab region: Managing scarcity and securing the future. 2013. 182 p.
[19] Darwish MA, Al Awadhi FM. The need for integrated water management in Kuwait. Desalin Water Treatment 2009;11:204–14.
[20] Ramadhan M, Hussain A. Kuwait Energy Profile for Electrical Power Generation. Strategic Plan Energy Environ 2012;32:18–25.
[21] IEA, 2016. CO2 emissions from fuel combustion, Statistics, International Energy Agency, Statistics, 166 p.
[22] Jaganathan, V.N, Mohamed, A.S. and Kremer, A. 2009. (Etd) Water in the Arab World: Management Perspectives and Innovations; MENA countries. World Bank, 554p.
[23] Hamoda MF. Desalination and water resources management in Kuwait. Desalination 2001;138:385–93.
[24] Amery HA. Water security for Kuwait: assessing the feasibility of water imports from Iran. Int J Hydrol Sci Technol 2012;2:292–305.
[25] Fadlelmawla A, AlOtaibi M. Analysis of the water resources status in Kuwait. Water Resour Manage 2005;19:555–70.
[26] Mukhopadhyay A, Quinn M, Al-Haddad A, Al-Khalid A, Al-Qallaf H, Rashed T, et al. Pollution of fresh groundwater from damaged oil wells, North Kuwait. Environ Earth Sci 2017;76:145. https://doi.org/10.1007/s12665-017-6457-4.
[27] Al-Rashed MF, Sherif MM. Water resources in the GCC countries: An overview. Water Res, Mgmt 2000;14:59–75.
[28] Hajeeh M. Water conservation in Kuwait: a fuzzy analysis approach. J Industr Eng 2010;6:90–105.
[29] Eltawil, M.A., Zhengming, Z. And Yuan, L. 2009. A review of renewable energy technologies integrated with desalination systems, RSER, 13, 2245-2262.
[30] MRMWR. Water resources in Oman Ministry of Regional Municipalities and Water Resources, Sultanate of Oman, Report 1991;140p.
[31] FAO. World agriculture: towards 2015/30. A FAO perspective. London, UK: Earth Scan; 2003.
[32] Rockstrom J, Falkenmark M, Karlberg L, Hoff H, Rost S, Gerten D. Future water availability for global food production: the potential of green water for increasing resilience to global change, Water Resour Res 2009;45. W00A12, https://doi.org/10.1029/2007WR006767.
[33] Mbaga MD. The prospects of sustainable desert agriculture to improve food security in OmanMsafiriDaudiMbaga. Consilience. J Sustain Develop 2014;13(2014):114–29.
[34] Fath H, Sadik A, Mezher T. Present and future trend in the production and energy consumption of desalinated water in GCC countries. Int J Therm Environ Eng 2013;5:155–65.
[35] MDPS. Water statistics in the state of Qatar 2013. In: Ministry of Development Planning and Statistics; 2016. 48 p.
[36] Ismail H. Kuwait: Food and Water Security, Future Directions International Pty Ltd.; 2015. www.futuredirections.org.au
[37] Ismail H. Qatar: Food and Water Security Part 2, Future Directions International Pty Ltd.; 2015. www.futuredirections.org.au

[38] Darwish M, Hassabou AH, Shomar B. Using Seawater Reverse Osmosis (SWRO) desalting system for less environmental impacts in Qatar. Desalination 2013;309:113–24.
[39] Ayoub N, Musharavati F, Pokharel S, Gabbar HA. Energy consumption and conservation practices in Qatar—A case study of a hotel building. Energy Build 2014;84 (2014):55–69.
[40] Lehane S. Bahrain Food and Water Security. Future Directions International Pty Ltd.; 2015, www.futuredirections.org.au.
[41] Alnaser NW. Building integrated renewable energy to achieve zero emission in Bahrain. Energy Buildings 2015;93:32–9.
[42] EIA 2016. How much carbon dioxide is produced per kilowatt hour when generating electricity with fossil fuels? US Energy Information Administration. https://www.eia.gov/tools/faqs.
[43] Mogaka, H., Gichere, S., Davis, R. and Hirji, R. 2006. Climate variability and water resources degradation in kenyaimproving water resources development and management. The World Bank Working Paper 69, 130p.
[44] WRMA. Project on the development of the National water master plan 2030, The Republic of Kenya Ministry of Environment, Water and Natural Resources. Final report, Water Resources Management Authority; 2013. 413p.
[45] IEA, 2015. CO_2 emissions from fuel combustion, International Energy Agency Pub, 136p.
[46] Omenda, P. and Simiyu, S. 2015. Country update report for Kenya 2010–2014. Proceed, World Geothermal Congress 2015, Melbourne, Australia April 19–25, 2015 (CD).
[47] VEGA. Priority geothermal direct use applications for Kenya: A prefeasibility study for greenhouses, In: VEGA/Power African Agriculture Project report. Land O Lakes and Winrock International; 2013, https://www.landolakes.org.
[48] MWR. Ethiopia: Strategic framework for managed groundwater development. Ethiopia. Report: Ministry of Water Resources; 2011. 60p.
[49] Tucho GT, Weesie PM, Nonhebel S. Assessment of renewable energy resources potential for large scale and standalone applications in Ethiopia. Renew Sustain Energy Rev 2014;40:422–31.
[50] Mosello B, Calow R, Tucker J, Parker H, Alamirew T, Kebede S, et al. Building adaptive water resources management in Ethiopia. London, Report: Overseas Development Institute; 2015. 100p.
[51] FDRE, 2011. Ethiopia's climate Resilient Green Economy, Green Economy Strategy. Federal Democratic Republic of Ethiopia Report, 200p.
[52] FDRE, 2014. Federal Democratic Republic of Ethiopia: Ministry of Finance and Economic Development. Growth and Transformation Plan, Annual Report, 2012/13, 119p.
[53] Yassin, A. 2014. Transnational Large Scale Agricultural Firms in Gambella Regional State, Ethiopia: Local Potentials, Opportunities and Constraints for Market Linkage and Contractual Farming Schemes. Research report, Future agriculture's early career fellowship programme, 32p (www.future-agricultures.org, accessed on 6 March 2017).
[54] FAO. Aquastat irrigation in Africa in figures—AQUASTAT Survey 2015;2005. 8p.
[55] Tripathi RP, Ogbazghi WAA. Rice Production Prospects in Eritrea. J Water Res Protection 2015;7:1429–34.
[56] Chandrasekharam D, Lashin A, Al Arifi N, Al-Bassam M. Red Sea Geothermal Provinces. CRC Press; 2016. 221p.
[57] World Bank, 2014. World Bank http://data.worldbank.org/indicator (viewed 20.01.14).
[58] IEA. 2014. International Energy Agency statistics. CO2 emission from fuel combustion. CO_2 emission from fuel combustion.

[59] Pacini N, Harper DM. Hydrological characteristics and water resources management in the NileBasin. Ecohydrol Hydrobiol 2016;16:242–54.
[60] ICARDA 2011. Water and Agriculture in Egypt, Technical paper based on the Egypt-Australia-Workshop on On-farm Water-use Efficiency, July 2011, Cairo-Egypt. 84p.
[61] Keith B, Enos J, Garlick B, Simmons G, Copeland D, Cortizo M. Limits to Population Growth and Water Resource Adequacy in the Nile River Basin, 1994–2100. Center for Nation Reconstruction and Capacity Development. West Point, New York: United States Military Academy; 2013. 50p.
[62] Ghaffer EA. In: Non-conventional water resources management. Proceed 10th Inter. Water Tech. Conference, Alexandria, Egypt; 2006. p. 549–59.
[63] Renault, D. 2002. Value of virtual water in food: Principles and virtues. FAO report, 25p.
[64] FAO 2016. Aquastst. FAO information system on water and agriculture: Libya. 14p.
[65] Abdudayem A, Scott AHS. Water infrastructure in Libya and the water situation in agriculture in the Jefara region of Libya. Afr J Econ Sustain Develop 2014;3:33–64.
[66] Mohamed AMA, Al-Habaibeh A, Abdo H. An investigation into the current utilisation and prospective of renewable energy resources and technologies in Libya. Rene Energy 2013;50:732–40.
[67] FAO. Water and agriculture: Djibouti AQUASTST; 2005. 9p.
[68] FAO. Groundwater Management in Yemen Draft Synthesis Report; 2009.13p.
[69] IMF. Issues in managing water challenges and policy instruments: Regional prospective and case studies. 2015 International Monitory Fund, Report. 22p.
[70] Ward C, Ueda S, McPhail A. Water Resources Management in Yemen Contribution to the CDR Yemen, http://siteresources.worldbank.org/INTYEMEN/Overview/20150274/YE-Water.pdf 12 March 2017.
[71] Al-Kebsi A, Chandrasekhara D. Saltwater contamination in the coastal aquifer, Tihama Plain, Wadi Surdud, Yemen Republic. In: Bjerg PL, Engesgaard P, Krom TD, editors. Groundwater 2000. Rotterdam: Balkema; 2000.
[72] Giesecke C. Yemen's water crisis: review of background and potential solutions. In: USAID report; 2012.
[73] Fiorenza G, Sharma VK, Braccio G. Techno-economic evaluation of a solar powered water desalination plant. Energy Convers Manage 2003;44:2217–40.
[74] Kalogirou SA. Seawater desalination using renewable energy sources. Prog Energy Combust Sci 2005;31(3):242–81.
[75] Gude VG. Energy storage for desalination processes powered by renewable energy and waste heat sources. Appl Energy 2015;137:877–98.
[76] Gude VG, Nirmalakhandan N, Deng S. Renewable and sustainable approaches for desalination. Renew Sustain Energy Rev 2010;14:2641–54.
[77] Gude VG. Desalination and sustainability—An appraisal and current perspective. Water Res 2016;89:87–106.
[78] Dawoud MA, Al Mulla MM. Environmental Impacts of Seawater Desalination: Arabian Gulf Case Study. Int J Environ and Sustain 2012;1:22–37.
[79] IRENA, 2012. Renewable energy technologies: Cost analyses series: Solar Photovoltaic. IRENA working paper Volume 1 Power Sector, Issue 4/5, 52p.
[80] Al-Dayel M. Geothermal resources in Saudi Arabia. Geothermics 1988;17(2–3):465–76.
[81] Lashin A, Chandrasekharam D, Al Arifi N, Al Bassam A, Chandrasekhar V. Geothermal energy resources of wadi Al-Lith, Saudi Arabia. J African Earth Sci 2014;97:357–67.
[82] Lashin A, Al Arifi N, Chandrasekharam D, Al Bassam A, Rehman S, Pipan M. In: Geothermal Energy Resources of Saudi Arabia: Country Update. Proceed. World Geother. Congress (CD), Melbourne, Australia, April 2015; 2015.
[83] Coleman RG, Gregory RT, Brown GF. Cenozoic volcanic rocks of Saudi Arabia. U S G S Open file report 1983;83–788.

[84] Koulakov I, El Khrepy S, Al Arifi N, Sychev I, Kuznetsov P. Evidence of magma activation beneath the Harrat Lunayyir basaltic field (Saudi Arabia) from attenuation tomography. Solid Earth 2014;5:873–82.
[85] Koulakov I, El Khrepy S, Al Arifi N, Kuznetsov P, Kasatkina E. Structural cause of a missed eruption in the Harrat Lunayyir basaltic field (Saudi Arabia) in 2009. Geology 2015;43:395–8.
[86] Bosworth W, Huchon P, McClay K. The Red Sea and Gulf of Aden Basins. J African Earth Sci 2005;43:334–78.
[87] Gettings ME, Showail A. Heat flow measurements at shot points along the 197X Saudi Arabian seismic deep-refraction line, part I: results of the measurements. US Geol Surv Open File Rept 82-793 1982; 98 pp.
[88] Gettings ME, Blank HR, Mooney WD, Healey JH. Crustal Structure of Southwestern Saudi Arabia. J Geophy Res 1986;91:6491–512.
[89] Rybach L. Radioactive heat production: a physical property determined by the chemistry of rocks. In: Strens RGJ, editor. The physical and chemistry of minerals and rocks. New York: Wiley-Interscience; 1976. p. 309–18.
[90] Lashin A, Al Arifi N. The geothermal potential of Jizan area Southwestern parts of Saudi Arabia. Int J Phys Sci 2012;7:664–75.
[91] Stoeser DB. Distribution and tectonic setting of plutonic rocks of the Arabian Shield. J. African. Earth Sci 1986;4:31–46.
[92] Stuckless, J.S., Van Trump Jr, G., Bush, C.A., Vaughn, R.B. and Bartel, A.J. 1987. Geochemistry and preliminary assessment of resource potential for postorogenic granites of the southwestern Arabian Shield, Kingdom of Saudi Arabia. USGS Open File report 87-509, 61p.
[93] Harris NBW, Marriner GF. Geochemistry and petrogenesis of peralkaline granite complex from Midian Mountains, Saudi Arabia. Lithos 1980;13:325–37.
[94] Cermak V, Huckenholz HG, Rybach L, Schmid R. Radioactive heat generation in rocks. In: Hellwege K, editor. Landolt-Bornstein numerical data and functional relationships in science and technology. New Series, Group V. Geophysics and Space Research, vol. 1, Physical properties of rocks, subvolume b. Berlin, Heidelberg, New York: Springer; 1982. p. 433–81.
[95] Lachenbruch AH. Preliminary geothermal model of the Sierra Nevada. J Geophys Res 1968;73:6977–89.
[96] Park Y, Nyblade A, Rodgers A, Al-Amri A. S wave velocity structure of the Arabian Shield upper mantle from Rayleigh wave tomography. Geochem Geophys Geosyst 2008;9:1–15.
[97] Girdler RW, Evans TR. Red Sea heat flow. Geophys J R Astr Soc 1977;51:245–51.
[98] Morgan P, Swanberg CA. Heat flow and the geothermal potential of Egypt. Pageophl 1978;17:213–26.
[99] Zaher MA, Saibi H, El-Nouby M, Ghamry E, Ehara S. A preliminary regional geothermal assessment of the Gulf of Suez, Egypt. J African Earth Sci 2011;60:117–32.
[100] Somerville M, Wyborn D, Chopra P, Rahman S, Don E, Van der Meulen T. Hot dry rock feasibility study. Energy Research and Development Corporation; 1994 unpublished report.
[101] Minissale A, Mattash MA, Vaselli O, Tassi F, Al-Ganad IN, Selmo E, et al. Thermal springs, fumaroles and gas vents of continental Yemen: Their relation with active tectonics, regional hydrology and the country's geothermal potential. Appl Geochem 2007;22:799–820.
[102] Mattash MA, Vaselli O, Minissale A, Ad-Dukhain A, Hazza M. The First Geothermal Resources Map of Yemen at a 1: 125 000 scale. Sana'a (Republic of Yemen): Ministry of Oil and Mineral Resources; 2005.

[103] Fouillac AM, Fouillac C, Cesbron F, Pillard F, Legendre O. Water rock interaction between basalt and high salinity fluids in the Asal Rift, Republic of Djibouti. Chem Geol 1989;76:271–89.
[104] Zan L, Gianelli G, Passerini P, Troisi C, Haga AO. Geothermal exploration in the Republic of Djibouti: Thermal and geological data of the Hanle and Asal areas. Geothermics 1990;19(6):561–82.
[105] Chandrasekharam D, Chandrasekhar V. Geothermal energy development of L. Abbe and CDM for Djibouti. Proceed. 3rd East African Rift Geothermal Conf (ArGeo-C3), Djibouti, Nov. 2010; 2010.
[106] Houssein B, Chandrasekharam D, Varun C, Jalludin M. Geochemistry of thermal springs around Lake Abhe, Western Djibouti. J Sustain Energy 2013; https://doi.org/10.1080/14786451.2013.813027.
[107] Moussa OA, Suleiman H. In: Country report, Geothermal development in Djibouti. Proceedings, World Geothermal Congress, April 19–29, 2015, Melbourne, Australia (CD); 2015.
[108] Duffield W.A., Bullen, T.D., Clynne, M.A., Fournier, R.O., Janik, C.J., Lanphere, M.A., Lowenstern, J., Smith, J.G., Giorgis, L., Kahsai, G., Mariam, K. and Tesfai, T. 1997. Geothermal potential of the Alid volcanic center, Danakil Depression, Eritrea. US Geological Survey Open File Report, 291, 62 p.
[109] Beyene AT. Multidimensional inversion of MT data from Alid geothermal area, Eritrea; comparison with geological structures and identification of a geothermal reservoir, 2012 Unpublished MS thesis, Univ. Iceland, 138 p.
[110] Yohannes, E., 2015. Geothermal exploration in Eritrea-country update, Proceed. WGC 2015, Melbourne (CD).
[111] Swanberg CA, Morgan P, Boulos FK. Geothermal potential of Egypt. Tectonophy 1983;96:77–94.
[112] Morgan P, Black'well DD, Fanis TG, Boulos FK, Salib PG. Preliminary temperature gradient and heat flow values for northern Egypt and the Gulf of Suez from oil well data. In: Proceedings International Congress on Thermal Waters. Geothermal Energy and Volcanism of Mediterranean Area. Geothermal Energy I,1976. p. 424–38.
[113] Zaher AM, Saibi H, Ehara S. Geochemical and stable isotopic studies of Gulf of Suez's hot springs, Egypt. Chin J Geochem 2012;31:120–7.
[114] Lashin A. A preliminary study on the potential of the geothermal resources around the Gulf of Suez, Egypt. Arabian J Geosci 2012; https://doi.org/10.1007/s12517-012-0543-4.
[115] Chandrasekharam D, Lashin A, Al Arifi N, Al Bassam A, Varun C, Singh HK. Geothermal energy potential of eastern desert region, Egypt. Environ Earth Sci 2016; https://doi.org/10.1007/s12665-016-5534-4.
[116] Ali S. Geochemical Studies of the Tendaho Geothermal Field. In: Proceed. World Geothermal Congress 2005, Antalia, Turkey; 2005.
[117] Didana YL, Thiel S, Heinson G. In: Magnetotelluric imaging at Tendaho high temperature geothermal field in North East Ethiopia. Proceed. World Geothermal Cong., 2015, Melbourne, Australia; 2015.
[118] MWE 2012. Scaling up renewable energy programme: Ethiopia investment plan. Ministry of Water and Energy, Ethiopia, 2012, report 69p.
[119] Ghaffour N, Lattemann S, Missimer T, Ng KC, Sinha S, Amy G. Renewable energy-driven innovative energy-efficient desalination technologies. Appl Energy 2014;136:1155–65.
[120] Al Mutaz IS. Potential of nuclear desalination in the Arabian Gulf countries. Desalination 2001;135:187–94.
[121] Aljohani MS. Nuclear desalination competitiveness in the western region of the Kingdom of Saudi Arabia. Desalination 2004;164:213–23.

[122] Mansouri NY, Ghoniem AF. Does nuclear desalination make sense for Saudi Arabia? Desalination 2017;406:37–43.
[123] Karytsas C, Mendrinos D, Radoglou G. The current geothermal exploration and development of the geothermal field of Milos Island in Greece. GeoHeat Centre Quart Bull 2004;25(2):17–21.
[124] Zarrouk SJ, Moon H. Efficiency of geothermal power plants: A worldwide review. Geothermics 2014;51:142–53.
[125] Kesieme U, Mlne N, Aral H, Cheng CY, Duke M. Economic analysis of desalination technologies in the context of carbon pricing, andopportunities for membrane distillation. Desalination 2013;323:66–74.
[126] Chandrasekharam D, Bundschuh J, editors. Geothermal energy resources for developing countries. Lisse, The Netherlands: A.A. Balkema; 2002.
[127] Aaheim A, Bundschuh J. The value of geothermal energy for developing countries. In: Chandrasekharam D, Bundschuh J, editors. Geothermal energy for developing countries. Lisse, The Netherlands: Balkema Publisher; 2002. p. 37–52.
[128] Bundschuh J, Birkle P, Aaheim A, Alvarado GE. Geothermal resources for development—valuation, present use and future opportunities. In: Bundschuh J, Alvarado GE, editors. Central America: geology, resources and hazards (2 Volumes). Lisse, The Netherlands: Balkema Publisher; 2007. p. 869–94.
[129] Bundschuh J, Coviello M. Geothermal energy: capacity building and technology dissemination. In: Chandrasekharam D, Bundschuh J, editors. Geothermal energy for developing countries. Lisse, The Netherlands: Balkema Publisher; 2002. p. 77–90.
[130] Clauser C. Geothermal energy. In: Heinloth K, editor. Landolt-Börnstein, Group VIII: Advanced Materials and Technologies, Vol. 3: Energy Technologies, Subvol. C: Renewable energies. Heidelberg-Berlin, Germany: Springer; 2006. p. 480–595.
[131] Bundschuh J, Suárez Arriaga MC. Introduction to the numerical modeling of groundwater and geothermal systems: fundamentals of mass, energy and solute transport in poroelastic rocks. Leiden, The Netherlands: Balkema/CRC Press/Francis & Taylor; 2010.
[132] Huenges E, Ledru P, editors. Geothermal energy systems: exploration, development, and utilization. Weinheim, Germany: John Wiley and Sons-VCH; 2010.
[133] MIT: The future of geothermal energy—Impact of enhanced geothermal systems (EGS) on the United States in the 21st Century. An assessment by an MIT-led interdisciplinary panel, Massachusetts Institute of Technology, 2006, http://www1.eere.energy.gov/geothermal/egs_technology.html (accessed April 2017).

FURTHER READING

[134] Shrestha E, Ahman S, Johnson W, Shrastha P, Batista R. Carbon footprint of water conveyance versus desalination as alternatives to expand water supply. Desalination 2011;280:33–43.
[135] Karagiannis IC, Soldatos PG. Water desalination cost literature: review and assessment. Desalination 2008;223:448–56.
[136] Miller, J.E. 2003. Review of Water Resources and Desalination Technologies, Sandia National Laboratories, SAND 2003-0800, 54p.

CHAPTER 6

Nuclear Energy Powered Seawater Desalination

Salah Ud-Din Khan*, Shahab Ud-Din Khan†, Syed Noman Danish*, Jamel Orfi*, Usman Ali Rana*, Sajjad Haider*
*King Saud University, Riyadh, Saudi Arabia
†COMSATS Institute of Information Technology, Attock, Pakistan

6.1 INTRODUCTION

It is estimated that the current freshwater reservoirs are only 2.5% apart from the major portion are locked up in glaciers and the remaining 30% are ground water and 0.4% as surface water [1]. This dictates the value of about 2.4 billion people suffering from freshwater resources across the globe [2]. This value is increasing at 7% per year and will be doubled in 2025 due to rapid population growth and considerable increase in urbanization [3]. One of the principal solutions to this problem is seawater desalination using fossil resources. The desalination technologies can be divided into two major techniques: thermal and membrane techniques. Multistage flash (MSF), multieffect distillation (MED), and vapor compression (VC) are the major conventional thermal technologies while reverse osmosis (RO) and electrodialysis (ED) are membrane technologies. The ratio of the deployed desalination technologies constitutes a share of RO as 60%, MSF as 25%, MED as 8%, and ED as 4% [4]. MSF is considered to be the proven technology for freshwater production but RO swapped the MSF technology recently [5]. The concept of nuclear desalination was first implemented in 1960s according to the International Atomic Energy Agency (IAEA) and it has already been proven with over 150 reactor years of experience. Also, the economic analysis showed that the nuclear energy could be a competitive choice when compared with fossil fuel. The assessment is presented in Table 6.1 [1,4].

While considering the energy production cost, the capital cost of nuclear energy is not always less and it is estimated that the advanced nuclear system costs 108 $/MWh and the advanced combined cycle and gas turbine costs around 66 and 104 $/MWh, respectively [6]. From an IAEA prospective,

Renewable Energy Powered Desalination Handbook
https://doi.org/10.1016/B978-0-12-815244-7.00006-4

Table 6.1 Techno-economic assessment of coupled units [1,4]

Coupled unit	**Water production cost (\$/m^3)**
Nuclear + reverse osmosis (RO)	0.75
Fossil + RO	0.85
Nuclear + multieffect distillation (MED)	0.90
Fossil + MED	1.4
Fossil (MED)	2.4

nuclear desalination plants are less costly as compared with fossil fuel desalination ones. There is also a concern of developing small modular reactors (SMRs) and according to IAEA there could be around 96 SMRs in operation by 2030 [7]. These SMRs are now building up under different purposes such as desalination and hybrid energy systems, access to remote areas, which are off-grid, and for developing countries and isolated populations.

In this chapter, different desalination technologies have been discussed along with integrated units in view of theoretical, computational, and experimental aspects. The possibility of small modular nuclear reactors for water desalination including the current status of all operated as well as planned desalination systems have been discussed in detail. Gulf Cooperation Council (GCC) region has been considered as special case for adopting nuclear desalination including pros and cons associated with it. Finally, technical and feasible assessments of coupled nuclear desalination units have been reported with the option of viable experimental design.

6.2 DESALINATION TECHNOLOGIES

The concept of water desalination is not new. However, it is gaining huge attention among masses with every passing day primarily because of the ever-increasing scarcity of potable water in most of the countries of the world, changed rainfall patterns, extended periods of droughts, and increase in population. It is expected that the world population will increase from 6.8 to 8.9 billion by 2050 [8]. Consequently, there has been an increasing trend in installing desalination plants. There were about 18,000 desalination plants globally by the end of 2015, with an overall installed production capacity of 86.55 million m^3/day or 22,870 million gallons per day (MGD) [9]. The major commercial technologies used on an industrial scale are generally classified into the categories as given in Table 6.2.

Table 6.2 Major desalination technologies

Thermal technologies	Vapor compression (VC) Multistage flash (MSF) Multieffect distillation (MED)
Membrane technologies	Electrodialysis (ED) Reverse osmosis (RO)

RO presently is dominant with 60% market share of all water desalination, followed by MSF at 27% and MED at 8% [10]. Nevertheless, if only seawater desalination is taken into account, the share from different technologies changes completely. In this case, the MSF become dominant with market share of 70% [11]. RO seawater desalination however is gaining its share mainly because of its lower cost, simplicity, and flexibility in operation due to its modular design. Following is a brief introduction of the major technologies used for desalination:

i. multistage flash (MSF)
ii. multieffect distillation (MED)
iii. vapor compression (VC) [mechanical vapor compression (MVC) and thermal thermal (TVC)]
iv. Reverse osmosis (RO)
v. Electrodialysis (ED)

6.2.1 MSF Desalination

According to an IAEA report MSF desalination systems yield 60% of all desalinated water in the world [12]. MSF has the most commercial experiences of the capacities over 100,000 tons/day compared with other desalination technologies [13,14]. The MSF comprises the use of distillation through multiple chambers called stages. In this process, each consecutive stage of the plant operates at relatively lower pressures. The process starts with heating of feedwater under high pressure. The heated water enters into first stage, at lower pressure triggering the sudden evaporation called flashing. The flashing of a percentage of the feedwater proceeds in each consecutive stage, because of the reduction of pressure in each successive stage. The heat exchanger tube bank that runs through each stage condenses the vapor and converts it into potable water.

Normally only a small proportion of the feedwater is converted into vapor and condensed. MSF distillation plants have been built since the late

1950s when the very first four-stage multiflash desalination plant was built in Kuwait [11]. Typically, MSF plants comprise 15–25 stages [15].

There are mainly two MSF desalination configurations which are either "once-through" or "brine recirculation." The main purpose of the once-through MSF system is to handle the major disadvantage of the single-stage flash unit that is low system performance ratio. This is very much accomplished by splitting the flashing range over a multiple number of stages which not only results in enhancing the performance ratio but also helps in reducing the stage temperature drop. In once-through configuration, the feedwater undergoes only one pass through the heater and flash chambers and disposed of. The main objective of "brine recirculation" configuration is to reduce the feedwater mass flow rate. This lowers the rate of consumption of chemical additives, reduces the volume of the pretreatment plant for the feedwater, and also enhances the thermal efficiency because of higher energy content of feedwater. MSF is the leading water desalination technology in the Middle East and North Africa region where fossil fuels are abundantly available at lower cost [16]. MSF is found to be a good choice of coupling with nuclear power plants. As an example, the Japan Atomic Energy Agency has installed a 600-MW thermal nuclear facility which generates 300 MW of electrical energy and uses the waste heat in MSF process.

6.2.2 Multieffect Distillation

The MED is considered to be the oldest desalination technology [17] and is found to be more efficient thermodynamically than MSF [18]. Similar to MSF, MED also consists of multiple stages or "effects" (4–16 effects) in which vapor from each effect is condensed in the proceeding stage and re-vaporized at gradually reducing vacuum conditions. Contrary to MSF, the MED does not require additional supply of heat for vaporization at each stage in order to process the feedwater. The evaporator surfaces in each stage are heated by the vapor produced in the preceding stage. This enhances the pressure and as a result the vapor condenses. The MED plants comprise numerous single units, each yielding between 100 and about 70,000 m^3/day. The starting temperature of the seawater in this process can be 70°C [19]. The use of low top brine temperature in MED results in less scaling and fouling problems compared with MSF technology. Many successful couplings of MED with nuclear plants are in operation. South Korea has developed a 330-MW nuclear reactor for cogeneration of electricity and potable water. The main concept has a small nuclear reactor coupled with four MED units. The plant is producing 40,000 m^3/day of freshwater with 90 MW of electrical energy.

6.2.3 VC Desalination

VC desalination is a process where the evaporation of feedwater is achieved by the application of heat delivered by compressed vapor. The process works on the principle of reduction of boiling point temperature with pressure and vice versa. Since, the compression process increases both the vapor pressure and temperature, therefore it enables to produce additional vapor thereby using the latent heat rejected during condensation.

There are two methods of vapor compression, which have found commercial application. The first one is called MVC in which a mechanical compressor is used to generate vacuum in the evaporator where the feedwater transforms into vapor. The compressor then compresses the vapor and condenses it inside the tubes of heat exchanger bundle. Feedwater is showered on the other side of the heated tube bundle where it partially boils and evaporates, producing additional vapor.

The second type of vapor compression is called TVC in which the principle of Venturi effect is used. An orifice extracts water vapor from the evaporator, creating a lower ambient pressure. The extracted water vapor is compressed by the steam jet. This mixture is condensed on the evaporator tube walls to provide the heat of condensation. This heat is utilized on the other side of the tube walls in the evaporator to evaporate the feedwater.

VC process is usually used in combination with other processes (MED or MSF) to improve overall system efficiency. Several plants based on the MED–TVC configuration are now widely used and show high energetic efficiencies compared with single MED. Single-vapor compression units are known to have low capacities that can vary between 20 and 2500 m^3/day. Compared with other desalination technologies, VC plants are usually smaller in size and therefore well suited for remote locations, resorts, industries, and drilling sites where freshwater is not readily available [20].

6.2.4 Reverse Osmosis

Membrane desalination uses membranes, which are becoming a common commodity to extract freshwater from saline feedwater. The feedwater is forced onto the surface of a membrane, which selectively passes water and retains salts. RO separates salts and small molecules from low molecular weight solutes (typically less than 10 MWCO [21]) at comparatively high pressures. In a typical RO plant, a pump increases the feedwater pressure above the osmotic pressure. Such pressures allow the freshwater to pass through the semipermeable membranes, leaving behind all the suspended solids, colloids, and dissolved solids. Typical osmotic pressures range is from 200 to 400 psi for brackish water and about 800–1000 psi for seawater. The quality of feedwater which includes salinity level, temperature, and turbidity plays a vital role in the selection of RO modules. Other desalination technologies however are not so sensitive to water quality [22].

The osmotic pressure increases with salinity which results in more pump work for high concentrated waters. If the feedwater has high turbidity, it will produce fouling and as a result membrane pores are clogged with suspended solids. RO is comparatively a new process, which was commercialized in 1970s [23].

Currently, RO is the most widely used method for desalination in the world. However, its performance expressed in terms of recovery ratio and salt rejection ratio decreases as the feed salinity increases. For instance, the RO recovery ratio can decrease from 30% to 40% for seawaters. In some regions especially the Middle East, the total salinity is in the range 41,000–45,000 ppm. In these regions, seawater has high fouling potential due to its high turbidity. High ambient temperatures in these regions also deteriorate the performance of RO plant. As a result, a suitable feedwater pretreatment is necessary before RO. Pretreatment may involve conventional methods such as chemical treatment followed by coagulation, sedimentation, sand filtration, flocculation, etc.

Currently, RO desalination is an ideal choice for a small-scale plant especially in rural areas or islands where there is no other source of freshwater is available. For example, most desalination plants in the Caribbean area use RO systems [24].

6.2.5 Electrodialysis

ED is another membrane process, which uses the high electrical potential energy to separate cations and anions from the stream. The electrical potential forces the salt particles to move through a membrane, leaving behind freshwater as a product. ED was originally conceived for seawater desalination, it has generally been used for brackish water desalination. This is due to the fact that the desired electrical energy for ED process is proportional to the salinity of feedwater that is why it is economical only for brackish water desalination. Like RO, ED sometimes needs pretreatment. Suspended solids with a diameter more than 10 μm need to be removed. Such large particles will clog the membrane pores.

6.3 APPLICATION WITH RENEWABLE ENERGIES

Desalination process is triggered by thermal or electrical energy with its source from renewable or sustainable energies. The dependency includes permeate water quality, plant capacity, economics, salinity, and availability of sources. The assessment showed that the geothermal and solar could produce low-temperature heat and electricity to drive MED, MSF, TVC, RO, and ED systems. The most up-to-date desalination systems integrated with renewable and sustainable energies are as follow:

- Coupled nuclear desalination system utilizes waste heat or low-pressure steam to operate RO/MED/MSF systems producing electricity and water at very reasonable cost in addition to maintenance and operating cost which drops significantly [25].
- Coupled solar photovoltaic (PV) system is a mature technology having life expectancy of more than 25 years providing both alternating current (AC) and direct current (DC) to desalination plant. For the purpose of desalinating brackish water seawater using PV with brackish water RO (BWRO) and seawater RO (SWRO) technology are currently being installed. Also, ED unit requires only DC power for electrodes while low-pressure pumps are driven by AC and DC, respectively [26–32].
- Coupling with wind turbine generating mechanical and electrical energy for RO, ED, and MVC desalination systems. The implementation capacity of RO/wind and MVC/wind is in the range of 50–2000 and 5–50 m^3/day, respectively [33–38].
- Geothermal energy coupled desalination units could drive RO, ED, and MVC both mechanically and electrically. Even, low-temperature geothermal sources can possibly drive MSF, MED, and TVC units [39–41].

It was observed that the output water cost is the highest for solar PV/RO system, while the lowest cost can be obtained when geothermal source is coupled with MED unit.

6.4 NUCLEAR DESALINATION

6.4.1 Nuclear Reactors

Nuclear reactor technology had been started since 1940s for the purpose of research applications and electricity was generated for the first time from nuclear energy in 1951 [1]. After that, many innovative nuclear reactors had been built with larger electricity output. With high energy density, nuclear power is more efficient and has extraordinary advantages compared with other energy resources. The fact is also of reducing the environmental degradation by emission of carbon. Under the economics crises and population growth the developing countries having no appropriate electricity grids solution intrigued toward the development of fast reactor as well small- and medium-sized nuclear reactors. The development of large nuclear reactors can be troublesome for the future goals of many under developed countries due to large space occupancy and economical risk. Considering these facts, the changing trend is to develop new technologies in reactor building including the construction in less ideal place, compactness in power plants, hydrogen production, electricity generation, freshwater production, and desalination systems. Moreover, innovative nuclear power reactors are capable of solving the current and upcoming crises including proliferation, security problems, global warming, fuel breeding, high-temperature process heat applications, and less hazardous affects.

6.4.2 Types of Nuclear Reactor

There are various types of nuclear reactors which can be summarized as follows [4].

Conventional Nuclear Reactors:

- Pressurized Water Reactor (PWR)
- Boiling Water Reactor (BWR)
- High-Power Channel Reactor (RBMK)
- Gas Cooled Reactor (GCR)
- Pressurized Heavy Water Reactor (PHWR)
- Advanced Gas Cooled Reactor (AGCR)
- Aqueous Homogeneous Reactor (AHR)
- Liquid Metal Fast Breeder Reactor (LMFBR)

Gen IV Nuclear Reactor Concepts:

- Very High-Temperature Reactor (VHTR)
- Gas-Cooled Fast Reactor (GCFR)
- Supercritical-Water-Cooled Reactor (SCWR)
- Sodium-Cooled Fast Reactor (SCFR)
- Molten-Salt Reactor (MSR)
- Lead-Cooled Fast Reactor (LCFR)

Advanced Nuclear Reactors (Gen III + Concepts):
- Fast Reactor Integral Type (IFR)
- The Pebble Bed Reactor
- Small, Sealed, Transportable Autonomous Reactor (SSTAR)
- The Clean and Environmentally Safe and Advanced Reactor (CAESAR)
- Subcritical Reactors
- Hydrogen Moderated Self-Regulating Nuclear Power Module (HPM)
- Advanced Heavy Water Reactors (AHWR)
- KAMINI (Reactor using U-233 Fuel)

6.4.3 Key Technologies of Nuclear Reactors

6.4.3.1 Floating Nuclear Power Plants

To supply electric power to far reaching areas, there is a need to develop ecological safety, radiation shielded, and high reliable and long-term operation of nuclear fuel without reloading. The concept appeared to develop floating-type nuclear power plant with low power levels of up to 10 MW having features of minimum construction time, at the same time capable of operating continuously for about 10–13 years without refueling and repairs, having full length total service of 40 years [42,43].

6.4.3.2 Seawater Desalination

About two-third of the world's population may suffer from water shortages by 2025 and seawater desalination is one of the most promising feature of supplying portable water. IAEA has already been assisting the production of water using nuclear energy [44,45]. Desalination plant have long fuel loading pattern and can easily be transported and are expedite to apply online in remote and difficult to access region where the supply of fossil fuel is impossible [46].

6.4.3.3 Space Reactors

The nuclear energy is considered to be the most appropriate to overcome the power limitation specifically in space mission as per the advent of international academy of astronautics (IAA). Many activities have been carried out for the development of strategies during the recent past and continue efforts are still underway. A number of impressive and developing strategies have been set up for the viability of space missions to meet the reliable and deployable under-controlled operations. Without any intervention, this technology extracted by-product for some applications as well. The impressive feature is to reduce the overall mass and volume of rocket propulsion, by carrying low power density, adopting less irradiation fuel at the launch, and by using low enriched uranium for propulsion. The other features include the subcritical core to reduce launching accidents, radiation protection systems, easy decommissioning, simple control of plant, simpler in maintenance and repairs to avoid any leakage of the constrained fluids. Two main problems experienced by the space reactors include heat sink and production of electricity as medium is space. Considering that the system is PWR, detailed neutornics and thermal hydraulic studies need to be conducted to define the preliminary scheme for the entire plant by the conventional generator. The fuel that can be perfect for these types of reactor is a composition of metal matrix and uranium [47].

6.5 SMALL MODULAR REACTORS

SMRs help to reduce carbon emission which is equally good for the environment. Owing to the smaller size, it does not require large area for construction making it safe and secure under economic limit. These reactors can be undergrounded with the purpose of reducing the hazardous effect with the addition of passive safety features. With reduced construction timing and less capital cost, SMRs also require less maintenance and thus have competitive electricity rates. There is also a possibility of simple refueling and little space for waste storage and can be remotely controlled between refueling. The heat can be recycled easily due to compact size of the reactor. Most important is the usage of less enriched uranium which cannot be further weapon graded.

There is a strong interest in developing SMRs across the globe due to enhanced affordability in technological advancements and ultimately ease pressure on grid. It is estimated that the global energy demand will increase by 50% over the next 35 years and considerably the carbon emission will also increase in the same way. Table 6.3 illustrates the current status of SMRs [48].

It has been observed that the capital costs of all types of reactor are nearly attributed to the same design characteristics while considering the economics aspects of reactors. However with the increase in energy demands, many countries are interested in the deployment of SMRs [49].

This situation affects the economy of nuclear power since there is a huge difference between large and small reactors. Also, the average investment and operating cost of reactors are contingent with the size of plant. SMRs are more flexible in siting, timing, and sizing than those of large reactors.

Table. 6.3 Current small modular reactors [48]

High-temperature gas-cooled reactors	Light water-cooled SMRs (iPWRs)	Liquid metal-cooled fast reactors	Heavy water-cooled SMRs
CEFR (China)	KLT-40 (Russia)	4S (Japan)	Pressurized heavy water reactor (PHWR) 220 (India)
HTR-10 (China)	SMART (Korea)	PFBR-500 (India)	EC-6/CANDU-6 (Canada)
HTR-PM (China)	CAREM-25 (Argentine)	Hyperion (USA)	AHWR300-LEU (India)
GTHTR300 (Japan)	IRIS (USA)	PRISM (USA)	–
PBMR (South Africa)	NuScale (USA)	SVBR (Russia)	
HTMR 100 (South Africa)	MPower (USA)	–	
EM2 (USA)	ACP 100 (China)	–	
SC-HTGR (USA)	VBER-300 (Russia)		
Xe-100 (USA)	ABV-6M (Russia)		
GT-MHR (Russia)	Flexblue (France)		
MHR-T/100 (Russia)	DMS (Japan)		
	IMR (Japan)		

The less construction time of SMRs make them viable for investment and can be postponed without any revenue losses giving the higher net values of the investment [50].

With over 9000 reactor years of operating experience accumulated with current nuclear energy systems, a thought for the development of small- and medium-sized nuclear reactors arouse. With added advantages of incorporated improved safety systems the protection against release of radioactivity into the environment increased. According to IAEA "Nuclear power less than 700 MWe are generally thought of having power outputs in the range of 100–400 MWe." The rough estimation shows that the about two-third of SMRs are in the ranges of 300–700 MWe and the rest would be in smaller range. The proportion of SMRs used to supply heat and electricity is about one-third. There is also interest in fast breeder reactors for nuclear power applications using liquid sodium as coolant and plutonium as fuel. There is also a high interest for the development of high-temperature GCR using helium as coolant for the purpose of attaining maximum thermal efficiencies. Many other concepts using heavy water as moderator and cooled by other fluids have also been developed and reached the stage of industrial prototype. The interest in the technology is due to advantages of reducing the heavy water inventory and associated risk of heavy water outflow.

6.5.1 Nuclear Power Suitable for Desalination Process

Production of potable water from sea by using nuclear reactors as source of energy is called nuclear desalination. In this process, thermal energy, electrical energy, or both can be used and the whole facility may be dedicated to the production of potable water as well as for the generation of electricity. The facility can be shared with some degree of common or shared facilities, operating strategies, planning, staff, and services and possibly control facilities adjacent with seawater intake and outfall structures.

Nuclear power as source of energy to power desalination systems has been proven with 150 reactor year experience mainly operated in India, Japan, and Kazakhstan. The main applicability is to run the reactor at full power capacity meeting the grid demand and a part of it is used to drive pumps for desalination purposes. There is an economical concern of commercialization and operation of large nuclear power plants. The ratio of different nuclear reactors by capacity factor are BWR as 20%, PWR as 68%, PHWR as 6%, LWGR as 3%, GCR as 2%, and FBR as 1%. The commercially available coupled desalination plants are PWR, PHWR, and LMFR. The optimized configurations of coupled nuclear desalination systems as headed by IAEA in different countries includes in Canada, Argentina, Japan, Pakistan, China, Russia, Morocco, Korea, Egypt, India, and Tunisia [51].

It is worth mentioning that desalination processes include MSF and MED processes for cogeneration while RO uses electricity only. These technologies can be coupled to nuclear power reactors. There is also a need to consider safety, economy, and operational factors which are also applicable for the prevention of radioactivity from entering into coupled desalination process or in case of accidental conditions. In addition, an isolation loop needs to be installed to prevent ingress of salinity into the turbine circuit. In some of the proven reactor design, two additional barriers exist between reactor coolant and desalination process. Some of the nuclear reactors are already operating in different countries such as Bulgaria, Hungary, Germany, Slovakia, and Switzerland. They operate under the configuration of the dual purpose of electricity generation and district heating. The reactors which are only used for district heating are considerably simpler in construction and operation but safety and other auxiliary systems cannot be compromised. These types of reactors are mostly operated in China, Canada, and in Russian Federation. The advantages of these nuclear heating reactors (NHRs) include smaller and more compact design with no special safety-related concerns.

The integration of nuclear reactors with desalination units has gained higher attention from the last few decades while improving the technological and safety concepts. Several studies have already been carried out on this concept of integration technology. For example,

- Coupling of NHR with MED process has been investigated by Wu and Dong in 2000 [52].
- Integration of MED + RO/VC with NHR was performed by Zhang and Shaorong in 2010 [53].
- Utilization of nuclear power for freshwater production and coupling of various desalination processes were reported by Misra and Kupitz [54].
- Feasibility assessment of KANUPP using RO technology has been performed on-site [55] and the same coupling scheme was also performed in Egypt as reported by Karameldin et al. [56].
- The feasibility of nuclear desalination options in UAE has been compared with fossil fuels. The results show that nuclear desalination is cost effective [57].
- The suitability of pressurized heavy water coupled with desalination technologies in Saudi Arabia has been investigated by Al-Johani [58].
- Similar studies were performed by Fried and Edlund [59] for economic and technological assessment of desalination plant in Middle East region.
- The safety of integrated desalination system from external events such missiles, aircraft warfare hazards has been examined by Abdul-Fattah [60].
- The adoption of small modular nuclear reactor for desalination purposes in UAE has been studied by Jung et al. [5].
- The potential of nuclear desalination in Saudi Arabia concluding Canadian-based nuclear reactor for desalination purposes as the best option [61].
- Hybrid technology has been adopted for the modification of important aspects pertaining to the economy of nuclear desalination plant [62].

- The history of nuclear desalination options were categorized by Belessiotis and Delyannis [63].
- The economic aspects of nuclear desalination were summarized by Senviratne [64] and the most recent scenarios were included in the world nuclear association report [65].
- Adoption of nuclear technology for desalination was discussed and studied by Darwish and Najem [66].
- A famous EURODESAL project covering the important aspects of utilizing nuclear power as heat source for desalination conducted by numerous scientists and engineers [67].
- Reliability of coupled nuclear reactors was investigated by Esen et al. [68].
- Calculations of the technical and economic analysis of the Russian design reactor ABV were performed by Khan et al. [69].
- Technical series report published by IAEA gives impressive and up-to-date content of nuclear desalination options [70].

6.6 STATUS OF NUCLEAR DESALINATION PLANTS

Nuclear energy is one of the main sources of energy for desalination purposes and is cost-effective as compared with fossil fuels. According to IAEA report "only nuclear reactors are capable of delivering the copious quantities of energy required for large-scale desalination projects." There is also a great concern about the desalination of brackish and seawater treatment plants in urban areas.

According to world energy report, there were about 19,000 desalination plants globally with an available production capacity of about 15 billion m^3/year and with increase in population, this value will increase considerably. The value of safe drinking water reaching out to world's population is estimated to be around one-fifth of total populations. The production of freshwater is a major hurdle in sustainable development and the statistics shows that 2.3 billion people live in water stressed areas and around 1.7 billion have less than 1000 m^3 of potable water per year. It is estimated that installed desalination capacity of United States, European Union, North Africa, and Middle East are 9%, 13%, 8%, and 50% while the rest contains 20% brackish water.

Today, most of the desalination plants run on fossil fuel thus triggering the greenhouse gas emission. As of 2016, the total installed capacity of 19,000 plants with potable water production is 88.6 million m^3/day. The ratio of membrane desalination and thermal desalination is 73:27 but mid-2016 added up 93% extra new capacity of membrane desalination. Middle East and North Africa (MENA) region contains majority of desalination plants and is economically advantageous for combined power generation and water production.

It is estimated that nuclear desalination plants can be identified as producing 80–100 or 200–500 km^3/day of desalinated water. The associated cost for RO is expected to be 50–94 cent/m^3, 60–96 cent/m^3 for MED, and 1.18–1.48 \$/m^3 for MSF plants [7]. A study reported in 2006 for competitiveness of nuclear against gas and oil plants at Tunisia indicated that the desalination cost was about half for the MED coupled with gas plant and about one-third less for RO coupled unit. Similarly, in 2015, IAEA report on nuclear desalination in Algeria revealed that the nuclear energy is cost competitiveness as compared to fossil fuels.

Nuclear desalination plants have been chiefly operated in Kazakhstan, India, and Japan with over 150 reactor years of experience confirming the feasibility of nuclear desalination system. This technology is cost effective and environmental-friendly with most important technologies being used are MED which uses thermal energy for auxiliary water system and RO which uses mixed energy systems. However, it is also noted that the performance ratio of the reactor is dependent on the type of the reactor used. Tables 6.4 and 6.5 present current status of nuclear desalination units mentioning the type of reactor and technology used along with the country of development.

The use of nuclear technology for desalination is a very attractive, cost effective, and environmental-friendly solution. The most important desalination processes that have been used are MSF, MED, and RO. In nuclear reactors, the MED uses thermal energy for the auxiliary water system whereas RO employs mixed energy (electrical or mechanical). The performance of desalination system integrated with nuclear reactor is mainly dependent on the type of nuclear reactor.

6.7 THEORETICAL–COMPUTATIONAL MODELS

6.7.1 Theoretical Modeling

For the mathematical modeling, first find the thermal performance model for the MED system, which is equally applicable to RO system. This model includes the gain output ratio (GOR) and the governing equation specific to the MED/MSF system [51,60,66] are given as:

Table 6.4 Current status of nuclear desalination units [7]

Reactor name	Power (MWth)	Desalination technology	Water ratio (m^3/day)	Country
SMART	330	4-MED unit with TVC	40,000	South Korea
CAREM	100	–	–	Argentine
NHR-200	200	MED	80–160,000	China
Abengoa	–	RO	330,000	China (from spain)
KLT-40S	–			Russia
ATETS-80	–	–	–	
ABV-6	–			
Volnolom floating NPP	–	–	40,000	
VBER-3000	–		–	
PAES-150	–		–	
VBER-300	–	–	–	
BN-350	750 (operating)	MED	80,000	Kazakhstan
Pressurized water reactor (PWR)	–	MSF	14,000	Japan
Pressurized heavy water reactor (PHWR)	170	MSF/RO	1800	India
Pressurized heavy water reactor (PHWR)	125	MED	4800	Pakistan
Bushehr NPP		RO	200,000	Iran
LMFB	150	MED/MSF	80–145,000	Russia
Pressurized water reactor (PWR)	566 890 3900 2600 1000 1000 1000	MSF RO MSF RO RO MED RO	2000 2000 3900 2600 1000 1000 1000	Japan
Pressurized heavy water reactor (PHWR)	6300 30	MSF/RO LTE	6300 30	India
Pressurized water reactor (PWR)	2180	RO	2180	USA

Table 6.5 Current status of new desalination projects

Country	Desalination technology	Water output (m^3/day)	Desalination cost (USD)	Status	Region
Algeria	RO	500,000	495 millions	2014	Magta
		120,000	185 millions	2011	Tripaza
	MSF	91,000	2.3 million	–	Arzew
Argentine	RO	3000	–	–	Puerto Deseado
Australia	RO	–	9.16 billions	2006–2012	-
		–	381,800	–	Sydney
		0.19	20.62 millions	2016	Melbourne
		200–280,000	–	–	South Antrala
		100,000	–	2011	Adelaide
		123,000	293.95 millions	2006–2022	Perth
Chile	Nine-operating plus 16 under construction	220,000	3.43 billion	2017	–
			10 billion	2026	–
	112 sea water desalination plant	972,000	–	2014	
China	RO	50,000	–	2011	Caofeiden
		1,000,000	1.02 billion	2019	Caofeiden
		2,200,000		2015	Shenzhen
		300,000	–	2017	Tianjin
India	MVC	10,200	0.9 million	2012	Kudankulam
Egypt	RO	24,000		2013	Marsa Matrouh
Ghana	RO	24,000	125 million		Nungua
Iraq	RO	199,000	–	2016	Basrah
Jordon	–	1,400,000	–	Planned	Aqaba

Continued

Table 6.5 Current status of new desalination projects—cont'd

Country	Desalination technology	Water output (m^3/day)	Desalination cost (USD)	Status	Region
Kenya		100,000	–	–	Mombasa
Kuwait	RO	140,000	1.4 billion		–
	RO	486,000	1.4 billion		Az-Zour
Mexico	RO	21,000		2013	San Quintin
		379,000	490 million	2024	Rosarito
Morocco	MED	8000	–	–	Tan–Tan
Abengoa	RO	100,000	102.1 million	2014	Agadir
Oman	RO	128,000		2009	Sur
		45,460		2013	Barka
		120,000		2014	Barka-2
		191,000		2016	BOO project
		69,000		2013	Salalah
		250,000		2007	Sohar
		225,000		2017	Qurayyat
		225,000		2016	Suwaiq
Pakistan	RO	10,000	3.0 million		–
		2000		2015	Karwat
		189,000		–	Gawader
Qatar	MSF	144,000	504 million	2015	Ras Abu Fontas
	RO	164,000	–	2016	–
		590,000		2015	Doha

Russia	RO	10,000		2011–12	Vladivostok
Saudi Arabia	MSF	11,000,000,000		2014	–
		4,600,000	–	–	–
		2,600,000			Jubail/Rabigh/Yanbu and Shuqaiq
		146,000			Yanbu-1
	MED	68,000	1.0 billion		Yanbu-2 (Doosan)
	MSF	550,000		2016	Yanbu-3
	RO	600,000			Rabigh
	MSF	1,025,000	7.2 billion		Ras Al khair, Jubail
		727,000			
	RO	880,000			Shoaiba-3
	RO	178,600	19.3 billion	2015	Sadara
		30,000		2014	Al-Khafji (Phase-1)
		300,000		2018	Al-Khafji (Phase-2)
				2020	Al-Khafji (Phase-3)
		2,500,000		2019	Makkah
Singapore	RO	136,000		2005	–
		3,185,000	700 million	2013	–
		228,000		2016	Changi
		137,000		–	Marina East

Continued

Table 6.5 Current status of new desalination projects—cont'd

Country	Desalination technology	Water output (m^3/day)	Desalination cost (USD)	Status	Region
South Africa	RO	5000			Lamberts bay
		450,000	1.23 billion		Cape town
Spain	20-RO plants	1,100,000			Canary islands
Tunisia	RO	150,000			Sfax
UAE	RO	68,000			Ras Al Kaimah
		136,000			Fujairah-1
	MSF	454,000		2011	Al Ruwais
	seawater RO (SWRO)–MED	4,546,000			Fujairah-2
	RO	385,000			Taweelah
		394,000			Umm al nar
		91,000		2014	Dubai
		136,400			Al Zawra
		136,380	1.5 billion		Abu-Dhabi
	MSF	34,095	–	–	–
		640,000		2013	Dubai
UK	RO	150,000	–	–	–
USA-Mexico	RO	375,000	–	–	Baja
USA		60,000	19 million	2016	San Antonio
		150,000		2026	San Antonio
		200,000	1.0 billion	2015	California
		415,000	3.5 billion	2035	San Diego

$$\mathrm{GOR(MSF)} = \frac{h_{fg}}{C_p\left(\Delta T_{bh} + \Delta T_{bpe}\right)}\left(1 - \mathrm{e}^{\left(-C_{vm}.\Delta T_{ao}/\lambda_m\right)}\right)$$

$$\mathrm{GOR(MED)} = \frac{\lambda_h}{\left(\lambda_m \dfrac{dT_{ae}}{dT_{do}} + c_h x\left(dT_{ph} + dT_{bpe}\right)\right)}$$

$$\mathrm{GOR(MED)} = \mathrm{GOR}(1 + R_{tvc})$$

$$\Delta T_{ae} = \frac{(T_o - T_n)}{n} h_{fg} = \frac{\dot{m}_f}{\dot{m}_s} C_p(T_o - T_1)$$

For an insulated system with assumption that the steam is only in saturated state, the energy balance is given by [71]

$$h_{fg} = \frac{\dot{m}_f}{\dot{m}_s} C_p(T_o - T_1)$$

The mass balance equation is given by

$$\dot{m}_f = \dot{m}_b + \dot{m}_d$$

For the salt mass balance,

$$s_f = \frac{\dot{m}_b s_b + \dot{m}_d}{\dot{m}_f}$$

For the salt concentration factor (CF),

$$\mathrm{CF} = \frac{W_p}{W_b} + 1$$

The lost shaft can be calculated as

$$Q_{ls} = \left(\frac{Q_{st}}{1 - \eta}\right)$$

The isentropic efficiency η for the condenser heat load case can be calculated as

$$\eta = \eta_{lpt}\left(\frac{T_{cm} - T_c}{T_{cm} + 273}\right)$$

Substituting the values of η in the above equation, we get the net effective heat load for the condenser as

$$Q_{ls} = \left(\frac{Q_{cr}}{1 - \eta_{lpt}\left(\frac{T_{cm} - T_c}{T_{cm} + 273}\right)} \right)$$

For the heat lost shaft for extraction case,

$$Q_{ls} = W_{st} h_{fg} \eta$$

In this case, η can redefined as

$$\eta = \eta_{lpt}\left(\frac{T_{st} - T_c}{T_{st} + 273}\right)$$

Substituting the values of η,

$$Q_{ls} = W_{st} h_{fg} \eta_{lpt}\left(\frac{T_{st} - T_c}{T_{st} + 273}\right)$$

6.7.2 Computational Techniques

There are some techniques used for the integration and assessment of nuclear reactor desalination systems as recognized and owned by IAEA. The subjected simulation tools are being used by number of researchers across the globe thus identifying the applicability and authenticity of the results obtained. These simulation tools are as follows:

a) desalination economic evaluation program (DEEP)
b) desalination thermodynamic optimization program (DE-TOP)
c) SEMER code
d) advanced process simulator (APROS) simulation facility

6.7.2.1 Desalination Economic Evaluation Program (DEEP)

Considering the importance of desalination from nuclear energy, IAEA has designed a software package named as DEEP [72]. This simulation tool helps to calculate and estimate technical, operational, and economic analysis of different desalination systems integrated with renewable and sustainable energies. The integrated desalination technologies include MED, MSF, RO separately as well as in hybrid configuration like RO + MSF, RO + MED, and RO + MED + MSF hybrid energy systems. This simulation technique helps to estimate the water production cost, the levelized cost, and the total cost and to evaluate the performance of various coupled desalination systems. The results obtained out of this code are helpful in calculating and analyzing the integrated desalination systems. The code can be used for the development of prototype desalination plants.

6.7.2.2 Desalination Thermodynamic Optimization Program

IAEA has developed a new simulation tool named as DE-TOP used for the thermodynamic assessment of desalination plants. This simulation tool is based on excel and is able to model generic water-cooled reactors coupled with desalination units. The program also compares the performances of various configurations. The DE-TOP has been extensively used by various educational and research purposes globally [73].

6.7.2.3 SEMER Code

Atomic energy of France (CEA) has developed a code named as SEMER for the purpose of developing process coupling schemes. The program is capable of process modeling and simulation, calculating economic results, and performing optimization of various integrated nuclear as well as fossil fuel power plants. The program works via input from other software packages named as power conversion cycle tools (CYCLOP), coupling scheme modeling (NDESL), and economic evaluation model [74].

6.7.2.4 Advanced Process Simulator

APROS facility has been developed by Fortum power division in Finland in cooperation with VTT technical research center in 1986. This simulator has the capability to simulate number of tasks for thermal and nuclear power plants. The application includes the design, process simulation, engineering, and safety performance of the plant. This facility contains wider applications while assessing the safety and accident analysis of nuclear power plants. The simulator is governed by boiler, chemical, and nuclear reactor models with technical specifications that include GUI interfaces, thermal hydraulic model, simulation database, and automatic system. The facility also works as thermal, chemical, and nuclear power plant analyzer. There is also a separate instructor console, training simulator, and control for the analysis purposes.

6.8 GCC COUNTRIES AS POTENTIAL DESALINATION SITES

The history of GCC dictates that the discovery of oil in the region aroused from the invaded Second World War foreign investors [75]. With passage of time and small population influx, the foreigners intruded the GCC buildup.

By that time, increasing urbanization compelled the GCC to adopt desalination technologies. The enclosed area of the sea contributed to about 240,000 km^2 across the Middle East and North Africa region.

The GCC countries are investing huge amount of money for water desalination to ensure environmental friendly and better usage of wastewater including high level of water conservation. Fig. 6.1 shows the projected water shortage and usage by the GCC up to 2025.

Online desalination capacity of the GCC, world and share to other countries are given in Fig. 6.2.

Table 6.6 illustrates the most recent figures of online and contracted plants and their capacities in the GCC countries [75,76]

The GCC countries are the top oil exporter countries but potable water shortage still remains a major problem [77,78]. The growing population in the GCC countries increases the demand of energy and water and it was estimated that natural freshwater per capita ranges from 60 to 370 m^3 across Middle East region [79]. The climate change decreases the freshwater production by 20% [80] and this situation became the backbone of many GCC countries that produce about 99% of potable water needs and representing

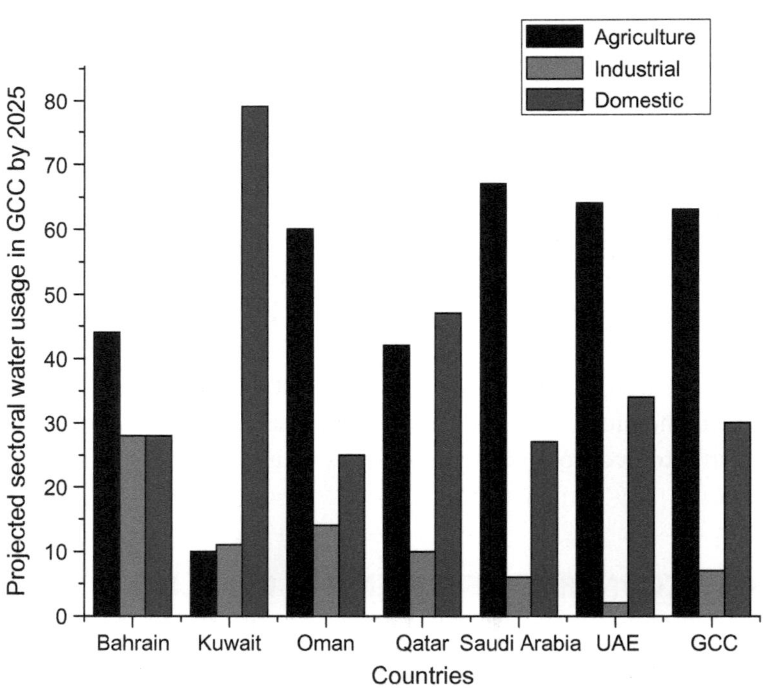

Fig. 6.1 Projected water usage plan up to 2025 in the GCC countries [75].

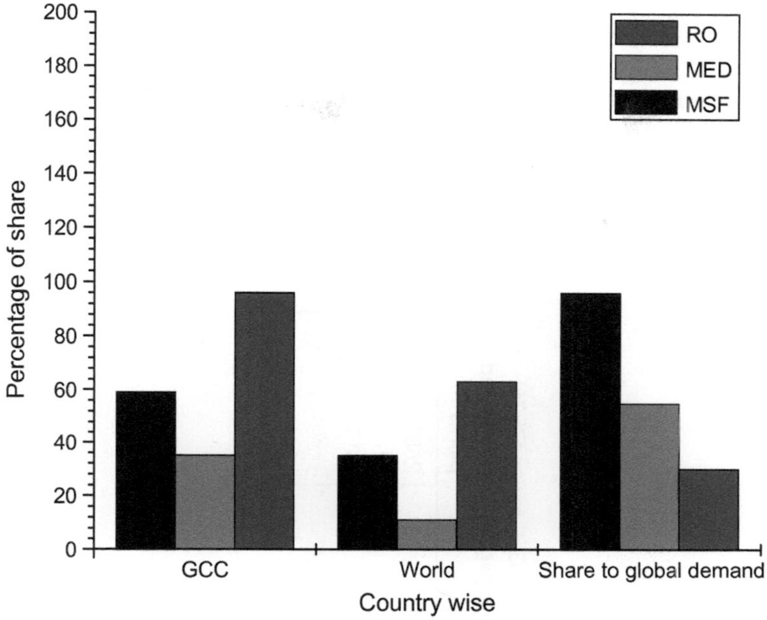

Fig. 6.2 Country wise share of desalination technologies [76].

about 57% of global desalination production [81]. This dependency on oil and consequently meeting the requirements of energy and water have led the GCC leaders to focus on water desalination with alternative energy resources (Fig. 6.3).

The major source of running thermal technologies is fuel which is then used to cogenerate both water and electricity across the GCC region as presented in Fig. 6.4. The need for adopting membrane technology is principally available in the region where feedwater is brackish. Membrane processes include essentially SWRO and BWRO but also ED. It is important to add that most of the new constructed and commissioned desalination plants in the GCC countries include RO units in single configuration or hybrid with MSF. For instance, MSF technology has been considered to be the most widely applied desalination technology but now RO is gaining higher attention. This is associated with several advantages of RO technology compared with thermal desalination including the fact that RO is cost effective, flexible in operation, and its efficiency can be improved further. Higher attention in hybrid desalination technologies, for example, RO + MED or RO + MSF is also observed.

Table 6.6 Recent figures of online and contracted plants with their capacities in GCC region [75,76]

Country	Population	Online capacity (m^3/day)	Online plants	Contracted capacity (m^3/day)	Contracted plants	Desalinated water production (million m^3)
KSA	28,290,000	13,080,497	2664	14,791,660	2688	1690.6
Oman	5,517,000	1,094,740	184	1,350,832	187	114.13
UAE	9,205,651	9,358,492	492	10,109,057	497	1357.1
Kuwait	3,250,500	3,023,369	88	3,477,969	89	749.34
Qatar	2,035,106	1,832,762	139	1,997,218	141	252.61
Bahrain	1,318,000	113,001	165	113,001	165	161.96

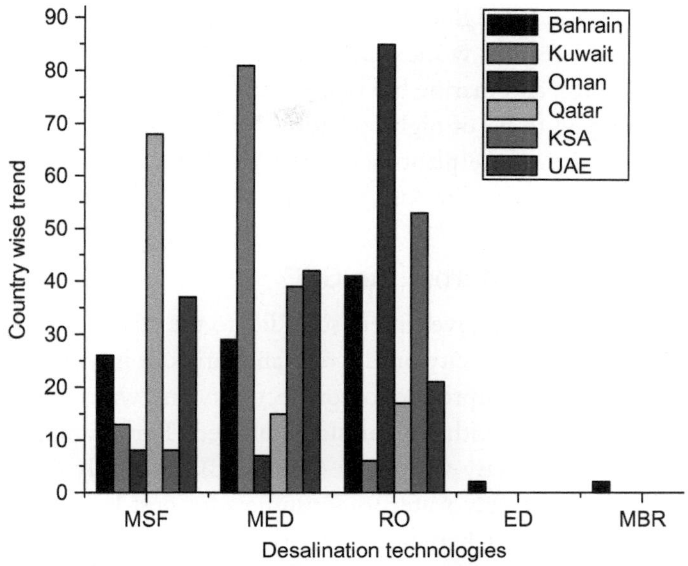

Fig. 6.3 Country wise trends with desalination technologies [75].

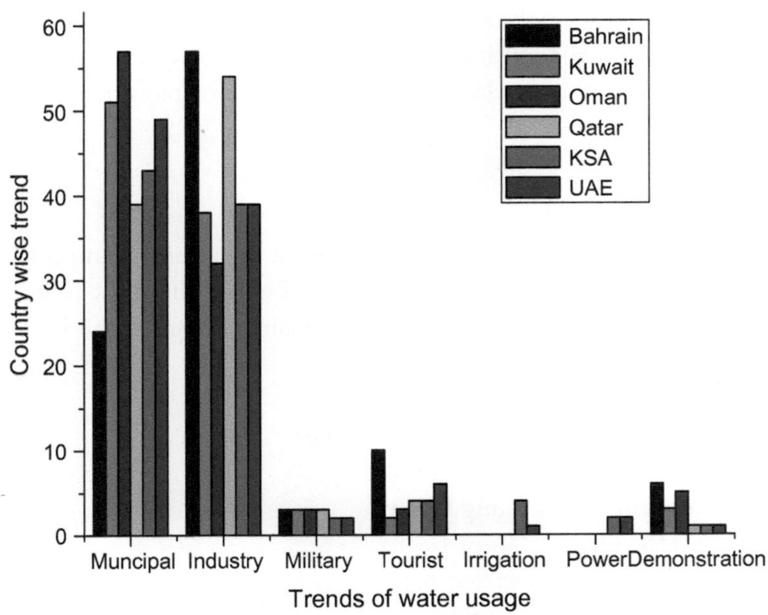

Fig. 6.4 Country wise trends with water usage [49].

The GCC region has higher concentration of desalination plants in comparison with other regions worldwide. Owing to high level of salinity, turbidity, temperature, and marine life in the Gulf water, thermal desalination units (MSF) proved to be of high operational reliability.

About two-third of the plants across the globe are processing seawater and the remaining process brackish water [81–85].

6.8.1 Financial Cost Across the GCC

There is also a concern of government subsidies to the residents of the GCC region including the electricity production and tariffs. It is worth mentioning that the actual costs of production of electricity and water are considerably high as compared with the subsidies offered. For example, in Abu Dhabi, Water & Electricity Company (ADWEC) financially compensates the distributors who supply water to consumers for cost less than the production cost [78]. Such schemes of low prices in water and energy result in higher usage per capita. According to IEA report, Qatar and UAE consume 25% more electricity per capita in comparison with KSA [78] due to the high electricity consumption for the purpose of air conditioning and cooling. Factors such as weather conditioning, virtual water imports, and wastewater treatment all contributed in energy intensity and water footprints of the GCC countries [79]. With the growing demands of nuclear power as promising option for desalination and electricity, many of the GCC countries are interested in adopting this technology. Among them, UAE has already started installing nuclear reactors while Saudi Arabia is planning to start in the near future.

6.8.2 Pros and Cons

For the purpose of well-defined technology, it is extremely important to look at the pros and cons associated with it. Few of the most prominent and easily understood pros [86] related to desalination are as follows:

- effective and proven technology
- scientifically proved technique
- ability to produce and preserve freshwater
- preserving ocean water as source
- not dependent on changing factors
- safely located plants

Similarly, some of the associated cons are as follows

- expensive and noneconomical

- huge energy to process
- environmental contamination
- brine extraction heavily damages environment
- production of contaminated water

6.9 POLICY ISSUES

Water as being the basic necessity of life subsequently becomes one of the most important issues of water shortage directly affecting the quality of life. In this context, seawater desalination is the most effective way to cope with this issue. However, the cost effectiveness makes this technology suitable to multiple uses including conservation of water and recycling purposes. In this criteria, less expensive options in line with policy issue establishment of nuclear desalination plant needs to be developed. Some of the main reasons for adopting this technology are as follows:

- Long-term fuel supply improving energy resources and self-sufficiency.
- Considerable economic value.
- Reduced environmental impact and public health.
- Saving fossil fuels needed as vital raw materials in industries.
- Expedites industrialization for developed countries with less fossil fuel resources.

Since applying nuclear desalination is a major issue likely developed by the government setting up the strategic policies on national and international relations, the governing policies are to be planned and developed by the national consensus including the role of international relations for nuclear energy development. The development of nuclear energy is one of the crucial issues likewise the nuclear desalination option needs to be addressed including technology, energy, water resources, and environmental protection.

The establishment of nuclear desalination program is a major step, which requires a wide range of participation including government involvement. If there is only consent of availability of electricity and water, then a minimum level of national participation may involve. This participation includes highly qualified and experienced participants considering all the design, construction, and operational aspect of plants. The development of nuclear desalination program is set to be an opportunity for national infrastructure, which can stimulate a country's long-term industrial development. In this way, policy makers consider the level of participation combined with the overall nuclear program. Moreover, the level of human resources is a critical factor for the success of nuclear desalination program. Once the policy issue regarding the establishment of nuclear desalination plant is set up, the long term and lead times need to be added for developing highly qualified human resources. Some of the major policy issues that should be considered [87] are as follows:

- The agreement between the mutual defense strategies for regional water supply among participating countries.
- Research and development in terms of regional water production and water reuse.
- Reduction in uncertainties and associated factors for improving the climate for financing and water management on governmental grounds.
- Nuclear desalination site preference as part of the national decision.
- Inducing a role of desalted water production.
- A well trained and dedicated professional staff applicable for safe and reliable operation.
- Establishing a structure for regional information sharing with environmental conditions and technological breakthrough.
- The strategic interaction over water consequently creating a coherent regional movement including the safety and availability of resources.
- Current initiatives need to be reexamined representing only the viable method of addressing required water sources.
- Progressive initiatives for domestic consumption of water through microscale dams and aquifers for the purpose of retaining rainfall.

6.10 FEASIBILITY AND TECHNICAL ASSESSMENT

As described in the previous sections, maximum number of desalinated plants (MSF) are located in the GCC countries. This massive increase in desalinated water production increases the power demand as well. Initially, there was no plan of adopting nuclear reactor technology in the GCC region but more recently, UAE have already planned four nuclear reactors with a total capacity of 5600 MW with first reactor to be operated in May, 2017. Jordan has already equipped with nuclear research reactor while Saudi Arabia will have its first nuclear reactor by 2022. This development in nuclear reactor technology produces a viable solution for utilizing the waste heat for water desalination. The working mechanism is such that the heat can be extracted at various temperature levels as steam, which is then used to run desalination units. Therefore, nuclear reactor can be a dual purpose plant capable of producing power and water at reasonable cost. The arrangements for coupling the NHR need to be on the same side so as to lessen the cost of transporting heat over long distances and to avoid unnecessary heat losses. A study conducted for the applicability of nuclear reactor option in the Arab region concluded that the CANDU-type nuclear reactor can be the most suitable type of reactor for desalination and power generation [88]. CANDU reactor uses heavy water unlike other nuclear reactors, which is simpler compared with enriching uranium. Some other studies have indicated that the CANDU reactor is attractive and appropriate for the GCC region because of the presence of natural uranium in some areas of the GCC countries [89].

Fourth generation high-temperature GCRs (Helium) can also be a good candidate for the GCC region in terms of technical and economic aspects. These reactors contribute to electricity production with other advantages including hydrogen production and seawater desalination. Many countries across the globe are looking forward to adopt these types of reactors. However, there is a need to conduct feasibility studies for technical and economic aspects of manufacturing these types of reactors. These reactors are inherently safe due to the type of fuel used, which cannot be further reprocessed for further use, thus obeying the nonproliferation treaty agreement. The main applicability for adopting these reactors for the GCC region is not using water as a coolant instead helium gas is used as coolant. In addition, the entire GCC region is water-scarce area and helium gas is inert which is transparent to neutron and cannot contribute to radioactive release into the environment. There is still a need to further assess the disadvantages of these reactors in terms of local perspectives (increase in energy demands and population growth) of the GCC region. Let us consider a case of Saudi Arabia, an increase in depletion of oil resources is mainly caused by domestic consumption and industrialization. It is estimated that the decline in oil resources is about 3.4 million barrel per day and it can further decrease to 8.3 billion barrel per day by 2028 [90–92]. The demand for electricity increased from 75 to 120 GWe during the year 2009–2028 and this definitely reduces the oil exportation [93]. This increase in demand of energy pulls the decision maker of the country to prepare and implement a roadmap for other sources including nuclear power. Under these criteria, freshwater production could be easily compensated with this technology. In KSA, the desalinated water production is about 3.3 million m^3/day including the consumption of 2.5 million barrel per day of oil [94]. The ratio of seawater–desalinated capacity is about two-third and the remaining one-third is due to brackish water [93,94].

The sudden increase in population growth resulted into the urbanization and industrialization is at the rate of 7% per year and will become double by 2025 [95]. The applicability and suitability of RO and MSF are already well understood and justified [96]. Like other renewable energy resources, nuclear energy has proved its potential for producing potable water in addition to electricity generation. Nuclear desalination system appears to be very interesting not only for the countries rich in fossil fuel but also equally applicable to agricultural areas [61]. It has been estimated that the ideal production of seawater desalination is in the range from 80–100 to about 200–250 km^3/day [5]. IAEA reported the cost of water desalination in the ranges of 1.18–1.48 USD per cubic meter for MSF processes while comparatively less in the case of MED and RO as 0.6–0.96 and 0.5–0.94 USD per cubic meter, respectively.

6.11 INTEGRATION OF NUCLEAR REACTOR WITH DESALINATION UNITS

Integration of nuclear reactors with desalination units is gaining higher attention across the globe and a great amount of work is already been published. Some of the very recent research studies are outlined below:

- Theoretical and computational evaluation of KANUPP reactor coupled with different desalination units is being evaluated by Salah [97].
- Techno-economic analysis was conducted by Esen and his team on different coupled systems, for example, performance analysis of appraisal ground source heat pump system [11] and coupled system of heat pump for the purpose of space cooling [60].
- Coupling of nuclear reactors with desalination technologies including hybrid system consisting of RO and MSF system were investigated by Ibrahim [59].
- All the concerned nuclear technologies including the coupled systems for seawater desalination are reported in IAEA technical document [94].
- IAEA introduced and launched very impressive and readable contents for nuclear desalination through the publication of technical series report [96].
- EURODESAL project headed by number of renowned scientist and engineers focused on other energy resources for effective use of seawater other than nuclear options [63].
- For the case study, feasibility studies for Kuwait to opt for nuclear desalination were conducted by Darwish and Najem [62].

- Most recently, nuclear desalination concepts and current scenarios were published by world nuclear association [61].
- Senviratne compiled the summary of research projects on nuclear desalination covering the economic indicators [5].
- Feasibility including technical and economic analysis of nuclear reactor coupled with water desalination across MENA region is reported by Khan et al. [7].
- Utilizing waste heat from nuclear reactor has been ideally compiled by Zhang et al. [60].
- The present and the past status of nuclear desalination were discussed by Belessiotis and Delyannis [59].
- A very important aspect is to deal with the improvement while considering the economy of desalination plant by using hybrid technologies [98].
- The effective assessment of nuclear desalination including the cost analysis for thermally coupled HRGRs and MED systems were investigated by Kim and No [99].
- A very important finding of the potable water needs in water-scarce areas were reported by Ibrahim [100].

In view of the above literature survey, it has been shown that the nuclear desalination installation is practical and has become a developed technology. This adopted coupled reactor technology will definitely contribute to meet the freshwater demands in the future in a better way. The applicability of adopting nuclear reactor technology coupled with desalination process is increasing at a higher rate thus producing the possibilities of cutting edge research.

As an example of nuclear power and desalination integration, Fig. 6.5 illustrates the schematic diagram of coupled nuclear power plant with desalination system. It is composed of a nuclear section with a hybrid desalination plant (MED and RO). It is worthy to remind that MSF technology can be easily used instead of the MED system. Other configurations of integrating nuclear power and desalination can be similarly drawn and discussed.

Low-pressure steam is condensed in the condenser and is sent back to feedwater pump for onward transmission to nuclear power plant. At the same time, seawater passes through a heat exchanger and then it exchanges heat and transfer to the MED unit where it passes through different stages and finally comes up with freshwater production. The brine may be collected from one side and the water is pumped to RO unit. There is connected energy recovery unit at RO system. The output is also the water product from the RO unit as shown in Fig. 6.5.

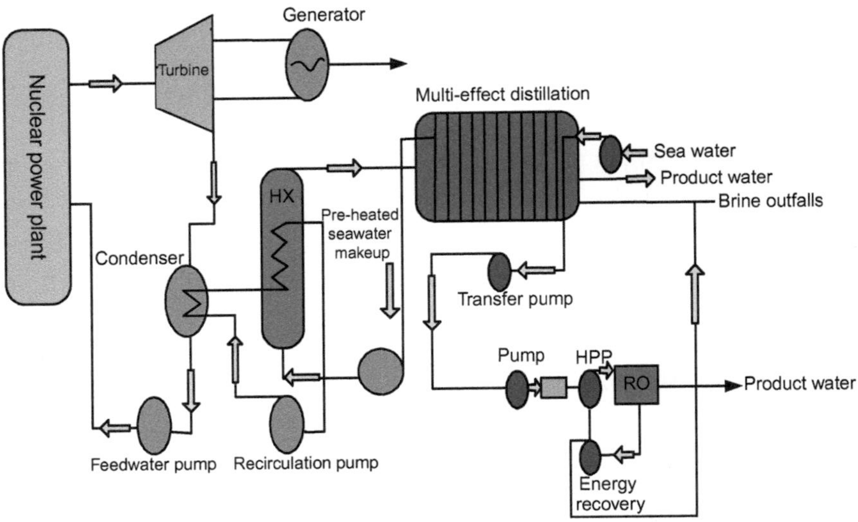

Fig. 6.5 Schematic diagram of a coupled nuclear power plant and desalination system [7].

6.12 FEASIBILITY OF NUCLEAR DESALINATION OPTION

The prime objective to setup and implement nuclear desalination project is to perform planning for nuclear desalination program. The main tasks needed for this planning study are as follows:

- Estimation of demand, need, and economic viability of nuclear desalination including water and energy sources.
- Need of the siting arrangement.
- Overall scheduling of the entire desalination planning program.
- Comparison of water and power supply options.
- Highly qualified professional required.
- Need of water consumption in different sectors.
- Cost involved in freshwater production, availability, and source potential.
- Need to conduct water demand forecast (short, medium, and long term).

For the planning of nuclear desalination systems, some specific measures need to be specified such as

- capacity of plant (single/double)
- heat- or electric power-driven system
- choice of large or small units
- licensing, construction, acquisition, and commissioning
- finance involvement

- infrastructure development and participation
- personnel resources and training programs
- siting considerations, public acceptance, and licensing considerations
- project management schemes
- benefits and constraints involved
- international and bilateral cooperation
- economic ground rules, operational experience, and reliability criteria
- choice of optimum power and water energy need
- technical and economical optimization.

The adoptability of nuclear desalination plant should consider various aspects and factors which include the technical, economical, institutional, and environmental ones. The possibility of the nuclear desalination plant includes the accommodation of electric grid system, addition of new electricity and water production, and long-range development program. This feasible assessment of nuclear desalination plant is primarily intended to provide relevant authorities with all necessary detailed information. The scope of feasibility of nuclear desalination includes the following:

- nuclear reactor type
- desalination technology and integration process
- capacity and main features of integrated unit
- site selection and associated specific problems
- detailed cost and economic evaluations
- financial viability
- organizational and human resource requirements
- overall project scheduled
- contractual approach for plant acquisition
- nuclear reactor safety criteria follow-up
- analyzing an international market for nuclear reactors
- selection of personal with water and energy experience
- foreign consultation and advisor from IAEA.

These aspects provide all the detailed information for establishing well-developed feasible studies. It is worth mentioning that 2–3 years is mandatory for conducting this entire feasible studies excluding the site selection and qualification required. Further, market potential is needed to determine the required desalination capacity, the plant capacity, type of the desalination technology used, operating characteristics including the reliability and operational concerns, and cost factor. However, expansion in the plant size may be considered for the overall cost benefit, water demand forecast, and integration process.

6.13 SUMMARY

Water desalination using nuclear power is growing very rapidly across the globe. The suitability of large reactors (LRs) to small reactors (SRs) for water production and energy has been reported in this chapter. The current status of all nuclear desalination plants planned, operated, or in progress has been summarized with the description of each. Three major desalination techniques (MSF, RO, and MED) are normally being employed for nuclear reactor desalination options worldwide. The integration of these desalination technologies with nuclear reactors has been investigated and included in this chapter, as there is a vital need to investigate the cost and the technical aspects of this integration process. Theoretical and computational techniques have been included for the purpose of calculating the different integration techniques. The chapter also explains useful details and analysis of the GCC nations for the water–energy consumption and production. The national policies and agendas over the water issue which can facilitate the preservation and protection of water resources have been reported. The advantages and disadvantages of the integration have been discussed including the technical assessment based on previous studies. The chapter also gathers literature survey of the most recent research and practicing notes on nuclear desalination integration. It highlights the fact that further studies need to be performed for such types of nuclear and desalination integration.

REFERENCES

[1] WEF & CERA. Thirsty energy: water and energy in the 21st century. Geneva: World Economic Forum and Cambridge Energy Research Associates; 2008.
[2] WHO/UNICEF. Progress on sanitation and drinking-water 2013 update. New York: WHO/UNICEF; 2013.
[3] Khalid F, Dincer I, Rosen MA. Comparative assessment of CANDU 6 and Sodium-cooled Fast Reactors for nuclear desalination. Desalination 2016;379:182–92.
[4] IDA. Desalination yearbook. International Desalination Association; 2012.
[5] Senviratne G. Research projects show nuclear desalination economical. Nuclear News 2007;60–3.
[6] Mansouri NM, Ghoniem AF. Does nuclear desalination makes sense for Saudi Arabia. Desalination 2017;406:37–43.
[7] Khan SU-D, et al. Development and techno-economic analysis of small modular nuclear reactor and desalination system across Middle East and North Africa region. Desalination 2017;406:51–9.
[8] Mekhilefa S, Saidurb R, Safari A. A review on solar energy use in industries. Renew Sust Energ Rev 2011;15:1777–90.
[9] Voutchkov N. Desalination—past, present and future. International Water Association; 2016. http://www.iwa-network.org/desalination-past-present-future/.

[10] J. Koschikowski, Water desalination: when and where will it make sense? 2011 Annual meeting of the American Association for the Advancement of Science, Fraunhofer Institute for Solar Energy Systems (ISE), 2011.
[11] Fiorenza G, Sharma VK, Braccio G. Techno-economic evaluation of a solar powered water desalination plant. Energy Convers Manag 2003;44(14):2217–40.
[12] IAEA. Introduction of Nuclear Desalination. IAEA Technical Report Series No. 400, ISSN 0074-1914, 2000.
[13] Noreddine G, Missimer TM, Amy GL. Technical review and evaluation of the economics of water desalination: current and future challenges for better water supply sustainability. Desalination 2013;309(15):197–207.
[14] Al-Mutaz IS. A comparative study of RO and MSF desalination plants. Desalination 1996;106(1):99–106.
[15] O.K. Buros, The ABCs of desalting, 2nd ed.. ASIN: B0006S2DHY, International Desalination Association, vol. 30, 2000.
[16] Water Desalination Using Renewable Energy—Technology Brief IEA-ETSAP and IRENA Technology Brief, March 12, 2012.
[17] Al-Shammiri M, Safar M. Multiple-effect distillation plants: state of the art. Desalination 1999;126(1–3):45–59.
[18] Ophir A, Lokiee F. Advanced MED process for most economical sea water desalination. Desalination 2005;182(1–3):187–98.
[19] S. Lattemann and T.Höpner, Desalination of seawater. Wissenschaftliche Auswertungen, Hamburg. Online: www.klima-warnsignale.uni-hamburg.de.
[20] Buros OK. The desalting ABC. Topsfield, MA, USA: International Desalination Association; 1990.
[21] Frenkel VS. Seawater desalination: trends and technologies. In: Schorr M, editor. desalination, trends and technologies. 2011, ISBN: 978-953-307-311-8 http://www.intechopen.com/books/desalination-trends-and-technologies/seawater-desalination-trends-andtechnologies.
[22] Goebel O. Co-generation of power and water-selection of desalination process. Germany: Lahmeyer International GmbH; 2003.
[23] Buros OK. The ABCs of desalting. International Desalination Association; 2000.
[24] Caribbean Environmental Health Institute (CEHI). The evaluation of the use of desalination plants in the Caribbean. United Nations Educational, Scientific and Cultural Organization (UNESCO); 2006.
[25] A.A. Karaghouli and L. Kazmerski, Comparisons of technical and economic performance of the main desalination process with and without renewable energy coupling, 2012.
[26] K. Perrakis, E. Tzen, and P. Baltas. PV-RO desalination plant in Morocco, JOR3-CT95-0066, CRES, Greece, 2001.
[27] Kuroda O, et al. An electrodialysis seawater desalination system powered by photovoltaic cells. Desalination 1987;67:161–9.
[28] M. Lichtwardt and H. Remmers. Water treatment using solar powered electrodialysis reversal. In: Proceedings of the Mediterranean conference on renewable energy sources for water production; 1996.
[29] Mohamed ES, Papadakis G. Design, simulation and economic analysis of a stand-alone reverse osmosis desalination unit powered by wind turbines and photovoltaics. Desalination 2004;164:87–97.
[30] Kershman SA, Rheinlander J, Neumann T, Goebel O. Hybrid wind/PV and conventional power for desalination in Libya—Gecol's facility for medium and small scale research at Ras Ejder. Desalination 2005;183:1–12.
[31] A. Maurel. Desalination by RO using RE (solar & wind): Cadarache Center Experience. Proceedings of the New Technologies for the Use of RE Sources in Water Desalination, Greece, 26–28 September 1991.

[32] Miranda M, Infield D. A wind powered seawater RO system without batteries. Desalination 2002;153:9–16.
[33] Pestana N, et al. Optimization of RO desalination systems powered by RE: part I. Wind energy. Desalination 2004;160:293–9.
[34] Carta JA, Gonzalez J, Subiela V. The SDAWES project: an ambitious R&D prototype for wind—powered desalination. Desalination 2014;161:33–48.
[35] Barbier E. Geothermal energy technology and current status: an overview. Renew Sustain Energy Rev 2002;6:3–65.
[36] Barbier E. Nature and technology of geothermal energy. Renew Sust Energ Rev 1997;1:1–69.
[37] Awerbuch L, Lindemuth TE, May SC, Rogers AN. Geothermal energy recovery process. Desalination 1976;19:325–36.
[38] Semiat R. Energy issues in desalination processes. Environ Sci Technol 2008;42(22).
[39] Wu S, Dong D. Seawater desalination plant using nuclear heating reactor coupled with MED process. Nucl Sci Tech 2000;11(1):6–12.
[40] Juslin K. A companion model approach to modelling and simulation of industrial processes. VTT Publications 574; 2005.
[41] Misra BM, Kupitz J. The role of nuclear desalination in meeting the potable water needs in water scarce areas in the next decades. Desalination 2004;166:1–9.
[42] Baranaev YD, Dolgov VV, Sergeev YA. Activities in the field of small nuclear power reactors. Nucl Eng Des 1997;173:159–66.
[43] V.I. Kostin et al, Small floating nuclear power plants with ABV reactors for electric power generation, heat production and sea water desalination, pub No.1NSAC-2004, 2004.
[44] Gowin JP, Konishi T. Nuclear seawater desalination—IAEA activities and economic evaluation for southern Europe. Desalination 1999;126:301–7.
[45] A.G. Grechko, A.A. Romenkov, V.A. Shishkin, Technical and economic evaluation of nuclear sea water desalination system. Pub No. XA9848809, 1998.
[46] Weimin M, Karbojian A, Sehgal BR. Experimental study on natural circulation and its stability in a heavy liquid metal loop. Nucl Eng Des 2007;237:1838–47.
[47] Cammi A, et al. Generation III + nuclear reactor for space needs. Prog Nucl Energy 2009;51:347–54.
[48] Baig H, Antar MA, Zubair SM. Performance evaluation of a once-through multi-stage flash distillation system: impact of brine heater fouling. Energy Convers Manag 2011; 52(2):1414–25.
[49] Kavvadias KC, Khamis I. The IAEA DEEP desalination economic model: a critical review. Desalination 2010;257:150–7.
[50] Cervera IGS, Kavvadias KC, Khamis I. DE-TOP: a new IAEA tool for the thermodynamic evaluation of nuclear desalination. Desalination 2013;321:103–9.
[51] IAEA. Optimization of the coupling of nuclear reactors and desalination systems. Vienna: IAEA; 2005.
[52] Karameldin A, Shamlou MM, Shaalan MR, Esawy MH. Dynamic of reverse osmosis in a standalone co-generative nuclear reactor, Part I: reactivity changes. Int J Nucl Desal 2007;2(3):219–33.
[53] M.S. Naighi, W.G. Tesfai, H.E. Fath, Economic feasibility of nuclear desalination for Abu Dhabi, Proceedings of the ASME 2011 international mechanical engineering congress & exposition, Denver, Colorado, USA, 2011.
[54] Aljohani MS. Nuclear desalination competitiveness in the western region of the Kingdom of Saudi Arabia. Desalination 2014;164:213–23.
[55] Fried JJ, Edlund MC. Desalting technology for Middle Eastern agriculture, an economic case. New York, NY: Pracger Publishers; 1971.
[56] Abdul-Fattah AF, Husseiny AA, Sabri ZA. Nuclear desalination for Saudi Arabia: an appraisal. Desalination 1978;25:163–85.
[57] Yung YH, Choi J, Wibisono AF, Lee JI, No HC. Feasibility study of a small-sized nuclear heat-only plant dedicated to desalination in the UAE. Desalination 2014;337:83–97.
[58] Al-Mutaz IS. Potential of nuclear desalination in the Arabian Gulf countries. Desalination 2001;34:187–94.

[59] Wu S, Zhang Z. An approach to improve the economy of desalination plants with a nuclear heating reactor by coupling with hybrid technologies. Desalination 2003;155:179–85.
[60] Belessiotis V, Papanicolaou E, Delyannis E. Nuclear desalination: a review on past and present. Desalin Water Treat 2010;20:45–50.
[61] Nuclear Desalination, World Nuclear Association, April, 2016.
[62] Darwish MA, Al-Najem NM. Nuclear desalination in Kuwait: a feasibility study. In: Globalization, Energy and Environment. Warsaw School of Economics; 2008.
[63] Nisan S, Caruso G, Humphries JR, Mini G, Naviglio A, Bielak B, et al. Sea-water desalination with nuclear and other energy sources: the EURODESAL project. Nucl Eng Des 2003;221:251–75.
[64] Esena H, Inallib M, Esena M. A techno-economic comparison of ground-coupled and air-coupled heat pump system for space cooling. Build Environ 2007;42(5):1955–65.
[65] Khan SUD, Danish SN, Haider S, Khan SUD. Theoretical calculation simulation studies of ABV nuclear reactor coupled with desalination system. Int J Energy Res 2015;39:1554–63.
[66] IAEA. Introduction to nuclear desalination: a guidebook. Technical Report Series Number 400Vienna: International Atomic Energy Agency; 2000.
[67] Subki MH. Advantage of nuclear plants in design and technology developments for near-term deployments and associated common issues. In: Proc. 69th General conference 2015 on small modular reactors: an option for safe nuclear power technology for near term deployments, 16 September 2015; 2015.
[68] A1-Mutaz IS. Coupling of a nuclear reactor to hybrid RO-MSF desalination plants. Desalination 2005;157:259–68.
[69] Salah U.D. Khan, Shahab U.D. Khan, Kingdom of Saudi Arabia's nuclear program: a future prospect, 22nd international conference nuclear energy for New Europe, Bled, Slovenia, September 9–12, 2013.
[70] Shropshire D. Economic viability of small to medium-sized reactors deployed in future European energy markets. Prog Nucl Energy 2011;53:299–307.
[71] Nisan S, Rouyer JL, Marcetteau P, Dufloc D. SEMER: a simple code for the economic evaluation of nuclear and fossil energy-based power production systems. Nucl Eng Des 2003;221:301–13.
[72] A.E. Safty, Implementation of sound policies to secure the future supply of water in the Arabian Gulf Region, Environment. UAE, 2016.
[73] Danyel R. Energy Policies of Gulf Cooperation Council (GCC) Countries—possibilities and limitations of ecological modernization in Rentier States. Energ Policy 2010;38(5):2395–403.
[74] B. Imen, Renewable energy in the GCC Countries. Friedrich-Ebert-Stiftung. Gulf Research Center, March 2012. Web. 13 September 2012.
[75] DesalData [www Document], 2013. URL http://desaldata.com/ (accessed 06.14.13).
[76] ESCWA [www Document], 2013. URL http://www.escwa.un.org/ (accessed 06.14.13).
[77] A.Z. Waleed, Use of simulation modeling in the prediction of the dammam aquifer system behavior in the Kingdom of Bahrain, 10th Gulf water conference, Grand Hyatt Doha Hotel, Doha, 2012. Lecture.
[78] A.M. Ali, Business opportunities in water industry in Qatar, Singapore International Water Week. KAHRAMAA, 2008. Web. 12 September 2012.
[79] El-Ghonemy AMK. Future sustainable water desalination technologies for the Saudi Arabia: a review. Renew Sust Energ Rev 2012;16:6566–97.
[80] J. Grooossen, Nuclear process heat desalination, GCEP-Fission energy workshop, November 29-30, 2007.
[81] IAEA Tec.doc No.1753, New Technologies for seawater desalination using Nuclear Energy Government of India, Dept. of atomic Energy, Desalination & Water Purification Technologies, Technical Information Document, 2010.
[82] Regulation & Supervision Bureau (RSB) (2014b). Overview. Retrieved on May 19, 2014 from http://rsb.gov.ae/en/sector/overview.

[83] IEA: World energy statistics. IEA World Energy Statistics and Balances (database), 2013, doi: https://doi.org/10.1787/data-00510-en.
[84] Enerdata: global energy statistical yearbook 2012: energy intensity of GDP at constant purchasing power parities. Data from http://yearbook.enerdata.net/energy-intensity-GDP-by-region.html.
[85] I. Berman, P.M. Wihbey, The new water politics of the Middle East, Strategic Review, 1999. Web link: http://www.ilanberman.com/6513/the-new-water-politics-of-the-middle-east.
[86] Hussies FM. Nucl Technol 1998;80:392.
[87] W.H. Abulfaraj et al., Technical features and economic comparison of major nuclear power reactor system, The Saudi symposium on energy utilization and conservation, King Abdul Aziz University, Jeddah, March 4–7, 1990.
[88] Yahya SA, Irfan MA. The techno-economic potential of Saudi Arabia's solar industry. Renew Sust Energ Rev 2016;55:697–702.
[89] IBP, Inc. Saudi Arabia energy policy, laws and regulations handbook Vol.1: strategic information and basic laws. USA: International Business Publications Inc.; 2015.
[90] A.A. Alshaikh, International conference on desalination and sustainability, Casablanca, 1–2 March 2012. Retrieved 23 February 2013, 2012.
[91] The report: Saudi Arabia 2014, Oxford Business Group.
[92] IAEA 2014, Advances in small modular reactor technology development, A supplement to: IAEA Advanced Reactors Information System (ARIS), http://aris.iaea.org.
[93] Report on the small secure transportable autonomous reactor (SSTAR)/lead-cooled fast reactor (LFR) and Supporting Research and Development, Nuclear Engineering Division, ANL-GenIV-089.
[94] IAEA.TecDuc.1527, Status of Nuclear desalination in IAEA member states, IAEA publication, 2008.
[95] Khan SUD, Khan SUD. Karachi Nuclear Power Plant (KANUPP): as case study for techno-economic assessment of nuclear power coupled with water desalination. Energy 2017;127:372–80.
[96] IAEA, Tecdoc No.1753 New Technologies for Seawater Desalination Using Nuclear Energy, 2016.
[97] Dong D, Wu S, Zhang D. Comprehensive utilization of nuclear heating reactor. J Tsinghua Univ (Sci Technol) 1994;34:32–8.
[98] Kim HS, No HC. Thermal coupling of HTGRs and MED desalination plants, and its performance and cost analysis for nuclear desalination. Desalination 2012;303(1):17–22.
[99] Mutaz ISA. Hybrid RO MSF: a practical option for nuclear desalination. Int J Nuclear Desal 2003;1(1).
[100] Esen M, Yuksel T. Experimental evaluation of using various renewable energy sources for heating a greenhouse. Energy Build 2013;65:340–51.

FURTHER READING

[1] IAEA.TecDoc.1444. Optimization of the coupling of nuclear reactors and desalination system. IAEA Publication; 2005.
[2] A water sector assessment report on the countries of the cooperation council of the Arab States of the Gulf. World Bank. World Bank, March 2005, 8 September 2012.

CHAPTER 7

Relevance of Nuclear Desalination as an Alternative to Water Transfer Geoengineering Projects: Example of China

Anne-Perrine Avrin*, Gang He†, Daniel M. Kammen*
*University of California, Berkeley, CA, United States
†Stony Brook University, Stony Brook, NY, United States

Nomenclature

ACPR1000	Advanced China Power Reactor, Generation III Chinese nuclear reactor design of 1000 MW electric
AP1000	Advanced Passive, Generation III Westinghouse Electric Company nuclear reactor design of 1000 MW electric
ARIS	Advanced Reactors Information System
CAP1400	Generation III Chinese nuclear reactor design of 1400 MW electric
CNP-600	China Nuclear Power, Generation II Chinese nuclear reactor design of 600 MW electric
CPR1000	Generation II Chinese nuclear reactor design of 1000 MW electric
DEEP	Desalination Economic Evaluation Program, code developed by the IAEA
EPR	European Pressurized Reactor, nuclear reactor design of 1700 MW electric
HTR-PM	High-Temperature gas-cooled Reactor-Pebble-bed Module, Generation IV Chinese nuclear reactor of 250 MW thermal
IAEA	International Atomic Energy Agency
MED	Multieffect Distillation
MSF	Multistage Flashing
NHR-200	Nuclear Heating Reactor, Generation IV Chinese nuclear reactor of 200 MW thermal
PRIS	Power Reactor Information System
RO	Reverse Osmosis
SNWTP	South to North Water Transfer Project
VVER-1000	Voda Voda Energo Reactor, Generation III Russian nuclear reactor design of 1000 MW electric

This chapter is an update of Anne-Perrine Avrin, Gang He, Daniel M. Kammen, Assessing the impacts of nuclear desalination and geoengineering to address China's water shortages. Desalination 360 (2015) 1–7, reprinted with permission from Elsevier.

Renewable Energy Powered Desalination Handbook
https://doi.org/10.1016/B978-0-12-815244-7.00007-6

7.1 INTRODUCTION

In China, several provinces, especially in the center and the northeastern parts of the country, experience moderate to severe water shortages, affecting municipal, industrial, and agriculture needs. While renewable freshwater resources per capita in China is about a third of the global average [1], the average residential price for cubic meter of water is only between 13% and 66% [2] of the average global price [3], depending on the quantity and location. The total national shortage in 2030 is forecasted to be nearly 200 billion m^3 including 16% for the domestic and municipal needs and 32% for industrial needs [4]. Water-use efficiency, recycling, and conservation programs have been implemented to save water, but they have not addressed the shortage issue in its totality.

Cross-region water diversion projects have been undertaken to address lack of water. In particular, the South-to-North Water Transfer Project (SNWTP), presented in Fig. 7.1, aims at diverting about 24 billion m^3 of water annually from Yangtze River—the Eastern route—and Danjiangkou reservoir—the Central route—to water-scarce Eastern provinces [9], and will require 530 MW of energy generation for pumping capacity. This project, whose costs have already reached $81 billion for the Eastern and Central routes, will not increase the total quantity of water available to the country.

Costs, scale, and impacts of the SNWTP warrant a framework to alternative approaches. One such alternative is the use of seawater to address freshwater needs through desalination. Desalination can play a major role for coastal cities, communities where access costs of other freshwater sources are high, or where the average value of prime energy is relatively low, for example, in areas with high levels of electricity curtailment. China already uses desalination, either using heat from coal-fired plants or electricity from a power grid dominated by coal. Although its desalination capacity has been expanding in the last decade, the country had 103 desalination facilities producing, in total, 900,830 m^3/day at the end of 2013—not even reaching the capacity of the Ras Al Khair desalination plant in Saudi Arabia [10]. At the same time, China is massively deploying nuclear power reactors [5]. While the country's electricity mix in 2030 is expected to remain dominated by fossil fuels [11], making electricity-based desalination technologies such as reverse osmosis (RO) polluting processes, hot steam from nuclear reactors can be extracted to enable low-carbon desalination. The purpose of this study is to determine if and under what assumptions, in the long term, nuclear thermal desalination can prove itself to be appropriate to China's

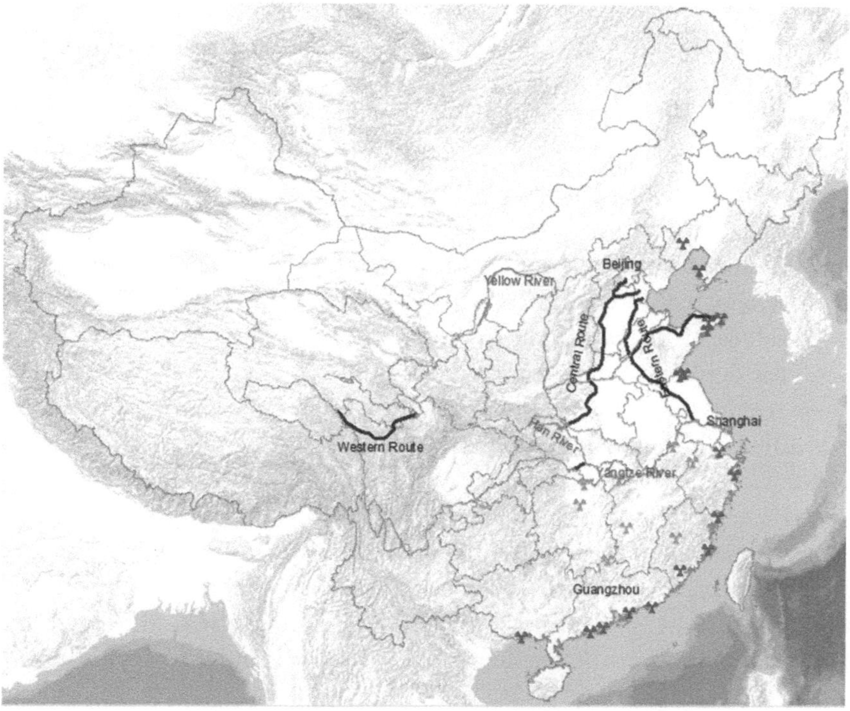

Fig. 7.1 Chinese nuclear fleet by 2030 [5,6] and Eastern, Central, and Western routes of the South-to-North Water Transfer Project (SNWTP) [7]. The three routes are drawn in dark blue. Red crosses represent nuclear plants hosting reactors appropriate to nuclear desalination in 2030, while yellow crosses indicate nuclear plants that, for geographic or age-related reasons, are considered not appropriate to desalination. *Source: [8].*

water needs and to compare it with water diversion projects and coal desalination on economic and environmental levels. The choice of a 2030 horizon results from a trade-off between the need of a period of time long enough to expect substantial evolution in the Chinese nuclear fleet, and the uncertainty of long-term planning.

7.1.1 Past Research

The literature addressing water scarcity in China is extensive but not comprehensive; several studies have been conducted on water desalination in China but with no detail on nuclear desalination [12–15], on nuclear desalination but not from a Chinese perspective [16–21], and on novel reactor technologies to perform desalination in China but not on how to develop a desalination industry to address shortage in its totality [22–26]. Zhang et al. [12] and Zhou and Tol [13] have described the technical and economic characteristics of different desalination processes and their implementation in China. However, the relationship between thermal water desalination and electricity production in China has not been evaluated. The present study focuses on one type of desalination process—steam extraction from nuclear reactors. Projections for China's water shortage and nuclear fleet in 2030 allow for a meaningful evaluation of the maximal production capacity of desalinated water, the distance between future water-scarce provinces and desalination plants, and the resulting costs and affordability of desalinated water for domestic and industrial needs in the 2030 Chinese context. This paper provides a specific framework to assess nuclear desalination impact: the affordability of freshwater supply in water-scarce provinces through the use of the 5% rule, as defined later in this chapter. Finally, previous studies mentioned other solutions to water shortage in China which could compete with nuclear desalination, particularly SNWTP, but no research was executed prior to the present study on the development of an interdisciplinary framework to evaluate and quantify nuclear desalination assets and drawbacks compared with its two main alternatives: SNWTP and coal desalination.

7.1.2 Data Sources and Methods

In this study, the technical configuration, performance, and economic assessments of building a desalination plant and coupling it to a nuclear reactor have been performed with the Desalination Economic Evaluation Program (DEEP) model developed by the International Atomic Energy Agency (IAEA) [27–29]. DEEP's calculation methods and inputs have been modified and adapted to the characteristics of the present study. The project's data come from the World Nuclear Association, DEEP, IAEA's, Power Reactor Information System (PRIS) [30], and Advanced Reactors Information System (ARIS) [31] for the nuclear and desalination technical and economic sections; from Zhou and Tol [32], Gleick [17], IAEA [33], and DEEP for the water transportation section; from the World Nuclear Association, Nicobar Group, University of South China (USC), and Google Maps for the nuclear facility location section; from China Water Risk [34] regarding China's current and future water situation; and from the Chinese Bureau of South-to-North Water Transfer Planning and Designing for the SNWTP. The costs and prices are given in 2017 US dollars, and the conversion rate between Chinese RMB and US dollars is 0.16.

7.2 METHODS

7.2.1 Selection of the Appropriate Desalination Technology

Desalination technologies differ on the energy consumption intensity, the form of energy required, the flexibility in the choice of the plant's location, and the quality of water produced. The existing technologies are divided into two broad categories: membrane separation and thermal processes [35]. The three predominant commercial seawater desalination processes, as identified by the IAEA, which are proven and reliable for large-scale production of water, are multistage flashing (MSF) and multieffect distillation (MED) for distillation processes; reverse osmosis (RO) for membrane processes [36].

Nuclear desalination relies on on-site coupling of a nuclear reactor with a desalination plant in order to produce freshwater through a thermal process. The coupling point is the cogeneration steam turbine, which used to produce electrical energy and bleeding steam for process heat. The electrical energy required by the desalination plant can also be supplied by the nuclear reactor [37]. As RO uses electricity from the grid, it is only as clean as the electricity grid it is powered from. MSF and MED can be coupled to a nuclear reactor to perform water desalination through steam extracted from the nuclear facility, therefore is free of air pollutants during normal operation.

To identify the most attractive technology among MSF and MED[1] for nuclear desalination purposes, a 500,000-m^3/day water plant powered by nuclear energy was modeled for the two thermal process cases using the DEEP model. The energy consumption has been evaluated in Table 7.1. The model shows that to process 1 m^3 of water, MED requires 28% less heat

Table 7.1 Energy consumption to desalinate one cubic meter of water depending on the thermal technology used

		Multieffect distillation (MED)	Multistage flashing (MSF)
Heat from nuclear to water plant	kWh/m^3	42.65	59.28
Electricity consumption for desalination	kWh/m^3	2.02	3.19
Loss of electricity due to steam extraction	kWh/m^3	4.08	10.9
Gain output ratio (GOR)	/	15	12

[1] A nomenclature is presented at the end of the document.

and 37% less electricity than MSF. The nominal gain output ratio (GOR), common measure of the efficiency of a thermal desalination plant representing the quantity of pure product produced per unit of driving steam, is slightly higher in the case of MED than MSF [35].

The advantage of MED is its low energy consumption compared with other thermal processes, because the vapor produced in each stage is used to heat up feed water in the next effect. Compared with MSF, this not only reduces the energy required for distillation but also the overall electric power consumption [35], as well as investment costs [38]. MED is more cost efficient and energy efficient than MSF, thus it is the technology considered in this study.

7.2.2 Technical Characteristics and Production Capacity of a Nuclear Desalination Fleet in China in 2030

In this study, we assume that water desalination in 2030 will be powered by China's nuclear reactors that are less than 30 years of age in that year. We use data for operating, under construction, and planned reactors from the IAEA, PRIS database [30], and from the World Nuclear Association [5]. We did not include reactor sites where construction has been considered but not confirmed. The Chinese nuclear fleet in 2017 consists of 36 reactors in operation on 11 different sites, for a total electricity net generation of 32.4 GWe. By 2030, the total nuclear net capacity could reach 91.5 GWe with 94 reactors—excluding nonPWR reactors and reactors smaller than 100 GWe (Fig. 7.2). The 58 additional reactors to be built by 2030 will be spread over 21 sites.

Inland and island nuclear plants, as well as plants whose decommissioning is scheduled before 2045, have not been considered in this study as calculations show that their use for desalination is not economically attractive. According to these criteria, 19 nuclear power plants[2] have been identified as appropriate to nuclear desalination by 2030 hosting 69 reactors in total, with a net total capacity of 71.9 GWe.

This study evaluates the desalination capacity as well as the construction and operation costs of 69 desalination plants. It is assumed that each desalination plant will be built as close as possible to a nuclear reactor and coupled with it. A desalination plant can be built either during or after the

[2] Bailong, Fangchenggang, Fangjiashan, Fuqing, Haixing, Haiyang, Hongyanhe, Huizhou, Ling Ao, Lufeng, Ningde, Putian, Qinshan, Sanmen, Shidaowan, Taishan, Tianwan, Xudabao, Yangjiang, and Zhangzhou.

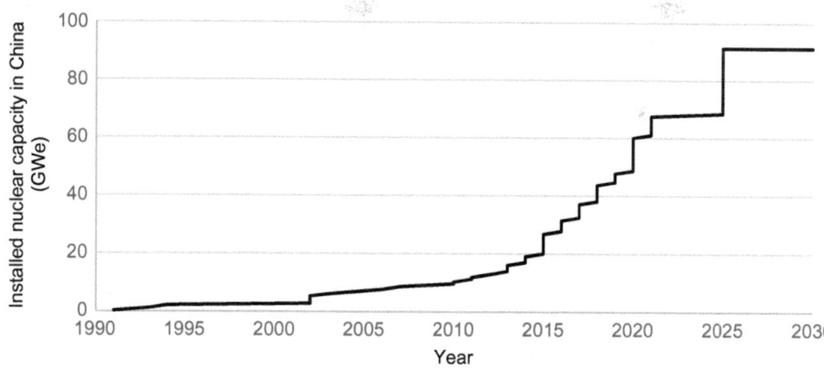

Fig. 7.2 Nuclear power capacity expansion in China until 2030 (Gigawatt-electric). For nuclear reactors whose starting year is not known yet but that are planned to be built before 2030, first grid connection is assumed to be 2025. This assumption does not impact final results. *Source: authors' calculations using data from [5].*

construction of a nuclear plant [39]. All desalination plants considered in this study are assumed to be built by 2030.

The main parameters that influence maximum desalination production capacity are the power capacity and thermal efficiency of nuclear reactors. They depend on the nuclear technology considered. The 69 reactors identified gather eight different nuclear technologies.

To determine the size of the desalination plants to be powered by the 69 nuclear reactors, available heat, depending on the nuclear reactor technology, is calculated. The theoretical quantity of water that can be produced by the maximum quantity of heat available is derated by 15%, rounded down to meet standard capacities of water plants. This margin is in line with the average reserve margin advocated by institutions' standards for power generation systems [40], as the current primary purpose of nuclear reactors planned by the government is power supply. To provide a meaningful framework for comparison purposes, we assume in our calculations that all desalination plants are built in 2030. Conversion between units of same nature (hour/day/year; kWh/MWh; etc.) are not mentioned in formulas to ease reading. Calculations of nuclear desalination economics are based on the DEEP model [41].

The nominal water production capacity of each reactor technology per unit of time is calculated using

$$W_y = \frac{P_{\text{th}} \times (1-\eta)}{L} \times GOR \times (1-M) \times CF_{nuclear} \times CF_{desalination}$$

where P_{th} is the thermal power specific to the nuclear reactor technology considered, η the thermal efficiency of the nuclear reactor technology, GOR the gain output ratio (*15 for the MED technology in this study* [35]), and L (*2334 kJ/kg* [41]) is the water latent heat at 70°C. M (*15%* [40]) is the percentage of maximum heat available that is used as a "reserve" for electricity purposes, therefore not used for desalination purposes.

The number of reactors and maximum total desalinated water production, classified by reactor technology, is provided in Table 7.2. A 9% water desalination plant's forced outage rate is additionally considered, assuming that scheduled outage for maintenance will be performed along with nuclear reactor's scheduled outages.

Water production capacity from the 69 MED desalination plants, to be built near the 19 nuclear sites, adds up 20.7 billion m^3 per year.

The change in temperature of a nuclear reactor's secondary loop induced by the desalination process—that is, the extraction of energy from steam—leads to a reduction in the reactor's power output. The resulting loss in electricity generation per unit of time is calculated as follows:

$$E_{reduc-y} = \frac{W_y \times L}{GOR} \times \left(1 - \frac{T_c}{T_{cm}}\right) \times \eta_{is} \times \eta_{turbine} \times \eta_{generator}$$

where T_c (*40°C* [41]) is the average condensing temperature, or lowest available temperature point and T_{cm} (*75°C* [41]) is the modified condensing temperature of the dual purpose plant. η_{is} (*85%* [41]) is the isentropic efficiency of the low pressure turbine, $\eta_{turbine}$ (*98.8%* [41]) the mechanical efficiency of the turbine, and $\eta_{generator}$ (*97%* [41]) is the generator efficiency.

Table 7.2 Composition, including maximum production capacity, of the fleet of nuclear reactors suitable for desalination purposes by 2030 [31]

Reactor technology	Number of reactors in 2030	Maximum annual water production (Billion m3/y)
ACPR1000	6	2.0
AP1000	18	5.7
CAP1400	2	0.7
CNP-600	2	0.4
CPR-1000	22	6.2
EPR	4	1.6
Hualong 1	13	3.5
VVER-1000	2	0.6
Total capacity	**69**	**20.7**

Over the entire nuclear fleet, the reduction in power production induced by desalination is 8.2 GW, assuming round-the-clock operating mode for both nuclear reactors and desalination plants.

7.2.3 Economics of Nuclear Desalination in China in 2030

7.2.3.1 Capital and Production Costs of Desalination Plants

The capital cost of a desalination plant is calculated using the following formula:

$$I_{desal} = SC_{base} \times W_y \times \left(F_{ow} + F_{contin} + F_{adj}\right) \times (1 + I)^{LT}$$

where SC_{base} is the addition of in and outfall specific base cost (*\$77/m³/day* [41]) and base unit cost (*\$900/m³/day* [23,28,42]). F_{ow} (*1* [41]), F_{contin} (*0.05* [41]), and F_{adj} (*0.105* [41]) are, respectively, the water plant's adjusted base cost factor, owners cost factor, and contingency cost factor. LT (*12 months* [41]) is the lead time of the desalination plants. The choice of the discount rate depends on the perspective of the study. A prescriptive approach, also known as social perspective which is what the analysis aims at, usually uses lower discount rate than a descriptive or "decision-maker's" perspective [43]. Here, discount rate I is *4%* [44].

The total capital cost of the 69 desalination plants is \$66 billion, which corresponds to a net fixed charge per year, FP_y of \$4.3 billion over the 25-year lifetime [39] of the water plants.

Total O&M costs, including human resources and equipment, represent \$7.7 billion per year:

$$OM_y = (S_{lab} \times P_{lab} + S_{man} \times P_{man}) \times W_y + \left(SC_{OM-spare} + SC_{replace} + SC_{OM-pre} + SC_{OM-post}\right) \times W_y$$

where S_{lab} (*\$23190/year* [45]) and S_{man} (*\$43481/year* [45]) are, respectively, the annual salary of production engineers and product manager. P_{lab} and P_{man} are the labor personnel and management personnel, respectively, (authors calculation based on data from the Carlsbad desalination facility [46] in the United States and the Victorian desalination plant [47] in Australia, and assuming that the share of labor and management personnel is 80% and 20%, respectively). $SC_{OM\text{-}spare}$ (*\$0.03/m³* [41]), $SC_{replace}$ (*\$0.01/m³* [41]), $SC_{OM\text{-}pre}$ (*\$0.03/m³* [41]), and $SC_{OM\text{-}post}$ (*\$0.02/m³* [41]) are, respectively, the specific O&M spare parts cost, tubing replacement cost, specific O&M chemicals cost for pretreatment, and specific O&M chemicals cost for posttreatment.

The resulting cost per unit of time of the construction and operation of a desalination plan over its 25-year lifetime is the sum of its fixed charges and variable charges:

$$C_y = FC_y + OMC_y$$

For the 69 desalination plants, C_y adds up to \$12 billion/year.

An additional cost, resulting from the reduction in electricity sales from the nuclear plants due to lost shaft work from steam extraction and water desalination plant electricity use, must be added to the total.

Total electricity cost, adding up to \$0.4 billion per year in this study, is calculated by

$$EC_y = (E_{sale} - E_{transm}) \times \left(W_y \times SE_{use} + E_{reduc-y}\right)$$

where E_{sale} is the average electricity sale price in China (*\$0.085*/kWh [48–50]), E_{transm} is the average electricity transmission cost (*\$0.010*/kWh [51]), and SE_{use} (*2.02 kWh*/m^3 [41]) is the specific electrical power use of the MED technology considered in this study.

The following formula is finally used to calculate the production cost of 1 m^{3}of desalinated water:

$$SC_{desal} = \frac{C_y + EC_y}{W_y}$$

The resulting specific water production cost from nuclear desalination is \$0.72/m^3. This cost is divided as follows: 29% (\$0.21) of fixed charge, 14% (\$0.10) of O&M costs, and 58% (\$0.41) due to loss of electricity generation resulting from steam extraction.

7.2.3.2 Economics of Water Transportation From Desalination Plants to Demand Centers

To the production cost, a cost for water supply must be added. The cost of transporting water from production plants to demand centers depends on the quantitative and geographical distribution of water needs, more specifically on three parameters: transport length, transport height, and water flow per unit of time. Applying the law of conservation of energy, and making the assumption that velocity stays constant as water flows along the pipe, the pumping energy to transport water on a given path is computable.

The average electric power cost per unit of time of transporting W_y of water in a pipeline is

$$EC_{transp-y} = \left(E_{flat} \times l + E_{lift} \times h\right) \times W_y \times E_{sale} \times \left(1 + \frac{L}{2}\right)$$

where E_{flat} ($0.018\ kJ/\mathrm{m}^3/m$ [52]) is the energy required by a pump to transport 1 m^3 of water over 1 m on a flat path, E_{lift} ($13.986\ kJ\ \mathrm{m}^3/m$ [52]) the energy required by a pump to lift 1 m^3 of water 1 m. l is the pipeline length and h is the total elevation of the pipeline, varying with the path considered (starting point: desalination plant; ending point: demand center). L (30%) is the percentage of water lost through leakage.

The corresponding fixed charge cost per unit of time is calculated as

$$FP_{transp-y} = l \times SC_p \times D \times (1+I)^{LT_p} \times \frac{I \times (1+I)^{L_p}}{(1+I)^{L_p} - 1}$$

$$\text{with } D = 2 \times \sqrt{\frac{W_y}{3600 \times \pi \times \mathrm{FS}}}$$

where SC_p (*\$160/m* [33]) is the cost of building 1 m of a 1-m diameter pipeline. D is the economically optimized diameter of the pipeline. FS (1 m/s) is the average water speed. LT_p (*48 months*) is the construction lead time of the pipeline and L_p (*25 years*) is the lifetime of the pipeline.

Operation and maintenance cost per unit of time is given by

$$OM_{transp-y} = FP_{transp-y} \times F_{OMp}$$

where F_{OMp} (0.07 [41]) is the operation and maintenance cost factor for pipelines.

Water transport specific cost (\$/m^3) is given by

$$SC_{transp} = \frac{FY_{transp-y} + OM_{transp-y} + EC_{\mathrm{transp}-y}}{W_y}$$

For fixed l and h, that is, for given desalination plant and demand center, SC_{transp} varies inversely as the square root of W_y. As a consequence, the larger the desalination capacity of a plant, the lower the specific water transportation cost.

7.3 RESULTS AND DISCUSSION

7.3.1 Relevance of Nuclear Desalination in 2030

7.3.1.1 Identification of a Metric

The average inflation rate in China over the past 10 years equals 2.83% [53]. Logic suggests that water production costs should be compared with forecasted water prices to assess the affordability of nuclear desalination by 2030. However, beyond the sole effect of inflation, water price in China has continuously been rising in the last 10 years due to political reforms [54]. In Beijing, the average water price has multiplied 30 times between 1991 and 2004, while the inflation raised prices by only a factor of 2. During the year 2005 alone, water price increased by 20% [55]. Nevertheless, government contributions are still too low to fill the gap between the true cost of service and the revenue from users [56], which suggests that additional reforms will be enforced in the near future, making the water price landscape even less predictable. Owing to the preponderance of political reforms over market laws to set water prices in China, it is almost impossible to forecast 2030 water tariffs.

Instead, high uncertainty on future water prices and unpredictable governmental reforms suggest that nuclear desalination be evaluated on the basis of its relative cost and benefits to population. The "5% rule" [57] advocates that water purchase must be equal to or lower than 5% of households' income to be affordable. Here, we evaluate the affordability of freshwater from nuclear desalination, including both production and supply costs, according to the 5% rule. This metric is more stable than the 2030 water tariffs projection as historical wage changes in China have shown to be more correlated to overall economic context than water price changes.

Using data for renewable water resources per capita at the provincial level from China Water Risk [34] and natural growth rate per province from China Statistical Yearbook [58], we calculate that, in 2030, 11 provinces—Beijing, Tianjin, Shanghai, Shandong, Hebei, Jiangsu, Henan, Anhui, Shanxi, Gansu, and Ningxia—will experience renewable water resources per capita per annum below 500 m^3, which is the threshold for *absolute scarcity* [59,60].

Below, we evaluate the quantity of water that can be produced and supplied from nuclear desalination plants to the capital cities of the 11 water-scarce provinces identified above while meeting the 5% rule in 2030. The calculation method is detailed in the case of Beijing below; same formulas with different parameters have been used for all cities.

7.3.1.2 Desalinated Water Supply Quantity Within the 5% Rule Threshold

The minimum monthly wage in Beijing in 2013 was \$275 [61], that is, \$3302 per year. The average annual wage in China is expected to double between 2013 and 2030 [62]. Therefore, according to the 5% rule, supply of freshwater from desalination plants to Beijing in 2030 will be affordable if and only if the production and supply costs to Beijing is below \$165 per person per year.

The cost of water production and supply from Chinese desalination plants to Beijing in 2030 (Fig. 7.3) depends on the quantity of supply per person per day and the desalination plants that supply water. Supplying water to each of the 33.6 million inhabitants of Beijing at a cost equal to about 5% of the minimal salary translates into an additional quantity of water of 164 m^3 per person per year.

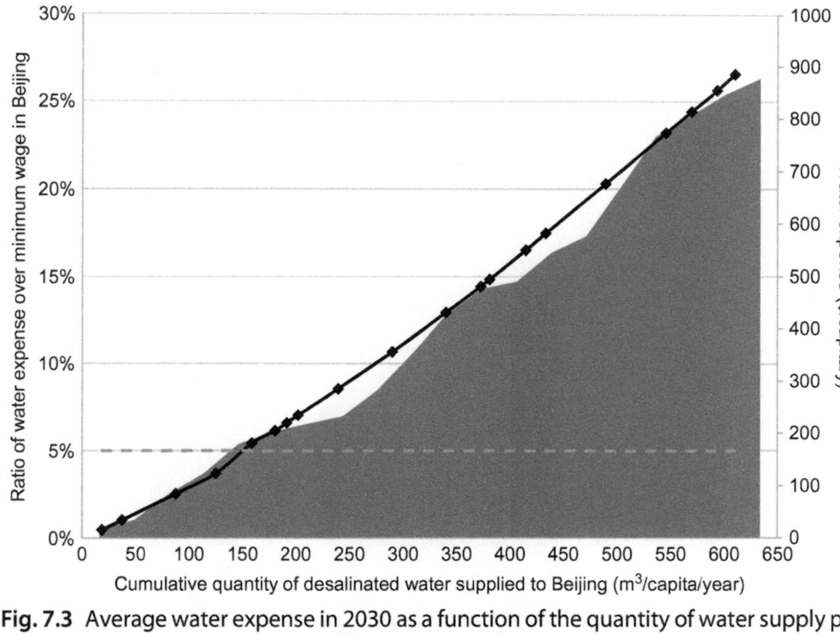

Fig. 7.3 Average water expense in 2030 as a function of the quantity of water supply per person per year in Beijing. The blue dotted line indicates the threshold for the 5% rule. Markers represent minimum quantities of water for which the capacity of an additional water plant is required (small quantities are supplied by nearby plants and cost increases as production from additional, further plants is required). The length in brackets indicates the distance in kilometers between the additional desalination plant considered and Beijing: the closest plant is Haixing, 230 km from Beijing, and the furthest plant that still meets the 5% rule is Tianwan, 670 km from the capital.

The maximum quantity of water that can be affordably supplied from desalination plants to Beijing is 159 m^3 per capita per year. On the basis of the data from China Water Risk, we find that renewable water resources in Beijing in 2030 will be 79 m^3 per capita per year. Water resources can therefore be doubled while remaining affordable through the use of nuclear desalination.

The same cost–supply simulation was performed to the capitals of the nine other water-scarce provinces; results are presented in Fig. 7.4.

Calculations show that, while water production cost remains around $0.72/$m^3$ across the nation, the supply cost varies significantly across regions between $0.87/$m^3$ and $1.96/$m^3$. In cities located relatively near to the coast such as Nanjing, Tianjin, Shanghai, Jinan, Hefei, Beijing, and Shijiazhuang, this cost is at or below $1/$m^3$. It significantly increases for inland cities such as Yinchuan and Lanzhou, which consequently limits the quantity of affordable water to be supplied according to the 5% rule. However, in cities such as Shijiazhuang, Hebei province and Yinchuan, Ningxia region, where the level of water scarcity is projected to be high in 2030, supplying freshwater from desalination plants multiplied the total amount of water resources by a factor of 2.76 and 2.22, respectively. On the contrary, despite low supply cost in Nanjing, affordable desalinated water multiplies the total projected water resources by a factor of 1.44. It is important to highlight that,

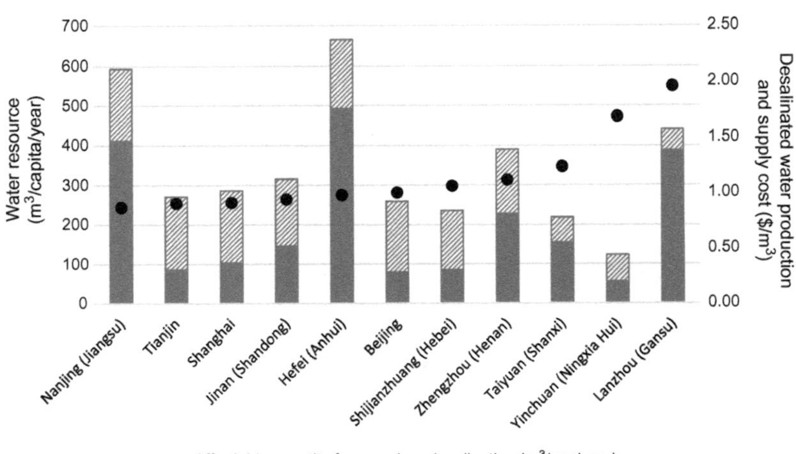

Fig. 7.4 Annual water resources as well as production and supply costs from nuclear desalination plants to the capitals of the 11 water-scarce provinces in 2030.

since the total nuclear desalination capacity is limited to current plans for 2030, there exists a trade-off between water supply allocations among those 11 places.

The 5% rule also shows that water tariffs by 2030 in all water-scarce capitals could be increased well above inflation's sole effect without jeopardizing Chinese households' ability to pay for freshwater.

In the case of pure water needs—for special industries, for example—water production cost in 2013 will be around $5/m^3 due to treatment. As MED produces pure water, nuclear desalination costs are already below water market price in China [33].

7.3.2 Comparison Between Nuclear Desalination and Two Other Projects With Similar Objectives: Coal Desalination and SNWTP

As of 2016, although China General Nuclear Power Group has commissioned a small 10,080 m^3/day desalination plant [63], there is no desalination plant thermally coupled with nuclear reactors in China. To address its water shortage, the country has conducted two main initiatives in addition to improving water-use efficiency: the SNWTP and water desalination using energy from coal-fired plants. Beijiang's desalination and power plant in Tianjin, with state-of-the-art MED and coal-fired technologies, is the most advanced and largest plant of its kind in China, aiming to reach a water production capacity of 400,000 m^3/day [64].

We evaluate CO_2 emissions and water production specific costs for nuclear desalination, SNWTP, and coal desalination. To enable meaningful comparison, we make the following hypotheses. First, the number of nuclear reactors considered for desalination purposes corresponds to the Chinese central government's nuclear reactor deployment plans, as considered in the previous sections of this study. Second, water production capacity from coal desalination is assumed to be the same as production capacity from nuclear desalination (20.7 billion m^3/year). Third, STNWTF's Western route is not considered here because no accurate cost estimations are available and no commitment has been made with regards to its construction[3]. As a consequence, the annual water transferred through the SNWTP, W_{SNWTP}, is 25 billion m^3 [9]. Third, coal plants used for desalination are assumed to use state-of-the-art technology, reflecting the characteristics

[3] First official economic evaluation for the Western route show that costs will likely be higher than for the two other routes as this route transfers water from far-West remote locations.

of Beijiang's ultra-supercritical coal-fired plant, with a thermal efficiency of 46% and CO_2 emissions of 700 g/kWh.

The specific water production cost of the SNWTP is \$0.51/m^3, calculated as follows:

$$SC_{SNWTP} = \frac{I_{SNWTP} \times CRF + OM_{SNWTP-y} + SP_{SNWTP-use} \times E_{csale}}{W_{SNWTP}}$$

where I_{SNWTP} (*\$81 billion* [65]) is the capital cost of the Eastern and Central routes of the SNWTP. CRF is the capital recovery factor using a discount rate of 4% and assuming that the SNWTP's lifetime is 50 years—twice a desalination plant. $SP_{SNWTP\text{-}use}$—including electricity used for pumping (0.11 kWh/m^3, author's calculation is based on Refs. [66–68]) and SNWTP water treatment (1.15 kWh/m^3 [69])—is the specific power use per cubic meter of water transferred.

The water diversion project is less energy intensive per cubic meter supplied than desalination. Coal and nuclear desalination use 2 kWh/m^3 with an additional 2 kWh/m^3 for water transportation on average, and reduce power plants' output by 4 kWh/m^3 produced, while the SNWTP requires 1.16 kWh/m^3 for water supply and treatment.

However, the construction of SNWTP requires 15 million m^3 of cement and 0.7 million tons of steel [70], while the construction of the 69 desalination plants, powered by coal or nuclear, would use 0.3 million m^3 of cement and 0.02 million tons of steel [71].

Carbon dioxide emissions from the construction and operation for the three alternatives over their respective lifetime add up to 3 million tons per year for nuclear desalination, 18 million tons per year for the SNWTP, and 182 million per year for coal desalination, using carbon intensity values for power plants from the IPCC [72] and carbon intensity projection for China's 2030 power grid from [73].

China's national emission trading scheme, announced by President Xi in September 2015, should be launched during the second half of 2017 [74]. While carbon price ranges are not known yet, the country has implemented seven cap-and-trade pilot projects since 2011 [75], with prices ranging between \$3 and \$12 per ton of CO_2 [76]. Water production and supply cost from the three alternatives are shown in Fig. 7.5 for three different hypothetical carbon prices.

Water production and supply costs from the SNWTP range from \$0.51/m^3 when no carbon price is implemented to \$0.58/m^3 when a ton of carbon costs \$100/m^3. This is not a significant variation compared

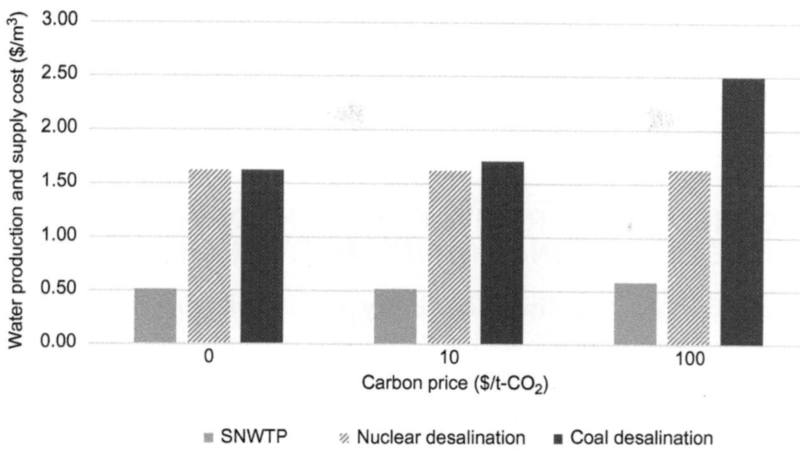

Fig. 7.5 Average specific desalinated water cost—including production and supply—for the SNWTP, nuclear desalination, and coal desalination in 2030, as a function of carbon price.

with coal desalination, whose costs vary between $1.62/m^3 and $2.50/m^3 across the same carbon price range. The price of carbon has no significant impact on the cost of nuclear desalination, which stays around $1.62/m^3.

The SNWTP therefore has lower energy requirements but larger carbon emissions than nuclear desalination, however, a carbon price does not challenge its economic attractiveness compared with nuclear and coal desalination.

The SNWTP presents several other challenges that are not quantified in the analysis presented above. In its current version, it is not able to supply several water-scarce places, including Taiyuan (Shanxi) and Yinchuan (Ningxia), accounting in total for more than 19 million people who will experience water scarcity by 2030. As a seasonal supply mean, the SNWTP's water availability will also be substantially reduced during some periods of the year. Water quality is high enough in the three water supply alternatives to be considered as *potable* and therefore meet domestic needs, but in the case of SNWTP it is not pure enough to meet needs of special industries. Therefore, in the latter case, a cost to purify water must be added to supply higher-quality water needs.

The social and environmental impacts of nuclear and coal desalination mostly lie in their potential effect on coastal marine's life. SNWTP will have a substantial impact on fluvial life, but the high social consequences of this project reside in the resettlement of 300,000 people [77], and the potential drying out of South China, which has already endured searing droughts in 2000, 2007, and 2009 [78], and could be accentuated because of water transfer toward Northern provinces.

7.4 CONCLUSION—RELEVANCE OF THE DEVELOPMENT OF A NUCLEAR DESALINATION INDUSTRY IN CHINA BY 2030

Through the use of an interdisciplinary framework, the affordability of nuclear desalination in China by 2030 to significantly increase projected water resources is confirmed through the use of a proxy: the capitals of China's 11 most-scarce provinces. The SNWTP has lower costs than nuclear desalination, however, its implementation is significantly more challenging than nuclear desalination from environmental—including carbon emissions—and social perspectives. While calculations show that coal desalination should not be part of the solution to China's water shortages, as it is economically not attractive and it is not consistent with the country's commitments toward sustainability, water diversion projects have an important role to play, although not as currently planned. Rather than providing water to coastal cities, SNWTP, rooted in river springs in the middle of the country, should be dedicated to supplying remote provinces such as Ningxia and Gansu, which are too far from the coast to receive significant amount of affordable freshwater from desalination plants.

In the longer term, desalination could be performed using high-temperature nuclear reactors such as the high-temperature reactor–pebble bed module (HTR–PM), whose efficiency will be higher than conventional reactors. The use of nuclear heating reactors such as the 200-MWt nuclear heating reactor (NHR-200) is designed to produce free steam, because the electricity-related cost would only result from power consumption by the desalination plant, relatively low, and not from the reactor's thermal efficiency decrease. This could reduce by half the water-production specific cost according to the calculations presented in this study. On the other hand, the costs induced by building an advanced reactor technology and adapt its fuel cycle consequently may be higher than well-known pressurized water reactor technologies that will still compose most of the Chinese nuclear fleet in 2030. Finally, desalination through wind and solar energies, although at its early days, may become predominant by 2030, considering the already favorable context for these energies in China in 2014. This could actually constitute a cost-efficient solution to current high levels of electricity curtailment, as desalination plants could operate only when the value of energy or electricity is low.

In addition, international working groups such as the Global Water Desalination Alliance [79] launched in Paris during the COP21 should be pushed further in order to favor the pooling of technical knowledge and the efficient use of available funds. To achieve this project, the public must be informed and the investors must be convinced. This requires the systematic and publicly available evaluation of assets and drawbacks presented by nuclear desalination and other types of energy as well as geoengineering projects—including risk assessment and cautious cost evaluation of each option in each country's particular context—as presented in this study. In addition to China, India, the Republic of Korea, and Pakistan, which all demonstrate simultaneously a lack of water and the ability to use nuclear energy for desalination [80], could represent a potential long-term market for nuclear desalination provided that such a methodological evaluation is performed in those countries.

REFERENCES

[1] Renewable internal freshwater resources per capita (cubic meters) | Data, 2017. [Online]. Available: http://data.worldbank.org/indicator/ER.H2O.INTR.PC.
[2] Renzetti S, Dupont DP. Water pricing experiences and innovations. Gewerbestrasse: Springer International Publishing; 2015.
[3] Price full cost of water—Opinion—Chinadaily.com.cn, 2017. [Online]. Available: http://www.chinadaily.com.cn/opinion/2014-04/04/content_17405783.htm.
[4] Addams L, Boccaletti G, Kerlin M, Stuchtey M. Charting our water future: economic frameworks to inform decision-making. 2030 Water Resources Group; 2009.
[5] World Nuclear Association, China Nuclear Power, 2017. [Online]. Available: http://www.world-nuclear.org/information-library/country-profiles/countries-a-f/china-nuclear-power.aspx.
[6] Google Maps. [Online]. Available: https://maps.google.com/.
[7] Bureau of South-to-North Water Transfer Planning Designing and management—Ministry of Water Resources, 欢迎浏览'中国南水北调'网站—South-to-North Water Diversion, 2017. [Online]. Available: http://www.nsbd.gov.cn.
[8] Avrin A-P, He G, Kammen DM. Assessing the impacts of nuclear desalination and geoengineering to address China's water shortages. Desalination 2015;360:1–7.
[9] Barnett J, Rogers S, Webber M, Finlayson B, Wang M. Transfer project cannot meet China's water needs. Nature 2015;527:295–7.
[10] Al-Nory MT, Brodsky A, Bozkaya B, Graves SC. Desalination supply chain decision analysis and optimization. Desalination 2014;347:144–57.
[11] He G, et al. SWITCH-China: a systems approach to decarbonize China's power system. Environ Sci Technol 2016;50:5467–73.
[12] Zhang L, Xie L, Chen H-L, Gao C-J. Progress and prospects of seawater desalination in China. Desalination Nov. 2005;182(1–3):13–8.
[13] Zhou Y, Tol RSJ. Implications of desalination for water resources in China - an economic perspective. Desalination 2004;164:225–40.

[14] Deng R, Xie L, Lin H, Liu J, Han W. Integration of thermal energy and seawater desalination. Energy Nov. 2010;35(11):4368–74.
[15] Zou Q, Liu X. Economic effects analysis of seawater desalination in China with input-output technology. Desalination 2016;380:18–28.
[16] Megahed MM. Feasibility of nuclear power and desalination on El-Dabaa site. Desalination 2009;246(1–3):238–56.
[17] Gleick P. Water and Energy. Annu Rev Energy Environ 1994;19(1):267–99.
[18] Methnani M. Influence of fuel costs on seawater desalination options. Desalination 2007;205(May 2006):332–9.
[19] Younos T, Tulou KE. Energy needs, consumption and sources. J Contemp Water Res Educ 2009;132(1):27–38.
[20] Kupitz J, Crijns MJ. Perspectives desalination of nuclear energy for seawater. Desalination 1994;99:299–327.
[21] Faibish RS, Konishi T. Nuclear desalination: a viable option for producing freshwater. Desalination 2003;157(1–3):241–52.
[22] Megahed MM. Nuclear desalination: history and prospects. Desalination 2001;135 (1–3):169–85.
[23] International Atomic Energy Agency. Status of design concepts of nuclear desalination plants. Vienna: IAEA; 2002.
[24] Wang D, Dong D, Zheng W, Zhang D, Wang L. The 200 MW nuclear heating reactor and its possible application in seawater desalination. Desalination 1994;99(2–3):383–99.
[25] Jia H, Zhang Y. Nuclear seawater desalination plant coupled with 200 MW heating reactor. Int J Nucl Desalin 2008;3(2):230.
[26] Tian L, Wang Y, Guo J. Economic analysis of a 2x200 MW nuclear heating reactor for seawater desalination by multi-effect distillation (MED). Desalination 2002;152:223–8.
[27] International Atomic Energy Agency. Desalination Economic Evaluation Program (DEEP-3.0): user's manual. Comput Man Ser 2006;19.
[28] International Atomic Energy Agency, "Examining the economics of seawater desalination using the DEEP code," IAEA, Vienna, 2000.
[29] International Atomic Energy Agency. DEEP 5 user manual; 2013.
[30] International Atomic Energy Agency, PRIS: power reactor information system—Country details (China), 2017. [Online]. Available: http://www.iaea.org/pris/CountryStatistics/CountryDetails.aspx?current=CN.
[31] International Atomic Energy Agency, "Advanced reactors information system (ARIS)," 2017. [Online]. Available: https://aris.iaea.org/ARIS/reactors.cgi.
[32] Zhou Y, Tol RSJ. Evaluating the costs of desalination and water transport. Water Resour Res 2004;41:1–16.
[33] International Atomic Energy Agency. Economics of nuclear desalination: new developments and site specific studies. Vienna: IAEA; 2007.
[34] China Water Risk and Unison Creative, Who's Running Dry?, 2013. [Online]. Available: http://chinawaterrisk.org/big-picture/whos-running-dry/.
[35] American Water Works Association. Desalination of seawater, manual of water supply practices. Denver: American Water Works Association; 2011.
[36] International Atomic Energy Agency. Optimization of the coupling of nuclear reactors and desalination systems. Vienna: IAEA; 2005.
[37] Darwish MA. Cogeneration power-desalination plants. Desalination 1988;69:27–46.
[38] Tonner J. Barriers to thermal desalination in the United States; 2008.
[39] International Atomic Energy Agency. Introduction of Nuclear Desalination: a guidebook. Vienna: IAEA; 2000.
[40] Energy Sector Management Assistance Program; The World Bank, Central America regional programmatic study for the energy sector: managing an electricity Shortfall, 2010.

[41] International Atomic Energy Agency. Desalination economic evaluation program (DEEP). Vienna: IAEA; 2011.
[42] University of Palermo. Seawater desalination: conventional and renewable energy processes. Palermo: Springer; 2009.
[43] Fridley D, Zheng N, Zhou N, Ke J, Hasanbeigi A, Price L. China energy and emissions paths to 2030; 2011.
[44] Zhang H, et al. China's green revolution; 2009.
[45] Kelly Services. Salary guide; 2016.
[46] Carlsbad Desalination Project, Carlsbad Desalination Project | FAQs, 2017. [Online]. Available: http://carlsbaddesal.com/faqs.
[47] Department of Environment and Primary Industries Victoria State Government, DEPI—Desalination jobs and contracts, 2017. [Online]. Available: http://www.depi.vic.gov.au/water/urban-water/desalination-project/desalination-background/desalination-jobs-and-contracts.
[48] Ni CC. China's electric power industry and its trends. Japan: Institute of Energy Economics; 2006.
[49] Beijing Internati, Guide to heating, electricity, water, and gas policies and procedures, 2012. [Online]. Available: http://www.ebeijing.gov.cn/feature_2/GuideToHeatingElectricityWaterAndGas/PriceGuide/t1107813.htm.
[50] China.org.cn, China's power consumption slows further, Xinhua, 2012.
[51] Delucchi MA, Jacobson MZ. Providing all global energy with wind, water, and solar power, Part II: reliability, system and transmission costs, and policies. Energ Policy 2011;39(3):1170–90.
[52] Anderson J. Potential impacts of desalination development on energy consumption. Brussels: DG Research, European Commission; 2008.
[53] Inflation.eu, Historic CPI inflation China, 2016. [Online]. Available: http://www.inflation.eu/inflation-rates/china/historic-inflation/cpi-inflation-china.aspx.
[54] Zhong L, Mol APJ. Water price reforms in China: policy-making and implementation. Water Resour Manag 2009;24(2):377–96.
[55] X. Yang, J. F. Donzier, C. Noel, Une Comparaison de la Politique de l'Eau en France et en Chine Deuxième Partie: Autorisation, Protection et Financement, 2012.
[56] Browder GJ, Xie S, Kim Y, Gu L, Fan ME. Improving the performance of China's urban water utilities. Washington, DC: World Bank; 2007.
[57] Water Engineering and Development Centre (WEDC), Water Sector Development Project (WSDP), and Department For International Development (DFID), China Water sector briefing note series: assessing ability and willingness to pay briefing, 2003, pp. 1–6.
[58] National Bureau of Statistics of China, China Statistical Yearbook, 2015. [Online]. Available: http://www.stats.gov.cn/tjsj/ndsj/2015/indexeh.htm.
[59] Falkenmark M, Lundqvist J, Widstrand C. Macro-scale water scarcity requires micro-scale approaches. Aspects of vulnerability in semi-arid development. Nat Resour Forum 1989;13(4):258–67.
[60] Food and Agriculture Organization of the United Nations—Water scarcity, 2014. [Online]. Available: http://www.fao.org/nr/water/topics_scarcity.html.
[61] 中华人民共和国人力资源和社会保障部 (Ministry of Human Resources and Social Security of the People's Republic of China), "全国各地区月最低工资标准情况." [Online]. Available: http://www.mohrss.gov.cn/ldgxs/LDGXqiyegongzi/LDGXzuidigongzibiaozhun/201509/t20150929_221690.htm (accessed 07.04.17).
[62] Hawksworth J, Lambe C. Global wage projections to 2030. PWC; 2013.
[63] World Nuclear Association, Nuclear desalination, 2016. [Online]. Available: http://world-nuclear.org/info/Non-Power-Nuclear-Applications/Industry/Nuclear-Desalination/.
[64] The Urban China Initiative, Xiao G, Xue L, Woetzel J. The urban sustainability index: a new tool for measuring China's cities; 2010.

[65] Photo slideshow: China completes second line of South-North Water Transfer Project—Circle of Blue, 2017. [Online]. Available: http://www.circleofblue.org/2015/world/photo-slideshow-china-completes-second-line-south-north-water-transfer/.
[66] Bureau of South-to-North Water Transfer Planning Designing and Management—Ministry of Water Resources, 南水北调工程总体规划内容简介, 2008. [Online]. Available: http://nsbd.mwr.gov.cn/nsbdgcjs/ztbj/201211/t20121129_333896.html.
[67] Bureau of South-to-North Water Transfer Planning Designing and management—Ministry of Water Resources, 南水北调东线工程规划(2001年修订)简介, 2008. [Online]. Available: http://nsbd.mwr.gov.cn/nsbdgcjs/ztbj/201211/t20121129_333895.html.
[68] Bureau of South-to-North Water Transfer Planning Designing and management—Ministry of Water Resources, 欢迎浏览 '中 国 南 水 北 调' 网站, 2017. [Online]. Available: http://www.nsbd.gov.cn/zx/gcgh/200308/t20030825_195174.html.
[69] Mallonée ML. The South-North Water Transfer Project: a cost-benefit analysis; 2016.
[70] Liu J, Chen S, Wang H, Chen X. Calculation of carbon footprints for water diversion and desalination projects. Energy Procedia 2015;75:2483–94.
[71] M. Fontseré, LCA software: practical example using SimaPro. Girona, 2012.
[72] Edenhofer O, Pichs Madruga R, Sokona Y. Renewable energy sources and climate change mitigation (special report of the Intergovernmental Panel on Climate Change). 2012;6(4).
[73] Cai W, Wang C, Wang K, Zhang Y, Chen J. Scenario analysis on CO2 emissions reduction potential in China's electricity sector. Energ Policy 2007;35(12):6445–56.
[74] China national ETS launch likely in second half of 2017—sources Carbon Pulse, 2016. [Online]. Available: http://carbon-pulse.com/17057/.
[75] Swartz J. China's National Emissions Trading System: implications for carbon markets and trade. Geneva: International Centre for Trade and Sustainable Development; 2016.
[76] Ge X. State and trends of Shenzhen carbon market. In: Proc. 14th IEA-IETA-EPRI Annual Workshop on Greenhouse Gas Emission Trading; 2014.
[77] Berkoff J. China: the South–North water transfer project—is it justified? Water Policy 2003;5:1–28.
[78] Jaffe A, Schneider K. A dry and anxious North awaits China's giant, unproven water transport scheme. Circle of Blue WaterNews; 2011.
[79] Global Clean Water Desalination Alliance, Global Clean Water Desalination Alliance—Concept Paper, 2015.
[80] International Atomic Energy Agency. Nuclear technology review. Vienna: IAEA; 2004.

CHAPTER 8

Desalination Costs Using Renewable Energy Technologies

Upeksha Caldera, Dmitrii Bogdanov, Christian Breyer
Lappeenranta University of Technology, Lappeenranta, Finland

8.1 INTRODUCTION

Renewable energy (RE) powered desalination is rapidly gaining interest as a means to augment the increasingly stressed global water supply [1,2]. The demand for seawater desalination is forecasted to grow at a compound annual increase of 8.1% from 2014 to 2020 [3]. This includes regions in Asia and sub-Saharan Africa where desalination has not typically been a contributor to the water supply [4]. However, this anticipated growth is not without concern. Desalination plants are more energy intensive than traditional water treatment methods [5]. The energy consumption for surface freshwater treatment is about 0.6 kWh/m^3 [5], whereas the least energy intensive seawater reverse osmosis (SWRO) desalination plants today consume around 3 kWh/m^3 [6]. Meanwhile, the demands of thermal desalination plants are much higher and range from 40 to 80 kWh/m^3, and additional electrical energy requirements range from 2.5 to 5 kWh/m^3, depending on the desalination technology [2]. Thus, conventional desalination plants are heavily dependent on diminishing and costly fossil fuel resources. For instance, the Kingdom of Saudi Arabia (KSA), the world's largest producer of desalinated water, is estimated to use nearly 25% of its oil and gas resource for desalination plants [7]. At the current rate of consumption, by 2030, Saudi Arabia is expected to utilize 50% of domestic fossil fuel resources for desalination plants [7]. At the other end of the spectrum, regions such as the islands of the South Aegean Sea resort to importing water while countries without sufficient domestic fossil fuel reserves, such as China and India, import fossil fuels for desalination plants [8,9].

In addition to the dependency on fossil fuels, desalination plants result in greenhouse gas emissions, only further contributing to one of the causes of water scarcity—climate change. RE enables desalination independent of fossil fuels, accessible to all countries regardless of domestic fossil fuel

Renewable Energy Powered Desalination Handbook
https://doi.org/10.1016/B978-0-12-815244-7.00008-8

resources, and eliminates emissions that lead to climate change [10]. Thus, a net zero carbon emission world cannot be achieved with fossil fuel based desalination.

8.2 CURRENT STATUS OF RE POWERED DESALINATION

The global installed desalination capacity as of 2016 is 79,220,166 m^3/day [11]. RE powered desalination is estimated to account for 1% of the global installed desalination capacity [9]. Most of the RE powered desalination plants are based on solar and wind energy [1].

Solar powered desalination is not a recent concept and, as explained by Ghermandi et al. [12], has been studied since the 1970s. Ghermandi et al. [12] list the solar powered RO plants in operation around 2009 and state that there were large-scale plants being investigated but none in operation. Most of the system capacities ranged from 0.1 m^3/day to 75.7 m^3/day. Fast-forward to 2017, and the RE-powered desalination market presents a different scenario.

The world's largest solar photovoltaic (PV)-based desalination plant is being built in Al-Khafji, Saudi Arabia and is expected to be completed by 2017 [13,14]. The SWRO plant will provide up to 60,000 m^3/day and utilize a 15 MW PV plant. When there is not enough PV generated electricity, the national grid will enable continual operation of the desalination plant. In addition, excess electricity from the PV plant will be injected back into the national grid [13].

Similarly, a 100,000 m^3/day SWRO plant is in the procurement stage in Abu Dhabi, United Arab Emirates (UAE). The plant is to be powered by a 22 MW PV plant. By 2030, Abu Dhabi, led by Masdar, the Abu Dhabi-based energy organization, aims to power the city's desalination plants by 100% RE [13].

Australia accounts for the largest capacities of wind powered desalination plants. There is about 393,700 m^3/day of SWRO plants whose energy consumption is offset by wind farms. Smaller SWRO plants powered by wind farms are also online in China and Greece [15].

Table 8.1 lists the RE powered SWRO plants >1000 m^3/day that have been contracted globally [13,15].

The main concerns with the use of RE systems are intermittency and energy storage requirements. Mentis et al. [8] analyzed the case of using completely RE powered desalination units in the South Aegean Islands. However, it was decided to opt for the mix of a PV-wind hybrid power

Table 8.1 Renewable energy powered SWRO plants globally [13,15]

Country	Plant	Capacity (m^3/day)	Contracted year	Operational year	RE power source
Saudi Arabia	Al Khafji	60,000	2015	2017	15 MW PV
Australia	Kurnell–Sydney	250,000	2007	2010	140 MW wind farm
Australia	Kiwnana–Perth	143,700	2005	2006	80 MW wind farm
China	Dafeng	10,000	2013	2014	2.5 MW wind farm
Greece	Milos	3360	Unknown	2008	0.6 MW wind turbine

plant and grid electricity to meet the energy demand of the desalination units. Novosel et al. [16] presented the case of using PV and wind energy for desalination in Jordan. Brine operated pump storage was used to allow for higher penetration of RE. However, fossil fuel balancing is necessary for optimal utilization of the desalination plants.

Today, RE powered desalination plants are said to result in higher costs of water production than fossil fuel powered plants due to higher capital costs [2,15]. Fthenakis et al. [14] present an analysis of SWRO desalination plants, located along the Red Sea in Saudi Arabia, powered by concentrated solar PV, fixed-tilted PV, and single-axis tracking PV. It was found that SWRO desalination plants of 190,000 m^3/day powered by a 116-MW single-axis tracking PV plant provided the least cost of water production of 0.85 €/m^3. The national grid served as backup at times of low PV electricity generation and additionally purchased excess electricity. Current water production costs of fossil powered SWRO plants in Saudi Arabia lie between 0.65 and 1.90 €/m^3 [17]. The low costs of fossil powered SWRO plants are due to the heavily subsidized fossil fuels and, therefore, electricity costs in the country. This is particularly due to not charging the world market price for the fuel but only the local production cost, which then constitutes a subsidy. As the cost of RE power plants and energy storage plummets [18,19], it is more cost-effective to power desalination plants with RE. This is highlighted in a recent presentation by the Managing Director of the Dubai Electricity and Water Authority (DEWA) [20]. During the presentation, it was announced that Dubai would generate 1,154,000 m^3 of desalinated water a day by 2030 through the use of low-cost solar energy. The switch from fossil fuels to solar energy, between now and 2030, is estimated to save DEWA 13 bUSD.

This chapter provides further insights into the concept discussed by DEWA. The aim of this chapter is to illustrate how the global desalination demand in 2030 can be met through 100% RE power plants in the most cost-effective manner. The subsections that follow discuss the estimated global desalination demand by 2030, technical and financial aspects of the optimal RE power and desalination plants, methodology to analyze water production costs in 2030, and results and conclusions with insights into future trends.

In this chapter, it is assumed that the 2030 desalination demand is met by SWRO plants. Reverse osmosis accounts for almost 67% of the total global installed capacity in 2016 [11]. In addition, 58% of the installed global desalination capacity used seawater as the feed water type. Ghaffour et al. [21]

explain that SWRO will continue to remain the dominant desalination technology, owing to the lower costs, energy consumption, and technological improvements. This view is also expressed by GWI Desal Data [22]. However, according to the group, the savings due to RO efficiency improvements may be negated by increasing energy and material costs [22].

8.3 GLOBAL DESALINATION DEMAND FOR 2030

As of 2016, the total online desalination capacity is estimated to be 79,220,166 m^3/day [11]. According to Global Water Intelligence Data, the most common global use of desalinated water today is for drinking water production. The municipal sector accounted for 55.2% of the total desalination capacity online in 2015, followed by the industrial sector at 42.5%. The municipal sector includes water for drinking consumption. The industrial sector includes the use of desalinated water for boilers in power generation. In such instances, co-generation power and water plants are utilized so that waste heat from the power plant can be captured for desalination [23]. The remaining 2.4% of the desalination capacity is used for other purposes including water for agricultural production [23].

What does the forecast for the global desalination capacity look like? Section 8.3.1 summarizes the GWI forecast up to 2020 while Section 8.3.2 illustrates the desalination demand up to 2030 based on global water stress.

8.3.1 Desalination Capacity Forecast

GWI Desal Data report that the growth in annual cumulative contracted capacity of desalination plants between 2004 and 2014 was 7.8% [23]. The anticipated annual cumulative growth from 2016 to 2020 is 6%, resulting in a total contracted capacity of 117,247,071 m^3/day by 2021 [24].

The GWI forecast for 2020 is based on the following methodology:

- knowledge of the already proposed desalination plants
- capacity of plants that may be additionally contracted by 2020
- anticipated project delays up to 2020

There is currently a lull in the desalination market attributed to low oil prices, commodity prices, and insufficient government budgets to invest in desalination projects [25]. However, after 2020, the desalination market is expected to pick up, driven by higher oil and commodity prices as well as an increasing demand for drinking water sources [25].

8.3.2 Desalination Demand for 2030 Based on Global Water Stress

The Aqueduct Water Risk Atlas illustrates the global water stress in 2010 (baseline water stress) and the projected global water stress for 2020, 2030, and 2040. The projections are based on climate change variables, used for the Coupled Model Intercomparison Project Phase 5 (CMIP5) project, and socioeconomic variables from the International Institute for Applied Systems Analysis (IIASA) [26].

Water stress is defined as the ratio of the water withdrawals to the mean available blue water [27]. The decadal water stress is presented for 15,006 global water catchments for the optimistic, business as usual, and pessimistic scenarios. In addition, the Atlas provides the respective changes in water supply and water demand for the water catchments in the respective decades. Change in climate change variables is used to gauge future water supplies while socioeconomic variables are used to analyze future water supplies [26]. Furthermore, based on combinations of climate change and socioeconomic variables, the water stress, supply, and demand data are presented for three scenarios: optimistic, business as usual, and pessimistic.

Fig. 8.1 presents global water stress for 2030 based on an optimistic scenario. High or extremely high water stress implies that water demand is much higher than the renewable water resource available in the region [28]. These are regions where nonrenewable groundwater is being extracted to meet water demands. Table 8.2 defines the water stress category labels in the figure.

Fig. 8.2 presents the total water demand projected for 2030, assuming an optimistic future scenario. The total water demand for any region is the sum of the water demand for the region's agricultural, municipal, and industrial sectors.

The industrial sector includes water demands for the manufacturing and power sectors. According to an OECD report, in 2010, 15% of global water withdrawals were used for the energy sector [29]. The energy sector accounts for both fuel production and electricity generation. The share is higher in developed countries. For instance, 40% of total freshwater withdrawals in the United States are used for the cooling of thermal power plants. Meanwhile, in China, 16% of the country's water demand is solely for the coal industry. The same OECD report states that by 2050, global water demand will increase by 55%, half the increase being driven by the energy sector [29].

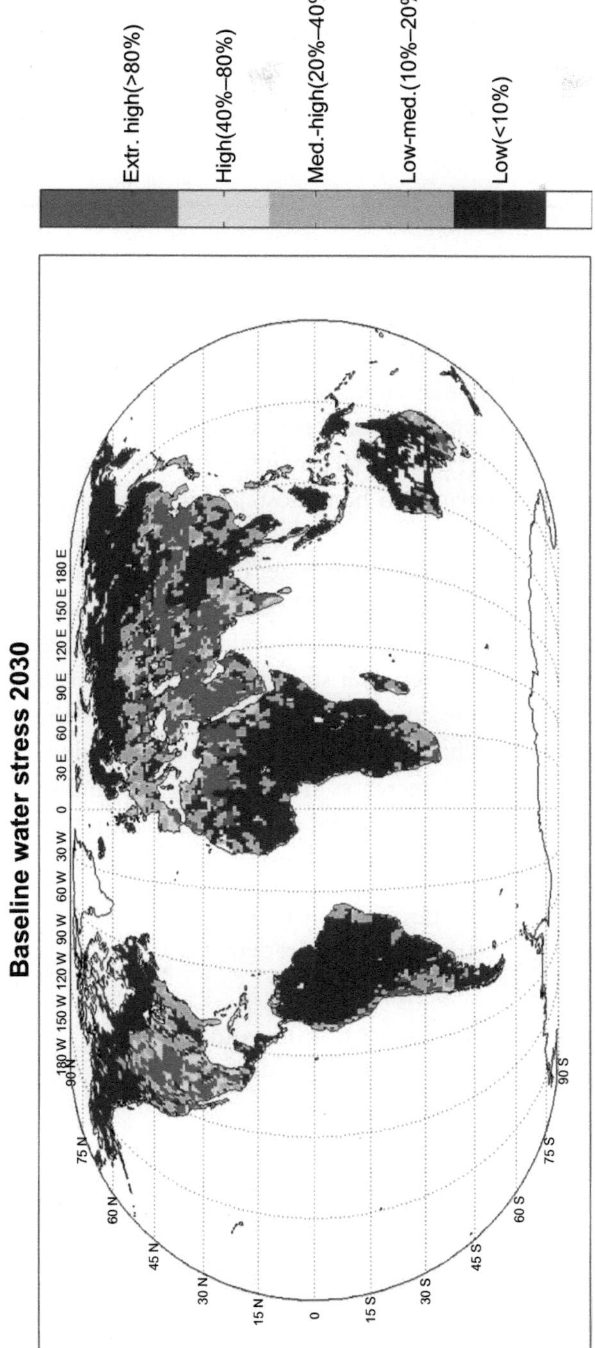

Fig. 8.1 Global Water Stress for 2030 based on the optimistic scenario.

Table 8.2 Water stress category definitions

Water stress category	% of available freshwater used
Low	<10%
Moderate	10%–20%
Medium high	20%–40%
High	40%–80%
Extremely high	>80%

Motivated by the global shift toward an RE powered future and urgency to curb climate change [30–38], the chapter assumes that by 2030 there will be no fossil powered thermal power plants providing electricity for desalination. As a result, the water demand for the industrial sector excludes the water demand for cooling of thermal power plants. The resulting total water demand in 2030 is calculated to be almost 15 billion m^3/day in this most optimistic scenario.

Based on the water stress and total water demand, the global desalination demand can be calculated via a logistic function. This is illustrated in Eq. (1) [26,39,40]. It is assumed that regions with water stress higher than 40% have a desalination demand. As the water stress increases, the demand for desalination increases. A logistic curve allows the expression of the variation in desalination demand with water stress: the desalination demand is lower for lower water stress and increases more steeply for higher water stress levels. At water stress levels >80%, the rate of desalination demand increase is lower and equal to the total water demand of the region.

$$\text{If } 40\% < WS < 50\% : \text{desalination demand} = (WS - 40\%) \times \frac{\text{Total water demand}}{WS} \times \text{logistic}(40\%) \quad (1a)$$

$$\text{If } 50\% < WS < 60\% : \text{desalination demand} = (WS - 50\%) \times \frac{\text{Total water demand}}{WS} \times \text{logistic}(50\%) \quad (1b)$$

$$\text{If } 60\% < WS < 70\% : \text{desalination demand} = (WS - 60\%) \times \frac{\text{Total water demand}}{WS} \times \text{logistic}(60\%) \quad (1c)$$

$$\text{If } 70\% < WS < 80\% : \text{desalination demand} = (WS - 70\%) \times \frac{\text{Total water demand}}{WS} \times \text{logistic}(70\%) \quad (1d)$$

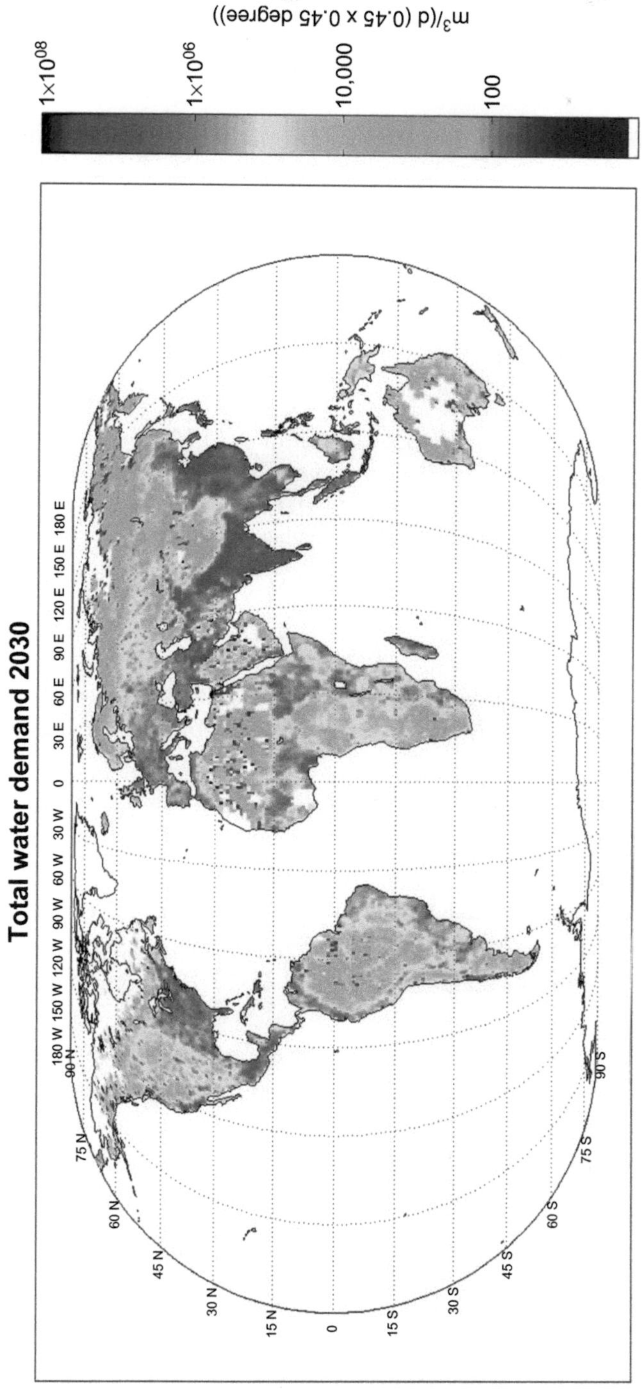

Fig. 8.2 Global Water Demand for 2030 based on the optimistic scenario.

$$\text{If WS} > 80\%: \text{desalination demand} = (\text{WS} - 80\%) \times \frac{\text{Total water demand}}{\text{WS}} \tag{1e}$$

$$\begin{aligned}\text{Total water demand} = {} & \text{Water demand of municipal sector} \\ & + \text{Water demand of industrial sector} \\ & + \text{Water demand of agricultural sector} \\ & - \text{Water demand of thermal power plants}\end{aligned} \tag{1f}$$

Equation 1

Desalination demand for a region based on a 2030 optimistic scenario [26,39,40]. Here, *WS* is the water stress for a region based on a 2030 optimistic scenario, and *Total water demand* is the total water demand for the region based on a 2030 optimistic scenario, excluding thermal power plants.

Based on Eq. (1), the total desalination demand is estimated to be 2.92 billion m^3/day, much higher than the current installed capacity of 79,220,166 m^3/day.

The estimate of 2.92 billion m^3/day of desalinated water assumes that by 2030, only renewable groundwater and surface freshwater will be used to meet global water demand. Given that groundwater is the world's highest withdrawn raw material, the source of drinking water for half the world's population, and main source of water for agriculture [41], it is unlikely that there will be no fossil groundwater usage by 2030. Reflecting this approach, the National Water Strategy (NWS) of Saudi Arabia has outlined specific amounts of nonrenewable water to be consumed from 2015 to 2040, gradually putting an end to the country's use of nonrenewable water [42]. The percentage of the total water demand to be met through nonrenewable water is summarized in Table 8.3 [42].

Such a scenario is assumed globally and 11% of the desalination demand is met through nonrenewable groundwater in 2030. The resulting daily global desalination demand is presented in Fig. 8.3. The total desalination demand is estimated to be 2.6 billion m^3/day, at least 3000% more than the current installed capacity.

Table 8.3 Percentage of Saudi Arabia's total water demand to be met through fossil groundwater

Year	% of water demand met by fossil groundwater
2015	42
2020	27
2025	18
2030	11
2035	10
2040	10
2045	0
2050	0

8.4 DESIGN OF 100% RE POWERED DESALINATION

To determine if it is viable to meet the 2030 desalination demand with SWRO desalination powered solely by RE, the corresponding unit cost of water production (€/m^3) or the levelized cost of water (LCOW) must be found. The costs can then be compared with costs of existing fossil-powered SWRO desalination plants to gauge the cost-effectiveness.

Fig. 8.4 presents the SWRO desalination system envisaged in this chapter. SWRO desalination plants are powered by hybrid RE power plants being cost optimized by storage. High-voltage direct current (DC) cables transport power to the desalination plants on the coast. The water produced is transported to meet the demands of the municipal, industrial, and agricultural sectors. Water storage at the site of demand ensures constant water supply. Hybrid RE plants comprise PV and wind power plants coupled with storage options such as batteries and power-to-gas (PtG). Hybrid RE power plants enable optimal full load hours (FLH) of the desalination plant. Higher FLH enables optimal water production costs.

In the subsections that follow, the economics of the proposed system in Fig. 8.4 are discussed, including water and electricity production costs, and the methodology used to apply on a global scale is described.

8.4.1 Levelized Cost of Water

The unit production cost of water (€/m^3) or LCOW is one of the key economic indicators used to analyze and compare desalination plants [18]. The LCOW is dependent on the capital expenditures (CAPEX) and the operational expenditures (OPEX) of the desalination plant. In this chapter,

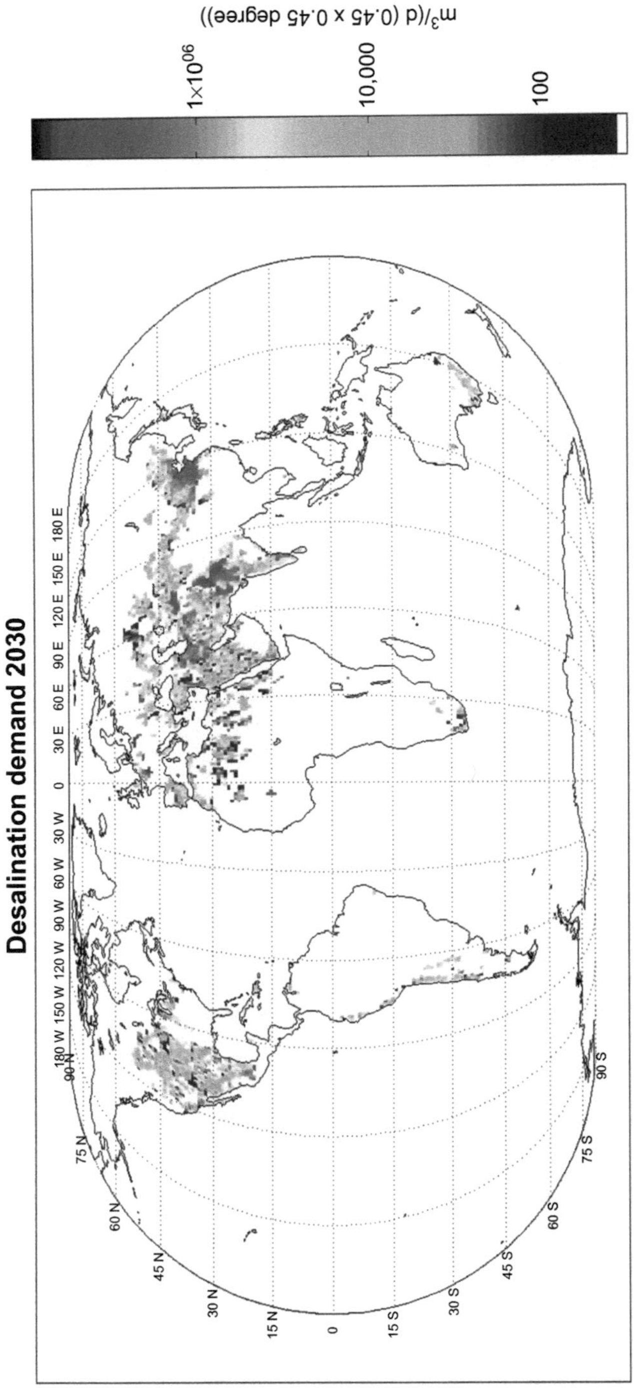

Fig. 8.3 Total desalination demand in 2030 for an optimistic scenario.

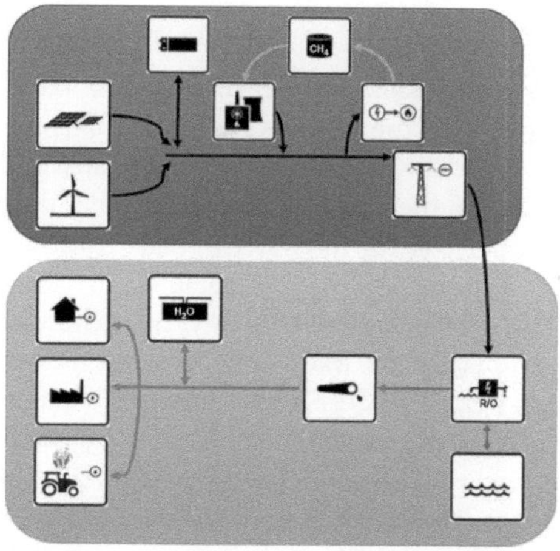

Fig. 8.4 100% RE powered SWRO system proposed to meet future desalination demand.

the LCOW is assumed to include the unit production cost of water at the site of the desalination plant, electricity costs, water storage at the site of demand, and the water transportation to the region of desalination demand.

Eq. (2) illustrates how the components contribute to the LCOW [44].

$$LCOW_{desal} = \frac{(capex_{desal} \times crf_{desal} + capex_{water\,storage} \times crf_{water\,storage}) + opex_{fixed}}{Total\ water\ produced\ in\ a\ year} + opex_{var\,desal} \times SEC \tag{2a}$$

$$opex_{fixed} = opex_{fixed\,desal} + opex_{water\,storage} \tag{2b}$$

$$opex_{var\,desal} = LCOE \tag{2c}$$

$$crf = \frac{WACC \times (1 + WACC)^N}{(1 + WACC)^N - 1} \tag{2d}$$

$$WACC = \frac{E}{E+D} \times k_E + \frac{D}{E+D} \times k_D \tag{2e}$$

$$LCOT_{desal} = \frac{\left(capex_{hpumps} \times crf_{hpumps} + capex_{vpumps} \times crf_{vpumps}\right) + capex_{pipes} \times crf_{pipes}) + opex_{fixedt}}{Total\ water\ produced\ in\ a\ year} + opex_{vart} \times SEC_t \tag{2f}$$

$$opex_{fixedt} = opex_{fixed\,pumps} + opex_{fixed\,pipes} \tag{2g}$$

$$opex_{vart} = LCOE \tag{2h}$$

$$LCOW = LCOW_{desal} + LCOT_{desal} \tag{2i}$$

Equation 2

LCOW for regions with desalination demand in 2030. Here, $capex_{desal}$ is the capital expenditures of the SWRO desalination plant in €/m^3, $capex_{water\ storage}$ is the CAPEX of the water storage tank at the desalination site in €/m^3, crf_{desal} is the annuity factor for the desalination plant, $crf_{water\ storage}$ is the annuity factor for water storage, *Total water produced in a year* is in m^3, and $opex_{fixed}$ is the total fixed annual operational expenditures of the desalination system in €/m^3. The total fixed OPEX is the sum of $opex_{fixed\ desal}$, the fixed OPEX of the desalination plant in €/m^3, and $opex_{water\ storage}$, the OPEX of the water storage tank in €/m^3. The value $opex_{\mathrm{vardesal}}$ is the variable OPEX of the desalination plant and is equal to the levelized cost of electricity (LCOE) of the plant in €/kWh. *SEC* is the specific energy consumption in kWh/m^3. The product of the *LCOE* and *SEC* is the energy cost of the desalination plant in €/m^3. *WACC* is the weighted average cost of capital, *N* is the lifetime of the desalination plant or the water storage, *E* is the equity in €, *D* is the debt in €, k_E is the return on equity, and k_D is the cost of debt. $LCOT_{desal}$ is the levelized cost of water transportation from the desalination plant to the demand site in €/m^3, $capex_{hpumps}$ is the CAPEX of the horizontal pumps in €/m^3,crf_{hpumps} is the annuity factor for the horizontal pumps, $capex_{vpumps}$ is the CAPEX of the vertical pumps in €/m^3, crf_{vpumps} is the annuity factor for the vertical pumps, $capex_{pipes}$ is the CAPEX for the vertical and horizontal pipes in €/m^3, $opex_{fixedt}$ is the fixed operational expenditures of the pipes and pumps in €/m^3, $opex_{fixed\ pumps}$ is the fixed operational expenditures of the horizontal and vertical pumps in €/m^3, $opex_{fixed\ pipes}$ is the fixed operational expenditures of the horizontal and vertical pipes in €/m^3, $opex_{vart}$ is equal to the *LCOE*, SEC_t is the specific energy consumption of the pumps in kWh/m^3, and *LCOW* is the resulting levelized cost of water in €/m^3.

The approach to calculate the LCOE of hybrid RE power plants is illustrated in Eq. (3) and is further discussed by Bogdanov and Breyer [34,45].

$$LCOE_r = LCOE_{prim,r} + LCOC_r + LCOS_r \tag{3a}$$

$$LCOE_{prim,r} = \frac{\sum_{t=1}^{REtech} \left((capex_t \times crf_t + opexfixed_t) \times cap_{t,r} + opexvar_t \times E_{gen,t,r} \right)}{Edemand_r + Eexp_r - Eimp_r} \tag{3b}$$

$$LCOC_r = LCOE_{prim,r} \times \frac{Ecurt_r}{Edemand_r + Eexp_r - Eimp_r} \tag{3c}$$

$$LCOS_r = \frac{\sum_{t=1}^{Stech}((capex_{st} \times crf_{st} + opexfixed_{st}) \times cap_{st,r} + opexvar_{st} \times E_{sdischarge,t,r}}{Edemand_r + Eexp_r - Eimp_r} \tag{3d}$$

Equation 3

LCOE for a region. Here, $LCOE_{prim,r}$ is the levelized cost of electricity for a primary generation source in the region r in €/kWh, $LCOC_r$ is the levelized cost of curtailment, and $LCOS_r$ is the levelized cost for energy storage in the region in €/kWh. RE_{tech} are the RE technologies utilized for the region, and t represents each of the RE technologies. The value $capex_t$ is the CAPEX for the respective RE technology in €/kW, crf_t is the capital recovery factor, $opexfixed_t$ is the fixed operational expenses of the technologyin €/kW, $cap_{t,r}$ is the installed capacity of the RE technologies in kW, $opexvar_t$ are the variable expenses of the RE technologies in €/kWh, and $E_{gen,t,r}$ is the annual energy generated in the region by the respective technology in kWh. $Edemand_r$ is the annual electricity demand of the region in kWh, $Eexp_r$ is the annual exported energy of a region in kWh, and $Eimp_r$ is the annual imported energy of a region in kWh. Both $Eexp_r$ and $Eimp_r$ are zero in this instance. $Ecurt_r$is the annual energy curtailed in the region in kWh. S_{tech} are the storage energy technologies utilized for the region, $capex_{st}$ is the CAPEX of the respective energy storage technology in €/kWh, crf_{st} is the capital recovery factor, $opexfixed_{st}$ are the fixed operational expenses of the storage technology in €/kWh,$cap_{st,r}$ is the capacity of the respective storage in the region in kWh, $opexvar_{st}$ are the variable operational expenses of the technology in €/kWh, and $E_{sdischarge,t,r}$ is the electricity from discharging the storage technology in the region in kWh.

8.4.2 Technical and Financial Parameters of SWRO Plants in 2030

The technical and financial parameters of the SWRO desalination plants utilized by 2030 can be based on the following literature and assumptions. Table 8.6 presents a summary of the technical and financial parameters that may be assumed globally for SWRO plants in 2030.

8.4.2.1 CAPEX and OPEX of SWRO Plants

Loutatidou et al.[46] present a model that estimates the CAPEX of RO plants located in the Gulf cooperation council (GCC) and Southern Europe. The model determines the CAPEX based on the nominal plant capacity, year of commissioning, type of RO plant (whether feedwater used is seawater or brackish water), and the region of the desalination plant. In addition, the research forecasts the CAPEX of SWRO plants located in the GCC region from 2014 to 2030. Loutatidou et al. [46] show that it is the capacity of the plant that most influences the CAPEX. With the increase in plant capacity, the CAPEX reduces, showing the trend of economies of scale. The region is comparatively less important. Thus, the forecasted CAPEX for SWRO plants in the GCC region in 2030 is assumed to be the same for global SWRO plants. In addition, the CAPEX is taken for large plants with a capacity of 20,000 m^3/day and is estimated to be 2.23 €/(m^3 a).

The OPEX of SWRO desalination plants, collected from the 2010 GWI database, is presented by Fichtner [18]. Energy consumption is the largest contributor to the OPEX of the SWRO plants. Excluding energy consumption, the fixed OPEX is estimated to be about 4% of the CAPEX. It must be noted that the fixed OPEX of SWRO plants may vary between regions. For example, labor and chemical costs may depend on the region. However, in this chapter, a constant value of 4% is assumed.

8.4.2.2 Energy Consumption of SWRO Plants

The SEC (kWh/m^3) of SWRO desalination plants varies with the salinity, temperature of feedwater, and the type of energy recovery device (ERD) used [43]. Current research explains that the most prevalent type of ERD used today is the hydraulic turbine [47]. Zahrani et al. demonstrate the variation of the SEC of SWRO plants with a hydraulic turbine ERD [43]. The data have been extracted and reproduced in Fig. 8.5.

The second-order polynomial $y = 0.0003x^2 + 0.0018x + 2.6049$, where x is the salinity in practical salinity unit (PSU), represents the trend line. The salinity, x, of the feedwater can be found for a specific region from the NCEI World Ocean Database [48]. Therefore, the minimum SEC for SWRO plants with hydraulic turbine ERDs, located in different regions, can be calculated.

8.4.2.3 Water Transportation Costs

Lamei et al. [49] have published a study comparing the production costs for RO desalination and long-distance piping to supply freshwater to tourist resorts in Egypt. The study concluded that after a certain distance and flow capacity, it is economically viable to use RO desalination as opposed to

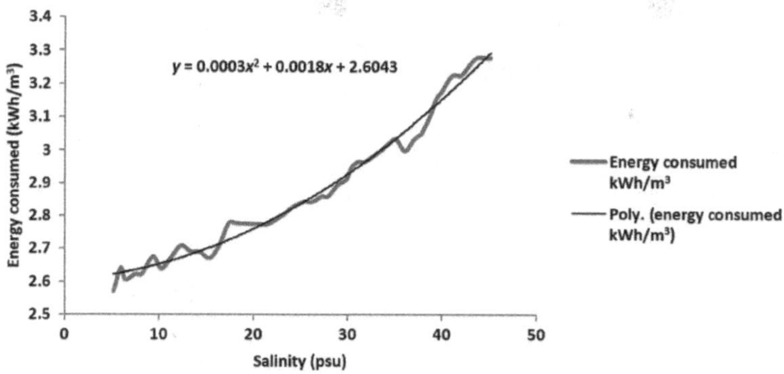

Fig. 8.5 SWRO energy consumption with a hydraulic turbine [43].

long-distance piping. This study highlights the significance of water transportation costs. However, Lamei et al. [49] explain that there is little in the literature about water transportation costs.

The economic and social commission for western Asia (ESCWA) [50] has published a report studying the role of desalination in the region. As part of the report, transportation costs were discussed. The data were based on the paper by Kally (1993) (referred in [50]) about water transport between the Suez and Negev. This is the most cited paper regarding water transportation costs. ESCWA separated the water transportation costs into piping and pumping, excluding the cost of water at the source, for both the horizontal and vertical distances. Table 8.4 summarizes the cost breakdown deduced by the ESCWA.

Missimer et al. [51] did an economic analysis of the cost of supplying desalinated water or treated wastewater to Wadi communities in the Kingdom of Saudi Arabia. As part of the study, Missimer et al. [51] provide the costs of water pumps for water transportation. The details can be summarized as follows:

- The CAPEX of a pumping station of 5000 m^3/day is 500,000 USD. It proportionately with the capacity of the pumping station.
- For a head loss of 120 m, at a specific flow rate, the pumping station can pump 40 km horizontally with an elevation of 70 m. Thus, the head loss due to friction of the pipes is 50 m.
- This implies that the same pump can transport water at the above flow rate horizontally only for 96 km.
- Similarly, the pump can raise water at the above flow rate vertically only for 120 m.

Table 8.4 Water transportation costs breakdown [50]

Horizontal piping (2008 costs in USD)		
CAPEX	USD/m^3/100 km	0.06
OPEX	USD/m^3/100 km	0.03
Horizontal pumping		
Energy consumption	kWh/(m^3/h)/100 km	0.04
Pumping costs	USD/m^3/100 km	0.01
Vertical piping		
CAPEX	USD/m^3/100 km	0.06
OPEX	USD/m^3/100 km	0.03
Vertical pumping		
Energy consumption	kWh/(m^3/h)/100 m	0.36
Pumping costs	USD/m^3/100 m	0.09

Based on the above, the CAPEX for a 5000 m^3/day pump can also be expressed as follows:

$$\text{For vertical pumping, the CAPEX} = 500{,}000\,\text{USD}/\left(5000\,\text{m}^3/24\,\text{h}\right)/120\,\text{m} = 20\,\text{USD}/\left(\text{m}^3/\text{h}\right)/\text{m}.$$

$$\text{For horizontal pumping, the CAPEX} = 500{,}000\,\text{USD}/\left(5000\,\text{m}^3/24\,\text{h}\right)/96\,\text{km} = 25\,\text{USD}/\left(\text{m}^3/\text{h}\right)/\text{km}.$$

8.4.2.4 Water Storage Costs

Table 8.5 lists the cost assumptions for water storage at the desalination site. These assumptions are based on the CAPEX values listed by Al-Nory and El-Beltagy [52] and the Government of Newfoundland and Labrador [53]. The operational expenses have not been specified and are therefore assumed to be 2% of the CAPEX.

Table 8.6 provides a summary of the technical and financial aspects of SWRO desalination plants assumed for 2030. Where necessary, all costs have been inflated to 2014 costs in €, with an exchange rate of 1.30 USD

Table 8.5 Water storage cost assumptions [52,54]

Water storage cost assumptions (2011 costs)		
CAPEX	USD/m^3	100
OPEX	USD/m^3	2% of CAPEX

Table 8.6 Key assumptions for the SWRO plant, water storage, and water transportation in the model for 2030 [43,46–53]

SWRO desalination system		
Capacity	m^3/a	Equal to desalination demand of region
CAPEX	€/(m^3 a)	2.23
Fixed OPEX	€/(m^3 a)	4% of CAPEX
Full load hours (FLH)	h	System optimum
Energy consumption	kWh/m^3	Calculated for desalination site Approximate range is 2.80–3.30. See Fig. 8.5
Lifetime	Years	30
Water storage		
CAPEX	€/m^3	65
Fixed OPEX	€/(m^3)	2% of CAPEX
Lifetime	Years	30
Horizontal pumping and piping		
Horizontal pipes CAPEX	€/(m^3 a km)	0.053
Horizontal pipes Fixed OPEX	€/(m^3 a 100 km)	0.023
Horizontal pump CAPEX	€/(m^3 h km)	19.23
Horizontal pump Fixed OPEX	€/(m^3 h km)	2% of CAPEX
Energy consumption	kWh/(m^3 h 100 km)	0.04
Lifetime	Years	30
Vertical pumping and piping		
Vertical pipes CAPEX	€/(m^3 a km)	0.053
Vertical pipes Fixed OPEX	€/(m^3 a 100 km)	0.023
Vertical pump CAPEX	€/(m^3 h m)	15.40
Vertical pump Fixed OPEX	€/(m^3 h m)	2% of CAPEX
Energy consumption	kWh/(m^3 h 100 m)	0.36
Lifetime	Years	30

to 1 €, which represents the long-term average. Water storage of a minimum of 7 days at the desalination site is assumed. A WACC of 7% is considered. Where specific data of the OPEX are not available, it is assumed to be 2% of the CAPEX.

8.4.3 Technical and Financial Parameters of Hybrid RE Power Plants in 2030

Hybrid RE power plants discussed in this chapter may be a combination of the following technologies:

- PV fixed-tilted plants
- PV single-axis tracking plants
- Wind power plants

In 2016, the newly installed global PV capacity amounted to 75 GW. This is an increase of 50% from the newly installed capacity in 2015. The total cumulative installed capacity by the end of 2016 reached 305 GW [54]. The 2030 CAPEX for PV can be determined on the basis of learning curves as described by Kersten et al. [19] and Liebreich [55]. The 2030 CAPEX is estimated to be 550 €/kW, while the fixed OPEX is 2%. Single-axis tracking PV follows the sun during the day, ensuring that more solar irradiation is captured than fixed-titled PV [56]. The global total installed capacity of single-axis tracking PV was higher than 10 GW in 2016 and is expected to grow sharply. As the learning curves in [19,55] show, with the improvement in efficiency and increase in installations, the CAPEX of single-axis tracking PV will decrease most likely to at least 620 €/kW by 2030. Werner et al. [57] demonstrate the rapid growth in solar energy markets globally, with most regions of the world experiencing at least a 50% increase in cumulative PV capacity in 2015 relative to 2014. The global PV installed capacity, by the end of 2015, was estimated to be 230 GWp where 23% of the total capacity was added in 2015 [57].

Meanwhile, the total installed capacity of wind power plants globally was approximately 433 GW at the end of 2015, an increase of 17% from the previous year [58]. Learning curves [55] show that the CAPEX decrease of wind onshore is rather low compared to PV up to the year 2030. The CAPEX of wind power plants by 2030 is assumed to be 1000 €/kW and the OPEX is 2% of the CAPEX.

Owing to the intermittency of the solar and wind resources, there is a need for energy storage to enable high FLH of the SWRO plants [59]. The Bloomberg New Energy Finances (BNEF) New Energy Outlook 2016 report forecasts that global storage capacities will increase to 25 GW by 2028 from the 1 GW installed today [53]. The storage technologies analyzed in this chapter are battery and PtG [60].

Lithium-ion batteries are currently experiencing by far the fastest growth of all storage options and are being used in small- and utility-scale applications [61]. Consequently, there has been a sharp decline in the CAPEX of batteries as presented by Liebreich from BNEF [55]. The price of the electric vehicle (EV) lithium-ion battery is estimated to have fallen from 770 €/kWh in 2010 to 243 €/kWh in 2015 [55]. The report forecasts that the cost will plunge even more sharply to 162 €/kWh by 2018, a 77% fall in cost between 2010 and 2018. Based on the discussed learning curve rate of 14%–19%, the capital cost of EV is expected to arrive at parity with internal combustion engine cars by 2022 [55]. The core technology of Li-ion batteries does not differ substantially between mobile and stationary applications; thus, this cost trend is thegeneral one for Li-ion battery systems.In this chapter, the CAPEX of battery storage is taken to be 150 €/kWh by 2030 with a fixed OPEX of 10 €/(kWh a).

PtG plants convert renewable electricity to renewable methane that can be stored in gas grids and used as per conventional natural gas. The plants utilize two processes, already used by industries: electrolysis and methanation. Sterner [60] and Lehner et al. [62] explain that the increase in RE systems results in growing demand for PtG storage. This will drive further improvements in PtG technology and reduce costs. PtG technology is already used in several countries [63], such as Germany, and has been commercially introduced by several technology providers. There are plants already at the MW-scale. The car manufacturer Audi invested in a 6.4-MW plant, which is fully integrated in the business model of sustainable methane for their compressed natural gas (CNG) cars in Germany—this plant has been operating regularly since 2013 [64]. Several countries, mainly in Europe, are currently following, such as Sweden, Denmark, Switzerland, Austria, the Netherlands, but also Canada and Japan [65]. The technology itself is based on fundamental processes and components well known in the chemical industry, which makes upscaling easier. A report by the Berlin-based think tank Agora [66] and a paper by Breyer et al. [67] describe the potential for the decrease in costs for PtG plants. PtG plants comprise a water electrolysis unit, a CO_2 scrubber, and a methanation unit. The costs assumed are provided in Table 8.7 below.

Table 8.7 presents the costs and technical parameters for the RE power plants assumed for 2030. A WACC of 7% is considered. Where specific data for the OPEX are not available, they are assumed to be 2% of the CAPEX. The high-voltage direct current (HVDC) transmission grid, illustrated in Fig. 8.5, is assumed to have a power loss of 1.6% per 1000 km. This is further explained in Bogdanov and Breyer [34].

Table 8.7 Costs and technical assumptions for PV, wind, battery, and PtG hybrid power plants in 2030 [19,55,66,67]

Fixed-tilted PV plant		
CAPEX	€/kW	550
OPEX	€/(kW a)	1.5% of CAPEX
Lifetime	Years	35
Single-axis tracking PV plant		
CAPEX	€/kW	620
OPEX	€/(kW a)	1.5% of CAPEX
Lifetime‘	Years	35
Wind plant		
CAPEX	€/kW	1000
OPEX	€/(kW a)	2% of CAPEX
Lifetime	Years	25
Battery system		
CAPEX	€/kWh	150
Fixed OPEX	€/(kWh a)	10
Lifetime	Years	15
Efficiency	%	90
PtG		
Water electrolysis CAPEX	€/kW,$_{H2}$	380
Water electrolysis Fixed OPEX	€/(kW,$_{H2}$ a)	13
Water electrolysis efficiency	%	84
Water electrolysis lifetime	Years	30
CO_2 direct air capture CAPEX	€/kW,$_{SNG}$	356
CO_2 direct air capture OPEX	€/(kW,$_{SNG}$ a)	14.24
CO_2 direct air capture efficiency	%	78
CO_2 direct air capture lifetime	Years	30
Methanation CAPEX	€/kW,$_{SNG}$	234
Methanation OPEX	€/(kW,$_{SNG}$ a)	5
Methanation efficiency	%	77
Methanation lifetime	Years	30
Gas storage	€/kWh	0.05
Gas storage efficiency	%	100
Gas storage lifetime	Years	50

8.4.4 Methodology

Based on the equations presented in Section 8.4.1 and the 2030 technical and financial parameters listed in Sections 8.4.2 and 8.4.3, it is possible to determine the LCOE and the corresponding LCOW for a region. However, to

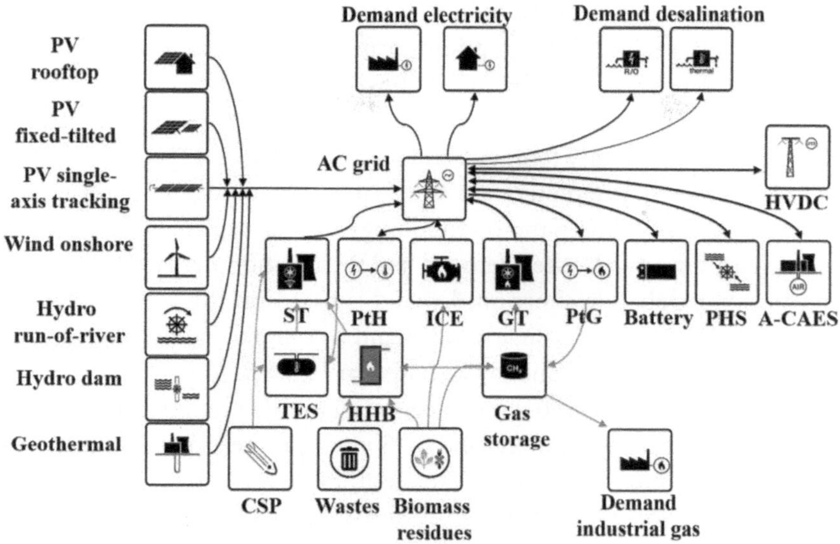

Fig. 8.6 Block diagram of the LUT energy systems model [34,68].

find the optimal combination of RE plants to power the SWRO desalination demand, on a global scale, an energy systems model is required.

For the design and analysis of the optimal RE power plants, the Lappeenranta University of Technology (LUT) energy systems model [34] is used. The energy model allows for the design of local, national, regional, or global energy systems. Fig. 8.6 is a block diagram of the energy systems model. It is composed of all relevant power generation and storage technologies, respective installed capacities, and different operation modes of these technologies. The three main components of the energy model can be listed as below:

- Technologies for the conversion of renewable resources into electricity
- Energy storage technologies
- Energy sector bridging technologies to provide more flexibility to the complete energy system

A key requirement of the LUT energy systems model is to match the power generation and demand for every hour of the applied year. The hourly modeling enables a more accurate system description including the synergy effects of different system components for power system balance [34]. The aim of the system optimization is to minimize the total annual cost of the installed capacities of the different technologies, cost of energy generation, and cost of generation ramping. The resulting system is then considered to be an optimal system design.

The full description of the model, as well as its input data including RE resources and technical assumptions, can be found in Bogdanov and Breyer [34], Breyer et al. [45], and Barbosa et al. [68].The energy model is based on the linear optimization method with interior point optimization and designed in an hourly temporal and 0.45 × 0.45 degree spatial resolution.

The following parameters are input into the model as explained in Caldera et al. [44]. The same methodology as that explained in Ref. [44] is used with improved water demand and desalination demand data:

1. The 2030 global desalination demand is based on the methodology in Section 8.3.2. The desalination demand is found for a spatial resolution of 0.45 × 0.45 degree and is illustrated in Fig. 8.3.
2. The solar irradiation and wind energy data for the year 2005 are used as a reference in hourly temporal and 0.45 × 0.45 degree spatial resolution as described in more detail in Breyer et al. [45].
3. The optimal location of SWRO desalination plants is based on the shortest distance from the demand node to the coast. In addition, the elevation of the demand node from sea level is taken to be the final water pumping height.
4. Energy consumption of SWRO plants is based on salinity of the plant's feedwater, as discussed in Section 8.4.2. The energy costs for water transportation to the demand site are found. This provides the total energy required to meet the desalination demand in a region.
5. Hybrid renewable power plant costs for 2030 comprise single-axis tracking PV, fixed-tilted PV, wind, batteries, and PtG power plant CAPEX and OPEX. SWRO desalination plant CAPEX and OPEX values for 2030, including water storage costs at the desalination site, are input as per Section 8.4.3.

Based on the described input parameters, the model determines the least cost hybrid RE power plant and SWRO plant system. The least cost system that meets the 2030 desalination demand of regions with water scarcity is the optimal system.

8.5 GLOBAL LCOW ESTIMATE FOR 2030

The model's results show that the least cost system to meet the 2030 desalination demand comprises a combination of fixed-tilted PV, single-axis tracking PV, wind energy plants, batteries, and PtG power plants. This energy system allows for a 11.6% reduction in the global average LCOW compared to the energy system without PtG power plants. The use of both batteries and PtG enables a higher fraction of usable FLH of the RE plant and therefore better utilization of the installed SWRO and water supply capacity. This results in lower LCOW. The figures that follow present an overview of the energy system, desalination system, and the resulting global LCOW.

Fig. 8.7 (*top*) presents the required installed capacities of hybrid PV–Wind power plants to meet a region's 2030 global desalination demand. A region is defined as an area of about 50 × 50 km (exactly 0.45 × 0.45 degree in units of latitude and longitude) and (*bottom*) presents the contribution of PV to the hybrid PV–Wind power plant energy generation.

PV power plants provide up to 85% of the total global energy demand with single-axis tracking accounting for 56% of this supplied energy. The higher contribution of PV power plants is attributed to the fact that there is higher desalination demand, and thus higher SWRO capacity, in regions where PV is more favorable than wind. Based on Fig. 8.7 (*top*), the global installed capacities of fixed-tilted PV, single-axis tracking PV, and wind power plants required are approximately 1060, 2072, and 598 GW, respectively. Most regions require total hybrid PV–Wind capacities of up to 5 GW. In addition, PV contributes toward at least 50% of the total generation in Asia, the west coast of America, and parts of Africa.

Fig. 8.8 (*top*) and (*bottom*) show the additional required capacities of batteries and PtG power plants. Batteries and PtG plants allow high FLH and therefore better utilization of the desalination capacity.

The required installed battery capacity is approximately 2011 TWh_{cap}. These batteries are cycled almost every day, reaching about 300 cycles per year and therefore providing up to 24% of the total global energy demand for the desalination demand. PtG contributes up to 15% of the total global energy demand with a required installed capacity of approximately 316 GW_e.

The results in Fig. 8.9 (*top*) show the optimal water storage required to obtain the lowest LCOW in all regions. The value ranges between 7 and 8 days. This can be explained through Fig. 8.9 (*bottom*). The figure illustrates the excess energy generated by the complete system. On a global average, the excess energy is about 6% of the total generated electricity.

The LCOE of the complete system to meet the 2030 desalination demand is presented in Fig. 8.10 (*top*). The LCOE of the system ranges from 0.06 to 0.15 €/kWh. The prevalent LCOE range is 0.06–0.12 €/kWh. Fig. 8.10 (*center*) presents the resulting global LCOW. The final LCOW range is between 1.0 and 4.5 €/m^3, mostly prevalent between 1.0 and 2.0 €/m^3.

Fig. 8.10 (*bottom*) explains the significance of the water transportation costs to the LCOW. The energy required for water pumping, and therefore

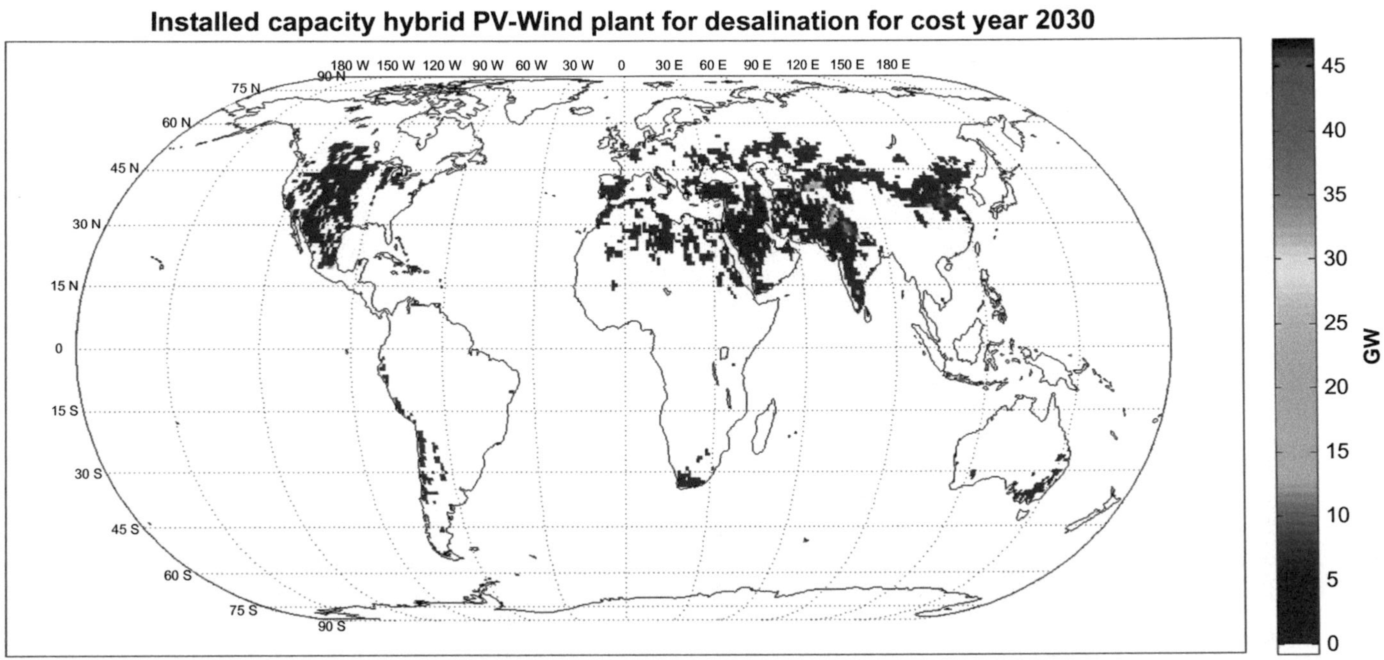

Fig. 8.7 *Top*: Installed hybrid PV–Wind power plant capacity for the 2030 desalination demand.

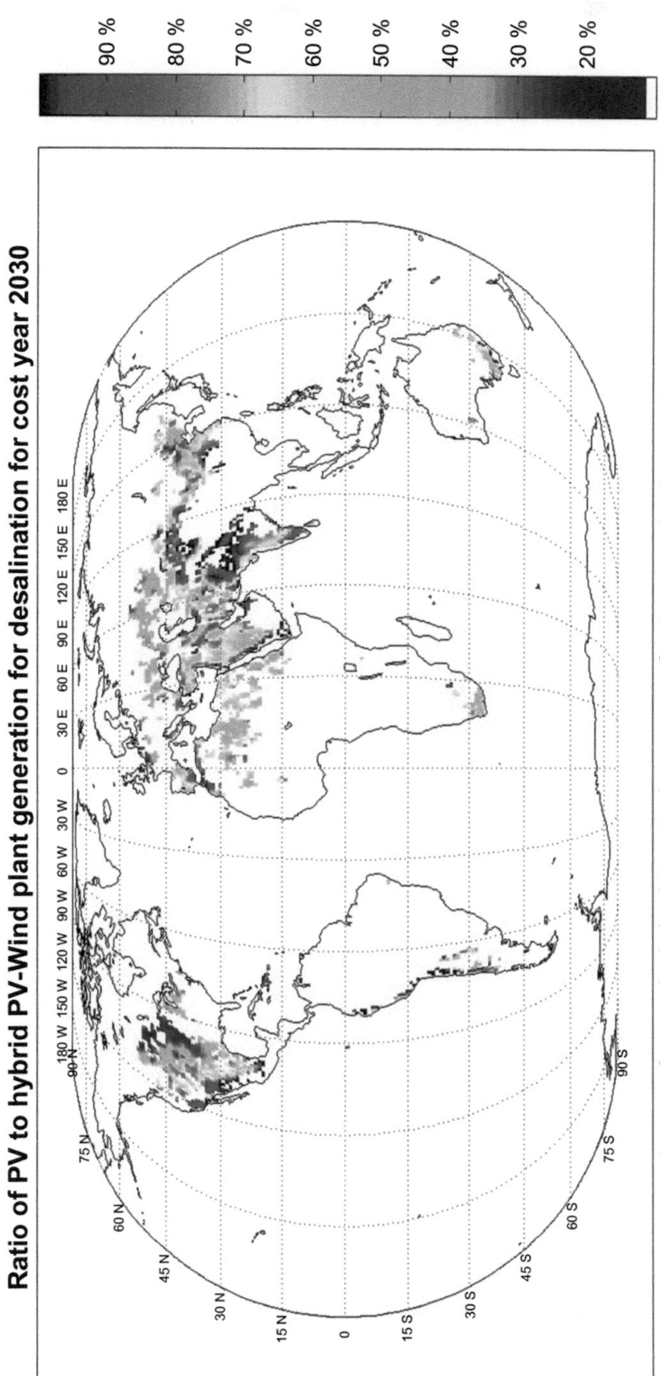

Fig. 8.7 *Bottom:* Contribution of PV to the hybrid PV–Wind generation for the 2030 desalination demand.

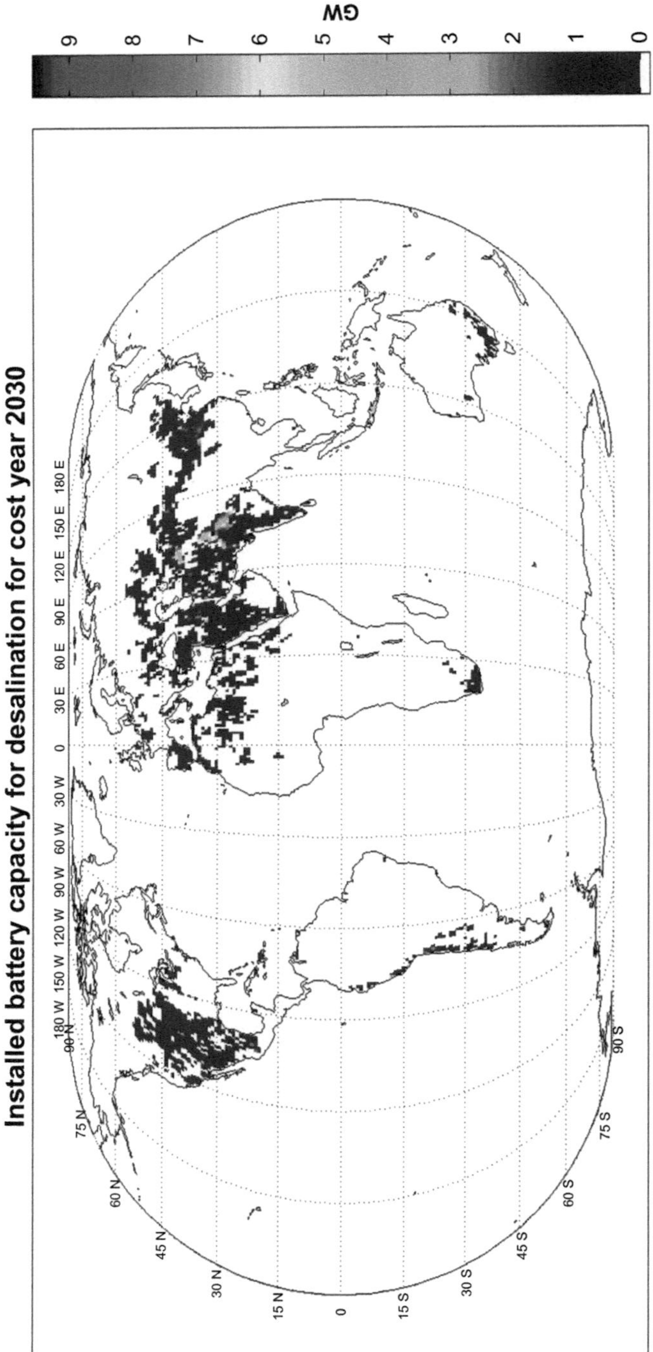

Fig. 8.8 *Top:* Installed battery capacity required for the2030 desalination demand.

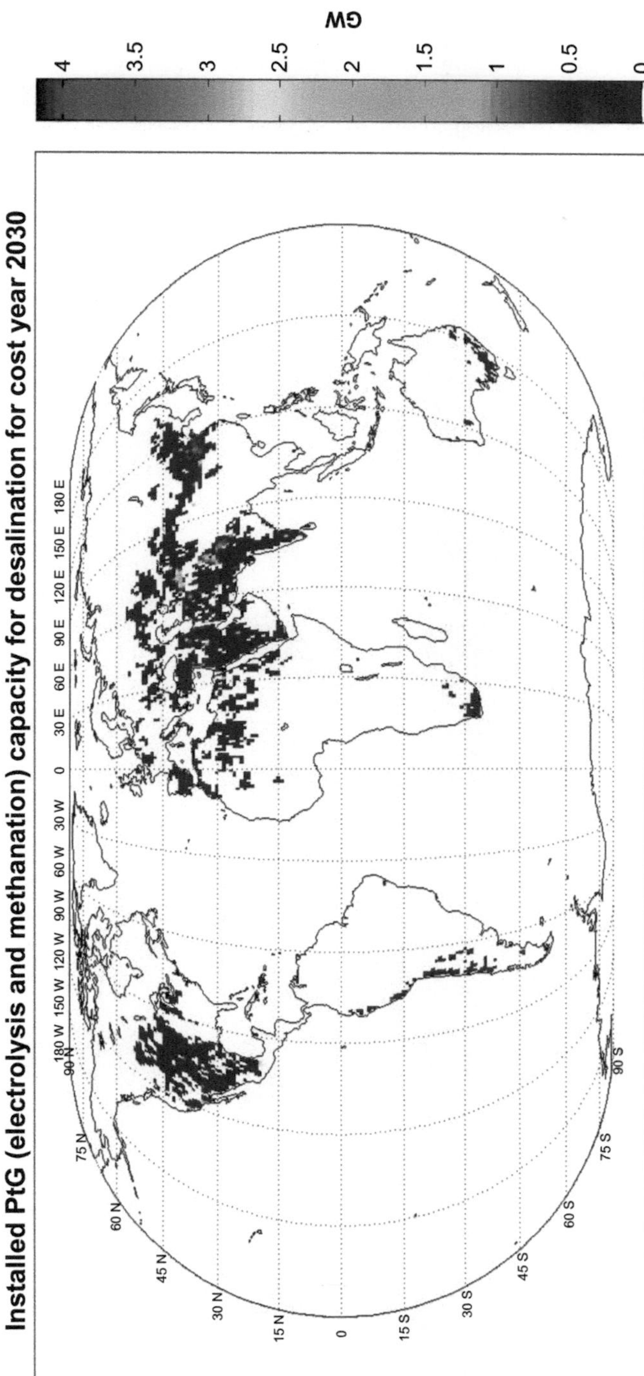

Fig. 8.8 *Bottom:* Installed power-to-gas (PtG) capacity required for the 2030 desalination demand.

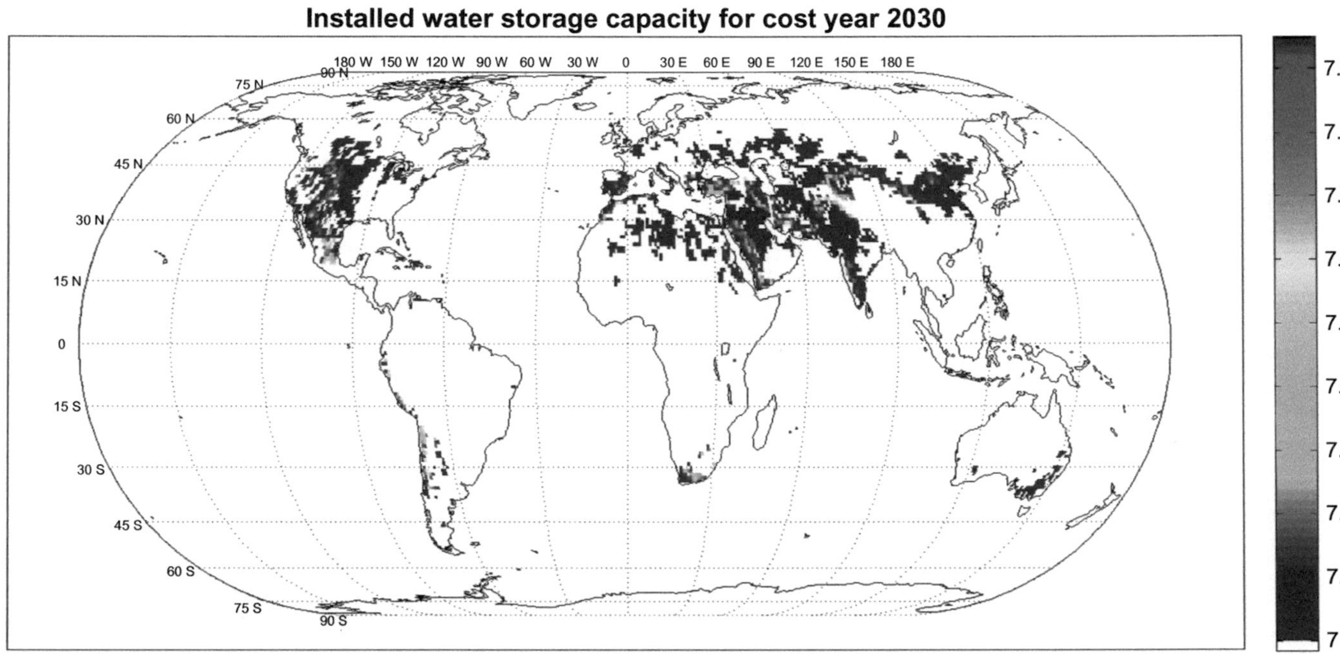

Fig. 8.9 *Top*: Optimal water storage capacity needed for the 2030 system.

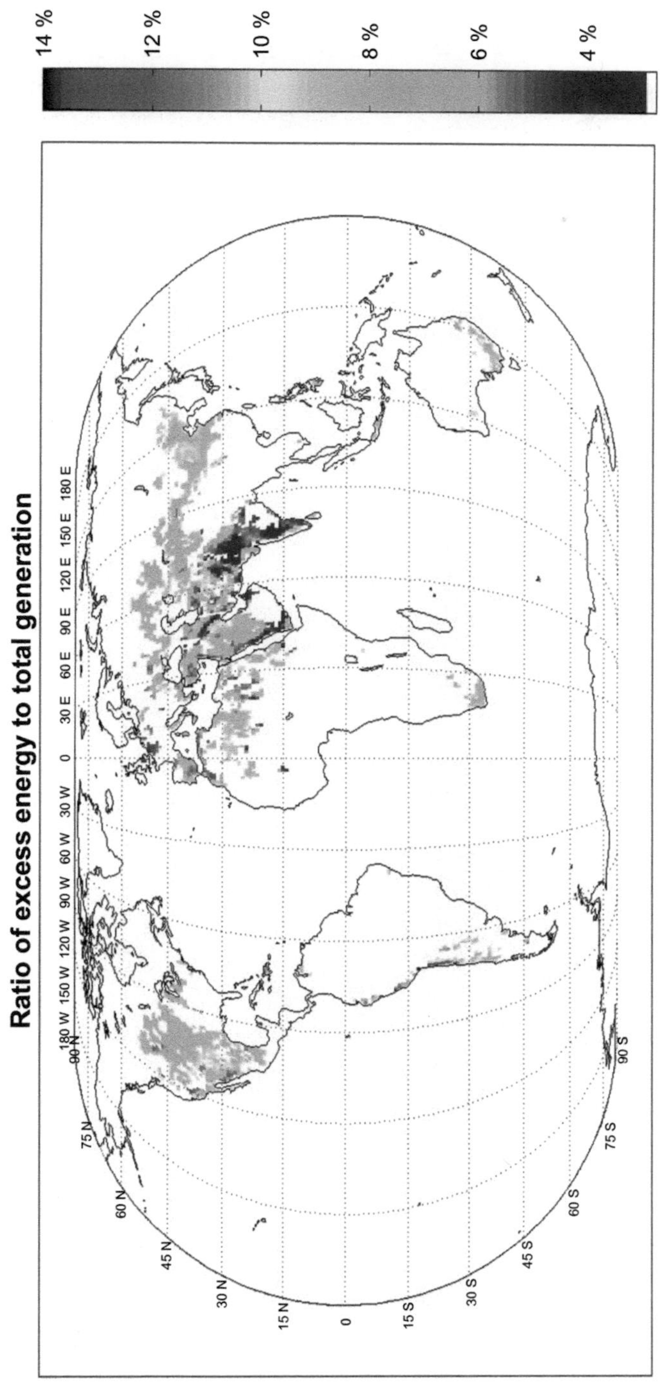

Fig. 8.9 *Bottom:* Ratio of the excess energy to the total energy generated for the 2030 system.

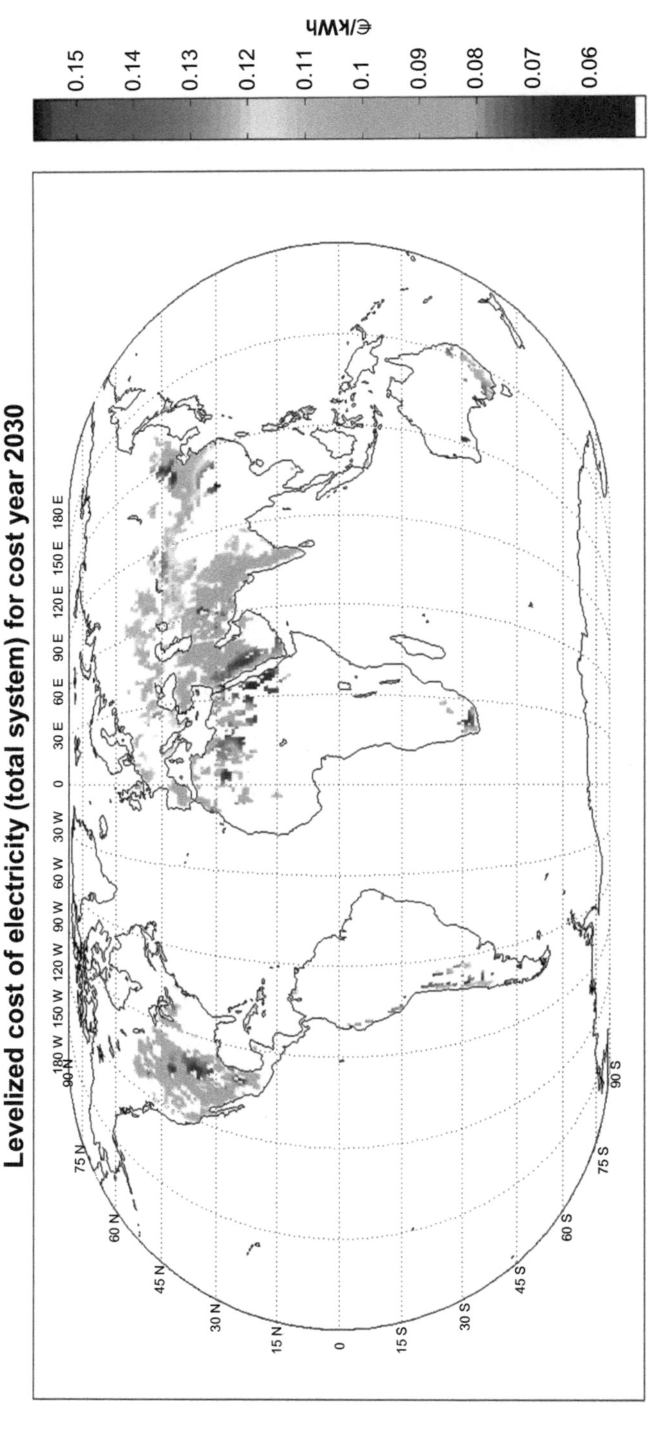

Fig. 8.10 *Top:* LCOE of the complete system for the 2030 system.

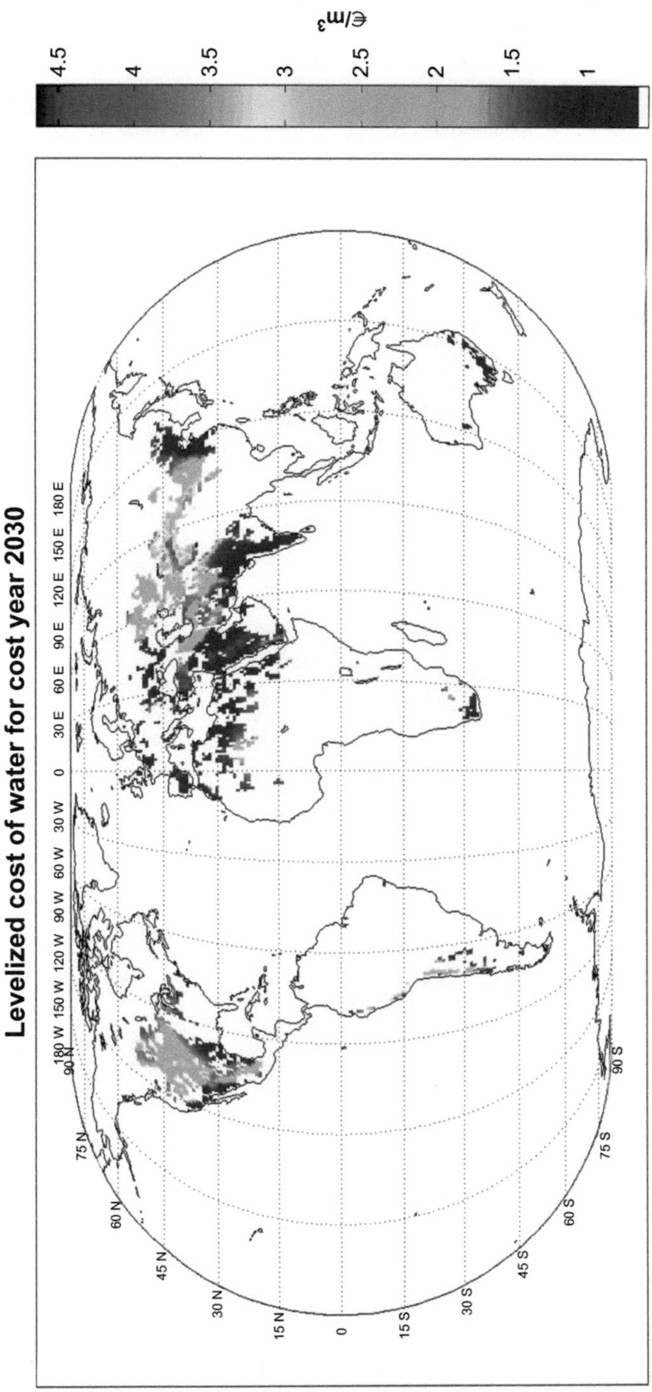

Fig. 8.10 *Center:* LCOW of the complete system for the 2030 system.

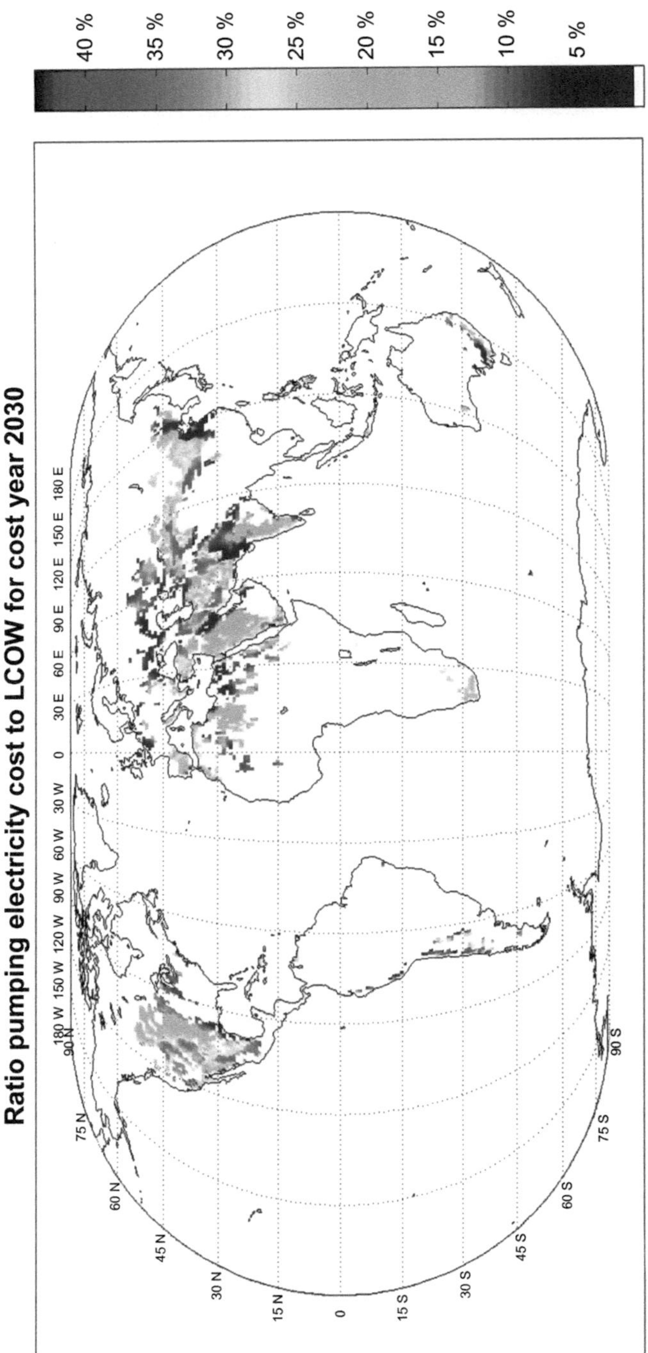

Fig. 8.10 *Bottom*: Pumping electricity cost contribution to the LCOW for the 2030 system.

the pumping electricity costs, are significant in regions where the demand site is distant from the desalination site. This includes both the vertical and horizontal distances. In regions like Central Asia, the electricity cost for pumping can contribute up to 40% toward the LCOW. The corresponding LCOW in this region is approximately 3.0–4.5 €/m^3. Globally, the average pumping electricity cost contribution to the LCOW is about 16% and the total OPEX contribution to the LCOW is approximately 39%.

To conclude, Fig. 8.11 illustrates the cost contribution to the final LCOW. Fig. 8.9 highlights the contribution of the total system CAPEX to the LCOW for regions with desalination demand in 2030. The CAPEX contributes between 55% and 80% to the final LCOW. The average global CAPEX contribution to the LCOW is approximately 62%. The closer the desalination demand is to the coast, the higher the contribution of the CAPEX. The further it is away from the coast, the contribution of electricity required for pumping increases, reducing the contribution of the CAPEX.

The LUT energy systems model was also used to analyze the energy system without PtG power plants as an option. After comparing the two results, it was found that the use of PtG reduced the demand for PV plants by 31%, battery storage by 21.9%, water storage by 36%, and excess energy share by 81%, all on a global average. Meanwhile, the use of PtG increased the wind power plants required by 15%. Thus, PtG enables a more optimal energy system, resulting in a 11.6% decrease in the total CAPEX and 11.4% decrease in annualized costs.

Table 8.8 summarizes the key parameters of the RE-powered SWRO system by 2030. The costs are based on a WACC of 7%. The total annualized costs required to meet the global desalination demand by 2030 is estimated to be 1260 b€, with a total cumulative CAPEX of 10,190 b€. According to Fichtner, the costs of water production at fossil powered SWRO plants today are between 0.6 and 1.9 €/m^3. The higher LCOW costs in Fig. 8.10 can be attributed to the high water transportation costs. The total required CAPEX seems to be remarkably high. However, this investment is equivalent to eradicating the lack of clean water available to humankind.

8.6 OPPORTUNITIES AND FUTURE DIRECTIONS FOR RESEARCH AND DEVELOPMENT

The results above illustrate that it is indeed possible to meet future global desalination demand with RE powered desalination, at costs competitive

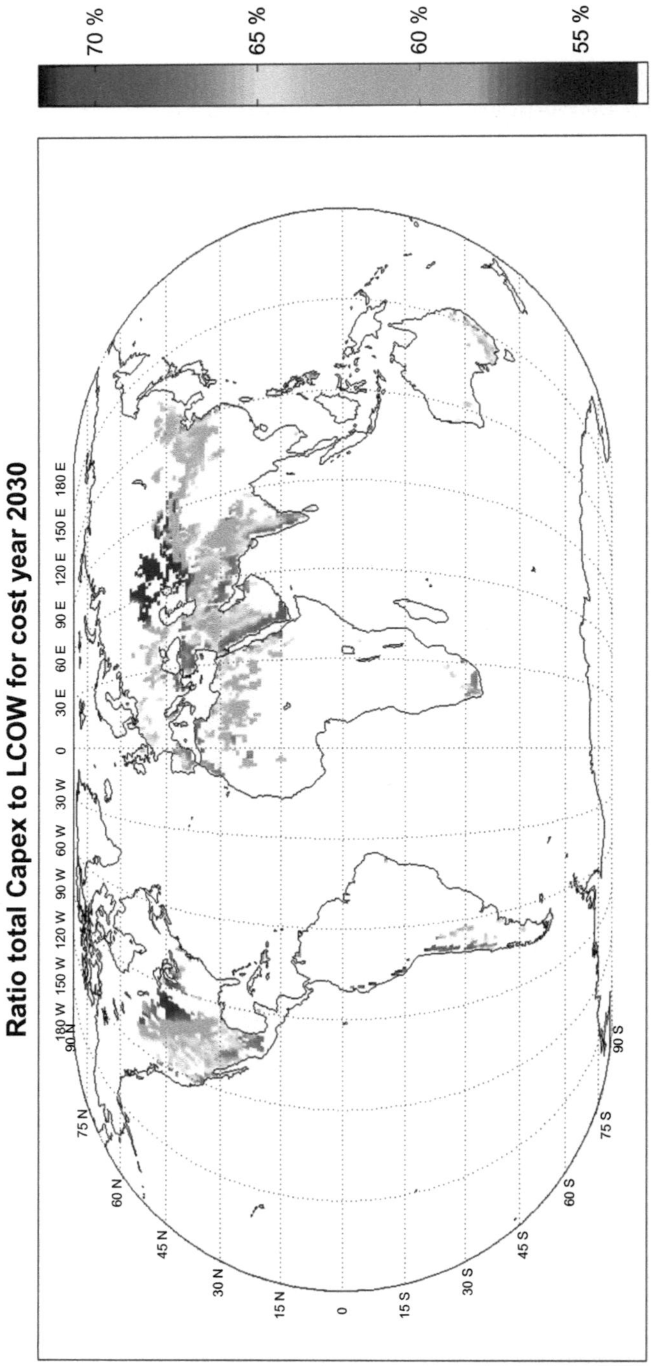

Fig. 8.11 Total system CAPEX contribution to the final LCOW.

Table 8.8 Summary of the key technical and cost data for the 2030 global system

		2030 global system
SWRO total global desalination demand per day	m^3/day	2.6 billion
Installed PV capacity	GW	3730
Fixed-tilted PV	GW	1060
Single-axis tracking PV	GW	2072
Installed wind capacity	GW	598
Installed battery capacity	TWh_{cap}	2011
Installed PtG capacity	GW_e	316
Average global excess energy generated	%	6
Total CAPEX	b€	10,190
Annualized costs	b€	1260
LCOW range	€/m^3	1.00–4.50
Prevalent LCOW range	€/m^3	1.00–2.00

with that of fossil powered plants today. With the continual decrease in CAPEX and increase in efficiency of SWRO and RE power plants, RE energy powered desalination is poised to become a lucrative solution to the global water crisis.

This chapter aims to provide the reader insights into how 100% RE powered SWRO can be optimally designed in all parts of the world facing water stress. However, there are certain gaps in the research that when answered will provide a more comprehensive analysis of how desalination and RE can work together to meet global water needs. These areas of research are summarized below:

1. There is currently no well-defined learning curve for SWRO desalination plants. This makes it difficult to project future SWRO costs in the same way as for PV plants or battery storage.
2. There is a lack of updated water transportation costs. The recent literature makes reference to the 1993 water transportation costs. In addition, the costs of water transportation will depend also on the piping used, which is in turn based on soil conditions. Thus, to get better water transportation costs, the study should take into account more accurate costs and the type of piping used.
3. There is incomplete knowledge of impacts of the integration of desalination on the power sector of a country. Initial results of this study, based on Saudi Arabia, are discussed in Refs. [69,70]. The integration of the desalination sector with the power sector increases the flexibility of

the power system by enabling better utilization of the hourly produced RE. The results in Ref. [70] show that the integration of the desalination sectors with the power and gas sectors results in a decrease varying from 1% to 6% of the annualized costs. Hence, the integrated energy system can be run in a more efficient way than the separated systems. Using a similar approach as that for Saudi Arabia, the authors aim to understand the effects of the integration of the desalination and power sectors on a global scale. The limited flexibility of SWRO desalination plants within the power system, relative to battery systems, due to the higher CAPEX requirements, is discussed in Ref. [59].

4. Multiple effect distillation (MED) desalination technology consumes less electrical energy than SWRO plants but requires additional thermal energy [2]. Thus, in the presence of very low-cost heat, MED offers a viable alternative to SWRO plants. In our research [70], the power sector and the industrial gas sector in Saudi Arabia did not offer much waste heat that can be captured by MED plants. However, if the heat sector waste was added too, perhaps the results may be different. In addition, in regions with geothermal potential, waste heat from geothermal electricity generation can be utilized for MED desalination [71]. Perhaps this might be an option for regions in China and Central Asia that have geothermal potential [72].

8.7 CONCLUSION

In this chapter, it is shown how and why it is possible to meet the increasing global water demand with SWRO desalination plants powered by hybrid RE power plants. Hybrid PV–Wind, battery, and PtG plants can power the global SWRO capacity by 2030. The LCOW range is found to be between 1.0and 4.5 €/m^3. The prevalent LCOW range is between 1.0 and 2.0 €/m^3.

Fichtner [18] presented water production costs for fossil powered SWRO desalination plants located in different regions. The costs range approximately between 0.6 and 1.9 €/m^3. The higher LCOW range of the model can be attributed to the high water transportation costs that contribute up to 40% toward the final LCOW. The costs presented by Fichtner do not include water transportation costs. The electricity costs for water pumping, in both horizontal and vertical distances, can make a substantial contribution to the final water production cost. On average, globally, electricity for pumping contributes up to 17% toward the final LCOW.

In the near future, the production cost of water from the hybrid system will be competitive with the cost of fossil powered SWRO plants today. This will be driven by the increase in desalination demand and the decrease in PV and storage costs due to learning curve effects. In addition, through the approaches discussed in Section 8.6, the costs of water production may be further optimized.

REFERENCES

[1] Global Water Intelligence Desal Data. Renewable desalination, www.desaldata.com/desalination-markets-2016/section-18, 2017 (accessed 02.02.2017).
[2] Al-Kharaghouli A, Kazmerski LL. Energy consumption and water production cost of renewable energy powered desalination processes. Renew Sust Energ Rev 2013;24:343–56.
[3] United Nations World Water Assessment Programme. The United Nations world water development report: water and jobs. Paris: UNESCO; 2016. 164 p.
[4] Virgili F. Desalination market update Third Quarter Assessment. Global Water Intelligence, www.desaldata.com/forecasts, 2016 (accessed 02.02.2017).
[5] Grubert AE, Stillwell AS, Webber ME. Where does solar-aided seawater desalination make sense? A method for identifying suitable sites. Desalination 2014;339:10–7.
[6] Water Online, Ghalilah Desalination Plant honored with distinction award for desalination plant of the year at 2016 Global Water Awards, www.wateronline.com/doc/ghalilah-desalination-plant-honored-plant-of-the-year-at-global-water-awards-0001, 2016.
[7] United Nations World Water Assessment Programme. The United Nations world water development report 2014: water and energy. Paris: UNESCO; 2014.
[8] Mentis D, Karalis G, Zervos A, Howells M, Taliotis C, Bazilian M, et al. Desalination using renewable energy sources on the arid islands of the south Aegean Sea. Energy 2016;94:262–72.
[9] International Renewable Energy Agency. Water desalination using renewable energy. Technology Brief, www.irena.org/DocumentDownloads/Publications/IRENA-ETSAP%20Tech%20Brief%20I12%20Water-Desalination.pdf.Abu Dhabi: IRENA, 2012 (accessed 02.02.2017).
[10] Dawoud MA, Al-Mulla MM. Environmental impacts of seawater desalination: Arabian gulf case study. Int J Environ Sustain 2012;1:22–37.
[11] Global Water Intelligence Desal Data. Plants, www.desaldata.com/projects, 2017 (accessed 02.02.2017).
[12] Ghermandi A, Messalem R. Solar-driven desalination with reverse osmosis: the state of the art. Desalin Water Treat 2009;7:285–96.
[13] Global water intelligence desal data. Desal knowledge solar desalination, www.desaldata.com/desalination-markets-2016/subsection-32, 2017 (accessed 02.02.2017).
[14] Water-technology, Al-Khafji solar saline water reverse osmosis (Solar SWRO) Desalination Plant, Saudi Arabia, www.water-technology.net/projects/al-khafji-solar-saline-water-reverse-osmosis-solar-swro-desalination-plant/, 2017 (accessed 02.02.2017).
[15] Global Water Intelligence Desal Data. Desal knowledge Wind desalination, www.desaldata.com/desalination-markets-2016/subsection-33, 2017 (accessed 02.02.2017).
[16] Novosel T, Cosic B, Puksec T, Krajacic G, Duic N, Mathiesen BV, et al. Integration of renewables and reverse osmosis desalination—case study for the Jordanian energy system with a high share of wind and photovoltaics. Energy Trans 2015;92:270–8.

[17] Fthenakis V, Atia AA, Morin O, Bkayrat R, Sinha P. New prospects for PV powered water desalination plants: case studies in Saudi Arabia. Prog Photovolt Res Appl 2016;24:543–50.
[18] Fichtner. MENA regional water outlook part 2 desalination using renewable energy. Final report, Stuttgart: Fichtner; 2011.
[19] Kersten F, Doll R, Kux A, Huljić D M, Görig M A, Breyer Ch, Müller J, Wawer P. PV-learning curves: past and future drivers of cost reduction, 26th EU PVSEC, Hamburg, September 5–9, 2011. doi: https://doi.org/10.4229/26thEUPVSEC2011-6CV.1.63. http://bit.ly/2iYu0OO.
[20] Gulf News. Solar energy to power Dubai desalination plants, http://gulfnews.com/news/uae/government/solar-energy-to-power-dubai-desalination-plants-1.1978217, Dubai: Al Nisr Publishing; 2017.
[21] Ghaffour N, Missimer MT, Amy LG. Technical review and evaluation of the economics of water desalination: current and future challenges for better water supply sustainability. Desalination 2013;309:197–207.
[22] Global Water Intelligence Desal Data. Desalknowledge. Forecast Trends, www.desaldata.com/desalination-markets-2016/subsection-245, 2017 (accessed 02.02.2017).
[23] Global Water Intelligence Desal Data. Desalknowledge Offtakers, www.desaldata.com/desalination-markets-2016/subsection-239, 2017 (accessed 02.02.2017).
[24] Global Water Intelligence Desal Data. Q4 2016 Market Update, www.desaldata.com/forecasts, 2016 (accessed 02.02.2017).
[25] Virgili F, Pankratz T. Desalination market update second quarter assessment. Global Water Intelligence, https://www.desaldata.com/forecasts, 2016 (accessed 02.02.2017).
[26] Landis M, Gassert F. Aqueduct water stress projections: decadal projections of water supply and demand using CMIP5 GCMs, www.wri.org/sites/default/files/aqueduct-water-stress-projections-technical-note.pdf. Washington DC: World Resources Institute; 2015 (accessed 09.09.2015).
[27] Reig P, Shiao T, Gassert F. Aqueduct water risk framework. Working paper, http://pdf.wri.org/aqueduct_water_risk_framework.pdf. Washington DC: World Resources Institute; 2013 (accessed 09.09.2015).
[28] iSciences. Freshwater sustainability analyses: interpretive guidelines, http://docs.wri.org/aqueduct/freshwater_sustainability_analyses.pdf, 2011 (accessed 09.09.2015).
[29] Biesheuvel A, Cheng I, Liu X, Koornneef J. Method and results report. Modelling water demand for coal based power generation, www.greenpeace.org/international/Global/international/briefings/climate/2016/Coal-Water/Technical-Report-Coal-Water-Methods-Results.pdf. St Petersburg/Amsterdam: Bos + Witteveen/Greenpeace 2016 (accessed 09.02.2017).
[30] Breyer C, Bogdanov D, Gulagi A, Aghahosseini A, Barbosa LSNS, Koskinen O, et al. On the role of solar photovoltaics in global energy transition scenarios, Prog Photovolt Res Appl 2017;25:727–45. https://doi.org/10.1002/pip.2885.
[31] Breyer Ch, Koskinen O, Bogdanov D. A low-cost power system for Europe based on renewable electricity. European Utility Week November 2016, Barcelona, Spain. http://bit.ly/2kZcNTB.
[32] Bogdanov D, Ch B. Eurasian super grid for 100% renewable energy power supply: generation and storage technologies in the cost optimal mix. ISES solar world congress, Daegu, Korea; 2015http://bit.ly/1IP8AeN.
[33] Gulagi A, Bogdanov D, Choudhary P, Breyer Ch. Electricity system based on 100% renewable energy for India and SAARC. PLoS ONE 2017;12:e0180611. https://doi.org/10.1371/journal.pone.0180611.
[34] Bogdanov D, Breyer C. North-east Asian super grid for 100% renewable energy supply: Optimal mix of energy technologies for electricity, gas and heat supply options. Energy Convers Manag 2016;112:176–90.

[35] Bogdanov D, Ch B. The role of solar energy towards 100% renewable power supply for Israel: integrating solar PV, wind energy, CSP and storages. Proceedings of the 19th Sede Boqer Symposium on Solar Electricity Production, Sede Boqer, Israel; 2015http://bit.ly/1JlaiDI.
[36] Connolly D, Mathiesen BV. A technical and economic analysis of one potential pathway to a 100% renewable energy systems. Int J Sustain Energy Plan Manage 2014;1:7–28.
[37] Center for Alternative Technology. Who is getting ready for zero? A report on the state of play of zero carbon modelling, http://zerocarbonbritain.org/en/ready-for-zero. Powys, Wales, 2016 (accessed 09.02.2017).
[38] Neo-Carbon Energy. Internet of Energy—online visualization tool, http://neocarbonenergy.fi/internetofenergy/#. Lappeenranta University of Technology, Lappeenranta, 2016 (accessed: 9.02.2017).
[39] Food and Agricultural Organisation of the United Nations, AQUASTAT database, www.fao.org/nr/water/aquastat/data/query/index.html?lang=en. Rome: FAO, 2015 (accessed 16.04.2015).
[40] Spang ES, Moomaw WR, Gallagher KS, Kirshen PH, Marks DH. The water consumption of energy production: an international comparison. Environ Res Lett 2014;9:14.
[41] The National Groundwater Association. Facts about global groundwater usage, www.ngwa.org/Fundamentals/Documents/global-groundwater-use-fact-sheet.pdf, Ohio, 2016 (accessed 02.02.2017).
[42] Ministry of Water and Electricity. National water strategy: the roadmap for sustainability, efficiency and security of the water future in the KSA National Water Strategy, www.sawea.org/pdf/waterarabia2013/Tue_Panel_Discussion/National_Water_Strategy.pdf, presentation for Water Arabia, 2013 (accessed 16.04.2015).
[43] Al-Zahrani A, Orfi J, Al-Suhaibani Z, Salim B, Al-Ansary H. Thermodynamic analysis of a reverse osmosis desalination unit with energy recovery system. Proc Eng 2012;33:404–14.
[44] Caldera U, Bogdanov D, Breyer C. Local cost of seawater RO desalination based on solar PV and wind energy: a global estimate. Desalination 2016;385:207–16.
[45] Breyer C, Bogdanov D, Komoto K, Ehara T, Song J, Enebish N. North-East Asian super grid: renewable energy mix and economics. Jpn J Appl Phys 2015;54. 08KJ0.
[46] Loutatidou S, Chalermthai B, Marpu PR, Arafat H. Capital cost estimation of RO plants: GCC countries versus southern Europe. Desalination 2014;347:103–11.
[47] Gude V. Energy consumption and recovery in reverse osmosis. Desalin Water Treat 2011;36:239–60.
[48] National Center for Environmental Information. World Ocean Atlas 2013, www.nodc.noaa.gov/OC5/woa13/woa13data.html. North Carolina, 2013 (accessed 15.06.2015).
[49] Lamei A, Zaag van der P, Münch von E. Basic cost equations to estimate unit productivity costs for RO desalination and long-distance piping to supply water to tourism-dominated arid coastal regions of Egypt. Desalination 2008;225:1–12.
[50] Economic & Social Commission for Western Asia. Role of desalination in addressing water scarcity, www.escwa.un.org/information/publications/edit/upload/sdpd-09-4.pdf, Beirut, 2009 (accessed 15.06.2015).
[51] Missimer MT, Maliva GR, Ghaffour N, Leiknes T, Amy GL. Managed aquifer recharge (MAR) economics for wastewater reuse in low population Wadi communities, Kingdom of Saudi Arabia. Water 2014;6:2322–38. https://doi.org/10.3390/w6082322.
[52] Al-Nory M, El-Beltagy M. An energy management approach for renewable energy integration with power generation and water desalination. Renew Energy 2014;72:377–85.
[53] Government of Newfoundland & Labrador Department of Environment and Conservation Water Resources Management Division. Evaluation of potable water storage tanks in Newfoundland and Labrador and their effect on drinking water quality,

www.env.gov.nl.ca/env/waterres/reports/drinking_water/Tank_Report_July_12_2011.pdf. Newfounland and Labrador, 2011 (accessed on 15.06.2015).
[54] The PV Market Alliance. PV Market Alliance announces the 2016 PV market at 75 GW and a stable market at 2017, www.pvmarketalliance.com/pv-market-alliance-announces-the-2016-pv-installations-at-75-gw-and-a-stable-market-in-2017/, 2017 (accessed 28.02.2017).
[55] Liebreich M. Bloomberg New Energy Finance Summit, in search of the miraculous. New York: Bloomberg New Energy Finance; 2016.
[56] Breyer C, Schmid J. Population density and area weighted solar irradiation: global overview on solar resource conditions for fixed tilted, 1-axis, 2-axis PV systems. 25th EUPVSEC, Valencia, Spain; 2010http://bit.ly/2kEpz8w.
[57] Werner C, Gerlach A, Breyer C, Masson G. Global photovoltaics in 2015—Analysis of current solar energy markets and hidden growth regions. 32nd EUPVSEC, Munich, Germany; 2016http://bit.ly/2lrc7Vy.
[58] Global Wind Energy Council. Global wind report. Annual market update, www.gwec.net/wp-content/uploads/vip/GWEC-Global-Wind-2015-Report_April-2016_19_04.pdf, GWEC, Brussel, 2015 (accessed 28.02.2017).
[59] Caldera U, Breyer C. Impact of battery and water storage on the transition to an integrated 100% renewable energy power system for Saudi Arabia. To be presented at 11th IRES March 2017, Energy Procedia, forthcoming, http://bit.ly/2mXiUbZ.
[60] Sterner M, Bioenergy and renewable power methane in integrated 100% renewable energy systems, PhD thesis, Faculty of Electrical Engineering and Computer Science, University of Kassel, 2009.
[61] PV Magazine. Energy storage market to grow to USD250 billion by 2040, www.pv-magazine.com/2016/06/13/energy-storage-market-to-grow-to-usd250-billion-by-2040_100024952/. Berlin, 2016 (accessed 01.02.2017).
[62] Lehner M, Tichler R, Steinmüller H, Koppe M. Power-to-gas: technology and business models. Heidelberg: Springer; 2014.
[63] Bailera M, Lisbona P, Romeo LM, Espatolero S. Power-to-gas projects review: lab, pilot and demo plants for storing renewable energy and CO_2. Renew Sust Energ Rev 2017;69:292–312.
[64] Energiespeicher. World's first industrial power-to-gas plant, http://forschung-energiespeicher.info/en/project-showcase/versorgungsnetze/projekt-einzelansicht/110/Weltweit_erste_industrielle_Power_to_Gas_Anlage/. 2016 (accessed 01.02.2017).
[65] Iskov H, Rasmussen BN. Global screening of projects and technologies for power-to-gas and Bio-SNG, A reference report, www.energinet.dk/SiteCollectionDocuments/Engelske%20dokumenter/Forskning/global_screening_08112013_final.pdf, Danish Gas Technology Center: Horsholm; 2013.
[66] Agora Energiewende. Stromspeicher in der Energiewende, www.agora-energiewende.de/fileadmin/Projekte/2013/speicher-in-der-energiewende/Agora_Speicherstudie_Web.pdf. Berlin, 2016 (accessed 01.02.2017).
[67] Breyer C, Tsupari E, Tikka V, Vainikka P. Power-to-gas as an emerging profitable business through creating an integrated value chain. Energy Procedia 2015;73:182–9.
[68] Barbosa LSNS, Bogdanov D, Vainikka P, Breyer C. Hydro, wind and solar power as a base for a 100% renewable energy supply for south and central America. PLoS One 2017;12:e0173820.
[69] Caldera U, Bogdanov D, Afanasyeva S, Breyer C. In: Integration of reverse osmosis seawater desalination in the power sector, based on PV and wind energy, for the Kingdom of Saudi Arabia. 32nd EUPVSEC, Munich, Germany; 2016, http://bit.ly/2iVKP97.
[70] Caldera U, Bogdanov D, Afanasyeva S, Breyer C, 2017, Role of seawater desalination in the management of an integrated water and 100% renewable energy based power sector in Saudi Arabia, submitted.

[71] Missimer TM, Ng CK, Thuw K, Shahzad WM. Geothermal electricity generation and desalination: an integrated process designed to conserve latent heat with operational improvements. Desalin Water Treat 2016;57:23110–8.
[72] Lund WJ, Boyd TL. Direct utilization of geothermal energy 2015 worldwide review. Geothermics 2016;60:66–93.

FURTHER READING

[1] Farfan J, Breyer C. Structural changes of global power generation capacity towards sustainability and the risk of stranded investments. J Clean Prod 2016;141:370–84.
[2] Luo X, Wang J, Dooner M, Clarke J, Krupke C. Overview of current development in electrical energy storage technologies and the application potential in power system operation. Appl Energy 2015;137:511–36.

CHAPTER 9

Integrated Planning of Energy and Water Supply in Islands

Raquel Segurado*, Mário Costa*, Neven Duić†
*Instituto Superior Técnico, Lisboa, Portugal
†University of Zagreb, Zagreb, Croatia

9.1 INTRODUCTION

Currently, about 18% of the world's population still lacks access to electricity, while about 38% relies on wood, coal, charcoal, or animal waste for cooking and heating [1]. This is a major barrier to eradicating poverty and building shared prosperity [2]. Moreover, where modern energy services are available, the problem is related to waste and pollution. Carbon dioxide (CO_2) and other greenhouse gases emissions from fossil fuels are contributing to climate change. The solution to both challenges is to provide sustainable energy for all, energy that is accessible, cleaner, and more efficient [2]. In Sep. 2011, the United Nations launched the "Sustainable Energy for All" initiative. This initiative is a partnership between governments, the private sector, and the civil society with three interconnected objectives to be reached until 2030:

- ensure universal access to modern energy services,
- double the global rate of improvement in energy efficiency,
- double the share of renewable energy in the global energy mix.

However, more importantly, these objectives together must act as a catalyst for the creation of conditions for the development of income generation activities as a development motor and an instrument of poverty alleviation. Achieving the three objectives together will maximize development benefits and help stabilize climate change over the long run [2,3].

Small islands face a unique challenge in achieving sustainable development and fighting climate change. They are on the front line of climate change and sustainable development. On the one hand, they are more vulnerable to climate change and sea level rise. On the other hand, their small size, remoteness, vulnerability to external shock, and fragile environments render more difficult the pursuit of sustainable development [4]. Most small islands that are not interconnected with the mainland depend mainly on the

Renewable Energy Powered Desalination Handbook
https://doi.org/10.1016/B978-0-12-815244-7.00009-X

importation of fossil fuels for energy production but, at the same time, present a considerable potential in renewable energy sources (RES). The use of this potential in the production of electricity and freshwater (in arid islands) could represent a large fraction of the total energy supply [5]. One of the major challenges of increasing the penetration of RES in a system is to integrate a high share of intermittent resources into the electricity supply system [6,7]. The intermittent nature of some RES, as well as the small energy systems of islands, introduces barriers to their penetration, like the struggle to match the demand with the supply and the problems related with the integration in the network, namely, the technical constraints of the conventional generating units (minimum load level of these units) and the dynamic penetration limit that is usually applied for grid stability [8]. The penetration of intermittent RES in the electricity supply system is therefore limited, even when there is a large RES potential.

The integration of RES in energy systems of small islands presents several advantages, namely, at the economic level, since their high investment cost is compensated by the high cost of the conventional energy sources. Indeed, the small dimension of the energy systems and the expensive importation of fossil fuels, whose transport costs make them even more expensive, cause a very expensive security of supply. Local RES provide a clean and sustainable approach to energy production, help to ensure security of energy supply, decrease energy import dependence, and contribute to the achievement of the Kyoto Protocol's objectives. The integration of RES in the power system, along with appropriate policies and regulations on the rational use of energy, is very important for the achievement of sustainable development [9]. The conventional electricity production technologies are rarely adapted to the conditions of isolated regions and can seriously damage the vulnerable ecosystems and natural habitats [10].

In arid islands, in addition to the energy problem, there is the water scarcity problem, with the need to provide freshwater to the population in regions where there are no natural sources of freshwater. The installation of desalination units is a common solution throughout the world in areas with water scarcity [11], especially in islands due to their proximity to the sea. However, desalination is a process that requires a significant amount of energy [12]; thus, renewable energy-driven desalination can play a vital role in the application of this technology.

The purpose of this study is to couple these two issues: the integration of RES in the electricity supply system and the water scarcity problem of islands. An integrated approach is used in energy and water supply systems' planning, with the aim of increasing the penetration of RES, taking into consideration the electricity and water production costs.

This study is applied to the island of S. Vicente, Cabo Verde. This island has significant problems regarding the power and water supply systems, like other islands of Cabo Verde. Cabo Verde's power prices are among the highest in Africa due to its dependency on fossil fuel-based plants, which in turn rely on the importation of expensive fuel. In 2012, the electricity tariff reached 0.283 €/kWh [13]. According to the Renewable Energies National Plan, the electricity production cost in Cabo Verde is about 70% higher than the European average [14]. Cabo Verde has by far the most expensive water tariffs in Africa and among the most expensive in the world. In 2012, the water tariff reached 3.313 €/m^3, which is more than double the average water tariffs in the Organization for Economic Co-operation and Development (OECD) major cities [15]. These high prices reflect the water scarcity problem that this country faces, where about 85% of its water production relies on desalination. Moreover, the cost of the energy-intensive desalination process is particularly high due to its dependence on power generation, whose high costs in turn reflect reliance on small-scale diesel generators and expensive imported oil [16].

The electricity supply system of S. Vicente is based on fossil fuel and wind power. Although this island has important wind resources, they are not fully used because of its intermittent nature. In addition, this island does not have any source of freshwater, being forced to desalinate seawater to produce water suitable for human consumption. On the one hand, this puts more pressure on the power supply system but; on the other hand, this can be an opportunity to implement renewable energy-driven desalination systems. Hence, this study proposes two scenarios to avoid the curtailment of this excess wind power, with two ways of storing excess wind power in S. Vicente. One way is to provide the excess wind power directly to the desalination units. The other way is to use both the desalination units and a pumped hydroenergy storage system to store this excess wind power. S. Vicente has the suitable topography to install a pumped hydrostorage (PHS) system and the water produced in the desalination units can be used in this system.

9.1.1 State of the Art

Several studies have been carried out on the feasibility of integrating RES in islands, and all of them rely on energy storage and/or demand side management strategies. Many analyses have been also carried out on the feasibility of using RES in desalination plants.

The use of excess wind power in the desalination units proposed in this study can be considered a demand side management strategy since the water cannot be turned back into electricity with a reasonable efficiency; water can, however, be stored.

Most of the studies undertaken so far have examined either the energy or the water supply system. When studies concentrate on the energy demand of the water supply systems, they usually do not analyze the overall power supply system of the region. However, there are some studies that begin to deal with this integrated approach.

In this section, a literature review is made concerning studies carried out on island energy systems with high penetration of intermittent RES and energy storage, namely, PHS; on the use of RES in desalination plants; and finally, on the energy–water nexus, focusing on the integrated supply of power and water.

9.1.1.1 Intermittent RES in Islands and Energy Storage

The intermittent nature of some RES and the small power systems of islands introduce barriers to their penetration. Wind power, for instance, has the largest disadvantage in its strong weather dependence. Since wind speed can have sudden and large changes, the power output of a wind turbine can have large fluctuations, with a significant impact on the operation of the power system and on the power quality. This impact increases with the increase of wind-power penetration [17,18]. The impact on power quality is primarily on the level and fluctuations of voltage and frequency. The system must ensure the stability of voltage and frequency within required limits [18]. The variations in the wind-power production can also cause disturbances between the power generation and power demand that can lead to dangerous operating conditions [19]. To avoid problems that can affect the safety and stability of the power supply system, wind power must be limited to a specific percentage of the system's load [19]. This limit is the maximum instantaneous wind power directly supplied to the power grid, and usually it is not higher than 30% [8,20]. In addition, the conventional generating units that are present in the power supply system have minimum loading levels, to avoid their operation in levels of lower efficiency. These two factors determine the maximum amount of wind power that can be injected in the power grid. For this reason, when there is large wind-power potential, there can be a significant amount of wind-power curtailed. The way to tackle this issue is to use energy storage. Research has shown that energy storage is crucial in the design of energy systems with high RES penetration. Energy storage allows the decoupling of production and demand, by transferring the energy surplus from one time to another where there is shortage, increasing the efficiency and the viability of such systems [21]. Energy storage is also advantageous for shorter duration applications, including grid stabilization, grid operational support, stable power quality and reliability, and load shifting [22]. Several solutions for wind energy storage are available, namely, flywheels, capacitors, superconducting magnetic energy storages, batteries, compressed air energy storages, pumped hydrostations, and hydrogen [17].

There is a significant amount of research carried out regarding the island energy systems with high penetration of intermittent RES.

Ntziachristos et al. [23] studied the coupling of a wind turbine with a fuel cell to improve the use of wind power in the noninterconnected Greek archipelago grid. Part of the wind power is stored in the form of hydrogen and then used to cover the demand through a fuel cell. This decoupling between the wind-power potential and the load is crucial to increase the contribution of RES to the small capacity grids of islands. The authors concluded that these types of installations can be used to complement or even replace conventional power stations, producing clean and renewable energy, without the limitations of the conventional wind-power systems [23].

Papathanassiou and Boulaxis [8] evaluated the energy yield of wind farms operating in isolated power systems. The authors developed a methodology and applied it to a Greek island and stated that wind farms installed on island systems are subject to significant restrictions, affecting their expected energy yield and thus the feasibility of investments. A significantly higher wind-power penetration requires fundamental changes in the structure and management of power supply systems, namely, the application of advanced control centers and the introduction of energy storage. For systems ranging from a few MW up to some hundreds MW, pumped storage seems to be the only viable alternative [8].

Kaldellis and Zafirakis [24] presented an integrated electricity production cost analysis for autonomous electrical networks based on RES and energy storage configurations and applied it to two Greek Aegean islands. The authors concluded that a properly sized RES-based power production system together with appropriate energy storage is a promising solution for this type of system, providing clean energy and contributing to a decrease in the important environmental problems resulting from the electricity generation sector [24].

Zoulias and Lymberopoulos [25] examined the technoeconomic aspects of replacing diesel generators of the Kythnos system by fuel cells, an electrolyzer, and a conventional hydrogen tank and presented sizing optimization and simulation results of both conventional and hydrogen-based power systems. Kythnos is an island of the Cyclades region of Greece, located in the Aegean Sea. The system proposed is technically feasible and increases RES penetration, decreasing the dumped excess energy, but to be economically viable the cost of the hydrogen-based system must significantly decrease [25].

Bağcı [26] has shown that it is technically feasible to produce all the required electricity of Peng Chau Island, Hong Kong, from RES, with almost no CO_2 emissions. The author concluded that solar, wind, and wave energy are the most suitable alternative and the model used can be applied to other outlying islands of Hong Kong or any other small island in other parts of the world with good RES potential [26].

Krajačić et al. [27] concluded that with an energy storage system based on hydrogen, the island of Mljet in Croatia could become a 100% renewable island concerning electricity and simulated transport needs and also could export additional power to the mainland power grid.

An autonomous wind/hydrogen energy demonstration system located on the island of Utsira in Norway was launched in Jul. 2004 [28]. The system is composed of a wind turbine, water electrolyzer, hydrogen gas storage, hydrogen engine, and a proton exchange membrane fuel cell. The system gives 2–3 days of full energy autonomy for 10 households on the island, and was the first of its kind in the world. This autonomous wind/hydrogen system at Utsira has demonstrated that it is possible to supply remote area communities with wind power using hydrogen as energy storage. However, further technical improvements and cost reductions need to be made before wind/hydrogen systems can compete with existing commercial solutions as, for example, wind/diesel hybrid power systems [28].

Parissis et al. [29] analyzed the introduction of hydrogen as a storage means and wind energy as an electricity production source into the power supply system of the Portuguese Island of Corvo in the Azores. The results indicate that a considerable RES penetration into the power system of Corvo Island is feasible and results in a significant reduction of the power generation costs [29].

The present study focuses on PHS because it is one of the most well-suited storage systems to achieve high wind-power penetrations in isolated systems [23] with a low maintenance cost. The main problem with this technology is the limited number of locations where it can be installed; however, the case study analyzed in this work has the suitable topography to implement this technology. Batteries are not considered due to their high maintenance costs and short lifetime expectancy [24] that result in significant environmental problems, especially in islands.

PHS is the most mature energy storage technology currently available [22]. Its operating principle is based on managing the gravitational potential energy of water. When the power demand is low and the intermittent RES production is high, this energy is used to pump water from a lower reservoir to an upper reservoir. When the power demand is high, and the RES production is low, this water is pumped back from the upper reservoir to the lower reservoir, activating hydroelectric turbines to produce electricity [30]. Generally, the lifetime of these systems is around 30–60 years, with a round trip efficiency of 65%–85% and capital costs of 500–1500 €/kW. It is an important instrument to control the electrical network frequency and as backup generation due to its fast response time (less than 1 min) [22,30]. The fact that PHS can quickly respond to energy demands also implies that these facilities can provide peak shaving, valley-filling, and spinning reserve capacity [31].

The use of PHS systems to increase the penetration of wind power has been widely analyzed. The prospect of creating a combined wind hydroenergy production station for Aegean Sea islands in Greece is analyzed by Kaldellis and Kavadias [32]. The main factors that delay the economic development of these islands are energy shortage and clean water deficit, especially during the summer. However, these islands present an outstanding wind potential, but there is a significant disharmony between wind-power production and electricity demand. Hence, the authors developed a methodology to estimate the most beneficial configuration of a wind-hydroenergy production station and applied it to several Aegean Sea islands. They concluded that RES penetration may exceed 85%, decreasing the fuel imports and minimizing the negative environmental effects related to the operation of the conventional units [32].

Bueno and Carta [33] presented a model for the technical and economic sizing of a medium-sized wind-powered PHS system that can be implemented in topographically suitable sites with enough wind resources. The authors [34] applied this model to the island of El Hierro, Canarian Archipelago. The results indicate that an annual renewable energy penetration of 68% can be achieved, with significant savings of diesel oil and CO_2 emissions reduction. In Jun. 2014, the installation of this wind-powered PHS system was completed and is currently operating on the island of El Hierro. The system is composed of a wind park with 11.5 MW, an upper and lower reservoir with 380,000 and 150,000 m^3 of capacity, respectively, a pumping station with 6 MW, a hydropower station with 11.32 MW, and three desalination plants [35].

Bueno and Carta [19] also presented a proposal to mitigate the problem of the restriction of wind-power penetration in the island grid of Gran Canaria, Canary Islands, through the installation of a wind-powered PHS system. The results obtained indicate that the penetration of RES can be increased by 2% (53 GWh/year) at a competitive cost, a fossil fuel saving of 13,655 ton/year, and a reduction in CO_2 emissions of 43,064 tCO_2/year. This system successfully combines a maximum exploitation of this RES with the maintenance of a power quality service [19].

Caralis et al. [36] state that the combined use of wind power with pumped storage systems is a means to exploit abundant wind potential, increase wind power installed, and substitute conventional peak supply, in autonomous islands where the wind penetration is restricted due to technical reasons related to the safe operation of the electrical systems. The authors concluded that another important advantage of the proposed system is that the production cost is, to a large extent, known in advance, contrary to the current cost, which depends strongly on the fossil fuel price, providing both financial and environmental benefits [36].

Duić et al. [37] proposed a wind-powered PHS system for the island of Corvo in the Azores. There are many places where there is already an excellent storage potential in the island's local water supply system, and by merging the energy and water supply systems, where there is sufficient elevation difference, it is possible to use pumped hydro to increase the penetration of wind power, even in cases where there is not much hydropotential [37].

Brown et al. [38] carried out an economic analysis of the inclusion of a PHS system in a small island with abundant RES available that cannot be integrated completely due to security criteria. Results showed that the inclusion of this technology can be an effective means of allowing larger penetration of intermittent RES, improving both the dynamic security and the economic operation of a test system [38].

Connolly et al. [39] investigated how PHS can support the integration of wind power in the Irish energy system. The authors focused on three key aspects of this technology: operation, size, and cost. PHS can increase the wind penetration on this energy system and reduce operating costs. These savings are, however, sensitive to changes in PHS capacities used, fuel prices, interest rates, and total wind power produced [39].

9.1.1.2 RES Desalination

The increasing scarcity of water is driving a growing implementation of desalination technologies. However, the energy needed to run the desalination plants is a disadvantage. Hence, many studies have been made for the use of RES in desalination [40].

This study will focus only on reverse osmosis (RO) desalination since it is the current technology installed in S. Vicente. In addition, this is the most efficient desalination technology, requiring about 3–10 kWh of electricity to produce 1 m^3 of freshwater from seawater. Sometimes this value is as low as 2.8 kWh/m^3 [41]. RO is a membrane separation process in which the water from a pressurized saline solution is separated from the dissolved material by flowing through a membrane. The major energy required is for pressurizing the feed seawater [42]. RO is well developed and has been in commercial use for decades. Moreover, this method provides significant flexibility in adding capacity and can be provided in various sizes of power consumption and amount of water produced. The modular nature of this technology and its flexibility makes this technology an ideal candidate for the combination of RES and desalination systems.

Wind-powered RO plants appear to be one of the most promising alternatives of RES desalination. This is because RO is the desalination process with lowest energy requirements and desalination plants are mostly located in coastal areas, where there is usually a high availability of wind resources [43].

Several studies are dedicated to the analysis of renewable energy-powered desalination systems.

Carta et al. [44] analyzed a prototype of an innovative, fully autonomous wind-powered desalination system, installed on the island of Gran Canaria in the Canarian Archipelago. This system was designed to determine experimentally the feasibility of a stand-alone desalination unit powered by wind power, to verify the operational feasibility of several desalination techniques when the energy source driving the system is intermittent, and to analyze the influence that various operational strategies have on the volume and quality of the desalinated water produced and on the effect on the main components of the desalination unit. The system can be applied to seawater desalination, both on a small and large scale, in coastal regions with a scarcity of water and with wind-power resource. RO technology seems to be the most suitable for coupling to a wind farm isolated from a conventional power grid, if it is operated with variation in capacity (number of desalination plants), but maintaining practically constant the operating parameters (pressure and flow) [44]. Later, Carta et al. [45] redesigned this small-scale prototype seawater RO desalination plant to adapt continuously its energy consumption to the variable power supplied by a wind turbine, dispensing with the use of batteries and proposing the use of a supercapacitor bank as a dynamic regulation system.

Paulsen and Hensel [46] analyzed how to combine a desalination plant with a fluctuating energy source like wind. The main problem for the combination of desalination processes and wind power is the fluctuation of power supply generated through renewable energies. To address this issue, an innovative RO desalination unit was designed and installed in the Island of Utsira, Norway. The ENERCON Energy Recovery System was designed for very low-energy consumption, high flexibility, and efficient combination with fluctuating energy sources. This system is supplied only by wind power, eliminating the need to use fossil fuels. The system was designed to maximize daily water production, while minimizing the costs [46].

Fadigas and Dias [12] proposed an alternative configuration to conventional RO desalination systems by incorporating the use of both gravitational potential energy and wind power. In general, RO plants use a high-pressure pump powered by electricity. The pump is used to send a flux of saline water to a group of semipermeable membrane modules, capable of filtering the dissolved salts. In this alternative configuration proposed, it is intended to achieve a flux at the inlet of the membrane modules with a pressure high enough for the desalination process, without using either electricity or fossil fuels. For this, a hybrid system that uses both gravitational potential energy and wind power is devised. This configuration was modeled and the authors concluded that it is technically and economically viable. After this theoretical analysis, a prototype of this system was constructed and proved that this mechanism is technically viable.

Spyrou and Anagnostopoulos [11] investigated the optimum design and operation strategy of a stand-alone hybrid desalination scheme, capable of fulfilling the freshwater demand of a Greek island. The scheme consists of an RO desalination unit powered by wind and solar electricity production systems and by a PHS unit. The proposed scheme can produce water at a competitive water production cost (about 1.5–3.0 €/m^3), in comparison with the current water supply in the Greek islands (5–8 €/m^3) [11].

The use of wind turbines for desalination has been proven to be economically feasible as wind technology is well advanced and coastal sites often have a good wind resource [47]. The main challenges associated with the use of wind turbines are the intermittency and fluctuations of the wind resource which occur due to turbulence and gusts over short periods of time (seconds to a few minutes) and mass air movements over long periods of time (tens to hundreds of hours) [47]. The direct connection of a wind turbine to an RO system with no form of energy storage will inevitably result in large fluctuations in pressure and flow rate. This presents a considerable challenge for wind-powered membrane (wind-membrane) systems as membranes are designed to operate at constant operating conditions with no abrupt pressure or cross-flow variations to minimize damage [47]. In Park et al. [47], a wind-powered RO membrane system without energy storage was tested using synthetic brackish water over a range of simulated wind speeds under both steady-state and fluctuating conditions. The main challenge associated with this operation is not the size of the fluctuations but the effect of the power switching off which causes reduced flux and permeate quality of the membrane. They concluded that this must be controlled with energy buffering, careful control of the water treatment load, or disposal of the poor-quality permeate [47].

The most suitable design for seawater RO desalination driven by off-grid wind energy systems is presented by Peñate et al. [48]. The authors address the design of two possible combinations useful for RES powered seawater RO desalination plants: with gradual capacity and with fixed capacity. In general, the designs proposed require a study of the annual local water needs and the wind availability to ensure the optimized operation and stability in the functioning of the system. The use of wind power to supply such a system, adapted to medium or large water demands, compels one to consider the water storage as the key energy storage system. This solution guarantees meeting the water demand in periods of lack of wind resource or serious breakdowns of the system [48].

When suitable water storage is not available, energy storage is crucial for the uninterrupted supply of freshwater from desalination technologies that depend on intermittent RES. Gude [49] carried out a critical review on current energy storage options for different desalination technologies powered by RES.

9.1.1.3 Energy-Water Nexus

Energy and water are inextricably linked. Water is needed to produce energy and energy is needed to extract, treat, and distribute water and to clean the used and polluted water [50]. This is called the energy-water nexus, and it is important to understand due to the increasing energy demands and decreasing freshwater supplies in many regions [51]. Owing to this close interrelationship, the design and operation of water and wastewater systems should take the energy aspect into consideration. Similarly, energy production should not be planned without taking water resources and water quality into consideration [50].

Siddiqi et al. [51] identified the water use in the energy sector and the energy use in the water sector in Jordan and concluded that integrated policy and planning is needed to meet the challenges of growing water and energy interdependencies in many regions. Joint consideration of both water and energy domains can identify new options for increasing overall resource use efficiencies. However, to identify such opportunities, a detailed knowledge of current and emerging water-energy couplings is needed along with an understanding of key actors and institutions engaged in decision-making [51].In another example of studies that consider the energy-water nexus, Gadhamshetty et al. [52] presented a novel application of sensible thermal energy storage to integrate a multieffect evaporation desalination system in a combined cycle power plant for simultaneous energy conservation and water desalination.

Power and water supply are two types of systems modeled and analyzed frequently, but often separately. Most studies focus either on the power or the water supply system. Some studies, as seen in Section 1.1.1, consider water as a means of storing energy in a power supply system (as in PHS). Moreover, some analyze the energy demand of the water supply system, as seen in Section 1.1.2, with the use of RES in desalination units. It is uncommon that these two supply systems are analyzed together in an integrated way.

There is a significant difference between the issues concerning the power supply and the water supply. Power production must meet demand always, while water can be easily stored. Hence, the studies that analyze the integrated water and power supply in some cases do not consider the water demand curve explicitly, but only determine the amount of water that can be supplied with the system proposed [53].

Corsini et al. [54] compared a hydrogen-based system and a desalinated water production system as two effective alternatives for renewable energy seasonal buffering in an island context (stand-alone system). Although the studied buffering systems have different purposes, both scenarios lead to fuel oil savings greater than 60% in relation to the current situation. The higher fuel oil saving values in the hydrogen-based scenario, compared to the water-based one, result from the fuel cell energy contribution, which allows a further reduction of fossil fuel-based unit operating hours [54].

Henderson et al. [55] studied the feasibility of a wind diesel hybrid system that also includes a desalination system component, on Star Island, New Hampshire, in the United States. The proposed system aimed to supply electricity during the peak demand summer months and to balance the seasonal mismatch between wind resource and electricity demand load via the production and storage of potable water during winter months. Although the addition of a desalination system is not economically viable, from a technical point of view, seawater desalination offers an interesting solution to energy storage or long-term load management for wind diesel hybrid systems [55].

Setiawan et al. [56] analyzed a scenario for supplying electricity and fulfilling demand for clean water in remote areas using RES and a diesel generator with an RO desalination plant as deferrable load. The authors simulated the performance of the proposed system in a remote area in the Maldives with about 300 inhabitants and concluded that the proposed system is economically and environmentally viable. They also set up a laboratory experiment to demonstrate the feasibility of providing reliable power and water supply to remote areas [56].

Currently, Jordan is one of the most water-deprived countries in the world [57]. On the other hand, it is also a country with high RES potential that is not used [57]. Novosel et al. [57] proposed a combination of desalination, pump storage that uses the produced brine and RES to tackle this issue. The results demonstrate that an increase in the flexibility of the desalination units and the use of the brine operated PHS system greatly benefit the reduction of excess RES, increasing the penetration of wind power and solar power. The case study also demonstrated that it is possible to increase greatly the water availability in Jordan [57].

The effects of large-scale desalination on the Jordanian energy system were analyzed by Østergaard et al. [58], with focus on the large-scale introduction of wind power into the energy system. The authors used desalination to decrease excess electricity production and concluded that water storage has some implication on the system's ability to integrate wind power [58]. Contrary to the previous studies mentioned, this one considers the hourly freshwater demand in Jordan. Two different desalination technologies were investigated: RO driven by electricity and multistage flash driven by cogeneration of heat and power. The two systems impact the energy systems in different ways due to the technologies' characteristics. The Jordanian primary energy supply with multistage flash is similar to the one with RO. The difference between the two technologies arises when considering the ability to integrate wind power in the power system. In this case, RO is more appropriate, especially if water storage is considered [58].

Traditionally, the infrastructure systems that deliver energy and water, the water distribution and power transmission networks, are thought of as separate, uncoupled systems. However, in reality, they are very much coupled in what is commonly known as the energy-water nexus. Although the energy-water nexus has recently caught the attention of several policy and regulatory agencies, it is rarely addressed in terms of an integrated engineering system for its management, planning, and regulation as an interdisciplinary concern [59]. Santhosh et al. [59] addressed this issue by focusing on the supply side of this integrated engineering system, developing a multiplant real-time simultaneous economic dispatch of power and water. The authors demonstrated that the coproduction minimum capacity limits and process constraints can lead to scenarios where the dispatch can choose the multiplant instead of the cheaper single product plants [59].

The alternative of using seawater desalination as a flexible load whenever excess RES power is present was analyzed by Bognar et al. [60]. The authors carried out numerical simulations of combined energy and water supply systems for the island of Petite Martinique, Grenada, located in the Caribbean. The authors determined that, for an optimal energy and water supply system on the island, the desalination plant's energy demand should be integrated as deferrable load [60].

Bognar et al. [61] analyzed the effects of integrating desalination into an island grid with a high share of renewable energies. The authors presented options to include an RO desalination unit into an optimized wind diesel energy supply system for the Island of Brava in Cabo Verde. The main benefits of using desalination as deferrable load in a microgrid are: better capacity use of the fossil fuel-based units, less fuel consumption and therefore less dependency on fuel imports, less unused excess electricity generated by RES technologies, saving of energy storages within the microgrid, and lower production costs of electricity and water [61].

From the examples mentioned, it is clear that the integrated planning of energy and water supply is crucial to tackle some of the problems arid islands face in providing quality power and water to their population.

9.1.2 Objectives

This work studies the way to increase the penetration of RES in islands by coupling the power and water supply systems. For this, an integrated supply system is proposed, where excess wind power is provided to desalination units that can produce freshwater that can be used in a PHS system and later be supplied to the population. The excess wind power can also be stored in that PHS system. The electricity and water production costs of the scenarios modeled are estimated to find the solution that minimizes the cost, while keeping RES penetration as high as possible. Increasing the wind power installed will increase the amount of wind energy curtailed, increasing the electricity production costs. To decrease the wind energy curtailed, there is the need to increase the desalination capacity of the system, so the desalination units can use this wind energy that otherwise would be curtailed. However, it is very important to ensure that the load of the desalination units is enough to guarantee the financial viability of the system. The main purpose of this study is to understand how the electricity and water production costs vary with the intermittent power curtailed and with the load of the desalination units to find an optimum solution.

Most previous research on the energy and water nexus focused on desalination with RES (RES to supply water) or the use of water to produce energy (water to supply energy) and not both issues at the same time. The novel aspect of this study is the coupling of these two supply systems to increase the integration of intermittent RES and minimize the electricity and water production costs, thereby increasing their efficiency.

9.2 CASE STUDY

Cabo Verde is an insular country, with about half a million inhabitants and a reduced and scattered territory (about 4000 km^2) [3]. It is composed of nine inhabited islands situated at about 450 km off the West African coast in the Atlantic Ocean. Cabo Verde has a semiarid climate, hot and dry, with scarce rainfall. The islands have a volcanic origin, mostly are mountainous, and with almost no vegetation and natural resources. The country is extremely dependent on importation; it imports about 80% of all it consumes [3].

9.2.1 Power and Water Supply in Cabo Verde

Power in Cabo Verde is supplied by the multiutility company ELECTRA, which is also responsible for the water supply in some of the islands. Cabo Verde has achieved a relatively high electrification rate. According to the National Census of 2010, 81% of the total population has access to electricity [62]. This value is above the average for African countries (43%), and above the average for developing countries (76%), but below the average for transition economies and OECD countries that reach 99.8% [1]. However, the increase in the electrification rate over the years in Cabo Verde was not accompanied by a proportional increase in the electricity production, with the power installed decreasing. Between 2002 and 2006, the power installed decreased about 4%, while connections rose by 40% and production by 10%. Hence, blackouts became more frequent and longer due to a tight demand–supply balance [13].

In Cabo Verde, energy production is largely dependent on fossil fuel-based plants, which in turn rely on expensive fuel imports. Fuel costs represent more than 70% of the power production costs [13]. A single power price is applied across all islands despite differences in costs due to a geographical cross-subsidization justified on the grounds of social equity. Cabo Verde's power prices are among the highest in Africa due to its dependency on the importation of expensive fuel [13]. According to ELECTRA, the average electricity tariff in Cabo Verde in 2012 was 0.283 €/kWh [13], about 45% higher than the average European electricity tariff for households [63]. The power production costs are about 70% higher than the European average [14].

In 2012, the water tariff of Cabo Verde was 3.313 €/m^3 [13], by far the most expensive water tariff in Africa, and among the most expensive in the world [16]. This value is more than double the average water tariffs in the major cities of OECD countries [15]. These high prices reflect the scarcity of water resources that has forced the country to rely on desalination for approximately 85% of the production. Moreover, the cost of the energy-intensive desalination process is particularly high due to its dependence on power generation, whose high costs in turn reflect the reliance on small-scale diesel generators and expensive imported oil [16].

The Government of Cabo Verde launched recently the National Plan for Renewable Energy and the National Plan for Energy Efficiency, as well as the Action Agenda in the scope of "Sustainable Energy for All" of the Economic Community of West African States [3]. In Cabo Verde, the objectives of the "Sustainable Energy for All" initiative are perfectly aligned with the developed strategies for the national energy sector and with its role in the development of the country. The focus on renewable energies is considered structuring for Cabo Verde and a way to ensure the sustainability of the target of universal access to energy. In the context of Cabo Verde, the use of endogenous energy sources allows higher energy independence and the access to energy at competitive prices for families and companies [3].

The foreseen strong penetration of RES in the power supply should be accompanied by measures to reduce losses in the power distribution system, which is currently quite high (about 30%), to about 8% until 2020 [14]. Regarding RES integration in Cabo Verde, one of the main barriers is the technical limitation of integrating variable wind resources in the grid. The existing grid is quite weak, which does not ensure that the power is in fact delivered to the consumers due to structural problems. The existing grid also restricts the addition of capacity, and thus the development of small- to medium-scale renewable energy projects [64].

9.2.2 The Island of S. Vicente, Cabo Verde

S. Vicente is the second most crowded Island in the Archipelago of Cabo Verde, with about 76,000 inhabitants in 2010 [62], mostly concentrated in its main city Mindelo. The island has about 227 km^2 of area. The electricity supply system of S. Vicente is based on fossil fuel and wind and, although this island has important wind resources, they are not fully used because of its intermittent nature. In addition, this island does not have any source of freshwater, being forced to desalinate seawater to produce water suitable for human consumption. This puts more pressure on the electricity supply system, since desalination requires a significant amount of energy. This can be an opportunity to implement renewable energy-driven desalination. However, it is always necessary to have energy storage. The topography of this Island is relatively uniform, with the exception of Mont Verde, a 774-m high mountain, which could be suitable for PHS. This island is a suitable case study to test the use of desalinated water in a pumping and hydrostation before being supplied to the population—a wind-powered desalination and PHS.

9.2.2.1 Electricity Supply System of S. Vicente

The electrification rate of S. Vicente reached 90% in 2010 [62], a value above the national average. The technologies used to produce electricity on the island are based on fossil fuel-fired plants and wind turbines. Furthermore, this island is not interconnected with any of its neighboring islands. There are two conventional thermal fossil fuel-based plants, the Matiota and the Lazareto plants. In 2012, the Matiota plant had four fuel oil and diesel generators, with an installed capacity of 10.9 MW. The Lazareto plant had two fuel oil generators, with an installed capacity of 6.44 MW.

In 2011, the installation of the Cabeólica Wind Park increased the wind-power penetration in the power system from 8% (0.9 MW) to 31% (6.85 MW). However, this great increase came with the increase of the wind-power curtailed. Owing to the hourly dynamic penetration limit, not all wind-power produced can be used to cover the load; hence, wind power is curtailed. This wind-power curtailed is wasted power, since it has no additional production cost; it is free power. Not using this wind power will increase the specific power production cost of S. Vicente, which is already quite high.

9.2.2.2 Wind Resources in S. Vicente

Wind power represents the greater contribution within the RES in the Cabo Verde Islands. This is due to good wind conditions in the islands where the trade winds prevail (winds of north-east direction) which are characterized by being constant and with medium-high speeds. Indeed, wind parks in Cabo Verde are the ones with the greatest productivity in the world, with capacity factors higher than 40% [65]. The wind resources in Cabo Verde present a clear seasonal asymmetry with two distinct periods: from Jan. to Jun. with high average wind speeds, and from Jul. to Dec. with a significant decrease in the average wind speeds.

S. Vicente is the island of Cabo Verde with the highest wind resource potential. Although it presents a rugged terrain, its structure presents favorable conditions to the acceleration of the wind resource, increasing the average wind speeds. This island presents vast areas with average wind speed higher than 8.5 m/s [66]. For this study, the hourly wind speed values are collected from the meteorological station of S. Pedro in 2005 [67].

9.2.2.3 Water Supply System of S. Vicente

All freshwater supplied to the population in S. Vicente is desalinated water [13,68]. In 2012, all the desalination units used RO technology, three with a production capacity of 1000 m^3/day and two with a production capacity of 1200 m^3/day. According to the ELECTRA 2012 report, to produce and pump 1 m^3 of water to be supplied to the population, about 5 kWh of electricity is needed [13]. The water reservoirs available in S. Vicente are distributed throughout the island, close to the population clusters. The total capacity of these reservoirs is about 14,680 m^3.

9.2.2.4 Future Electricity and Water Demand

The evolution of the electricity and water demand considered in this study is the one estimated in the National Energy Plan for Cabo Verde [69]. Table 9.1 shows the electricity and water production in 2010 [68] and the forecast for 2020 along with the peak power production for each year.

Table 9.1 Electricity and water production in 2010 [68] and the forecast for 2020

Year	2010	2020
Electricity production (MWh)	65,029	88,518
Peak production (MW)	10.9	16.3
Water production (m^3)	1,252,665	1,736,061

Several other studies present values for the forecasted power production in S. Vicente for the year 2020. For instances, in the Renewable Energy Plan of Cabo Verde, the baseline scenario considers a power production of 107,659 MWh in 2020 for S. Vicente [66]. With this in mind, a sensitivity analysis is done, analyzing the results obtained considering an increase of 21.6% of the power production for 2020.

9.2.2.5 Economic Data

The discount rate used to annualize the investment costs of the system is 10% [27,66], which is the value usually used in the region for the analyses of the financial viability of this type of project. Table 9.2 shows the costs and lifetime of the installed equipment. The yearly operation and maintenance (O&M) costs can be variable or fixed, depending on whether the costs are dependent on the energy (or water) produced or not. However, in some studies, only the fixed O&M cost is mentioned, because the average load factor of the plant is considered. The yearly fixed O&M costs are a given percentage of the total investment cost.

The investment and O&M costs of the wind parks and of the fossil fuel-fired units are based on the costs for projects foreseen for Cabo Verde stated in the Renewable Energy Plan of Cabo Verde published in 2011 [66].

Table 9.2 Costs and lifetime of the installed equipment [39,66,70].

Technology		Investment cost	Fixed O&M cost (%)	Variable O&M cost (€/MWh)	Lifetime (years)
Wind turbines [66]		2000 €/kW	3	–	20
Fossil fuel-based units [66]		1200 €/kW	1.5	–	20
Desalination [70]		1000 €/(m^3/day)	10	–	20
PHS [39]	Hydro	500 €/kW	1.5	1.5	40
	Pump	500 €/kW			
	Storage	7.5 €/kWh			

The fuel cost per kWh of electricity produced by the fossil fuel technologies is estimated for the years from 2007 to 2012 based on the information on the fuel specific consumption (g/kWh) of the existing fossil fuel-fired units, percentage of use of each fuel to produce electricity [13,68], and the current prices of each fuel (€/kg) in Cabo Verde, according to the National Economic Regulatory Agency [71]. In 2010, fuel costs reached 0.129 €/kWh. On the basis of these values, the estimations for the following years are made by linear regression, which lead to 0.268 €/kWh in 2020. Owing to the volatile nature of the fuel costs, a sensitivity analysis is carried out.

The estimation of the investment costs of PHS is very difficult to accomplish because it is very site specific, largely dependent on the geography of the location, on labor costs, etc. [17]. Hence, a sensitivity analysis is done, analyzing the results obtained with a higher investment cost of the PHS system.

9.2.3 Energy Scenarios for S. Vicente

Five different scenarios are considered in this study, all having 2010 as the base year. The first is the baseline scenario, as it only considers the projects already foreseen for the island, and it is established to allow the comparison between the results of the proposed solutions and the current energy and water supply systems, including electricity and water production costs and total yearly costs [72,73].

The second scenario considers the supply of wind power to the desalination plants already installed on the island. Although S. Vicente has several reservoirs spread through the island, this scenario considers only one reservoir in the island with the capacity of all of them, about 14,680 m^3, where the water that comes out of the desalination plant is stored before being supplied to the population. When the excess wind power is not enough to desalinate all the water needed, the fossil fuel-based generators are used to supply the remaining required electricity.

The third scenario is established by finding which installed wind power and desalination capacity minimizes the total annual costs of the electricity and water supply systems for 2020, while keeping the wind-power penetration the highest possible.

The fourth scenario considers the storage of the excess wind-power production through pumping of the desalinated water. This scenario contemplates the construction of two water reservoirs, one at low altitude and another at

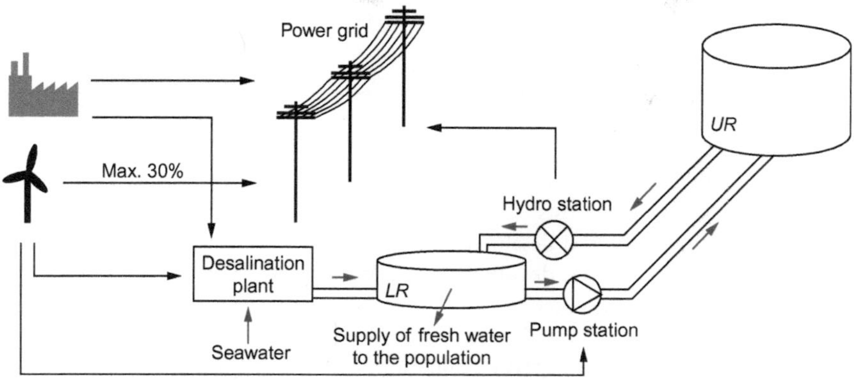

Fig. 9.1 Schematic diagram of the wind-powered desalination and pumped hydrostorage (PHS) system [74].

500 m of altitude. The wind park would supply electricity to a desalination plant and to a pumping station that pumps desalinated water from the lower reservoir into the upper reservoir. When it is necessary to supply water and electricity to the population, the water is retrieved from the upper to the lower reservoir, passing through the hydroturbine, as shown schematically in Fig. 9.1. The cycle efficiency of the PHS is about 69%.

The fifth scenario is established by finding which installed wind power, desalination capacity, hydropower, and pump power minimizes the total annual costs of the electricity and water supply systems for 2020, while keeping the wind-power penetration the highest possible.

These five scenarios are modeled limiting to 30% the hourly intermittent energy penetration, which means that only 30% of the load of 1 h can be covered by wind power [8,20].

For Scenarios 1, 3, and 5, alternative scenarios are made taking into consideration an hourly intermittent energy penetration of 100% (Scenarios 1a, 3a, and 5a), to compare the results with the remaining scenarios. It is important to notice that these scenarios present great risk of power instability on the grid; hence, it is necessary to consider that the wind turbines installed possess some degree of frequency and voltage control. There is no need to simulate Scenarios 2 and 4 with an hourly intermittent energy penetration of 100% because, with this limit, the wind-power curtailed is not enough to supply the desalination units and the pumps. For the other three scenarios, it is interesting to examine the consequences of an increase in the installed wind power.

Table 9.3 —Scenarios considered

Scenario	Description	Hourly wind-power penetration
1	Baseline	30%
2	Only wind-powered desalination	
3	Only wind-powered desalination with minimum costs	
4	Wind-powered desalination and PHS	
5	Wind-powered desalination and PHS with minimum costs	
1a	Baseline	100%
3a	Only wind-powered desalination with minimum costs	
5a	Wind-powered desalination and PHS with minimum costs	

In Scenarios 1, 1a, 2, and 4, the current installed wind power and desalination capacity in S. Vicente is considered. In Scenarios 3, 3a, 5, and 5a, the installed wind power and desalination capacity are those that minimized the total costs, while keeping the RES penetration the highest possible. Table 9.3 summarizes all scenarios considered in this analysis.

For the first part of this study, the tool used to model these scenarios (H2RES model [75]) does not allow performing optimization; hence, it is necessary to run all potential configurations and verify their technical feasibility (i.e., if they are able to supply the required electricity and water demand at all hours) and identify the one with lower total annual costs. The optimization performed in this study is an investment and operational optimization. On the one hand, each iteration has a specific potential configuration (capacity of the installed equipment) and, on the other hand, certain operational conditions could be changed in order to avoid the overflow of the reservoirs, namely, the maximum amount of wind-powered desalinated water in each hour. For the second part of this study, a mathematical optimization is performed where the sizing and operational strategy of the proposed system is optimized.

The power technologies considered are the ones already installed in S. Vicente (fossil fuel-based units and wind turbines) and a system for energy storage (pumped hydro). The technology considered to produce water is the only one used in S. Vicente, i.e., desalination of seawater using RO.

9.3 METHODS

9.3.1 Integrated Power and Water Supply Systems' Modeling

The proposed integrated system is modeled firstly with the H2RES model [72], which simulates the integration of RES and hydrogen in the energy systems of islands or other isolated locations. It is based on hourly time series analysis of demand (water, electricity, hydrogen, heat); storage (pumped hydro, batteries, hydrogen, heat); and resources (wind speed, solar radiation, precipitation) [27]. The wind-power produced is used firstly to cover the load, according to the dynamic (hourly) penetration limit allowed. The wind power that surpasses this limit (excess) is used in the desalination units. The desalination units use this wind power to fill the reservoir used to supply water to the population (lower reservoir). Afterward, if there is still wind power available, it is stored as pumped water into an upper reservoir. The energy stored can be retrieved later and supplied to the system as electricity. The remaining energy needs are covered by fossil fuel-based systems. The desalination module uses excess wind power to supply the desalination units, that produce drinkable water and put it on the lower reservoir; this reservoir is then used to supply the population. This module takes into account the desalination capacity of these units, their electricity consumption, and the maximum level of the reservoir to be filled with desalinated water (n_{desal}). This module also allows the use of fossil fuel-based units to supply electricity to the desalination in a given condition. This condition is determined by h_{FF}, the number of hours of average water demand that must be available in the reservoir always. If the amount of water in the reservoir is less than the amount of water of h_{FF}, the desalination units can be supplied by the fossil fuel-based units [72].

9.3.1.1 Financial Analysis

The total annual costs are estimated using the simplified levelized cost of energy method. The term levelized cost of energy emphasizes the fact that this cost is determined over a certain time (time of life of a specific technology). In practice, the objective is to find the price of energy that sets the sum of all future discounted cash flows to zero [58]. This method is adapted to calculate also the water production cost. Each production cost includes the cost of the components used to produce the specific output (electricity and water) [73].

9.3.1.2 Internalizing the CO_2 Emissions Cost

To internalize environmental problems related to CO_2 emissions, namely, climate change, the cost of a ton of CO_2 emitted is included in the total production costs. The cost of CO_2 considered in this study is the cost of the certified emission reduction (CER). CER are emission reduction certificates issued by the United Nations Framework Convention on Climate Change and the Kyoto Protocol for the successful completion of Clean Development Mechanism climate protection projects. This mechanism allows countries or companies to acquire CER that can be used to meet their own commitments by investing in projects in developing and newly industrializing countries (without themselves having to reduce emissions) [76].

The assumption is that if in the future CO_2 emissions are to be charged, i.e., if a carbon price would be charged to a company or country that has CO_2 emitting systems, one starting value for this price can be the current value of CER. The CER value is extremely volatile; hence, the proposed system is analyzed with two different values of CER: the average value of CER since this market was created until Feb. 2015—6.96 €/tCO_2, and the maximum value of CER ever reached—22.60 €/tCO_2 [76].

9.3.1.3 Sensitivity Analysis

To analyze the influence of the inputs on the results of the modeling, a sensitivity analysis is accomplished. The inputs varied are the ones considered more volatile and/or more unpredictable, i.e., electricity and water demand, fuel costs, and investment costs of the PHS.

9.3.2 Optimization

To optimize the sizing and the operational strategies of this integrated system and facilitate the use of optimization methods, the system is modeled separately from H2RES. The proposed system is modeled for the year 2020, taking into account the electricity and water demand forecasted for that year. The size and the operational strategy of this system are translated into the variables of the optimization problem. The proposed integrated system is modeled based on the hourly wind-power production potential (provided by the H2RES Model). On the basis of these hourly values and on the hourly load and hourly water load, it is possible to calculate the hourly wind-power excess, the hourly wind power used to produce water and to pump water, as well as the water produced and pumped. In each hour of the year, the wind-power excess not used to produce or pump water is the wind-power curtailed. The water turbinated is also calculated and the corresponding hydroproduction. The water produced from fossil fuel-based units is also calculated. These calculations allow the estimation of the level of the upper and lower reservoirs in each hour of the year.

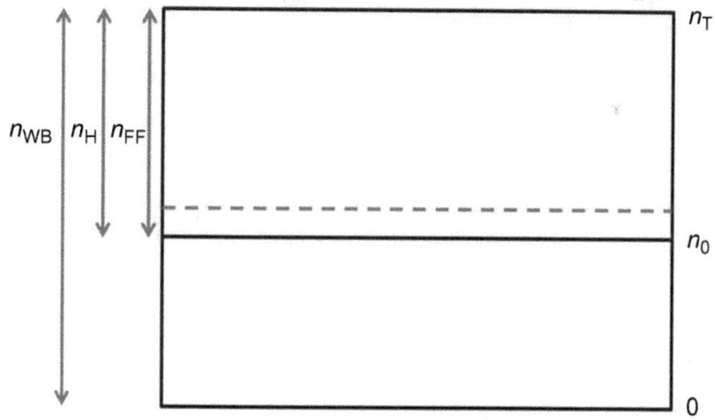

Fig. 9.2 Definition of variables n_{WB}, n_H, and n_{FF} [74].

On the basis of these hourly values, it is possible to determine the annual load covered by the fossil fuel-based units and the total annual costs of the system [74].

Fig. 9.2 presents the scheme of the lower reservoir (LR in Fig. 9.1), as well as the definition of the three variables related to the operational levels of the wind-powered desalination and PHS system (n_{WB}, n_H, and n_{FF}). The variable n_{WB} is the level of the lower reservoir that determines the balance between the excess wind power used to desalinate and to pump water to the upper reservoir. This variable ranges from zero to the maximum level of the reservoir (n_T). The variable n_H is the level of the lower reservoir in which the hydroproduction stops. This variable ranges from n_0, the minimum level of the reservoir, to n_T. When the level of the lower reservoir is less than n_0, and it is not possible to turbinate water from the upper reservoir, the fossil fuel-based units supply the desalination units to produce water until the level of the lower reservoir reaches n_{FF}. This variable also ranges between n_0 and n_T [74].

The proposed integrated system is optimized considering three objective functions: f_1, the minimization of the total annualized costs of the integrated energy and water supply system; f_2, the maximization of the percentage of RES in the total power production; and f_3, the minimization of the curtailed wind power [74].

To solve the optimization problem, a derivative-free method is used: the multiobjective optimization method DMS [77,78]. The Pareto optimal set is obtained; from this set, several solutions are chosen and analyzed in more detail [74].

9.4 RESULTS AND DISCUSSION

9.4.1 Integrated Power and Water Supply Systems' Modeling

The integrated power and water supply system proposed for S. Vicente is modeled with the H2RES model for all scenarios mentioned in Section 2.3. The base year is the year 2010, and the system is modeled for 2020, according to the forecasted electricity and water demand. The optimization scenarios consider only the electricity and water demand for 2020 [73].

9.4.1.1 Scenario 1: Baseline

This scenario considers the currently installed wind power and fossil fuel-based generators in S. Vicente. The results show that wind power will have a strong presence in the power generation of S. Vicente in the upcoming years, but the fossil fuel will remain very important, promoting a considerable increase in the electricity production cost and consequently in the water production cost. The results also reveal that the wind-power curtailed will be significant; specifically, in 2020, it will reach 36.5% of all potential wind power. Scenario 2 is built to use this wasted wind power.

9.4.1.2 Scenario 2: Wind Desalination

A good way to decrease the wind-power curtailed is to supply directly the desalination units with this excess wind power. However, when modeling this scenario, the first conclusion reached is that this excess wind power is not enough to desalinate all the water needed. Hence, the supply of electricity produced by the fossil fuel-based units to the desalination units is considered.

Scenario 2 considers the construction of a medium voltage (MV) power line from the wind parks to the desalination units. This would enable the supply of wind power directly to the desalination units. The distance between these two locations is about 12 km, and the cost per km of the construction of a MV power line is considered to be 20,000 €/km [79]. This cost is annualized considering the discount rate and the lifetime of 20 years and is added to the total yearly costs of the water supply system. Besides the power line, this scenario considers the equipment already installed in the island. Although S. Vicente has several reservoirs spread through the island, this scenario considers that there is only one reservoir in the island with the capacity of all of them.

In 2020, the electricity production costs are about 4% lower than those in the baseline scenario. This is because there is more wind power used, lowering the electricity needed from fossil fuel. The wind-power curtailed decreases from 36.5% to 24.7% in 2020. The water production costs, which include the investment in the new power line mentioned before, are slightly lower, and the electricity used to desalinate water is mostly wind power. This percentage of wind-powered desalination can increase if the capacity of the reservoir increases. This, however, would involve the construction of additional reservoirs, which is not considered in this scenario.

9.4.1.3 Scenario 3: Wind Desalination With Minimum Total Costs

Since the H2RES model does not allow performing optimization, it is necessary to run all proposed solutions and verify their technical feasibility (i.e., if they are able to supply the required electricity and water demand at all hours) and identify the one with lower total annual costs for the year 2020. The optimal configuration found has 6.85 MW of wind power installed and 5400 m^3/day of desalination capacity, which are the capacities currently installed on the island.

During the optimization process, it is possible to conclude that the addition of more desalination capacity only increased the costs. Even if the desalination capacity increases dramatically, the wind-power curtailed does not decrease much. This is due to the limited capacity of the reservoirs in which the desalinated water is stored, and the relatively low water consumption. If the reservoirs are full, the desalination stops and wind power is curtailed. Hence, from an economic point of view, it is possible to conclude that it is better to keep the load of the desalination units higher than to minimize the wind-power curtailed by adding more desalination units.

For the current desalination capacity installed, it is not possible to have 100% wind-powered desalination. Only if the installed desalination capacity doubles, it would be possible to achieve such a goal, and even then, a reservoir of about 300,000 m^3 would be needed, which corresponds to about 60 days of water demand in 2020. These values are disproportionate to the water consumption of this island. This is the first difficulty of coupling the electricity and water supply systems. The construction of new reservoirs is not accounted for in this scenario, as it is considered that the existing ones are suitable for the current and forecasted water consumption of the island.

9.4.1.4 Scenario 4: Wind Desalination and PHS

The previous scenario still resulted in a certain amount of wind-power curtailed. To use this curtailed wind power, a PHS is included in Scenario 4. Apart from the PHS, the existing equipment on the island is also considered (wind power and desalination capacity). It is also considered the construction of an upper reservoir with a capacity of 50,000 m^3, and a lower one with a capacity of 35,000 m^3 (about one week of average water demand in 2020).

To implement this system, there is the need to transport the desalinated water from where the desalination units are installed to the lower reservoir located at the bottom of Mont Verde. It is considered that the current water distribution system can be used in this case; however, it is necessary to confirm if that is the case and if it is not necessary to install a new dedicated penstock to transport the desalinated water. This would increase the total costs and the electricity consumption (due to pumping). Owing to the difficulty in assessing the costs of constructing a penstock, this case is not considered. Only the total costs of the installation and maintenance of the PHS system are considered, although with some degree of unpredictability. The value considered for the electricity needed to produce each m^3 of potable water already considers its pumping to the reservoirs spread throughout the island, as it is based on real data provided by ELECTRA and based on the water supply system present in S. Vicente. Hence, it is considered that this value already includes the extra energy needed to pump the desalinated water through this penstock.

It is also considered that the electricity produced in the PHS system is injected in the existent power grid. Since the exact location of the PHS station is not determined, it is not possible to estimate the distance between this facility and the power grid. This also can have some impact on the investment cost of the PHS system. Owing to this and the fact that the investment cost of a PHS is very difficult to estimate since it depends greatly on the location, a sensibility analysis is carried out, where the investment cost of the PHS is doubled.

According to Kaldellis and Kavadias [32], the most theoretically disturbed energy management scenario is based on the hypothesis that there is complete disharmony between electricity demand and wind-power harnessing. Hence, the rated power of the hydrostation is determined to cover the peak production. The pump station should have capacity to absorb the rated wind power minus the minimum consumption of the grid. With these hydro and pump rated powers, the production of RES increases significantly, but the load of the hydroturbines is very low. To face this issue, the rated power of the pumps and the hydroturbine is decreased step by step, always checking if the percentage of RES did not decrease.

It is noticed that the capacity of the upper reservoir does not influence the results, since all that is pumped is immediately retrieved back through the

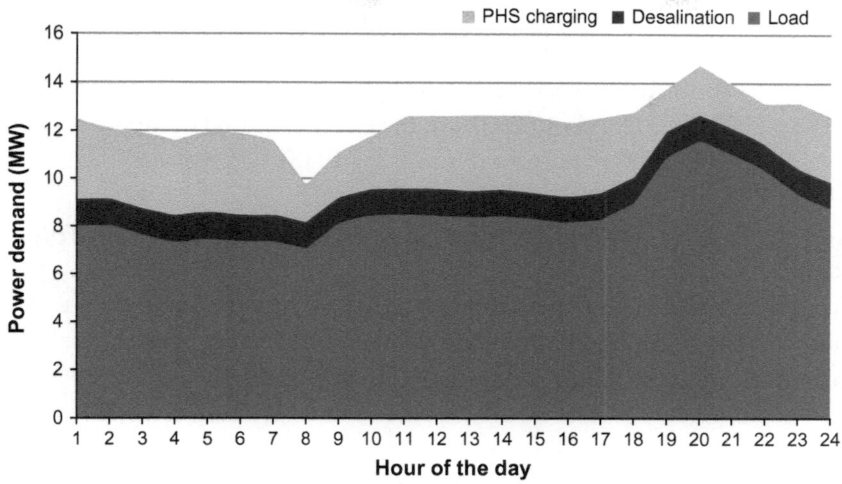

Fig. 9.3 Power demand of 1 day in 2020 for Scenario 4 [73].

hydroturbine. Hence, the capacity of this reservoir is kept as low as possible (10,000 m^3) to decrease the costs. Fig. 9.3 shows the power demand (load, desalination, and PHS charging) of one day in January 2020 in this scenario, and Fig. 9.4 shows the power production (wind, fossil fuel, and PHS) of the same day.

The total annual costs of Scenario 4 are lower than those for Scenario 2 because there is more wind power used, lowering the electricity needed

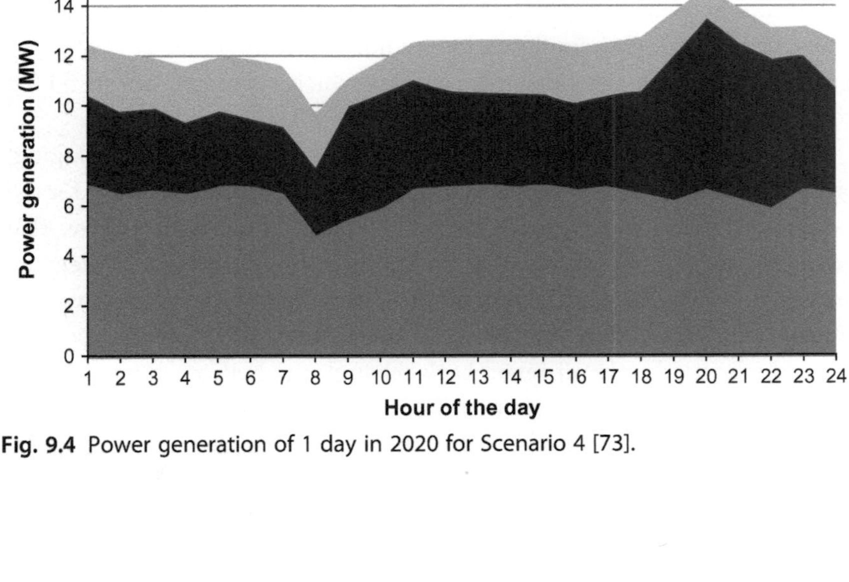

Fig. 9.4 Power generation of 1 day in 2020 for Scenario 4 [73].

from fossil fuel-based units. The wind-power curtailed decreases from 24.7% to 1.0% in 2020, which means that almost all wind-power potential is used. In comparison with the baseline scenario, the costs decrease about 7.3% in 2020. In 2020, about 28.0% of the electricity used to supply demand is RES based (22.0% wind power and 6.1% PHS).

9.4.1.5 Scenario 5: Wind Desalination and PHS With Minimum Total Costs

To establish Scenario 5, and having Scenario 4 as a starting point, several values for the wind, hydro, pump power, and desalination capacity are evaluated with the H2RES model.

The configuration that minimizes the total yearly costs is 18.75 MW of installed wind power, 7400 m^3/day of desalination capacity, 8 MW of hydropower, and 10.5 MW of pump power. This results in a load of the desalination units of 64.8%, a load of the hydroturbine of 42.3%, and a load of the pumps of 46.7%. The RES production reached 75.8% (53.3% wind power and 22.5% PHS) and the wind-power curtailed is about 15.0%. Regarding the electricity used to supply demand, about 61.4% is RES based (24.4% wind power and 36.9% PHS).

Again, it is noticed that the capacity of the upper reservoir has a small influence on the results, since all water that is pumped is immediately retrieved back through the hydroturbine. Hence, the capacity of this reservoir is kept as low as possible, in this case 35,000 m^3, to decrease costs.

The higher penetration of wind power with the wind-powered desalination and the PHS results in a reduction of the total yearly costs of about 21.4% in relation to the baseline scenario.

Considering an emission factor of 0.66 $kgCO_2$/kWh for the fuel oil power plants [66], this configuration avoids the emission of 24,782 tCO_2 in comparison with the baseline scenario, which represents about 54.0% of the total CO_2 emissions foreseen for 2020.

Fig. 9.5 shows the power demand (load, desalination, and PHS charging) for each month of 2020 in this scenario, and Fig. 9.6 shows the same for the power production (wind, fossil fuel, and PHS). These figures show that there are not significant changes in the electricity and water demand throughout the year. It also shows that the PHS charging follows the wind-power production, as the PHS production also, because what is pumped is immediately retrieved back through the hydroturbine, as mentioned before. Since the system that minimizes the costs is not large enough to store electricity over a longer term, the month that has the lower wind-power production is the month with higher fossil fuel production (Jul.).

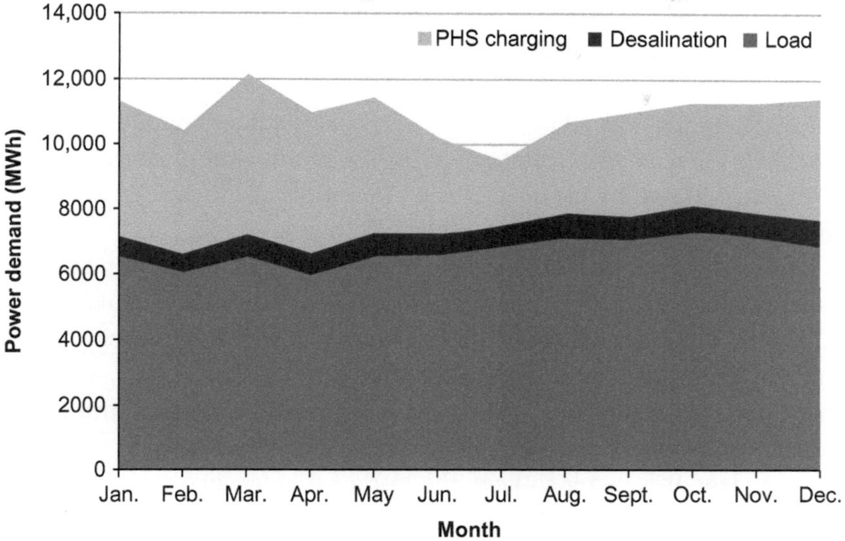

Fig. 9.5 Power demand in each month of 2020 for Scenario 5.

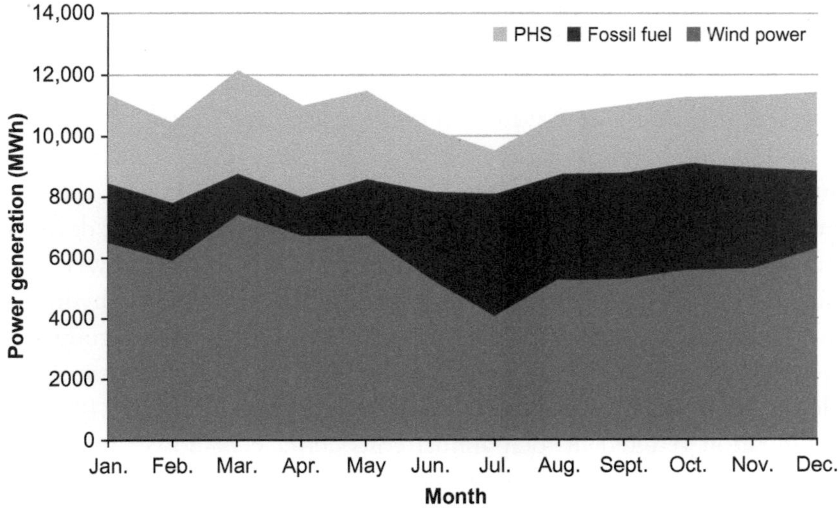

Fig. 9.6 Power generation in each month of 2020 for Scenario 5.

9.4.1.6 100% Hourly Intermittent Energy Penetration Scenarios

As referred earlier, Scenarios 1, 3, and 5 are modeled one more time, but now allowing the hourly intermittent energy penetration rate to reach 100% (see Table 9.3). For this, it is necessary to consider that the wind turbines installed possess some degree of frequency and voltage control, with a corresponding investment cost rise of 15% [65].

The baseline scenario with this condition (Scenario 1a, Table 9.3) decreases dramatically the wind-power curtailed; specifically, 0.5% in 2020. The percentage of wind power in the production of energy reaches 33.6% in 2020, which means that the current installed wind power is not sufficient to cover the load at all hours of the year.

To establish Scenario 3a, the method described earlier to set up Scenario 3 is repeated for 2020. The configuration that minimizes the annual costs is 17.05 MW of wind power and 5400 m^3/day of desalination capacity, which results in 65.7% of total electricity produced from wind power, 49.9% of wind-powered desalination, 22.3% of wind-power curtailed, and a load of the desalination units of 88.4%. Again, it is noticed that from an economic point of view, it is better to keep the load of the desalination units higher than to minimize the wind-power curtailed by adding more desalination units. In this scenario, the electricity production cost is 24.9% lower than in Scenario 3, the water production cost is 16.4% lower, and the total costs decrease 23.8%. In this case, although these wind turbines are more expensive, the extra amount of wind-power produced compensates the higher costs. Moreover, increasing significantly the desalination capacity does not reduce much the wind-power curtailed, due again to the capacity of the reservoirs and the low water demand.

To establish Scenario 5a, the method described earlier to establish Scenario 5 is repeated for 2020. The configuration that minimizes the total yearly costs is 17.9 MW of installed wind power, 6400 m^3/day of desalination capacity, 1 MW of hydropower, and 1.5 MW of pump power. This resulted in a load of the desalination units of 74.7%, a load of the hydroturbine of 15.3%, and a load of the pumps of 14.9%. The RES production reached 69.8% (68.3% wind power and 1.5% PHS), the wind-power curtailed is about 21.4%, and the wind-powered desalination is 61.1%. This configuration resulted in total annual costs about 6.7% lower than those of Scenario 5, but with a slightly lower RES production. It seems that Scenario 5a does not take advantage of the PHS system as much as Scenario 5, and this is expected since there is less need for storage when it is possible to deliver 100% of wind power directly to the grid. The fact that the configuration that minimizes the total yearly costs in this case has a lower installed wind power than in Scenario 5 is related to the higher cost of the wind turbines.

9.4.1.7 Comparison Between Scenarios

Fig. 9.7 shows the power generation in 2020 for five different scenarios. Scenario 3 (only desalination) does not differ much from the baseline scenario, although the costs are lower. In Scenario 5 (desalination and PHS), there is a significant increase in the penetration of RES. Comparing the scenarios that only consider desalination but have different intermittent limits, a 100% intermittent limit allows a significantly higher penetration of wind power. However, for the scenarios that consider desalination and PHS, the penetration of RES is almost the same.

Fig. 9.8 shows the desalinated water production in 2020 for four different scenarios. The desalination and PHS scenarios allowed for a higher percentage of water desalinated from wind power than the scenarios with only desalination for both intermittent limits. This is due to two factors. On the one hand, the reservoirs used to store the desalinated water in Scenarios 3 and 3a are smaller than the ones used on Scenarios 5 and 5a. On the other hand, since Scenarios 5 and 5a consider PHS, the water stored in the lower reservoir can be pumped to the upper reservoir, resulting in more space to put wind-powered desalinated water in the lower reservoir.

9.4.1.8 Internalizing the Cost of the CO_2 Emissions

To internalize environmental problems related to CO_2 emissions, the scenarios are modeled again considering the cost of CO_2. It is assumed that if in

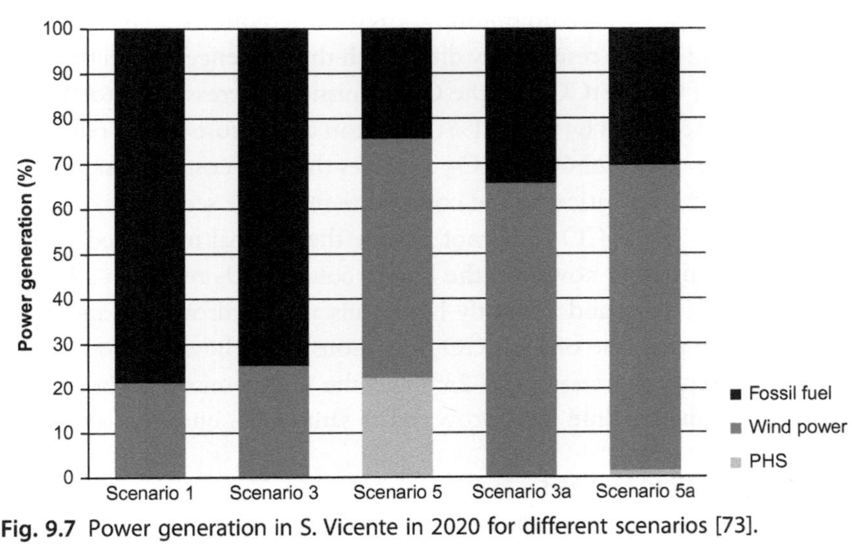

Fig. 9.7 Power generation in S. Vicente in 2020 for different scenarios [73].

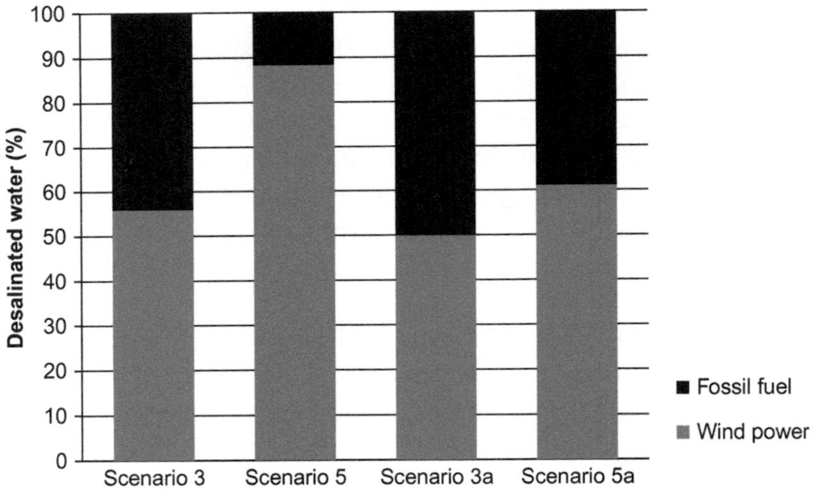

Fig. 9.8 Production of desalinated water in S. Vicente in 2020 for different scenarios [73].

the future CO_2 emissions are to be charged, the value could be the current CER cost. Since the CER cost is extremely volatile, two different values are considered: the average value of CER since this market was created until today—6.96 €/tCO_2, and the maximum value of CER ever reached—22.60 €/tCO_2 [76]. The minimum value ever reached was 0.01 €/tCO_2, which is not much different than the reference scenarios of this study, where this cost is not considered.

In the scenarios where no optimization is carried out (baseline scenario, Scenarios 2 and 4), the results only differ with the reference case in the total costs. A cost of 6.96 €/tCO_2 for the CO_2 emissions increases the total costs of these three scenarios by about 1% in relation to the costs of the reference scenarios. The cost of 22.60 €/tCO_2 increases the total cost of these scenarios by about 4% in relation to the cost of the reference scenarios.

The lower cost of CO_2 does not change the optimal integrated system reached in Scenario 5; however, the higher cost of CO_2 results in a higher installed wind power and a slightly higher installed hydropower and RES production. In this case of high CO_2 emissions cost, the total costs of the proposed systems decrease about 23% and the CO_2 emissions about 56% in relation to the baseline scenario with the same CO_2 emissions cost.

9.4.1.9 Sensitivity Analysis

Naturally, the increased electricity and water demand results in an optimal integrated system larger than in the reference case. The increase of 21.6% of electricity and water demand results in an increase of 19.3%, 20.8%, and 24.2% of the total costs of Scenarios 1, 2, and 4, respectively, in relation to the reference case.

To assess the influence of the fuel cost in the baseline scenario, it is seen that the decrease of 53% in the fuel cost results in a reduction of 38.6% of the total costs of the baseline scenario in comparison with the reference case. The scenarios that consider the proposed system also verify significant reductions in the total costs, due to the decrease in the fuel cost. The savings in the total cost verified for the proposed system in relation to the baseline scenario are due to the fuel cost; hence, with the decrease in fuel cost, it is natural that the proposed system allows for lower savings; nevertheless, these savings still exist. Even with a significant decrease in fuel cost, the proposed system is more economically viable than the current one installed in S. Vicente, with the advantage of being more environmentally friendly since its yields lower CO_2 emissions.

An increase in the PHS investment cost of 100% results in an increase in the total costs of 1.7% and 5.6% for Scenarios 4 and 5 in relation to the reference case. The proposed system still allows for significant savings in relation to the baseline scenario.

9.4.2 Optimization

To optimize the integrated electricity and water supply system proposed for S. Vicente, the same is modeled outside of the H2RES model. A tool has been designed to model this particular system for the year 2020, estimating its hourly operation. The proposed integrated system is optimized considering three objective functions: f_1, the minimization of the total annualized costs of the integrated energy and water supply system; f_2, the maximization of the percentage of RES in the total power production; and f_3, the minimization of the curtailed wind power. The method DMS, a derivative free multiobjective method, is used. With this optimization, 5910 nondominated solutions (Pareto optimal set) are obtained. The solution that minimizes the total cost is solution 1501, and it is the same one that minimizes f_1 in relation to f_2 and in relation to f_3. The solution that maximizes the share of RES in total power production is solution 4429, and it is the same one that minimizes f_2 in relation to f_1 and in relation to f_3 [74].

Table 9.4 presents the results of these solutions together with the baseline scenario foreseen for 2020. The baseline scenario considers the system installed in S. Vicente and the forecasted electricity and water consumption

Table 9.4 Results of solutions 1501 and 4429 compared with the baseline scenario

Solution		1501	4429	Baseline
Installed equipment	Wind power (MW)	21.30	28.10	6.85
	Desalination capacity (m^3/day)	6400	7400	5400
	Pump power (MW)	14.5	19.5	–
	Hydropower (MW)	8.5	19.5	–
	LR (m^3)	100,000	100,000	14,980
	UR (m^3)	150,000	280,000	–
Operational strategy	n_{WB}	77.0%	5.0%	–
	n_H	99.2%	100%	–
	n_{FF}	21.4%	25.3%	–
f_2—Total RES	83.5%	89.7%	21.4%	
f_3—Wind power curtailed (MWh)	12,126	30,418	10,898	
Wind power curtailed	12.9%	24.5%	36.5%	
f_1—Total costs (M€)	20.74	22.08	28.40	
CO_2 emissions ($ktCO_2$)	15.4	10.4	45.9	

for 2020. The total costs of this scenario also consider the CO_2 emissions cost of 6.96 €/tCO_2.

Both solutions of the optimization problem present a significant reduction in total annualized costs, increase in the share of RES in total power production and decrease of CO_2 emissions in comparison with the baseline scenario. Naturally, solution 1501 presents the greatest savings regarding total costs and solution 4429 presents the greatest increase in the share of RES and decrease in CO_2 emissions. The adoption of the proposed integrated system can reduce the total annualized costs by 22%–27% in 2020, can decrease the percentage of wind-power curtailed in relation to the wind-power potential between 33% and 65% (i.e., a decrease of 24pp and 12pp, respectively), increase the share of RES four times, and reduce the CO_2 emissions between 67% and 77%.

It is important to refer that the optimal capacity of the lower reservoir in these solutions is the maximum allowed. Despite this, the range of this variable is not increased because this capacity is considered the maximum suitable for construction on this island. Regarding the solution that maximizes the share of RES production, the same occurs for the wind power installed. A wind power of 28.10 MW is the equivalent of installing 25 more wind turbines, with the same capacity, than the ones already installed on S. Vicente. Hence, the range of this variable is also not increased.

9.4.3 Summary

From the results of the modeling of the system proposed for S. Vicente, it is possible to conclude that the direct supply of wind power to the desalination units can somewhat decrease the wind-power curtailed, improving the electricity and water production costs and reducing the CO_2 emissions. However, only with the addition of a PHS system is it possible to cancel out all wind-power curtailed. With this storage system, it is even possible to increase the installed wind power and desalination capacity to decrease the total costs, resulting in 75.8% of RES in the total electricity production and in a reduction of the total production costs of 21.4% in relation to the baseline scenario.

The optimal configuration found for the proposed system practically does not change with the addition of the CO_2 emission cost, and the total costs increase very slightly.

The sensitivity analysis revealed that even with a large decrease in the fossil fuel cost, the proposed system still brings economic savings in relation to the baseline scenario, and that a high increase of the PHS investment cost results in a small increase in the total production costs, i.e., the proposed system still allows for significant savings in relation to the baseline scenario.

The optimization carried out for the sizing and the operational strategy of the proposed integrated system for power and water supply revealed that, considering that the most important objective function is the minimization of the total costs, the optimized solution is the one listed in Table 9.5. This table also includes the results of Scenario 5 considering a CO_2 emission cost of6.96 €/tCO_2 for comparison purposes. Since the capacities of the reservoirs of the PHS can vary in the optimization process, it is expected that the results from this optimization reach higher values for the installed wind, pump, and hydropower than in Scenario 5. Nevertheless, the desalination capacity reached is smaller. The share of wind power and PHS is of the same order of magnitude, although it increases slightly for the optimized solution, for the same reasons. This implies a decrease in the CO_2 emissions by 27% in relation to Scenario 5. Although the optimized solution presents a higher installed wind power, the percentage of wind power curtailed in relation to the total wind-power production potential decreases slightly. Even though the desalination capacity installed is smaller in the optimized solution

Table 9.5 Comparison between the results of the modeling and of the optimization

		Scenario 5 with 6.96 €/tCO_2		Optimized solution	
Installed equipment	Wind power (MW)	18.75		21.30	
	Desalination capacity (m^3/day)	7400		6400	
	Pump power (MW)	10.5		14.5	
	Hydropower (MW)	8.0		8.5	
	LR (m^3)	35,000		100,000	
	UR (m^3)	35,000		150,000	
Operational strategy		n_{desal}	30%	n_{WB}	77.0%
				n_H	99.2%
		h_{FF} (h)	36	n_{FF}	21.4%
RES production	Wind power	53.3%		58.0%	
	PHS	22.5%		25.5%	
	Total RES	75.8%		83.5%	
Wind powered desalination		88.5%		99.6%	
Wind power curtailed		15.0%		12.9%	
Total cost (M€)		22.21		20.74	
CO_2 emissions (ktCO_2)		21.1		15.4	

than in Scenario 5, the wind-powered desalination is higher. This is a result of the higher capacity of the lower reservoir.

9.5 CONCLUSIONS

In this study, an integrated power and water supply system is proposed to tackle the issues of RES integration and water scarcity of the Island of S. Vicente in Cabo Verde.

Based on the existing load, water load and meteorological data, different scenarios are built and modeled using the H2RES model. The proposed wind-powered desalination and PHS system is then optimized regarding its size and operational strategy.

The financial analysis is carried out using the simplified levelized cost of energy method that finds the price of energy that sets the sum of all future discounted cash flows to zero. This method is adapted to consider the power and water production costs. Each production cost includes the investment cost of the components used to produce the specific output (power and water). The costs are annualized considering the capital recovery factor that uses the discount rate and the lifetime of the equipment. The discount rate considered in this study is 10%.

The results reveal that, with the current installed wind power and desalination capacity on S. Vicente, it is possible to have, by 2020, more than 25% of wind-power production, together with more than 56% of the water supplied to the population produced from wind power, with slightly lower costs than the ones foreseen for this year in the baseline scenario. With a PHS system, the RES production can reach 36% (31% wind power and 5% PHS), with 56% of wind-powered desalinated water and with about 7% decrease in costs in relation to the ones foreseen for 2020 in the baseline scenario. If the installation of more wind power and desalination capacity is considered, RES production can reach 76% (53% wind power and 23% PHS), with 89% of wind-powered desalinated water, with about 21% decrease in costs in relation to those predicted for 2020. This configuration avoids about 54% of CO_2 emissions forecasted for 2020.

The consideration of the CO_2 emissions cost in the production costs of the integrated system has a small influence on the results. This is because fossil fuel is already quite expensive in S. Vicente.

In relation to the sensitivity analysis carried out, the increase in electricity and water demand results in an increase in the production costs of the same magnitude. It also results in a larger optimal wind-powered desalination and PHS system, as expected. Also, as expected, the decrease of fuel costs results in a smaller optimal integrated system. However, even with a significant decrease in fuel cost, the proposed system is more economically viable than the current one installed in S. Vicente, with the advantages of being more environmentally friendly since it produces lower CO_2 emissions, using mostly endogenous resources, and is less subjected to fossil fuel cost oscillations that are difficult to predict. The large increase in the PHS system investment cost results in a relatively low increase in the production costs and in an optimal integrated system very similar to the reference case.

An optimization analysis is carried out applying the Direct Multisearch method for the sizing and operational strategy of the proposed integrated system. The optimized solution, considering the minimization of the total costs of the integrated power and water supply system, indicates that the RES production can reach 84% (58% wind power and 26% PHS), with 99.6% of wind-powered desalinated water, with about 27% decrease in costs in relation to those predicted for 2020. This configuration avoids about 67% of CO_2 emissions forecasted for 2020.

An integrated power and water supply system can indeed increase the penetration of RES in an arid island, while minimizing the production costs. In the case of S. Vicente, an island with important wind resources and a topography suitable for PHS, the proposed wind-powered desalination and PHS system can significantly increase the RES penetration and decrease the electricity and water production costs in relation to the ones foreseen for 2020.

Generally, the total annualized costs increase with the wind-power curtailed; however, the relation between these two values is not linear. The results reveal that in some situations it is economically viable to have a little more wind-power curtailed.

The load of the desalination units is extremely dependent on the capacity of the reservoirs of the PHS system.

The optimal operational strategy of this integrated system depends greatly on the sizing of the system installed. Considering the results of the optimization analysis regarding the minimization of the total costs, the optimal operational strategy consists in only producing water from the fossil fuel-based units when the lower reservoir has less water than the minimum allowed, producing electricity from the hydroplant as long as there is water in the upper reservoir and space for water in the lower reservoir and finally balancing the wind power that goes to the desalination units and to the pumps from about 77% of capacity of the lower reservoir.

ACKNOWLEDGMENTS

This work was supported by Fundação para a Ciência e Tecnologia through the scholarship SFRH/BD/31663/2006 and through IDMEC, under LAETA Pest-OE/EME/LA0022 and UID/EMS/50022/2013.

REFERENCES

[1] International Energy Agency. World Energy Outlook; 2014.
[2] Sustainable energy for all initiative. http://www.se4all.org, accessed on April 22nd, 2015.
[3] Direção Geral de Energia. Agenda de Ação para a Energia Sustentável para todos-Cabo Verde, Fevereiro. (in Portuguese).
[4] United Nations. Trends in sustainable development. Small Island Developing States Report.
[5] Duić N, Krajačić G, Carvalho MG. RenewIslands methodology for sustainable energy and resource planning for islands. Renew Sust Energy Rev 2008;12(4):1032–62.
[6] Mathiesen BV, Lund H, Karlsson K. 100% Renewable energy systems, climate mitigation and economic growth. Appl Energy 2011;88(2):488–501.

[7] Lund H, Mathiesen BV. Energy system analysis of 100% renewable energy systems—the case of Denmark in years 2030 and 2050. Energy 2009;34(5):524–31.
[8] Papathanassiou SA, Boulaxis NG. Power limitations and energy yield evaluation for wind farms operating in island systems. Renew Energy 2006;31(4):457–79.
[9] Lund H, Duić N, Krajačić G, Carvalho MG. Two energy system analysis models: a comparison of methodologies and results. Energy 2007;32(6):948–54.
[10] Rei P, Fonseca JP, Duić N, Carvalho MG. In: Integration of renewable energy sources and hydrogen storage in the Azores archipelago. Proceedings of the international conference on new and renewable technologies for sustainable development. Ponta Delgada, Azores; 2002.
[11] Spyrou ID, Anagnostopoulos JS. Design study of a stand-alone desalination system powered by renewable energy sources and a pumped storage unit. Desalination 2010;257(1-3):137–49.
[12] Fadigas EAFA, Dias JR. Desalination of water by reverse osmosis using gravitational potential energy and wind energy. Desalination 2009;237(1-3):140–6.
[13] ELECTRA. Annual report 2012 (Relatório e Contas—Exercício de 2012), 2013 (in Portuguese).
[14] Direção Geral de Energia. Plano nacional de Ação para as energias renováveis—período 2015-2020/2030—Cabo Verde, April 2015 (in Portuguese).
[15] OECD. Environment at a Glance. In: OECD Indicators. Paris: OECD Publishing; 2013.
[16] Briceño-Garmendia CM, Benitez DA. Cape Verde's infrastructure: a continental perspective. The World Bank Africa Region Sustainable Development Department. Policy research Working Paper, 2011; 5687.
[17] Spahić E, Balzer G, Hellmich B, Münch W. Wind energy storages—possibilities. Proc IEEE Power Technol 2007;615–20.
[18] Hansen JC. In: Sleeter T, Begeman P, editors. Wind power and small islands: ideas, theories and practical realities. Proceedings. St. George's: Bermuda Biological Station for Research, Inc; 2002. p. 39–48 [Bermuda Biological Station for Research Special Publication, 36].
[19] Bueno C, Carta JA. Wind powered pumped hydro storage systems, a means of increasing the penetration of renewable energy in the Canary Islands. Renew Sust Energy Rev 2006;10(4):312–40.
[20] Weisser D, Garcia R. Instantaneous wind energy penetration in isolated electricity grids: concept and review. Renew Energy 2005;30(8):1299–308.
[21] Krajačić G. The role of energy storage in planning of 100% renewable energy systems. Doctoral Thesis, Zagreb: University of Zagreb; 2012.
[22] Evans A, Strezov V, Evans TJ. Assessment of utility energy storage options for increased renewable energy penetration. Renew Sust Energ Rev 2012;16(6):4141–7.
[23] Ntziachristos L, Kouridis C, Samaras Z, Pattas K. A wind-power fuel-cell hybrid system study on the non-interconnected Aegean islands grid. Renew Energy 2005; 30(10):1471–87.
[24] Kaldellis JK, Zafirakis D. Optimum energy storage techniques for the improvement of renewable. Energy 2007;32(12):2295–305.
[25] Zoulias EI, Lymberopoulos N. Techno-economic analysis of the integration of hydrogen energy technologies in renewable energy-based stand-alone power systems. Renew Energy 2007;32(4):680–96.
[26] Bağcı B. Towards a zero energy island. Renew Energy 2009;34(3):784–9.
[27] Krajačić G, Duić N, Carvalho MG. H2RES, energy planning tool for island energy systems—the case of the island of Mljet. Int J Hydrog Energy 2009;34(16):7015–26.
[28] Ulleberg Ø, Nakken T, Eté A. The wind/hydrogen demonstration system at Utsira in Norway: evaluation of system performance using operational data and updated hydrogen energy system modelling tools. Int J Hydrog Energy 2010;35(5):1841–52.

[29] Parissis OS, Zoulias E, Stamatakis E, Sioulas K, Alves L, Martins R, et al. Integration of wind and hydrogen technologies in the power system of Corvo island Azores—a cost-benefit analysis. Int J Hydrog Energy 2011;36(13):8143–51.
[30] Díaz-González F, Sumper A, Gomis-Bellmunt O, Villafáfila-Robles F. A review of energy storage technologies for wind power applications. Renew Sust Energy Rev 2012;16(4):2154–71.
[31] Deane JP, Ó Gallachóir BP, McKeogh EJ. Techno-economic review of existing and new pumped hydro energy storage plant. Renew Sust Energy Rev 2010;14 (4):1293–302.
[32] Kaldellis JK, Kavadias KA. Optimal wind-hydro solution for Aegean Sea Island's electricity-demand fulfilment. Appl Energy 2001;70(4):333–54.
[33] Bueno C, Carta JA. Technical–economic analysis of wind-powered pumped hydrostorage systems. part I: model development. Sol Energy 2005;78(3):382–95.
[34] Bueno C, Carta JA. Technical-economic analysis of wind-powered pumped hydro storage systems. Part II: model application to the island of El Hierro. Sol Energy 2005;78(3):396–405.
[35] Gorona del Viento El Hierro, S.A. http://www.goronadelviento.es, accessed on May 6th, 2015.
[36] Caralis G, Zervos A. Analysis of the combined use of wind and pumped storage systems in autonomous Greek islands. IET Renew Power Gen 2007;1(1):49–60.
[37] Duić N, Fonseca JP, Carvalho MG. Increasing the penetration of renewable energy by merging energy and water supply systems: case study Corvo Island, Azores. In: Proceedings of the 3rd Dubrovnik conference on sustainable development of energy, water and environment systems. Dubrovnik, Croatia, 2005.
[38] Brown P, Peças Lopes JA, Matos MA. Optimization of pumped storage capacity in an isolated power system with large renewable penetration. IEEE Trans Power Syst 2008;23(2):523–31.
[39] Connolly D, Lund H, Mathiesen BV, Pican E, Leahy M. The technical and economic implications of integrating fluctuating renewable energy using energy storage. Renew Energy 2012;43:47–60.
[40] Charcosset C. A review of membrane processes and renewable energies for desalination. Desalination 2009;245(1-3):214–31.
[41] Gude VG. Energy consumption and recovery in reverse osmosis. Desalin Water Treat 2011;36(1-3):239–60.
[42] Koroneos C, Dompros A, Roumbas G. Wind energy technologies integrated with desalination systems: review and state-of-the-art. J Clean Prod 2007;15(1-3):449–64.
[43] Ma Q, Lu H. Wind energy technologies integrated with desalination systems: review and state-of-the-art. Desalination 2011;277(1-3):274–80.
[44] Carta JA, González J, Subiela V. The SDAWES project: an ambitious R&D prototype for wind-powered desalination. Desalination 2004;161(1):33–48.
[45] Carta JA, González J, Cabrera P, Subiela V. Preliminary experimental analysis of a small-scale prototype SWRO desalination plant, designed for continuous adjustment of its energy consumption to the widely varying power generated by a stand-alone wind turbine. Appl Energy 2015;137:222–39.
[46] Paulsen K, Hensel F. Design of an autarkic water and energy supply driven by renewable energy using commercially available components. Desalination 2007;203(1-3):455–62.
[47] Park G, Schäfer A, Richards B. Renewable energy powered membrane technology: the effect of wind speed fluctuations on the performance of a wind-powered membrane system for brackish water desalination. J Membr Sci 2011;370(1-2):34–44.
[48] Peñate B, Castellano F, Bello A, García-Rodríguez L. Assessment of a stand-alone gradual capacity reverse osmosis desalination plant to adapt to wind power availability: a case study. Energy 2011;36(7):4372–84.

[49] Gude VG. Energy storage for desalination processes powered by renewable energy and waste heat sources. Appl Energy 2015;137:877–98.
[50] Olsson G. Water and energy nexus. chapter in: encyclopedia of sustainability science and technology. Germany: Springer Verlag; 2011.
[51] Siddiqi A, Kajenthira A, Anadon LD. Bridging decision networks for integrated water and energy planning. Energy Strat Rev 2013;2(1):46–58.
[52] Gadhamshetty V, Gude VG, Nirmalakhandan N. Thermal energy storage system for energy conservation and water desalination in power plants. Energy 2014;66:938–49.
[53] Market applications for energy storage methods and RES units: case studies. In: Deliverable 2.1 of Renewable Energy Storage in Islands - STORIES Project; 2009.
[54] Corsini A, Rispoli F, Gamberale M, Tortora E. Assessment of H2- and H2O-based renewable energy-buffering systems in minor islands. Renew Energy 2009;34(1): 279–88.
[55] Henderson CR, Manwell JF, McGowan JG. A wind/diesel hybrid system with desalination for Star Island, NH: feasibility study results. Desalination 2009;237(1-3): 318–29.
[56] Setiawan AA, Zhao Y, Nayar CV. Design, economic analysis and environmental considerations of mini-grid hybrid power system with reverse osmosis desalination plant for remote areas. Renew Energy 2009;34(2):374–83.
[57] Novosel T, Cosić B, Krajačić G, Duić N, Puksec T, Mohsen MS, et al. The influence of reverse osmosis desalination in a combination with pump storage on the penetration of wind and PV energy: a case study for Jordan. Energy 2014;76:73–81.
[58] Østergaard PA, Lund H, Mathiesen BV. Energy system impacts of desalination in Jordan. Int J Sust Energy Plan Manag 2014;01:29–40.
[59] Santhosh A, Farid AM, Youcef-Toumi K. Real-time economic dispatch for the supply side of the energy-water nexus. Appl Energy 2014;122:42–52.
[60] Bognar K, Blechinger P, Behrendt F. Seawater desalination in micro grids: an integrated planning approach. Energy Sust Soc 2012;2:14.
[61] Bognar K, Pohl R, Behrendt F. Seawater reverse osmosis (SWRO) as deferrable load in micro grids. Desalin Water Treat 2012;51(4-6):1190–9.
[62] National Statistics Office of Cabo Verde, Census 2010 - National Statistics Institute of Cabo Verde. Demographical predictions by geographical code and year (Projecções Demográficas por Código Geográfico e Ano) (in Portuguese).
[63] Eurostat - Energy price statistics - http://ec.europa.eu, accessed on July 15th, 2015.
[64] UNIDO, ECREEE, GEF. Report Cape Verde: energy analysis and recommendations, in the scope of the project promoting market based development of small to medium scale renewable energy system in Cape Verde, 2011. Available on http://www.ecreee.org.
[65] Description of system configurations and model design results and analysis for selected integrated RES/H2 systems and H2 end-uses. In: Deliverable 2 of Renewable Energy Solutions for Islands—RENEWISLANDS Project; 2005. February.
[66] Gesto Energia SA. Plano energético renovável de cabo verde; 2011 [in Portuguese].
[67] Instituto Nacional de Meteorologia e Geofisica de Cabo Verde. Velocidade do vento na estação meteorológica de S. In: Pedro; 2005.
[68] ELECTRA. Annual report 2010 (Relatório e Contas—Exercício de 2010), 2011 (in Portuguese).
[69] Ministry of Economy, Growth and competitivity of Cabo Verde. National energy plan—energy policy plan for Cabo Verde (Plano Energético Nacional—Plano de Política Energética da República de Cabo Verde), May 2003 (in Portuguese).
[70] Cost-benefit analysis. Deliverable 2.3 of renewable energy storage in islands—STORIES project. December; 2009.
[71] Economic Regulatory Agency of Cabo Verde. http://www.are.cv, accessed on March 24th 2014.

[72] Segurado R, Krajačić G, Duić N, Alves L. Increasing the penetration of renewable energy resources in S. Vicente, Cape Verde. Appl Energy 2011;88(2):466–72.

[73] Segurado R, Costa M, Duić N, Carvalho MG. Integrated analysis of energy and water supply in islands. Case study of S. Vicente, Cape Verde. Energy 2015;92(3):639–48.

[74] Segurado R, Madeira JFA, Costa M, Duić N, Carvalho MG. Optimization of a wind powered desalination and pumped hydro storage system. Appl Energy 2016;177:487–99.

[75] H2RES - Energy planning of islands and isolated regions. http://h2res.fsb.hr, accessed on September 9th, 2009.

[76] Carbon dioxide emission allowances electronic trading system—SENDECO2. http://www.sendeco2.com, accessed on February 18th, 2015.

[77] Custódio AL, Madeira JFA, Vaz AIF, Vicente LN. Direct multisearch for multiobjective optimization. SIAM J Optim 2011;21(3):1109–40.

[78] Custódio AL, Emmerich M, Madeira JFA. Recent developments in derivative-free multiobjective optimisation. Comput Technol Rev 2012;5:1–30.

[79] Ramos JS, Ramos HM. Sustainable application of renewable sources in water pumping systems: optimized energy system configuration. Energy Policy 2009;37(2):633–43.

PART 2

Thermodynamics

CHAPTER 10

Energy Storage for Desalination

Veera Gnaneswar Gude
Mississippi State University, Mississippi State, MS, United States

Nomenclature

A	area (m^2)
AC	alternate current
Ah	ampere hour
c_p	specific heat (kJ/kg K)
DC	direct current
I	solar insolation (kJ/h m^2)
m	mass of storage medium (kg)
MDOD	maximum depth of discharge
P	pressure (atm), power (kW)
PV	photovoltaic
q	evaporation energy (kJ/h)
Q	energy or useful energy from solar collectors (kJ/h)
S	entropy (kJ/K)
t	time (h, s)
T	temperature (K)
TCM	thermochemical reaction material
U	heat loss/transfer coefficient (kJ/h m^2 K)
U	internal energy
v	volume of the saline water or storage tank (m^3)
V	wind speed (m/s)

Special characters

ρ	density (kg/m^3)
λ	latent heat (kJ/kg)
η	energy (first law) efficiency (%); panel efficiency
Δh	latent heat for PCM storage

Subscripts

1	TES temperature
2	evaporation chamber temperature
a	ambient
ac	alternate current
c	collector
Comb	combined

Renewable Energy Powered Desalination Handbook
https://doi.org/10.1016/B978-0-12-815244-7.00010-6

CR	chemical reaction
dc	direct current
DES	desalination
EC	evaporation chamber
h	hot stream
in	inlet, supply
e	evaporation, electrical
l,ls,L	latent, losses, load
out	energy supplied by TES
PCM	phase change material
s	solar, collectors, TES
sensible	sensible heat
STC	standard conditions
t	total
TES	thermal energy storage
th	thermal

10.1 INTRODUCTION

Desalination is an energy-demanding process commonly powered by non-conventional energy sources which are unsustainable [1,2]. Renewable energy sources (RES) such as solar, wind, and geothermal sources can serve as alternative energy sources for desalination processes. Desalination processes are mainly based on thermal and membrane technologies. Thermal desalination processes include multieffect distillation (MED), multistage flash (MSF) desalination, vapor compression (mechanical and thermal, MVC or TVC), and solar desalination (SD). Membrane-based desalination processes include reverse osmosis (RO) and electrodialysis (ED). The potential renewable energy and desalination technology combinations are shown in Fig. 10.1 [2,4]. The RES should be integrated with the relevant desalination technology that has capability to utilize the energy in the most effective manner. Some RES-dependent desalination technologies must be placed on the same site (colocation) and some do not have this requirement. Colocating desalination and power generation systems can be cost effective [5]. Accordingly, the following thermal desalination-renewable energy combinations require colocation (located on same site): (a) wind-shaft-MVC, (b) solar thermal-TVC, (c) solar thermal-MSF, (d) solar thermal-MED, (e) solar thermal-SD, (f) geothermal-TVC, (g) geothermal-MSF or MED. The other electricity-driven combinations that do not require colocation are (a) wind-MVC, (b) wind-RO, (c) solar PV-RO, (d) solar PV-MVC, (e) geothermal-MVC, and (f) geothermal-RO.

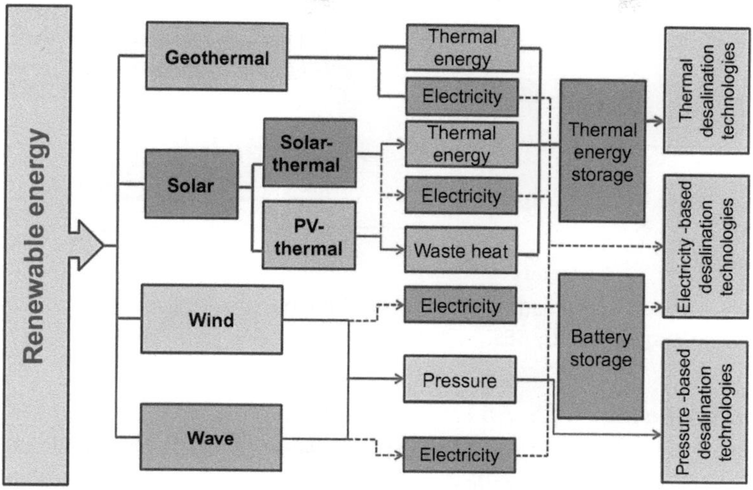

Fig. 10.1 Desalination technologies coupled with renewable energy and storage systems.

10.2 ENERGY STORAGE OPTIONS FOR VARIOUS DESALINATION PROCESSES

Desalination technologies that utilize thermal energy and thus require thermal energy storage (TES) for uninterrupted process operation are MED, MSF, low-temperature MED (LTMED) low-temperature desalination (LTD), humidification-dehumidification (HD), and membrane distillation (MD). TES technology requires a suitable medium for the storage and circulation for heat transfer while the photovoltaic (PV)/wind-generated electricity needs to be stored in batteries for later use as shown in Fig. 10.1. Similar to TES, the battery energy storage (BES) is essential to store electrical energy for ED, RO, and MVC technologies.

10.3 TES FOR DESALINATION

Energy absorption-accumulation, storage, and release are the principal elements of the TES concept which result in improved economics, better resource management, and lower environmental emissions of a variable energy source-powered desalination system (DES), for instance, solar energy. TES units function in three important stages: (1) charging period, (2) storing period, and (3) discharging period [6]. Thermal energy accumulation and storage can be accomplished by three different principal methods: (1) sensible heat TES which includes both solid-state and liquid-state materials for storage; (2) phase change TES, also known as latent heat storage; and

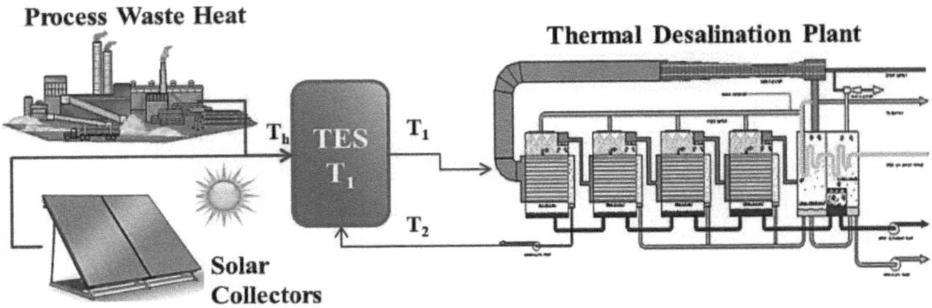

Fig. 10.2 Thermal desalination system powered by solar collectors or waste heat sources and augmented by TES.

(3) thermochemical TES. An example of the sensible heat TES unit for desalination application powered by process waste heat or solar collectors is shown in Fig. 10.2. The following equation expresses the amount of heat stored in a sensible heat TES in general which depends on the amount of heat transferred from a heat source:

$$Q_{sensible} = m \int_{T_1}^{T_2} c_p(T)dT \tag{10.1}$$

where $Q_{sensible}$ is the amount of sensible heat stored, m the flow rate of the storage medium, c_p the specific heat capacity of the storage medium, and T_1 and T_2 are the outlet and inlet temperatures of the heat transfer fluid (HTF).

Thermodynamic efficiency of a TES augmented desalination process can be illustrated using the first and second law efficiencies [7]. The first law (energy) efficiency of the TES can be expressed as (Fig. 10.2)

$$\eta_{TES} = \frac{Q_{out}}{Q_{in}} = \frac{m_s c_s (T_1 - T_2)}{m_h c_h (T_h - T_1)} \tag{10.2}$$

where Q_{in} is the energy supplied to the TES by a hot stream of mass flow rate m_h, specific heat c_h, and inlet and outlet temperatures of T_h and T_1, respectively. Q_{out} is the energy supplied by the TES to the DES with a mass flow rate m_s, specific heat c_s, and inlet and outlet temperatures of T_2 (evaporation temperature T_{EC}) and T_1, respectively.

First law (energy) efficiency of the DES can be found from

$$\eta_{DES} = \frac{m\lambda_{(T_{EC})}}{m_s c_s[(T_1 - T_2)]} \tag{10.3}$$

where m is the mass of the freshwater produced and λ the latent heat of the freshwater at the evaporation temperature T_{EC} which is same as T_2.

First law (energy) efficiency of the combined system can be found from

$$\eta_{Comb} = \frac{m\lambda_{(T_{EC})}}{m_h c_h[(T_h - T_1)]} \tag{10.4}$$

Details on the thermodynamic analysis (first and second law efficiencies) of TES systems augmenting the desalination application at different TES and desalination temperatures can be found elsewhere [7].

Since thermal energy can be stored and transported by fluids in different phases, a generalized expression can be written to represent various fluids and associated operational parameters such as temperature and pressure. Even when a system is exposed to its environment, as long as it is in its equilibrium state so that the temperature T and the pressure P of the system are equal to those of the environment (i.e., $T = T_{environment}$ and $P = P_{environment}$), its internal energy has a well-defined value on the macroscopic level so that the internal energy is a state variable. Being the sum of the kinetic and the potential energy of the constituent atoms or molecules, the internal energy must be proportional to the total number N of these atoms or molecules, which is in turn proportional to their mole number (n): $U \alpha N \alpha n$, which implies that the internal energy is extensive so that we can express it as $U = nu\ (T, P)$, where u is the molar internal energy defined by $u = U/n$. To simplify the illustrations, the previously defined equations will be used in further discussions.

The quality of thermal energy that is stored by the storage medium determines its appropriate application which in turn depends on the availability of heat source. Sensible heat TES systems with water as storage medium store thermal energy below 100°C, therefore their application could possibly be limited to LTD systems such as solar stills, low-temperature multieffect evaporation (MEE) systems, MD, HD, and other novel DESs. Solid-state or liquid-state materials can store high-grade thermal heat in the range of 200–500°C which would be ideal for power generation combined with MSF, MEE or MED, and MVC/TVC desalination processes. Table 10.1 provides a comparison of the various storage mediums for TES and their potential for desalination and power generation applications [8]. Briefly, sensible heat TES technologies can be used in desalination process applications while the phase change and thermochemical TES technologies are ideal for cogeneration purposes.

Table 10.1 TES materials and technology applications in desalination and power industries

Technology	Molten salt	Concrete	Phase change material	Water/steam	Hot water
Capacity range (MWh)	500–>3000	1–>3000	1–>3000	1–200	1–3000
Annual efficiency	98%	98%	98%	90%	98%
Heat transfer fluid (HTF)	Synthetic oil	Synthetic oil, water/steam	Water/steam	Water/steam	Water
Temperature range (°C)	290–390	200–500	Up to 350	Up to 550	50–95
Investment cost ($/kWh)	40–60	30–40	40–50	180	2–5
Advantages	High storage capacity at relatively low cost Experience in industrial applications Well suited for synthetic oil HTF	Well suited for synthetic oil heat transfer fluid Easily available material Well suited for preheating and superheating in direct steam generating collectors	Latent heat storage allows for constant temperature and heat transfer Low material requirements Well suited for evaporation/condensation process in direct steam generating collectors	Latent heat storage allows for constant temperature at heat transfer Experience in industrial applications Well suited for evaporation/condensation process in direct steam generating collectors	Very low cost storage for processes heat below 100°C Experience in industrial applications

Disadvantages	Sensible heat storage requires temperature drop at heat transfer Molten salt freezes at 230°C	Not suited for evaporation/ condensation in direct steam generating collectors Recent development	Not suitable for preheating and superheating in direct steam generating collectors Very early stage development	Not suitable for preheating and superheating	Sensible heat storage requires temperature drop at heat transfer Not applicable to power generation
Desalination and/or power generation application	Cogeneration, power production, and desalination Large-scale applications for MSF, MED, MVC, and RO processes	Cogeneration, power production and desalination Large-scale applications for MSF, MED, MVC, and RO processes	Cogeneration, power production, and desalination Large-scale applications for MSF, MED, MVC, and RO processes	Cogeneration, power, and desalination Large-scale applications District heating and cooling	Solar still, humidification–dehumidification, membrane distillation, other low-temperature desalination processes

10.3.1 Sensible Heat TES for Desalination

The most commonly used sensible heat storage medium is water. In water-based TES systems, water serves both as the energy storage and the heat transfer medium. In addition, various solid-state and liquid/molten-state sensible heat TES mediums can be used depending on the desalination application. Tables 10.2 and 10.3 present the heat storage capacity and costs per unit of thermal energy ($/kWh$_t$) for different solid-state and liquid-state sensible heat TES mediums [9,10]. The economics of the energy storage systems vary significantly with different geographical locations, economic packages, and the material costs. The economics of these systems are discussed in the later sections. Sensible heat TES technology has been utilized in numerous solar desalination applications. The applications include solar stills, solar ponds, and in cogeneration systems with concentrated solar collectors for power generation and water production. In this section, we focus on TES applications in solar stills and solar ponds.

Table 10.2 Solid-state sensible heat storage materials

Storage materials	Working temperature (°C)	Specific capacity ($kWh_t/m^3 °C$)	Cost per kg ($/kg)	Cost per kWh ($/kWh)
Sand-rock minerals	200–300	0.61	0.15	4.2
Reinforced concrete	200–400	0.52	0.05	1.0
NaCl	200–500	0.51	0.15	1.5
Cast steel	200–700	1.30	5.00	60.0
Silica fire bricks	200–700	0.51	1.00	7.0
Magnesia fire rocks	200–1200	0.96	2.00	6.0

Table 10.3 Molten salts and high temperature oils

Storage materials	Working temperature (°C)	Specific capacity ($kWh_t/m^3 °C$)	Cost per kg ($/kg)	Cost per kWh ($/kWh)
Mineral oil	200–300	0.56	0.30	4.2
Synthetic oil	250–350	0.58	3.00	43.0
Silicone oil	300–400	0.53	5.00	80.0
Nitrite salts	250–450	0.76	1.00	12.0
Nitrate salts	265–565	0.83	0.5.	3.7
Carbon salts	450–850	1.05	2.40	11.0

10.3.1.1 Solar Stills

In solar stills, the solar radiation is reflected and lost to the ambient. The sinusoidal trend of the solar radiation and the ambient temperatures do not favor freshwater production during nonsunlight hours. During the night time, the solar still productivity could be higher owing to the lower ambient temperatures if the heat source is available during this period. A simple way to increase the heat storage capacity of a solar still is to increase the saline water depth in the still. By increasing the depth (TES volume) of the saline water, the heat storage capacity of the device can be increased which would result in continued evaporation during cloudy hours and nonsunlight hours. An optimum water depth should not decrease the water temperature during sunlight hours which may result in lower evaporation rates. Solar stills incorporating TES have shown significant improvement in the overall freshwater production rates. To describe this effect a few examples can be discussed. Tabrizi and Sharak [11] investigated a basin solar still integrated with a sandy heat reservoir. The productivity of this configuration was almost twice that of a conventional solar still at the end of a 14-hour test. Murugavel et al. [12] tested different TES materials such as quartzite rock, red brick, cement concrete, washed stones, and iron scraps in a single-basin double slope solar still to improve the nonsunlight hour distillate production. Among these materials, 3/4 in. sized quartzite rock was found to be the most effective, with a productivity of $\sim$2.1 kg/m^2/day and an enhancement of 6.2% compared with the still with same amount of water but without any energy storage medium. These studies confirm that TES increases the total freshwater productivity albeit at a slight increase in operational costs in practical applications.

A sensible heat TES was employed in a solar (three flat-plate solar collectors) desalination process with three stages of multieffect humidification (MEH)–dehumidification process. The energy storage unit provided the desalination chamber with additional heat for three nonsunlight hours. The experimental results showed that higher seawater mass flow rate through the system (from 0.1 to 0.13 L/s) increase the productivity of the system by 10%. This study concluded that the use of energy storage increased the productivity by 13.5% and the productivity of the experimental system was about three times the productivity of a conventional solar still [13].

10.3.1.2 Solar Pond

A solar pond can be considered as a large solar collector with huge storage volume where the solar energy trapped into the stored brine solution is circulated as a heating medium in the desalination process unit. Solar ponds

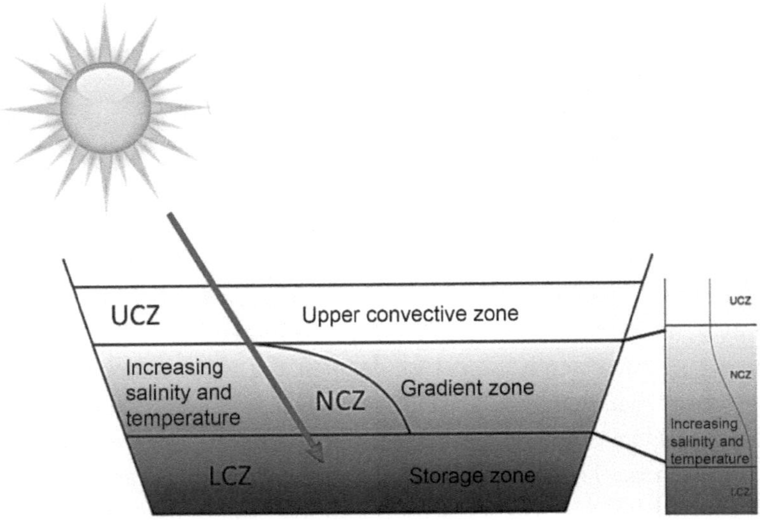

Fig. 10.3 Solar pond and its convective zones. *(After Busquets E, Kumar V, Motta J, Chacon R, Lu H. Thermal analysis and measurement of a solar pond prototype to study the non-convective zone salt gradient stability. Sol Energy 2012;86:1366–77.)*

provide the most convenient and least expensive option for heat storage for daily and seasonal variations [14–16] in solar energy including cloud effects while simultaneously providing for brine management. The solar pond consists of three distinct zones as shown in Fig. 10.3 [17]. The first zone, located at the top of the pond, contains the low-density saltwater mixture. This zone is called the upper convective zone (UCZ) which is the absorption and transmission region. The second zone, which contains a variation of saltwater densities increasing with depth, is called the gradient zone or nonconvective zone (NCZ). This zone acts as an insulator to prevent heat from escaping to the UCZ, by maintaining higher temperatures at lower zones. The bottom zone is the heat storage zone or lower convective zone (LCZ) with uniform salt density [18]. Sodium chloride (NaCl) is the commonly used storage medium in solar ponds. Magnesium chloride ($MgCl_2$), sodium nitrate ($NaNO_3$), sodium carbonate (Na_2CO_3), sodium sulfate (Na_2SO_4), ammonium nitrate (NH_4NO_3), fertilizer salts such as urea ($NH_2CO.NH_2$) satisfy the thermal stability criterion and are considered suitable for solar pond applications [19,20].

The temperature differences between the bottom and top layers of a solar pond are adequate to drive a generator. An HTF piped through the bottom layer harvests the heat for direct desalination application. Thermal energy may be circulated through a closed-loop Rankine cycle (RC) system to rotate a turbine to generate electricity. The annual collection efficiency of useful thermal energy for desalination is in the order of 10%–15% with larger ponds being more efficient than smaller ones. Solar ponds produce relatively low-grade thermal energy (<100°C) and are generally considered suitable for thermal distillation processes. Brine stream generated from the desalination processes can be stored in solar ponds to serve as storage medium. Brine utilization as storage medium in solar ponds provides a convenient method and inexpensive source to maintain solar pond salinity [21].

A study on a single-slope basin solar still integrated with a shallow solar pond (SSP) found that the average productivity and thermal efficiency of this system were higher than those obtained without the SSP by 52.4% and 43.8%, respectively, over a year [22]. Extensive research conducted for more than 20 years at El Paso (Texas, USA) solar pond demonstrated the reliability of a salinity-gradient solar pond in desalination application. This research focused on testing various operating conditions for a multieffect, multistage (MEMS) flash desalination unit. This research provided operation and maintenance procedures of the salinity-gradient solar pond coupled with the desalination [23]. An experimental solar pond with a surface area about 830 m^2 and a depth of 2.5 m was studied in Tajoura, Libya [24]. This solar pond was coupled with a 5 m^3/day MSF desalination plant. Tahri studied the possibility of combining an MSF desalination plant with a solar pond to recover the waste heat from exhaust gas of a thermal plant [25]. Posnansky [26] presented computer simulations and experimental results on a small-sized solar pond for the performance data of the coupled MSF unit [27]. Szacsvay and coworkers developed a DES with autoflash MSF unit consisting of a solar pond as the heat source. Performance and layout data were obtained from both computer simulation and experimental results with a small-sized solar pond coupled with desalination subsystem in Switzerland in operation for 9 years. The authors concluded that the cost of distillate could be reduced from \$5.48/m^3 for small DES with a capacity of 15 m^3/day to \$2.39/m^3 for DESs with a capacity of 300 m^3/day [28]. Solar-powered MEH studied by Muller-Host et al. used 2 m^3 TES increased the production rate to 500 L/day with 38 m^2 collector area (about 13 L/m^2). Two different humidification units were tested in this study [29]. A multiobjective mixed-integer nonlinear programming model was developed to model an integrated system of RO, a RC, parabolic trough solar collectors with TES. This study optimized the design and operating conditions with a focus on economic and environmental metrics. A molten-salt thermocline TES was integrated in the system to increase the solar energy utilization. Molten salt was used as the HTF to transport thermal energy between the TES and the relevant components of the power system (e.g., collector field and RC boiler). The method was applied to a case study of an RO plant coupled with a solar RC and a TES located in Tarragona. This study reported the benefits of TES as improved process performance and efficiency with reduced carbon dioxide emissions up to 55.6% albeit at a slightly higher cost, 14% [30].

10.3.1.3 Waste Heat Utilization and LTD

The TES can be used to store the process waste heat to be utilized for desalination purposes as shown in Fig. 10.2. The LTD is beneficial from many perspectives which include less resource (thermal) losses, lower operation, maintenance, and capital costs. A LTD process tapping the reject heat from the condenser of a domestic air-conditioning system was studied previously. In this study, evaporation of saline water occurred at near vacuum pressures created by exploiting the principles of local barometric head. The evaporator operates in a temperature range of 40–50°C with the heat supplied by a TES unit. The energy requirements for the system were less than that required for a MSF distillation process. It was shown that the thermal energy rejected by an absorption refrigeration system (ARS) of cooling capacity of 3.25 kW (0.975 tons of refrigeration) along with an additional energy input of 208 kJ/kg of desalinated water was sufficient to produce desalinated water at an average rate of 4.5 kg/h. This energy consumption was only 60% of a typical MSF distillation process (338 kJ/kg) [31,32]. A TES unit volume of 10 m^3 with a solar panel area of 25 m^2 was required for this application.

In another recent study [33], an integrated process model was developed for a novel application of a sensible TES system for energy conservation and water desalination in power plants. In this configuration, a cold TES was designed to alleviate the negative effects of high ambient temperatures on the performance of an air-cooled condenser that cools a 500-MW CCPP (combined cycle power plant). The cold TES also provided for the cooling requirements of a final water cooled condenser in a MED unit with a capacity of 950 m^3/day. Stack gases from CCPP were used to drive an ARS which maintains the chilled water temperature in a TES tank. A process model integrating CCPP, ARS, TES, and MED was developed to optimize the volume of the TES. Preliminary analysis of this integrated process showed that a cold TES tank volume of 2950 m^3 could meet the cooling requirements of ACC and MED in both hot and cold seasons. A potential savings of 2.5% of the power loss in a CCPP were realized on a hot summer day for this TES system along with an estimated desalination capacity of 950–1600 m^3/day for top brine temperatures between 100°C and 70°C of the MED.

10.3.1.4 Sizing of TES Systems

The size of the TES system depends on several factors: (a) the desalination technology (heat load), (b) the heat source whether solar energy or process waste heat (availability and duration), (c) the heat capacity of the storage material, (d) the required storage period, and (e) the expected standby loss (heat losses to ambient). Optimal sizing of the TES system is important to maximize the integrated system efficiency (see Eqs. 10.1–10.4). Without a supplemental or auxiliary heat source, undersized TES systems are incapable of meeting the energy demands. Oversized TES systems may result in higher capital costs as well as operation and maintenance costs, and can waste energy through standby losses. Sizing the system is even more critical as even optimally sized systems can occupy a large space and require high installation costs. Two important examples on the TES sizing are described below.

TES in Solar Ponds: Effect of the Thickness of the LCZ

Similar to solar stills (where excess solar energy is stored in the depth of the saline water medium), the lower zone (LCZ) in solar ponds stores the useful heat which is extracted for heating applications. A recent study reported the sizing of salinity gradient solar pond (SGSP) connected to a MED desalination plant of 5 m^3/day capacity [27]. It was shown that the thickness of the lower zone affects the amplitude of temperature fluctuations in response to daily solar irradiation (Fig. 10.4). The temperature variation in the LCZ was

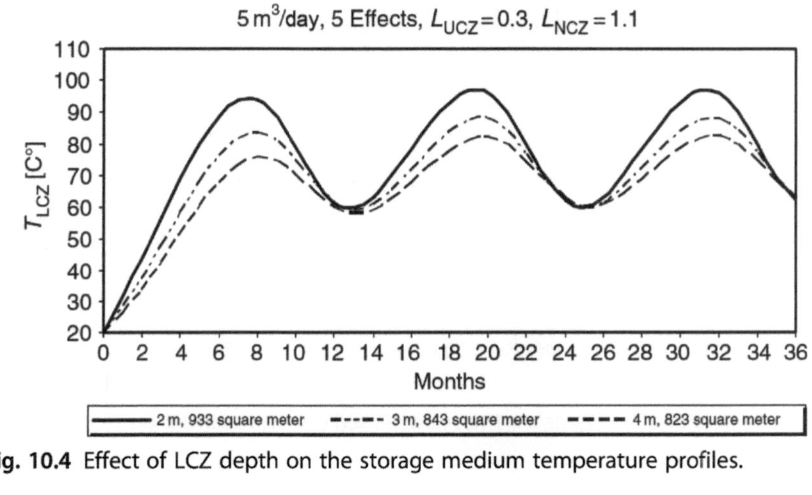

Fig. 10.4 Effect of LCZ depth on the storage medium temperature profiles.

between 60°C and 80°C which was found to be suitable for the desalination application. It was reported that a deeper LCZ zone (4 m) results in the smallest surface area required for the solar pond but with increased storage volume and increased capacity. This study also investigated the effects of UCZ and NCZ. The study concluded that the optimum thickness of LCZ zone is of the order of 4 m with 0.3 m thickness for the UCZ and 1.1 m for the NCZ. Solar pond with these specifications will provide working temperatures of the medium suitable for desalination application and to meet the extracted load requirement from the storage zone. Excellent details of heat extraction methods for SGSP and other solar ponds are discussed elsewhere [34,35].

TES in Solar Desalination

Storage volume is critical to the performance of the TES tank. In desalination application powered by solar collectors, a TES tank is required to mitigate the effects of cloudy hours and nonsunlight hours. In a recent study, the effect of TES tank volume was simulated for a LTD process (at a capacity of 100 L/day) supported by solar collectors [7]. The temperature profiles over a week of operation for the TES tank at two different volumes (1 and 6 m^3) are shown in Fig. 10.5. As expected, the TES temperature profiles followed the

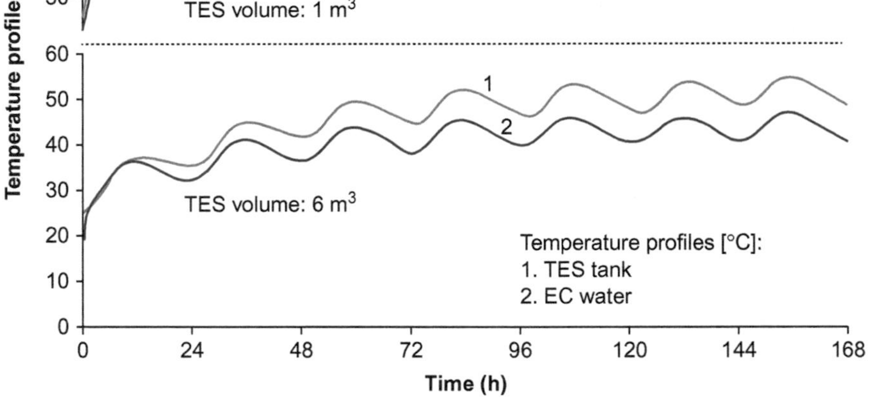

Fig. 10.5 Effect of TES volume on the storage medium temperature profiles.

sinusoidal nature of the solar irradiation and the ambient temperature. Smaller TES volume ($1\,m^3$) temperatures were more responsive to these variations than higher TES volume ($6\,m^3$). The temperatures in smaller TES tank (30–35°C) were not suitable for desalination application in the nonsunlight hours whereas the temperatures in the higher TES tank (45–48°C) were adequate for desalination application during nonsunlight hours. Further, simulations on 21 days of continued operation have shown that the storage medium temperature increases with storage time and that higher TES volume increases the quality and quantity of the heat source available for nonsunlight hour operation increases [7]. As a comparison, in DESs without energy storage, the desalination unit remains idle during nonsunlight hours requiring a higher desalination production capacity during the sunlight hours. Instead, a TES system can reduce the footprint of the DES with reduced downtime. In short, without a TES system, the desalination unit operates in a batch mode but with a TES the desalination can occur in a continuous mode which may result in lower capital and operation and maintenance costs [7]. In some other novel configurations where adsorption cooling and MED cycles are integrated, the evaporation of freshwater can be achieved at heat source temperatures as low as 35°C [36].

Finally, to gain optimum storage dynamics and longevity important factors need serious consideration: high energy density in the storage materials; good heat transfer between the HTF and the storage medium; mechanical and chemical stability of the storage materials; chemical compatibility between HTF, heat exchanger, and storage medium; complete reversibility for a large number of charging/discharging cycles; low cost; low environmental impact [37].

10.3.2 Phase Change TES

Phase change materials (PCMs) store thermal energy as a result of phase change phenomenon, i.e., as heat of fusion (solid-liquid transition), heat of vaporization (liquid-vapor), or heat of solid-solid crystalline phase transformation. Since the heat released during the phase change of a material (solid to liquid; liquid to vapor or gas) is much higher than sensible heat of the materials, PCMs reduce the energy storage unit size significantly. For example, latent heat of water is rather high compared with the sensible heat that can be stored as sensible heat [38]. Some PCMs can absorb heat beyond the phase change. Eq. 10.5 shows the amount of heat that can be stored in a PCM TES unit.

$$Q_{PCM} = m\int_{T_1}^{T_L} c_p(T)dT + m\Delta h|_{T=T_L} + m\int_{T_L}^{T_2} c_p(T)dT \qquad (10.5)$$

where Q_{PCM} is the energy stored in the phase change TES medium with a mass flow rate m, specific heat c_p, and initial and latent heat temperatures of T_1 and T_L, respectively. Δh is the latent heat and T_2 is the final temperature of the PCM.

The PCMs can be incorporated into TES units in two ways. PCMs can be suspended in the storage medium in nano- or microscale particles to increase the heat storage capacity. These PCMs flow with the storage medium which remains in liquid state in all stages of TES operation (charging-storing-discharging). The storage system can be a two-tank system with the PCM and storage fluids both contributing to the storage capacity of the system. The PCMs can also be used as a stationery medium over which the HTF circulates to extract and transfer the heat to the load unit (desalination process). Commonly used PCMs are encapsulated as spheres or other shapes to fill a fixed bed. In this geometry, the HTF passes through the void volume of the fixed/stationery bed. Alternatively, a pipe design such as a heat exchanger within a PCM storage system with charging and discharging cycles can be considered [38].

The PCM storage systems have a basic limitation of low power densities caused by the low thermal conductivity of the solid phase of these materials. Metals and metal alloys have conductivities in the range of 100–400 W/m K and do not present this problem when used as PCMs. Salts, however, have conductivity of around 0.5 W/m K and limit heat transfer and power density as the storage system discharges. This issue can be addressed by introducing a higher thermal conductivity material while increasing the heat transfer area of the heat exchanger. A large difference in the PCM solid- and liquid-phase densities also presents a design issue and the tank must be sized to accommodate the volume of the lower density phase and also ensure that the heat transfer surfaces are contacted by both solid and liquid phases. Other issues of consideration are the selection of heat transfer design and media since some PCMs exhibit reduced performance after a number of cycles of operation. In addition, phase change storage materials cost more than the sensible heat storage materials. Sensible heat storage medium usually costs in the range of 0.05–5.00 \$/kg, compared with the high cost of latent heat or PCM storage which usually ranges from 4.28 to 334.00 \$/kg. Thermal properties of some PCM storage materials are illustrated in Table 10.4 [9,39]. Fig. 10.6

Table 10.4 Phase change thermal energy storage materials

Storage materials	Working temperature (°C)	Specific density (kg/m³)	Latent heat (kJ/kg)	Latent heat (MJ/m³)
RT100 (paraffin)	100	880	124	Unavailable
E117 (inorganic	117	1450	169	245
$NaNO_3$	307	2260	172	389
KOH	380	2044	149.7	306
$MgCl_2$	714	2140	452	967
Na_2CO_3	854	2533	275.7	698
$MgCl_2$–KCl–NaCl	380	2044	149.7	306

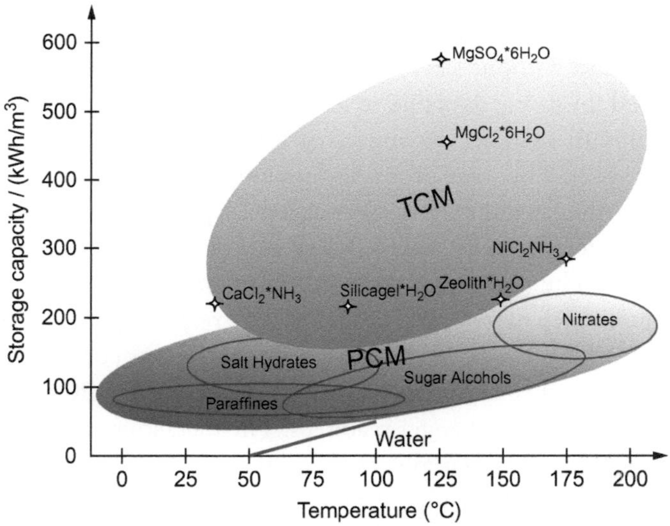

Fig. 10.6 Storage capacity vs. temperature for sensible, latent and thermochemical TES.

shows a comparison of the TES capacity of different storage materials. It can be noted that most of the phase change and thermochemical storage materials have much higher storage capacities than water which is commonly used as sensible heat TES medium [40].

10.3.2.1 PCM Energy Storage Applications in Desalination

Phase change or latent heat storage mediums are as commonly used as sensible heat storage medium in desalination applications. The latent heat TES systems have many advantages over sensible heat storage systems.

The benefits are a larger energy storage capacity per unit volume, and nearly constant temperature for energy charging and discharging cycles [41]. Table 10.4 presents the main properties of some PCM energy storage materials. The PCM technology is more suited for high-temperature desalination technologies such as MSF distillation, MED, and TVC processes which require very high heat source temperatures (>100°C) and for cogeneration plants where the power plants are colocated with the DESs.

The PCMs have been widely used in solar still applications. In addition, PCMs can be integrated within the still design. Built-in latent heat thermal energy storage (LHTES) was incorporated in a weir-type cascade solar still and the performance was compared with a still without storage. The total productivity of the still without LHTES was slightly higher than the still with LHTES during sunny days. A significant difference in the productivity of stills with and without LHTES was reported. The still without LHTES was preferred for sunny days due to its simplicity and low construction cost and the still with LHTES was suggested for cloudy areas due to its higher productivity [42]. The same group also investigated the effect of water flow rate on internal heat and mass transfer and daily productivity. Higher water flow rates resulted in lower internal heat and mass transfer rates and daily productivities. The maximum and minimum water flow rates resulted in daily productivities of 7.4 and 4.3 kg/m^2 day, respectively [43]. Since the storage medium has the potential to corrode the process components, Farell and coworkers studied the corrosive effects of salt hydrate PCMs on aluminum and copper heat exchanger materials. The metallographic examination of copper and aluminum samples revealed that aluminum has lower pitting corrosion compared with copper heat exchanger [44]. Thermal cycling tests on a few inorganic and organic PCMs have shown that inorganic PCMs were not found suitable after some cycles while thermal cycling for organic PCMs can be done up to 1000 thermal cycles with gradual change in melting temperature and latent heat of fusion [45]. El-sebaii et al. [46] tested a single basin solar still with PCM as a storage medium in Saudi Arabia. The daily productivity of the still increased significantly with an increase in the mass of PCM due to the increased heat storage. During discharging period of the PCM, the convective heat transfer coefficient from the basin liner (3.3 cm of stearic acid) to water was doubled which increased the evaporative heat transfer coefficient by 27%. This study reported a high daily productivity of 9 kg/m^2 day with 85.3% energy efficiency compared with a still productivity of ~5.0 kg/m^2 day without the PCM in summer season. The PCM was also found effective for lower masses of basin water in winter season.

Another group studied the effect of shape on a PCM heat storage system for rapid heat supply [43]. Four different encapsulated (sphere, cylinder, plate, and tube) PCMs were considered in numerical simulations. Among the different shapes considered, the sphere capsules showed the best heat release performance while the tubular capsule with low void fraction was not suitable for the heat release application [47]. Ramasamy and Sivaraman evaluated the performance of a solar still with Paraffin wax as LHTES material. In this study, the still with LHTES material showed higher efficiency (60.11%) compared with a still without LHTES (52.62%). This study also reported a disadvantage of PCM as corrosion when in direct contact with metal piping or housing [48]. Another study on the weir-type cascade solar still used a heat storage system of paraffin wax (18 kg over 2 cm thickness) beneath the absorber plate. As a result, the daily productivity of the still with PCM was 31% higher than that of the still without PCM [49]. These results are quite different from those reported in Ref. [42]. The variations in the still efficiencies suggest that the still efficiencies depend on the water flow rate, solar intensity, and the climatic conditions.

10.3.3 Thermochemical TES

Thermochemical storage has the advantage of very high stored energy densities compared with PCM or sensible only thermal storage [37]. Comparing Eqs. (10.1) and (10.6), Eq. (10.6) shows an additional enthalpy term associated with the chemical reaction at temperature, T_{CR}. The first term represents the enthalpy associated with heating the reactant(s) from T_1 to the chemical reaction temperature, T_{CR}. The third term in the equation accounts for the optional sensible enthalpy for increasing the temperature of the product(s) beyond the reaction temperature, T_{CR}, to temperature, T_2. The reaction occurs in two heat exchangers with chemical reaction occurring in the forward direction which is endothermic reaction (absorbs solar energy) in one heat exchanger and chemical reaction occurring in the backward reaction which is exothermic in another heat exchanger. The reaction occurs at the ambient temperature. The basic principle can be explained as: AB + heat $\Leftrightarrow$ A + B; using heat, a compound AB is broken into components A and B which can be stored separately; bringing A and B together AB is formed and heat is released. The storage capacity is the heat of the reaction or free energy of the reaction. A few examples and properties of the thermochemical TES materials that could be utilized in desalination and cogeneration applications are presented in Table 10.5 [50–54].

Table 10.5 Chemical reaction thermal energy storage materials

Storage materials	Working temperature (°C)	Enthalpy change during chemical reaction	Chemical reaction
Iron carbonate	180	2.6 GJ/m^3	$FeCO_3 \leftrightarrow FeO + CO_2$
Metal hydrides	200–300	4.0 GJ/m^3	Metal $xH_2 \leftrightarrow$ metal $yH_2 + (x-y)H_2$
Ammonia	400–500	67 kJ/mol	$NH_3 + \Delta H \leftrightarrow 1/2\ N_2 + 3/2\ H_2$
Hydroxides	500	3.0 GJ/m^3	$Ca(OH)_2 \leftrightarrow CaO + H_2O$

$$Q_{TCM} = m \int_{T_1}^{T_{CR}} c_p(T)dT + m\Delta h|_{T=T_{CR}} + m \int_{T_{CR}}^{T_2} c_p(T)dT \qquad (10.6)$$

where Q_{TCM} is the energy stored in the thermochemical TES medium with a mass flow rate m, specific heat c_p, and initial and chemical reaction temperatures of T_1 and T_{CR}, respectively. Δh is the heat released at the chemical reaction and T_2 is the final temperature of the TCM.

10.3.4 Current Status of TES Technologies

Among the available TES options, water storage, molten salt for concentrated solar power applications, ice storage for cooling applications, solid sensible storage, and PCM for temperature-sensitive products have been implemented at commercial scales (Fig. 10.7). Some underground storage, PCM storage technologies for cooling, and high-temperature applications are in demonstration stage. While other TES systems for power plants and industrial applications, microencapsulated and slurry PCMs for heating and cooling are in early stage of development and industrialization;

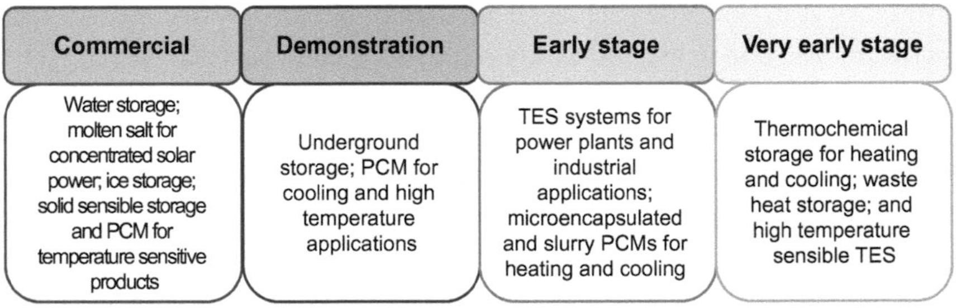

Fig. 10.7 Current status of thermal energy storage technologies.

thermochemical storage for heating and cooling, waste heat storage, and high-temperature sensible TES are in very early stage of development. At present, the sensitive TES technologies appear to be more practical for the desalination applications. The molten salt and phase change technologies will suit the power generation and desalination cogeneration schemes but further developments are required. In cogeneration schemes, power production with MSF and MED processes may benefit from TES whereas LTD systems may take advantage of the solar and process waste heat sources. For cogeneration schemes using concentrating solar power technologies such as parabolic trough, solar tower, Linear Fresnel, and Dish-Stirling technologies; the molten salts, high-temperature oils, phase change, and chemical reaction TES materials will best suit due to the need for high process fluid temperatures (Tables 10.3–10.5). Parabolic trough and solar tower require a two-tank molten salt or PCM storage while the Linear Fresnel and Dish-Stirling methods require short-term pressurized storage. For Dish-Stirling technology, the chemical reaction for TES is under development [37].

10.3.5 Economics of TES in Desalination

A comprehensive economic analysis of a DES supported by the thermal energy is not yet available. Fig. 10.8A shows process schematic of an MSF process supported by a solar pond [55]. The MSF process operates in 28 evaporation stages and three heat rejection stages with a performance ratio of 10 and a storage zone temperature of 90°C. The effect of the storage zone temperature and the salt cost on the produced thermal energy cost ($/kWh) is shown in Fig. 10.8B. It shows that the thermal energy costs are minimum when the salt is available at free of cost which is $0.0138/kWh at 70°C and increased to $0.0225/kWh when the cost of the salt is $80/ton. Thermal energy cost increased with increasing storage zone temperature due to higher losses to the ambient and lower quantities of useful energy derived. The oversizing of the pond may lead to higher heat losses during summer months which is referred as peak clipping. This affects the heat utilization factor of the desalination plant. The sensitivity analysis of various factors affecting the overall costs showed that the capital costs comprise about 66% of the total freshwater costs. It was concluded in this study that about 1% increase in interest rate for capital costs increases solar pond thermal energy costs by about 13%–15% and desalinated water costs by about 10%–13%. The break-even analysis shown in Fig. 10.8C illustrates the water costs for

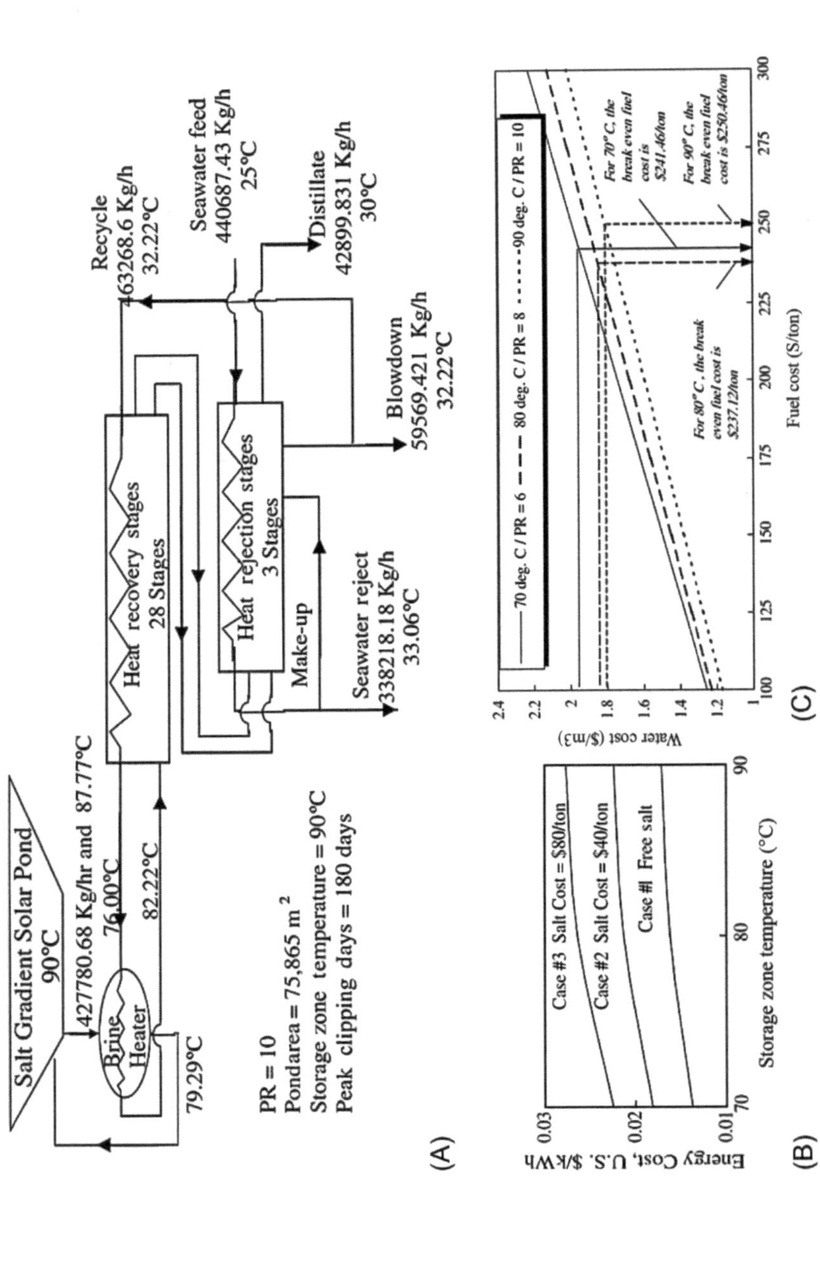

Fig. 10.8 Economics of an MSF process integrated with solar pond.

different fuel costs ($/ton). Although the cost of product water for the case of 90°C with a PR of 10 was the lowest, the break-even fuel cost was the highest compared with 70°C with a PR of 6°C and 80°C with a PR of 8. This is because the lowest cost of product water from a SP/MSF combination system occurs at a higher performance ratio than those of 70°C and 80°C, and therefore, the fraction of energy cost for the case of 90°C represents a smaller portion of the total product water cost when using a dual purpose (power and desalination) plant. More details on the economic analysis can be found elsewhere [55]. In another study where an MEMS coupled with a solar pond at the University of Texas, El Paso, the economic analysis of the system has reported the estimated desalination costs of 1.95–2.33 $/kgal (0.52–0.62 $/m^3), which depends on the cost of the solar pond liner [56].

Economic analysis on a LTD process powered by solar collectors augmented by TES unit has shown the desalination costs of 1.4 ¢/L or 14 $/m^3 and 1.32 ¢/L or 13.2 $/m^3, respectively, for DES with and without TES tank system [7]. The following assumptions were made in the economic analysis: interest rate: 5%, lifetime: 20 years, and plant availability of 90%. When lifetime of the system is considered as 25 years, the costs for the desalinated water are determined as 1.17 ¢/L or 11.7 $/m^3 and 1.1 ¢/L or 11.0 $/m^3, respectively, which are well-accepted values for a small-scale DES. This system was designed to produce 100 L/day of freshwater. This study concluded that the advantages of continuous and uninterrupted water supply provided by TES system regardless of weather conditions (cloudy days and passing clouds) outweighs the small difference in the desalinated water costs. However, the desalination costs for large desalination plants integrated with TES are yet to be reported. It is clear that this will involve high initial capital costs but with continued operation and in the long-term operation, the cost factors will become acceptable with payback.

10.4 ELECTRICAL ENERGY STORAGE OR BES FOR DESALINATION

DESs that primarily or partially depend on electrical energy sources are membrane-based technologies such as RO, ED, capacitive deionization (CD), electrodialysis-reversal (EDR), and MD. These technologies can be supported by the electrical energy generated by either PVs or wind turbine units. Similar to thermal energy produced by solar thermal collectors, these technologies also suffer from varying source of energy for electricity harvesting and need storage devices to meet the demand-supply logistics and to improve reliability and performance [57]. The solar and wind sources fluctuate very significantly during the course of a day resulting in potential excess generation or under supply. Excess electrical energy can be stored in batteries for the desired periods before it is discharged in process applications while the electrical energy that was already stored can be released during the periods of low power generation. Electrical energy storage, in large-scale applications, can be divided into three main functional categories such as (i) *power quality*: stored energy applied only for seconds or less to ensure continuity of power quality; (ii) *bridging power*: stored energy applied for seconds to minutes to assure continuity of service when switching from one source of energy to another (uninterrupted power supply, UPS); and (iii) *energy management*: stored energy used to decouple the timing of energy generation and consumption especially in the application of load leveling. Load leveling involves charging of storage in low demand time and use in peak time which enables consumers to minimize the total energy cost [58]. The third category is the ideal mechanism for desalination processes. The following sections discuss the energy requirements, energy recovery, energy storage, and sizing methods for membrane-based desalination processes requiring electricity.

10.4.1 Energy Consumption in RO

The energy consumed per unit freshwater (kWh/m^3) production is defined as specific energy consumption (SEC). This energy is supplied to the pumps in the form of electrical energy which is converted into mechanical energy to provide a pressure impact higher than the osmotic pressure of seawater on the membrane to extract freshwater. The specific energy requirements for RO process can vary between 1 and 10 kWh/m^3 which depends on the type of feedwater, level of pretreatment, and the membrane module and energy recovery scheme. The highest SEC is observed in the systems without an energy recovery device (ERD). The SEC for seawater desalination by RO process with ERDs such as Pelton turbine, pressure or work exchangers varies between 3 and 5 kWh/m^3. In an RO process, energy costs may represent up to 50% of the total water production cost [59]. In seawater reverse osmosis (SWRO) plants, the high-pressure feed pumps consume 50%–75% of the total energy supplied contributing to 35% of the total operating costs [60]. The energy costs ($/m^3) are very significant in RO process due to the high-pressure requirements (80–100 bar for SWRO and 15–40 bar for BWRO) [61]. Energy consumption in RO can be reduced by incorporating ERDs; and by developing high permeability membranes or low-energy membranes. More details on energy consumption and recovery schemes can be found elsewhere [3].

10.4.2 PV and Wind-Generated Electricity Storage

The process diagram for PV-wind-battery operated RO systems is shown in Fig. 10.9. A PV system consists of a PV array (a number of PV modules arranged in series), which converts solar radiation into direct current (DC) electricity. The system accessories include a charge controller, battery storage, inverter, and other components needed to provide the output electric power suitable to operate the systems coupled with the PV system. The PV supported RO and ED processes are the most cost competitive for small-scale systems where other thermal-based technologies are less competitive other than those that use direct solar energy which are less common [62]. The RO process uses alternating current (AC) for the pumps while the ED process uses DC for electrode stack operation. A DC/AC inverter is required for RO process [62]. However, ED can use the energy directly from the PV panels without major modifications. Battery storage systems are used for PV output power to ensure reliable system operation and improved process performance when solar radiation is inadequate. Sizing the battery charging system requires consideration of many physical and operating factors as discussed below.

Electrical storage sizing involves the following steps: (1) estimation of desalination load requirement, (2) estimation of required PV or wind turbine rating for the desalination load, (3) estimation of the daily power output from the PV array or the wind turbine, (4) estimation of PV array size and wind turbine rotor diameter, and (5) battery storage capacity requirements in ampere-hour (Ah) for the desalination load.

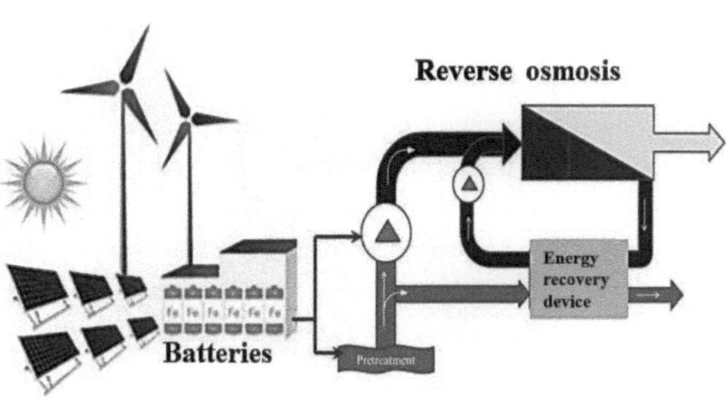

Fig. 10.9 Reverse osmosis system combined with PV or Wind or hybrid energy source.

10.4.3 PV-Battery System Sizing

Basis: $100\,m^3/day$ of freshwater production from seawater source with SEC of $5\,kWh/m^3$. The total electrical energy demand for this application will be 500 kWh. The PV array capacity for this application can be calculated using a solar window of 10 h as

$$P_{dc}(kW) = \frac{\text{Load\_(kWh/day)}}{\text{Solar\_window\_(h/day)}} = \frac{500}{10} = 50\,kW$$

For RO process unit, the equivalent AC capacity can be calculated from the efficiency of the PV unit. Considering a typical PV system efficiency (η) of 85% which includes inverter efficiency, dirt and other losses:

$$P_{ac,STC} = \frac{P_{dc}(kW)}{\eta}$$

The size of the PV array should be more than 1.2 times the desired load to charge the battery while supporting the load [63]. Therefore, the PV array capacity should be adjusted to the equivalent DC load as

$$P_{ac,STC(Adjusted)} = 1.2 \times P_{ac,STC} = 1.2 \times 60 = 72\,kW$$

Therefore, for the RO desalination application, 72 kW capacity of PV array with an optimum storage capacity (battery) will be required to support the desalination load for 24 h a day. For a known PV panel efficiency and for an estimated $1\,kW/m^2$ rated PV module, the required surface area of the PV array can be calculated. The efficiency of crystal silicon PV module is around 12.5%–15% [64], which gives PV array area as: $P_{ac,STC} = (1\ kW/m^2)$ insolation $\times A \times \eta$

$$A = \frac{P_{ac,STC}}{(1\,kW/m^2) \times \eta} = \frac{72}{1 \times 0.125} = 576\,m^2$$

Thus, a PV module area of $576\,m^2$ is adequate to support the desalination load with battery storage.

The required battery storage for the desalination application depends on the total energy generated from the PV array. Battery storage capacity can be estimated by multiplying the daily load on battery by number of days it should support the system to provide power for continued operation. Considering a system voltage of 24 V and 1 day battery storage [65], the Ah rating of the battery bank can be calculated as follows [66]:

$$\begin{aligned}\text{Daily load in Ah@system voltage} &= \frac{\text{Load (Wh/day)}}{\text{System Voltage}} = \frac{72 \times 10^3}{24} \\ &= 3000\,\text{Ah/day}\end{aligned}$$

The total energy that the battery storage system should store for 1 day demand at a maximum depth of discharge (MDOD; for lead-acid battery—80%) would be [64]

$$\text{Battery storage} = \frac{\text{Load (Ah/day)} \times \text{No of days}}{\text{MDOD}} = \frac{3000}{0.80} = 3750\text{Ah}$$

The final battery storage capacity for battery (C/20) bank with a discharge rate of 96% at 25°C would be

$$\begin{aligned}\text{Required battery storage}\,(25°C,\,20\text{h}-\text{rate}) &= \frac{\text{Battery storage}}{\text{Rated capacity}} = \frac{3750}{0.96}\\ &= 3906\,\text{Ah}\end{aligned}$$

Therefore, 3906 Ah of battery storage is required for a 500 kWh/day of load at a system voltage of 24 V with a 72 kW solar PV array system.

The above design calculations are based on known PV panel efficiency and electricity generation rate. However, proper design must consider the solar insolation, altitude, and the geographical and weather conditions of a particular location. These factors need consideration when designing a PV-based DES and a battery storage unit. The solar power map can be used as a guideline in this design.

10.4.4 Photovoltaic/Thermal Collectors

The photovoltaic/thermal (PV/T) collectors are hybrid collectors which integrate solar collectors with PV modules. A thermal energy harvesting system (heat exchanger) is incorporated to harvest the heat absorbed by the PV module. The circulating fluid cools the surface of the PV module which increases the electrical efficiency while providing a heat source for desalination application [67]. The energy efficiency of these hybrid systems can be higher than the sum of individually operated PV and solar thermal collectors (SC). As a comparison, 1 m^2 of the solar thermal collector and 1 m^2 of PV together would yield 520 kWh thermal energy and 72 kWh electrical energy annually, whereas 2 m^2 of PVT collector would alone yield 700 kWh thermal energy and 132 kWh electrical energy [68]. The PVT system also costs about 25% lower than the combined cost of solar collectors and PV panels [69]. The economic and environmental payback times are much shorter for the hybrid collectors compared with PV modules [68,70]. The amount of heat that can be extracted from these collectors depends on the collector type and design, cooling fluid circulation rate, and the site-specific weather conditions.

Two recent studies evaluated the possibility of combining the electricity generation with desalination by using PV/T and concentrated PV/T collectors. One study evaluated modification of PV modules with a heat exchanger attached to the absorber plate to generate thermal energy that supported a LTD system operating under natural vacuum for domestic applications. Simulation studies and economic analysis of the integrated process were reported [71]. Another study by Mittelman et al. investigated the performance of MEE DES integrated with concentrated PV/T collectors for large-scale application and the economics of the process were reported. Details on these studies can be found elsewhere [72].

10.4.5 Wind Power Battery Sizing

Similar to PV-battery system sizing, wind power battery storage can be designed for a 100 m^3/day of freshwater from seawater source with SEC of 5 kWh/m^3. The total energy demand for this system will be 500 kWh. Wind turbines convert the kinetic energy of the wind into electrical energy and a maximum theoretical conversion of 59.3% was reported which is known as Betz limit (also power coefficient, $C_p = 0.593$). Available power from wind turbine is expressed [73] as $P = \frac{1}{2} C_p \rho A V^3$

where P is the power output from wind turbine (Watts, W), ρ the air density (1.225 kg/m^3 at 15°C at sea level), A the area swept by the rotor (m^2), and V is the wind speed (m/s) which varies with different geographic conditions.

The required wind turbine capacity for the desalination load can be calculated as

$$P_{ac}(\text{kW}) = \frac{\text{Load (kWh/day)}}{\text{Windwindow (h/day)}} = \frac{500}{20} = 25\,\text{kW}$$

If an inverter (from AC to DC) is required to support the desalination process, the DC capacity can be calculated using an inverter efficiency which is typically 90%

$$P_{dc,\text{STC}} = \frac{P_{ac}(\text{kW})}{\eta} = \frac{25}{0.9} = 27.8\,\text{kW}$$

Similar to PV capacity calculation, 1.2 times the wind turbine capacity will be required for desalination application to charge the batteries while supporting the desalination process. Therefore, the new or adjusted wind turbine capacity for the equivalent DC load will be

$$P_{\mathrm{dc,STC(Adjusted)}} = 1.2 \times P_{\mathrm{dc,STC}} = 1.2 \times 27.8 = 33.3\,\mathrm{kW}$$

Considering the DC system voltage as 24 V, load on battery in Ah can be calculated for 1 day as Daily load in Ah@system voltage = $\frac{\text{Load (Wh/day)}}{\text{System voltage}} = \frac{33.3 \times 10^3}{24} = 1375\,\text{Ah/day}$

Considering MDOD of 80% for commonly used lead-acid batteries, the battery storage for 1-day operation is

$$\text{Battery storage} = \frac{\text{Load (Ah/day)} \times \text{No.of days}}{\text{MDOD}} = \frac{1375 \times 1}{0.80} = 1719\,\text{Ah}$$

The final required battery storage capacity will be (considering a discharge rate of C/20 batteries of 96% [64])

$$\text{Final battery storage } (25°C, 20\text{h} - \text{rate}) = \frac{\text{Battery storage}}{\text{Rated capacity}} = \frac{1719}{0.96} = 1790\,\text{Ah}$$

The above calculations illustrate how to size a battery system to match the energy demand in DESs powered by PV or wind energy that require AC or DC power supply. The battery sizing calculation was based on 1 day of uninterrupted supply but the battery storage will increase proportionately with increase in number of storage days required. The wind energy source will change with seasonal variations, latitude, and the location of the desalination unit. Wind source availability should be used as a guideline to design the desired turbine blade area, height, and turbine dimensions. These again depend on the air density which is a function of temperature which in turn affects the wind velocity and finally the energy generated from the swept area by the wind turbine blades. A proper design will consider these factors when sizing the blades and the turbine height and capacity.

10.4.6 Desalination Applications for BES Technology

Al-Karaghouli et al. reported recent updates on the installed PV-RO and wind-RO desalination plants in 2009 [74]. Zezli et al. reported a few geothermal, wind, and PV-RO desalination plants worldwide. These include a geothermal-driven MED plant of 80 m^3/day in Kimolos island, a wind energy-driven MVC plant (50 m^3/day) in Gran Canaria, PV/RO plants of 120, 6, and 1.5 m^3/day in Lampedusa island, Brazil, and Nevada, respectively, and a wind-PV-driven RO of (3 m^3/day) for seawater desalination in Lavrio, Greece [75]. Table 10.6 illustrates the details of PV-RO installations with battery storage details. Some hybrid configurations integrating PV and

Table 10.6 PV-Wind-RO desalination plants with battery energy storage systems

Capacity (m^3/d)	Specific energy consumption (SEC) (kWh/m^3)	Energy source and storage details	Refs.
PV-RO desalination systems			
3.2	16.1–19.7	8 kWp PV, 46.56 kWh battery	[71,72]
0.4–0.7	4–5.8 (BWRO)	1.2 kWp PV, 4.3 kWh battery	[73,74]
12	8	24.5 kWp (pump); 1.22 kWp (control); PV, 132 kWh (pump) and 4.8 kWh (control) battery	[75]
53	0.89 (BWRO)	19.84 kWp (pump); 0.64 kWp (control equipment) PV, 208 kWh battery	[76]
24	1.38–2.77 (BWRO)	2.59 kWp PV, 60 kWh battery	[77]
5–7.5	2.45	3.25 kWp PV, 9.6 kWh battery	[78]
0.8–4.2	18–19	4.8 kWp PV, 59.52 kWh battery storage	[79]
6	3 (BWRO)	1.1 kWp PV, 9.6 kWh battery	[72]
10	2.54	5.6 kWp PV with tracking, 41 kWh battery	[62]
36	4.86	125 kWp PV, 160 kVA diesel generator, 1236 kWh battery	[80]
0.9	4.3–4.6	0.846 kWp PV, 7.56 kWh battery	[81]
50	8 (BWRO)	10.5 kWp PV, 72 kWh battery	[82]
0.192	2.7 (BWRO)	0.136 kWp PV, 0.744 kWh battery	[83]
2	15	6 kWp PV, 4.8 kWh battery	[82]
PV-Wind-RO desalination systems			
3	–	3.5 kWp PV, 0.6 kWp wind, 36 kWh battery	[84]
3.12	16.5	3.96 kWp PV, 0.9 kWp wind, 44.4 kWh battery	[85,86]
2.2	3.3–5.2	0.846 kWp PV, 1 kWp wind, 7.56 kWh battery	[82,86]
1	3.74	0.6 kWp PV, 0.89 kWp wind, 21 kWh battery	[62]
300	4.3	50 kWp PV, 275 kWp wind, grid back up	[87]
30	– (BWRO)	7.6 kWp PV, 5 kW wind, 5 kVA diesel generator	[88]

wind energies (PV-wind-RO) are also installed. This configuration might have an additional benefit of requiring less battery storage since the wind generation continues round the clock. Other configurations include integration with the grid power to assure reliable power supply and stable performance [62,74,76–92].

10.4.7 Economics of BES in Desalination

Energy storage generally increases the capital costs, thereby affecting the freshwater costs. Evaluation of a few recent comparative studies with and without BES shows varying conclusions. In a study conducted at the Agricultural University of Athens, Greece, unit freshwater costs were reported as 7.8 $\$/m^3$ without using BES and 8.3 $\$/m^3$ with a BES system. The SEC of this system was between 4.3 and 4.6 kWh/m^3 by using a Clark pump ERD. However, the freshwater production was higher for the system with BES (0.9 m^3) than without BES (0.8 m^3) which could be very valuable for remote communities [85]. In another study in Jordan, for a desalination capacity of 0.192 m^3/day, the freshwater costs for without BES and with BES were 10 $\$/m^3$ and 13 $\$/m^3$, respectively [87].

10.5 FUTURE OF ENERGY STORAGE IN DESALINATION

The existing energy storage technologies suitable for desalination application can be primarily divided into two categories as thermal storage and electrical energy storage. Among the TES technologies, sensible heat storage technologies are considered mature with many successful demonstrations throughout world in a variety of process application. The high-temperature sensible TES technologies are still in very early stage of development. Water storage (sensible TES) technology can be considered most simple and economic alternative for LTD technologies, followed by molten storage and PCMs for high-temperature applications including power generation. PCMs involving microcapsules and nanoparticles still need some demonstration to gain more experience and to overcome the barriers of this technology. For electrical energy storage, the only available option is battery storage with various chemical electrolytes, lithium-ion and lead-acid batteries being the most commonly used. The main foreseeable objection for BES implementation in desalination applications would be their prohibitive costs. With increasing penetration of the renewable energy technologies (PV and wind) into large-scale desalination applications, current storage capacity or energy density of the BES technology does not prove to be cost competitive. Alternate large-scale energy storage technologies such as

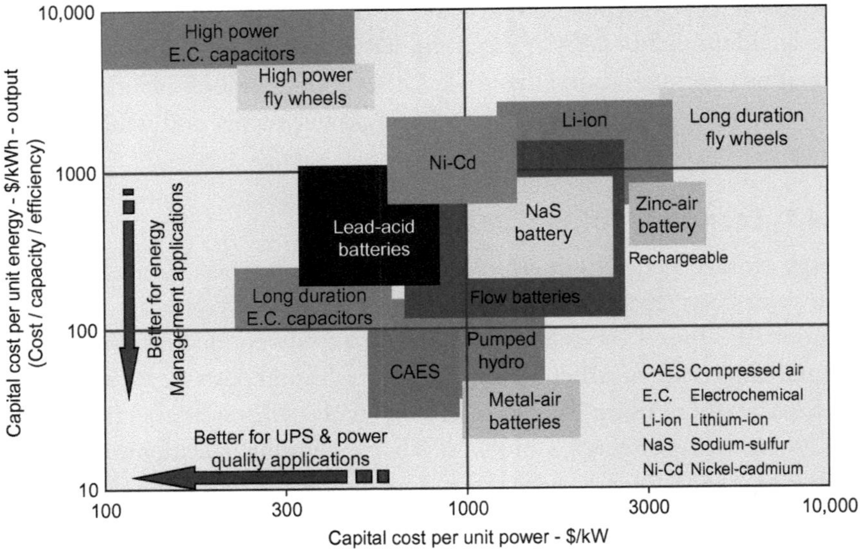

Fig. 10.10 Capital costs for different battery energy storage technologies.

chemical energy storage (CES), compressed air energy storage (CAES), and pumped hydrostatic energy storage need further consideration and development. A comparison of the capital costs for different energy storage technologies is shown in Fig. 10.10 [93]. The CES and CAES technologies can be used in power generation schemes to produce electrical energy suitable for desalination applications due their low cost and ability to utilize waste heat from other sources [94–96]. Use of low-grade heat for power generation using RCs is also viable for low-cost desalination. Pumped hydrostatic energy, on the other hand, does not require energy conversion but provides desalination directly by virtue of the hydrostatic effect. The CAES and pumped hydro energy storage technologies have lower capital costs compared with others due to their large-scale application feasibility (see Fig. 10.10). A novel RO desalination operating under this principle was proposed by Al-Kharabsheh [97]. This system requires a saline water storage tank positioned at a height to create a hydrostatic pressure equivalent to or higher than the osmotic pressure of the saline water from which freshwater is to be produced. The specific energy requirement for this process was 0.85 kWh/m^3 which is very close to thermodynamic limit (0.70 kWh/m^3). The drawback with this process is the geographical height difference to provide for the required hydrostatic pressure with segregated system components. Ocean communities may utilize the seawater column as a potential hydrostatic head to perform desalination on-site while eliminating the feed collection and brine disposal issues. A submarine RO desalination plant was proposed to take advantage of the hydraulic pressure head of the seawater [98]. In this configuration, the RO unit is submerged and located at the bottom of the seabed and the produced freshwater was supplied to the surface via mechanical pumping. The feedwater to the RO process also requires some mechanical pumping which depends on the height of the water column (available hydrostatic pressure) above the RO unit. The energy requirement for this configuration was 1.88 kWh/m^3, much less than the conventional RO (5–8 kWh/m^3) process. However, the RO process unit in this configuration has to be located on the seabed which requires a different operational environment and process skill set to operate the process with additional process engineering and safety concerns. Another recent development in this area is a hybrid system which utilizes the wind energy and gravitational potential energy to supply energy required for RO process for sweater desalination. The estimated SEC was 2.81 kWh/m^3 for this process [99].

10.6 SUMMARY

Energy storage is critical for uninterrupted supply of freshwater sources from desalination technologies that depend on variable energy sources. The type and size of energy storage depends on the combinations of the desalination process and renewable energy technology as discussed earlier. While the benefits of energy storage systems can be realized from energy, environmental, and economic perspectives, the energy storage option may not be ideal or economical in all cases since its feasibility depends on the location, type, and size of the desalination application and the available RES. Whether an energy storage option is feasible for a given application has to be determined by careful evaluation of the aforementioned factors. Apart from the cost issues, other barriers for TES implementation include material properties and stability issues. The design of TES systems needs several considerations and varies considerably for different applications. Future research efforts should focus on the development of storage materials for different temperature ranges, containers, and thermal insulation. Phase change and thermochemical storage technologies require additional efforts to improve material reactions along with process parametric optimization for desalination applications. BES technologies also suffer from similar issues such as TES technologies. Significant research efforts need to be devoted to study the electrochemistry of the materials to produce low cost, high density, and low environmental impact batteries. Environmental conditions (humidity, dirt, solar insolation, and ambient temperature) play an important role in the design and sizing of PV-wind-RO-battery storage systems. With increasing desalination capacity worldwide, these technologies need to be developed in parallel to provide solutions to the energy supply-demand mismatch and to improve the economics of the renewable energy powered DESs.

REFERENCES

[1] Gude VG, Khandan NN, Deng S. Low temperature process to recover impaired waters. Desalin Water Treat 2010;20:281–90.
[2] Gude VG, Khandan NN, Deng S. Renewable and sustainable approaches for desalination. Renew Sust Energ Rev 2010;14:2641–54.
[3] Gude VG. Energy consumption and recovery in reverse osmosis. Desalin Water Treat 2011;36:239–60.
[4] Kalogirou SA. Seawater desalination using renewable energy sources. Prog Energy Combust Sci 2005;31:242–81.
[5] Voutchkov N. Seawater desalination costs cut through power plant co-location. Filtr Sep 2004;41(7):24–6.
[6] Dincer I, Dost S. A perspective on thermal energy storage systems for solar energy applications. Int J Energy Res 1996;20:547–57.
[7] Gude VG, Nirmalakhandan N, Deng S, Maganti A. Low temperature desalination using solar collectors augmented by thermal energy storage. Appl Energy 2012;91:466–74.
[8] Thermal energy storage—technology brief. IEA-ETSAP and IRENA© Technology Brief E17 January 2013
[9] Gil A, Medrano M, Martorell I, Lazaro A, Dolado P, Zalba B, et al. State of the art on high temperature thermal energy storage for power generation. Part 1—concepts, materials and modellization. Renew Sust Energ Rev 2010;14:31–55.
[10] Tian Y, Zhao CY. A review of solar collectors and thermal energy storage in solar thermal applications. Appl Energy 2013;104:538–53.
[11] Tabrizi FF, Sharak AZ. Experimental study of an integrated basin solar still with a sandy heat reservoir. Desalination 2010;253:195–9.
[12] Murugavel KK, Sivakumar S, Ahamed JR, Chockalingam KKSK, Srithar K. Single basin double slope solar still with minimum basin depth and energy storing materials. Appl Energy 2010;87:514–23.
[13] Abd Elkader M. Solar seawater desalination using a multi-stage multi-effect humidification (Meh)-dehumidification system with energy storage. Int J Water Res Arid Environ 2011;1(2):116–22.
[14] Lu H, Walton J, Hein H. Thermal desalination using MEMS and salinity-gradient solar pond technology. El Paso: US Department of the Interior; 2002.
[15] Dickinson WC, Clark AF, Day JA, Wouters LF. The shallow solar pond energy conversion system. Sol Energy 1976;18(1):3–10.
[16] Tabor H. Non-convective solar ponds. Philos Trans R Soc *Lond Ser A* 1980;295:423–33.
[17] Hassairi M, Safi MJ, Chibani S. Natural brine solar pond: an experimental study. Sol Energy 2001;70(1):45–50.

[18] Busquets E, Kumar V, Motta J, Chacon R, Lu H. Thermal analysis and measurement of a solar pond prototype to study the non-convective zone salt gradient stability. Sol Energy 2012;86:1366–77.
[19] Kurt H, Ozkaymak M, Binark AK. Experimental and numerical analysis of sodium-carbonate salt gradient solar-pond performance under simulated solar-radiation. Appl Energy 2006;83:324–42.
[20] Saifullah AZA, Shahed Iqubal AM, Saha, Solar pond and its application to desalination A. Asian Trans Sci Technol 2012;2(3):1–25.
[21] Qiblawey HM, Banat F. Solar thermal desalination technologies. Desalination 2008;220:633–44.
[22] El-Sebaii AA, Ramadan MRI, Aboul-Enein S, Salem N. Thermal performance of a single-basin solar still integrated with a shallow solar pond. Energy Convers Manag 2008;49:2839–48.
[23] Lu H, Walton JC, Swift AHP. Desalination coupled with salinity gradient solar ponds. Desalination 2001;136:13–23.
[24] Mohammed Ramadan A-G. Design methodology and maintenance strategy for a salt gradient solar pond coupled with an evaporation pond. M.Sc. Thesis, Tripoli, Libya: Alfateh University; 1999.
[25] Tahri K. The prospects of fresh water supply for Tan Tan City from non-conventional water resources. Desalination 2001;135:43–50.
[26] Posnansky M. Technical and economical aspects of solar desalination with particular emphasis on solar pond powered distillation plants. Desalination 1987;67:81–95.
[27] Garman MA, Muntasser MA. Sizing and thermal study of salinity gradient solar ponds connecting with the MED desalination unit. Desalination 2008;222:689–95.
[28] Szacsvay T, Hofer-Noser P, Posnansky M. Technical and economic aspects of small-scale solar pond powered seawater desalination systems. Desalination 1999;122:185–93.
[29] Müller-Holst H, Engelhardt M, Herve M, Scholkopf W. Solar thermal seawater desalination systems for decentralized use. Renew Energy 1998;14:311–8.
[30] Antipova E, Boer D, Cabeza LF, Guillén-Gosálbez G, Jiménez L. Multi-objective design of reverse osmosis plants integrated with solar Rankine cycles and thermal energy storage. Appl Energy 2013;102:1137–47.
[31] Gude VG, Nirmalakhandan N. Combined desalination and solar-assisted air conditioning system. Energy Convers Manag 2008;49:3326–30.
[32] Gude VG, Nirmalakhandan N. Desalination using low-grade heat sources. ASCE J Energy Eng 2008;134:95–101.
[33] Gadhamshetty V, Gude VG, Nirmalakhandan N. Thermal energy storage system for energy conservation and water desalination in power plants. Energy 2014;66:938–49.
[34] Leblanc J, Akbarzadeh A, Andrews J, Lu H, Golding P. Heat extraction methods from salinity-gradient solar ponds and introduction of a novel system of heat extraction for improved efficiency. Sol Energy 2011;85:3103–42.
[35] El-Sebaii AA, Ramadan MRI, Aboul-Enein S, Khallaf AM. History of the solar ponds: a review study. Renew Sust Energ Rev 2011;15:3319–25.
[36] Ng KC, Thu K, Shahzad MW, Chun W. Progress of adsorption cycle and its hybrids with conventional multi-effect desalination processes. IDA J Desalin Water Reuse 2014;6(1):44–56.
[37] Kuravi S, Trahan J, Goswami DY, Rahman MM, Stefanakos EK. Thermal energy storage technologies and systems for concentrating solar power plants. Prog Energy Combust Sci 2013;39(4):285–319.
[38] Glatzmaier G. Developing a cost model and methodology to estimate capital costs for thermal energy storage technical report NREL/TP-5500-53066, 2011.
[39] Pilkington Solar International GmbH. Survey of thermal storage for parabolic trough power plants. NREL, Report. NREL/SR-550-27925.

[40] Laevemann E. In: Thermische energiespeicher, theoretische grenzen und beurteilung-skriterien. Experten-Workshop "Thermische Speicher: Potentiale und Grenzen der Steigerung der Energiespeicherdichten", DFG/PTJ, Berlin, Germany; June 2010.
[41] Radhwan AM. Transient performance of a stepped solar still with built-in latent heat thermal energy storage. Desalination 2004;171:61–76.
[42] Tabrizi FF, Dashtban M, Moghaddam H. Experimental investigation of a Weir—type cascade solar still with built—in latent heat thermal energy storage. Desalination 2010;260:248–53.
[43] Tabrizi FF, Dashtban M, Moghaddam H. Effect of water flow rate on internal heat and mass transfer and daily productivity of a weir-type cascade solar still. Desalination 2010;260:239–47.
[44] Farell AJ, Norton B, Kennedy DM. Corrosive effects of salt hydrated phase change materials used with copper and aluminium. J Mater Process Technol 2006;175:198–205.
[45] Shukla A, Buddhi D, Sawhney RL. Thermal cycling of few selected inorganic & organic phase change materials. Renew Energy 2008;33:2606–14.
[46] El-Sebaii AA, Al-Ghamdi AA, Al-Hazmi FS, Faidah AS. Thermal performance of a single basin solar still with PCM as s storage medium. Appl Energy 2009;86:1187–95.
[47] Wei J, Kawaguchi Y, Hirano S, Takeuchi H. Study on a PCM heat storage system for rapid heat supply. Appl Therm Energy 2005;25:2903–20.
[48] Ramasamy S, Sivaraman B. Heat transfer enhancement of solar still using phase change materials (PCMs). Int J Eng Adv Technol 2013;2(3):597–600.
[49] Dashtban M, Tabrizi FF. Thermal analysis of a weir-type cascade solar still integrated with PCM storage. Desalination 2011;279:415–22.
[50] Shiizaki S, Nagashimga I, Iwata K, Hosoda T, Kameyama H. Development of plate fin reactor for heat recovery system using methanol decomposition. Proceedings of the 8th international conference on thermal energy storage (Terrastock Stuttgart, Germany 2000); 2000.
[51] Hahne E. In: Thermal energy storage some view on some problems. Proceedings of the 8th international heat transfer conference San Francisco, USA; 1986.
[52] Lovegrove K. In: A solar driven ammonia based thermochemical energy storage system. Proceedings of ISES'99 solar world congress Jerusalem, Israel; 1999.
[53] Kubota M, Yokoyama K, Watanabe F, Hasatani M. In: Heat releasing characteristics of $CaO/CaCO_3$ reaction in a packed bed for high temperature heat storage and temperature up-grading. Proceedings of the 8th international conference on thermal energy storage (Terrastock Stuttgart, Germany 2000); 2000.
[54] Steinfeld A, Sanders S, Palumbo R. Design aspects of solar thermochemical engineering – a case study: two-step water splitting cycle using Fe_3O_4/FeO redox system. Sol Energy 1999;65:43–53.
[55] Agha KR. The thermal characteristics and economic analysis of a solar pond coupled low temperature multi stage desalination plant. Sol Energy 2009;83:501–10.
[56] Lu H, Walton JC, Hein H. Thermal desalination using MEMS and salinity-gradient solar pond technology 98-FC-81-0047. Desalination Research and Development Program Report No, August 2002. p. 80.
[57] Divya KC, Ostergaard J. Battery energy storage technology for power systems—an overview. Electr Power Syst Res 2009;79:511–20.
[58] Arif MT, Oo AMT, Shawkat Ali ABM. Investigation of Energy Storage Systems, Its Advantage and Requirement in Various Locations in Australia. J Renew Energy 2013;835309.
[59] Khawaji AD, Kutubkhanah IK, Wie J. Advances in seawater desalination technologies. Desalination 2008;21:47–69.

[60] Farooque AM, Jamaluddin ATM, Al-Reweli AR, Jamaluddin PAM, Al-Marwani SM, Al-Mobayed AA, et al. Parametric analyses of energy consumption and losses in SWCC SWRO plants utilizing energy recovery devices. Desalination 2008;219:137–59.

[61] Ibanez-Mengual JA, Garcia-Gamuz JA, Valerdi-Perez RP. Basic equations for desalination by reverse osmosis: a deduction based on the thermodynamics of irreversible processes. Desalin Water Treat 2009;2:287–94.

[62] Al-Karaghouli A, Renne D, Kazmerski LL. Technical and economic assessment of photovoltaic-driven desalination systems. Renew Energy 2010;35:323–8.

[63] IEEE. IEEE recommended practice for sizing lead-acid batteries for stand-alone photovoltaic (PV) systems. IEEE Standard 1013.

[64] Masters GM. Renewable and efficient electric power systems. New York, NY, USA: JohnWiley and Sons; 2004.

[65] IEEE. IEEE guide for optimizing the performance and life of lead-acid batteries in remote hybrid power systems. IEEE Std 2007;1561–2007.

[66] Energy Matters, "SMA Sunny Boy 3800W Grid-connected Inverter," http://www.energymatters.com.au/sma-sunny-boy-3800watt-grid-connect-inverter-p-412.html.

[67] Effendi Y. In: Three years experiences for PVRO-desalination. Photovoltaic Specialists Conference, 1988. Conference Record of the Twentieth IEEE; 2002. p. 1194–9.

[68] van Helden WGJ, Ronald JCZ, Herbert AZ. PV thermal systems: PV panels supplying renewable electricity and heat progress in photovoltaics. Res Appl 2004;12:415–26.

[69] Jaikra S, Nualboonrueng T, Sichanugrist P. In: Development of PVIT system in Thailand. 31st IEEE Photovoltaic Specialists Conference Proceedings; 2005. p. 1687–9.

[70] Tripanagnostopoulos Y, Souliotis M, Battisti R, Corrado A. Energy, cost and LCA results of PV and hybrid PV/T solar systems. Prog Photovolt Res Appl 2005;13:235–50.

[71] Mittelman G, Kribus A, Mouchtar O, Dayan A. Water desalination with concentrating photovoltaic/thermal (CPVT) system. Sol Energy 2009;83(8):1322–34.

[72] Gude VG, Nirmalakhandan N, Deng S. Integrated PV-thermal system for desalination and power production. Desalin Water Treat 2011;36:129–40.

[73] Burton DST, Jenkins N, Bossanyi E. Wind energy handbook. New York, NY, USA: JohnWiley and Sons; 2001.

[74] Al-Karaghouli A, Renne D, Kazmerski LL. Solar and wind opportunities for water desalination in the Arab regions. Renew Sust Energ Rev 2009;13:2397–407.

[75] Zejli D, Ouammi A, Sacile R, Dagdougui H, Elmidaoui A. An optimization model for a mechanical vapor compression desalination plant driven by a wind/PV hybrid system. Appl Energy 2011;88:4042–54.

[76] Boesch WW. World's first solar powered reverse osmosis desalination plant. Desalination 1982;41(2):233–7.

[77] de Carvalho PCM, Riffel DB, Freire C, Montenegro FFD. The Brazilian experience with a photovoltaic powered reverse osmosis plant. Prog Photovolt Res Appl 2004;12(5):373–85.

[78] Harrison DG, Ho GE, Mathew K. Desalination using renewable energy in Australia. Renew Energy 1996;8(1–4):509–13.

[79] James WL. In: Design, construction and operation of a 1.2kW photovoltaic reverse osmosis desalination plant. Presented at the solar energy coming of age: International Solar Energy Society; 1982.

[80] Sadi A, Kehal S. Retrospectives and potential use of saline water desalination in Algeria. Desalination 2003;152(1–3):51–6.

[81] Al Suleimani Z, Nair VR. Desalination by solar-powered reverse osmosis in a remote area of the Sultanate of Oman. Appl Energy 2000;65(1–4):367–80.

[82] Herold D, Horstmann V, Neskakis A, Plettner-Marliani J, Piernavieja G, Calero R. Small scale photovoltaic desalination for rural water supply-demonstration plant in Gran Canaria. Renew Energy 1998;14(1–4):293–8.
[83] Subiela VJ, de la Fuente JA, Piernavieja G, Pe˜nate B. Canary Islands Institute of Technology (ITC) experiences in desalination with renewable energies (1996–2008). Desalin Water Treat 2009;7:220–35.
[84] Scrivani A. Energy management and DSM techniques for a PV-diesel powered sea water reverse osmosis desalination plant in Ginostra, Sicily. Desalination 2005;183 (1–3):63–72.
[85] Mohamed ES, Papadakis G, Mathioulakis E, Belessiotis V. A direct coupled photovoltaic seawater reverse osmosis desalination system toward battery based systems—a technical and economical experimental comparative study. Desalination 2008;221 (1–3):17–22.
[86] Papapetrou M, Mohamed ES, Manolakos D, Papadakis G, Subiela VJ, Penate B. Operating RE/desalination units. Seawater Desalin 2009;247–72.
[87] Qiblawey H, Banat F, Al-Nasser Q. Laboratory setup for water purification using household PV-driven reverse osmosis unit. Desalin Water Treat 2009;7:53–9.
[88] Weiner D, Fisher D, Moses EJ, Katz B, Meron G. Operation experience of a solar- and wind-powered desalination demonstration plant. Desalination 2001;137(1–3):7–13.
[89] Tzen E, Theofilloyianakos D, Kologios Z. Autonomous reverse osmosis units driven by RE sources experiences and lessons learned. Desalination 2008;221(1–3):29–36.
[90] Mohamed ES, Papadakis G. In: A stand alone sea water reverse osmosis desalination unit powered by hybrid wind-photovoltaic system for rural development and small villages. Collection of Extent Abstracts of 2004 CIGR International Conferencevol. 2. ; 2004.
[91] Kershman SA, Rheinländer J, Neumann T, Goebel O. Hybrid wind/PV and conventional power for desalination in Libya—GECOL's facility for medium and small scale research at Ras Ejder. Desalination 2005;183(1–3):1–12.
[92] Ali MT, Fath HES, Armstrong PR. A comprehensive techno-economical review of indirect solar desalination. Renew Sust Energ Rev 2011;15:4187–99.
[93] Market Evaluation for Energy Storage in the United States. Fairfax, Virginia: KEMA, Inc.; 2012.
[94] Safaei H, Keith DW, Hugo RJ. Compressed air energy storage (CAES) with compressors distributed at heat loads to enable waste heat utilization. Appl Energy 2013;103:165–79.
[95] Madlener R, Latz J. Economics of centralized and decentralized compressed air energy storage for enhanced grid integration of wind power. Appl Energy 2013;101:299–309.
[96] Marano V, Rizzo G, Tiano FA. Application of dynamic programming to the optimal management of a hybrid power plant with wind turbines, photovoltaic panels and compressed air energy storage. Appl Energy 2012;97:849–59.
[97] Al-Kharabsheh S. An innovative reverse osmosis desalination system using hydrostatic pressure. Desalination 2006;196:210–4.
[98] Colombo D, de Gedoni M, Reali M. An energy-efficient submarine desalination plant. Desalination 1999;122:171–6.
[99] Fadigas EAFA, Dias JR. Desalination of water by reverse osmosis using gravitational potential energy and wind energy. Desalination 2009;237:140–6.

CHAPTER 11

Energy Recovery Devices in Membrane Desalination Processes

Eric Kadaj, Rolando Bosleman
Energy Recovery Inc., San Leandro, CA, United States

The use of energy recovery devices (ERDs) in membrane desalination continues to become a more common practice as the cost of electricity continues to increase throughout the world. System designers are more frequently being asked to minimize the specific energy consumption (SEC) even in the areas where the cost of electricity is relatively low. By far the largest contributor to the decrease in SEC over the past three decades has been the advancements made in energy recovery technologies. All ERDs used in the water treatment industry reduce energy consumption by harnessing the energy in the concentrate (or brine) waste stream and transferring it to the feed side via various methods.

11.1 HISTORY OF ERDs IN DESALINATION

Historically, the Achilles heel of seawater reverse osmosis (SWRO) systems, brackish water reverse osmosis (BWRO) systems, and industrial water systems has been the energy intensive nature of the membrane separation process. Over the past 30 years, the industry has seen a decrease in SEC from SWRO installations without ERDs operating close to 8 kWh/m^3 (for the RO portion of the process only) to 2.5 kWh/m^3 in today's state-of-the art facilities (Fig. 11.1). Currently, all medium- to large-scale SWRO facilities have adopted ERDs into their process designs and have benefited from the reduction in SEC [1].

11.1.1 ERDs Pre-1980

11.1.1.1 Francis Turbine/Reverse Running Pump

The Francis turbine or reverse running pump was the earliest used ERD for desalination (Fig. 11.2). The Francis turbine is a reaction turbine, which

Renewable Energy Powered Desalination Handbook
https://doi.org/10.1016/B978-0-12-815244-7.00011-8

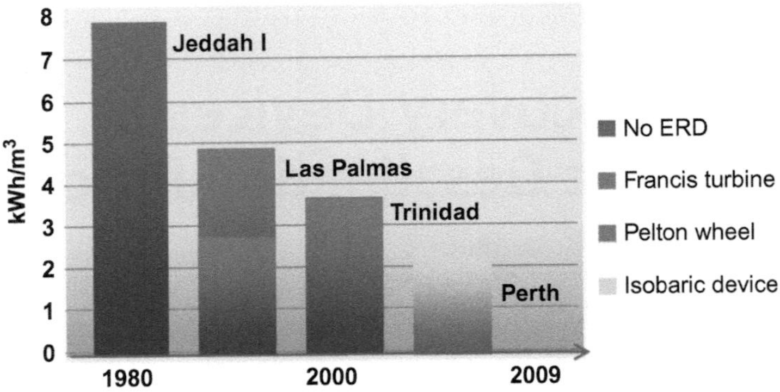

Fig. 11.1 Specific energy consumption (SEC) trend over the past three decades [reverse osmosis (RO) membrane unit only].

Fig. 11.2 Francis turbine with generator *(From the English Wikipedia. Taken By 13:14, 28 Mär 2004 Stahlkocher.)*

means that the working fluid changes pressure as it moves through the turbine, giving up its energy. The turbine is located between the high-pressure water source and the low-pressure water exit, usually at the base of a dam.

The inlet is spiral shaped. Guide vanes direct the water tangentially to the turbine wheel, known as a runner. This radial flow acts on the runner's vanes, causing the runner to spin. The guide vanes (or wicket gate) may be adjustable to allow efficient turbine operation for a range of water flow conditions.

As the water moves through the runner its spinning radius decreases, further acting on the runner. For an analogy, imagine swinging a ball on a string around in a circle; if the string is pulled short, the ball spins faster due to the conservation of angular momentum. This property, in addition to the water pressure, helps Francis and other inward-flow turbines harness water energy efficiently.

At the exit, water acts on cup-shaped runner features, leaving with no swirl and very little kinetic or potential energy. The turbine's exit tube is shaped to help decelerate the water flow and recover the pressure.

In summary, the water flow entering a turbine impeller where the hydraulic energy of the fluid is converted into mechanical energy via a shaft which is then connected to a generator. The generator then converts that mechanical energy into electrical energy which can then be used to power other equipment in the plant.

The disadvantage of the Francis turbine is its low peak efficiency and inability to operate optimally when conditions vary. Since desalination plants are subjected to varying conditions due to changes in the salinity and temperature, Francis turbines were frequently operated off of their original design which decreases performance. The majority of these Francis Turbines were replaced by Pelton wheels.

11.1.1.2 Pelton Wheel

The Pelton wheel was invented during the 1850s. The Pelton wheel has an input nozzle through which high-pressure feed is directed onto the buckets of the wheel (Fig. 11.3). In the case of desalination, the brine or waste stream from the membranes provide the fluid to the Pelton wheel. The nozzle is designed so that all of the hydraulic energy of the pressurized feed is converted into mechanical energy which makes the wheel rotate. The buckets, or vanes, of the wheel are arranged in series around the shaft. As the fluid impacts each bucket, the energy forces wheel to turn. The Pelton wheel is typically connected to a dual-shafted motor with one shaft connected to the Pelton wheel. The motor is sized to produce full power required by the high-pressure pump (HPP) should the Pelton wheel fail. The second motor shaft is then coupled to the HPP. The mechanical energy generated by the Pelton wheel supplements the power input from the motor in order to provide 100% of the required power to the HPP.

Fig. 11.3 Assembly of a Pelton wheel at Walchensee Hydroelectric Power Station, Germany (Wikipedia).

While the Pelton wheel has higher efficiency and wider operating range than the Francis turbine, there still are some disadvantages. Pelton wheel installations require a very large area due to the dual-shafted motor and large wheel designs. Specifically in desalination, the metal components of the Pelton wheel are susceptible to corrosion due to the corrosive nature of the fluid. After the brine flow has gone through the Pelton wheel, it is discharged at atmospheric pressure to a sump. Therefore, another pump must be installed in order to transport this fluid for disposal.

The relatively low efficiency, large footprint due to the pump–motor–Pelton arrangement, required sump pump, etc. all lead to a less than ideal solution for energy recovery in desalination.

11.1.2 ERDs 1980 to Present

11.1.2.1 Hydraulic Turbochargers

The turbochargers were first used in the late 1980s and more widely adopted in the 1990s. The turbocharger device recovers hydraulic energy from the high-pressure brine (concentrate) stream in the reverse osmosis (RO) process and transfers that energy to a feed stream. The feed stream may be seawater going into a single-stage RO membrane block, or it may be first-stage

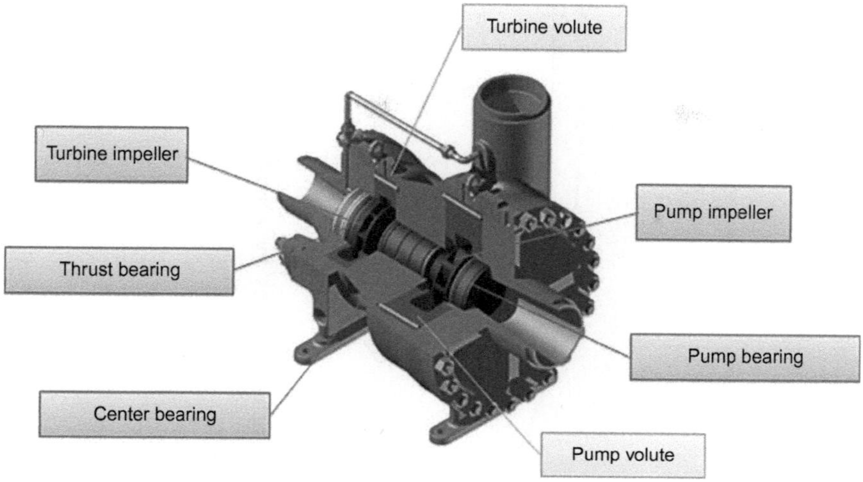

Fig. 11.4 Turbocharger cross-section.

brine stream being boosted in pressure for a second-stage membrane block for further recovery of permeate or flux balancing.

The turbocharger device consists of a pump section and a turbine section (Fig. 11.4). Both pump and turbine sections each contain a single-stage impeller. The turbocharger incorporates a number of features designed to ensure simple operation, long operating life, reliable performance, and high availability. A nozzle directs the flow of water into the turbine side which then energizes the turbine impeller. The turbine impeller extracts hydraulic energy from the brine stream and converts it into mechanical energy. The pump impeller converts the mechanical energy produced by the turbine impeller back to pressure energy in the feed stream.

The process fluid provides all required bearing lubrication eliminating any external oil or grease requirements.

The turbocharger provides partial boost of the full feed flow to the membrane system. It can be thought of as a second pump in parallel with the HPP driven by a turbine instead of an electric motor (Fig. 11.5A).

Brine pressure control: The turbocharger can discharge brine against a back pressure. The unit does not require a brine disposal pump as is required by the Pelton wheel.

As is typical for all centrifugal devices, the turbocharger efficiency increases as flow capacity increases. The relative efficiencies between the turbocharger and the Pelton wheel can be seen in Fig. 11.5B. It is important to keep in mind that the gap can be even larger since turbochargers are

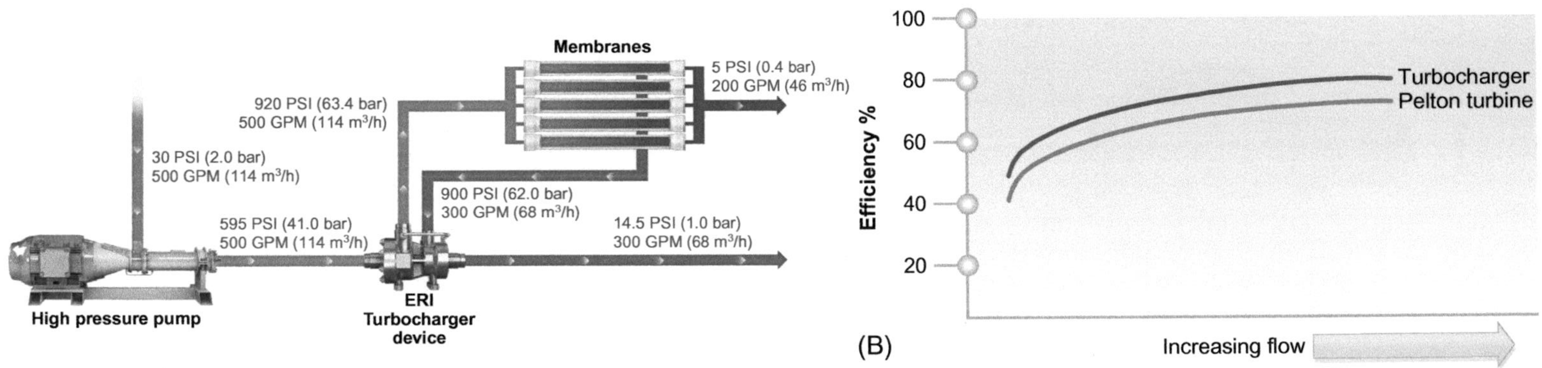

Fig. 11.5 (A) Typical turbocharger installation for seawater reverse osmosis (SWRO). (B) Typical efficiency curves for centrifugal energy recovery devices (ERDs).

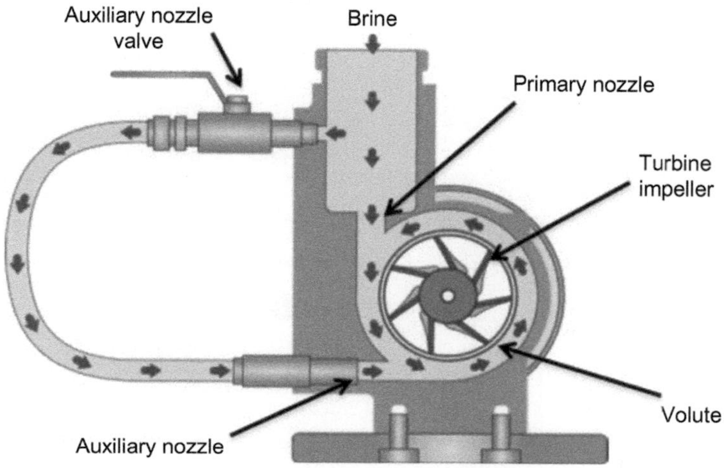

Fig. 11.6 Auxiliary Nozzle (AN) Line and Control Valve Arrangement.

customer machines to optimize efficiencies while Pelton wheel use standard hydraulic designs.

The turbocharger device is equipped with a primary nozzle, a secondary auxiliary nozzle (AN), and an auxiliary nozzle control valve (ANCV) as shown in Fig. 11.6. The primary nozzle is sized to provide a concentrate system resistance (concentrate pressure) equal to the maximum design pressure at the design concentrate brine flow rate. The ANCV controls flow to AN in the turbine casing. The ANCV can accommodate a 5%–15% brine pressure variation at a constant brine flow, should the system conditions require it.

The ANCV does not bypass flow around the turbocharger. It is a unique way to achieve variable area nozzle flow without the energy wasting throttling and bypassing valve arrangements needed by Francis turbines/reverse running pumps.

11.1.2.2 Isobaric ERDs

Isobaric ERDs transfer the hydraulic energy in the brine directly to the feedwater without an intermediate conversion to mechanical energy (Fig. 11.7). Efficiencies are kept very high, above 95%, because of this direct transfer of energy.

Rotary isobarics, such as the PX Pressure Exchanger, manufactured by Energy Recovery, Inc., use a rotor with a series of ducts to convey the seawater into direct contact with the pressurized brine.

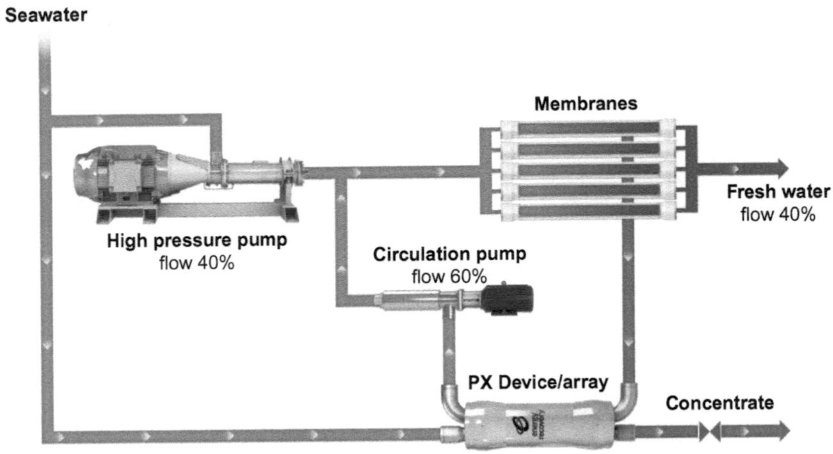

Fig. 11.7 Typical isobaric energy recovery device (ERD) configuration for seawater reverse osmosis (SWRO).

Piston-type isobarics, such as the dual work exchanger energy recovery (DWEER), manufactured by Flowserve use large cylinders with pistons and valves to pressurize and inject the seawater into the membrane using the brine energy.

While there are numerous differences in materials of construction, dimensions, number of moving parts and other working characteristics of rotary- and piston-type isobaric devices, the parameters used to measure how well they recover energy in a plant are identical. These parameters are: mixing and efficiency in energy transfer.

Piston-Type Isobaric ERD

An engineering team at Aqua Design (later Ionics and in turn GE) developed and used their own piston isobaric device using membrane pressure vessels and control valves in the 1980s, but discontinued the product. Then in 1996, Shumway filed a patent for a spool valve to operate large-chamber work exchanger devices, a design that evolved into the current DWEER device currently sold by Calder/Flowserve [2].

There have been several piston-type ERD designs manufactured over the years. They all operate via the same basic principal, however, the exact design and method of operation has been constantly evolving. At the time of writing the only commercially viable piston-type ERD is the DWEER device manufactured by Flowserve. Owing to this, most of the technical

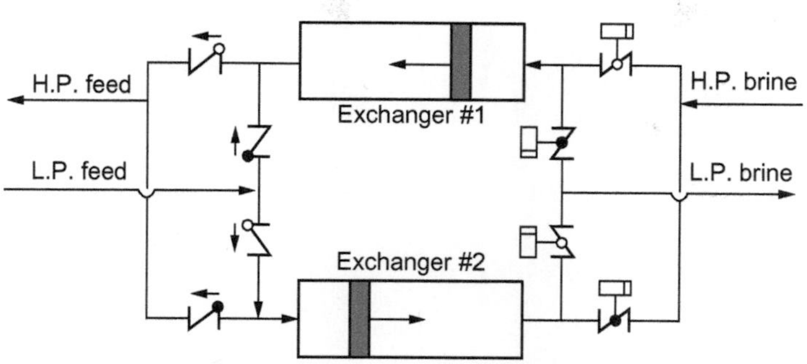

Fig. 11.8 Basic piston-type isobaric energy recovery device (ERD) scheme.

details will be based on the various DWEER designs produced through the years though most of the concepts disclosed apply to all piston-type ERDs.

Refer to Fig. 11.8 for a basic schematic of a piston ERD. Two or more large cylinders are usually fabricated out of pipe. A piston is fitted to the inside of the vessels and can travel the length of the vessel. The purpose of this piston is to provide a physical barrier between the seawater and the brine stream, however, this separation is not perfect and a very small amount of mixing between the two different fluids occurs which increments the salinity of the seawater feeding the RO membranes.

The top vessel is pressurized via the high-pressure brine and the flow of brine into the vessel is expelling seawater into the high-pressure feed. At the same time the bottom vessel is operating under low pressure and the low-pressure feed is pushing the depressurized brine out of the vessel. An array of actuated and check (nonreturn) valves direct the flows. When the pistons reach the end of their stroke the valves are quickly reversed (Fig. 11.9). This causes the flows to stop and reverse as the vessels are either pressurized or depressurized. In this way, the continuous flow of high-pressure brine is converted into a batch process for energy recovery [3].

As the process is a batch process, dedicated programmable logic controller (PLC) and control equipment are necessary to keep the pistons synchronized. The sensors on the vessels must detect the piston position and velocity in order to actuate the valves at precisely the correct moment before the piston hits the end of the vessel. This dedicated PLC is also connected to the plant supervisory control and data acquisition (SCADA) system. On the basis of the measured conditions the PLC sends a signal to a hydraulic pumping system. The hydraulic pumping system then sends pressurized oil to the piston system which actuates a large spool valve referred to as the linX valve.

Fig. 11.9 Typical piston-type isobaric energy recovery device (ERD) installation.

When the linX valve changes position one vessel is depressurized as the other is pressurized.

This pressurization causes the vessels to grow both axially and radially. Along with the inherent compressibility of water this expansion causes measurable lubrication losses. The movement of the vessels must be accounted for in the design of the plant as it can put stresses on pipe couplings and supports.

Rotary Isobaric ERD

The most commonly used rotary isobaric ERD in SWRO plants is the PX Pressure Exchanger device manufactured by Energy Recovery, Inc. (Fig. 11.10).

In 1987, Leif Hauge filed his first patent for the pressure exchanger, the basis of Energy Recovery, Inc.'s (ERI's) current line of rotary isobaric ERDs [2].

The PX Pressure Exchanger device consists of a ceramic cartridge where the pressure exchange occurs, a small number of other components which convey the water into the cartridge, and a vessel which contains all the components under high pressure.

The ceramic cartridge is comprised of four components, two end covers which direct the water into and out of the rotor, the rotor, and a ceramic sleeve which forms the bearing where the rotor spins (Fig. 11.11). The rotor is the only moving part in the rotary isobaric device.

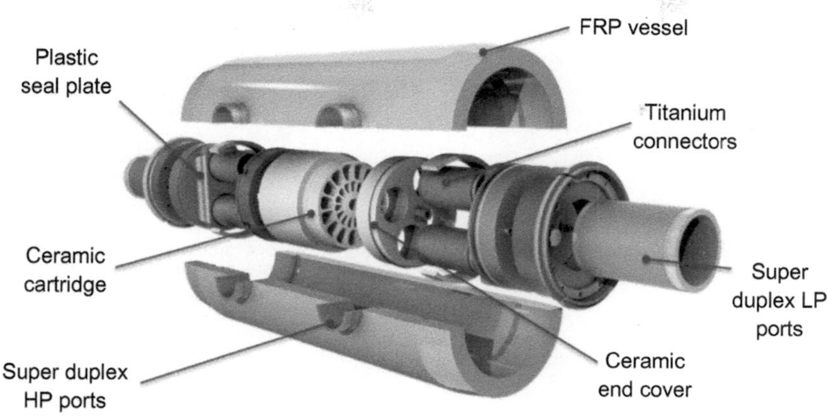

Fig. 11.10 Exploded view of a rotary isobaric PX Pressure Exchanger device.

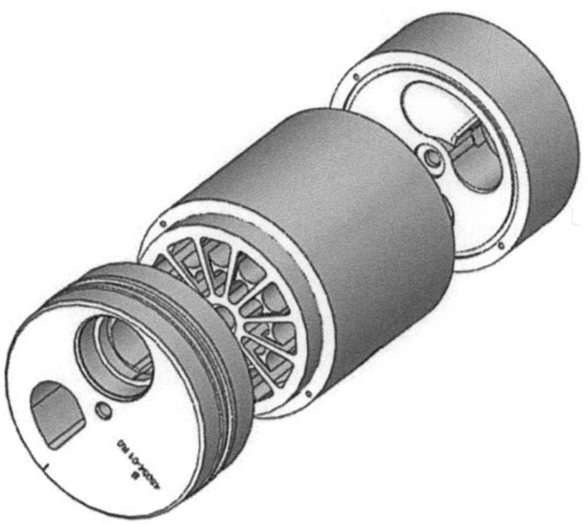

Fig. 11.11 Isobaric device cartridge showing end covers adjacent to the rotor and sleeve.

As can be seen in Fig. 11.12 there are four steps in the operation of the rotary isobaric device. In the first step low-pressure seawater enters the duct and pushes the brine out. Next the duct now full of seawater moves through a seal area. In the third step the duct opens to the high-pressure side and the high-pressure brine pushes the seawater out of the duct. The duct, now full of brine goes through another seal area in the fourth step. Finally as the duct enters the low-pressure area of step 1 it is depressurized and the whole process starts again. The rotor of an isobaric device has about 12 ducts and this process happens sequentially and continuously on all ducts at the same time.

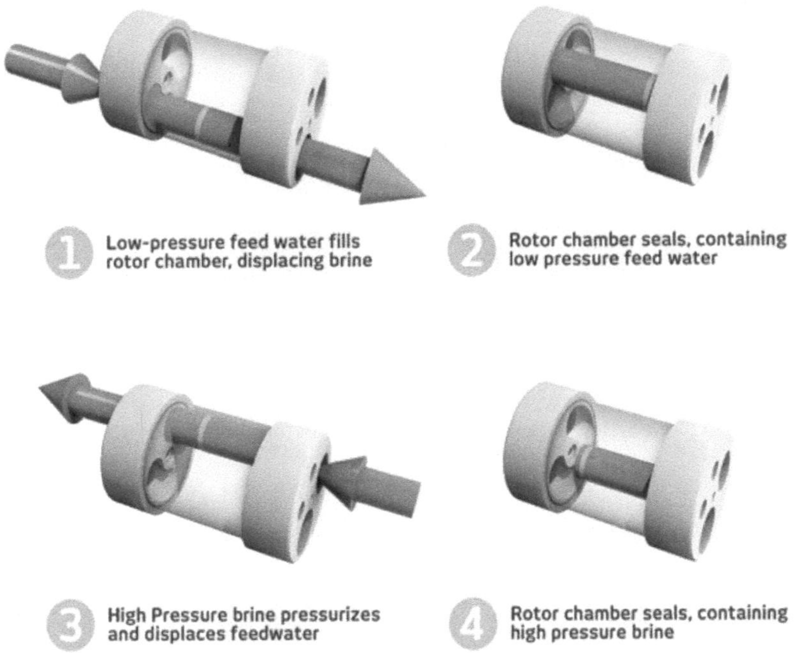

Fig. 11.12 Four-step pressure exchange process in a rotary isobaric device.

The flow entering and leaving the unit is continuous and there are no significant pressure fluctuations in these flows as this pressure exchange happens hundreds of times per second.

The rotor turns on an almost frictionless bearing at a rotational speed that is set by the flow rate of water into the device. In this way, the rotor is self-regulated. The isobaric rotary device does not require electric motors, actuators, sensors, or electronic control systems. The plant operator sets the high- and low-pressure flows using standard flow control technology such as valves or variable speed controlled pumps (Fig. 11.13).

The simplicity of this rotary isobaric device design is evident in the fact that there is only one moving part; however, the design does have some drawbacks. As there are no piston barriers between the brine concentrate and the seawater there is a small amount of mixing that occurs when the two fluids are in momentary contact. This mixing usually produces a small salinity increase at the membranes and incrementally higher feed pressure. A thorough energy analysis can take this into account and predict the resultant efficiency decrease.

Fig. 11.13 Typical rotary isobaric energy recovery device (ERD) installation.

11.2 RECENT DEVELOPMENTS IN ERDs

The isobaric ERDs have become the state-of-the-art technology for energy savings in SWRO plants, consequently all new developments have been mostly made in isobaric technology and therefore several new products are being introduced to the market.

11.2.1 iSave

This technology was developed by the HPPs division of the Danfoss group. In summary it is a device that incorporates a rotary isobaric device (similar to the PX Pressure Exchanger) and a booster pump in one single device, including the motor driver of the booster pump (Fig. 11.14). All the main components are made out of super duplex and the booster pump is a sliding vane pump, which is a volumetric pump with lower efficiency compared with piston volumetric pumps.

11.2.2 Saltec

Manufactured by KSB, it is a piston isobaric device consisting of one or more pairs of cylinders, pistons, and valves, similar to the DWEER device. This technology features electrical actuators and a rotary spool valve in comparison with previous piston isobaric devices.

Fig. 11.14 iSave energy recovery device (ERD).

11.2.3 Osmorec

It can be classified as a piston-type device, with the difference that it does not require a booster pump, the necessary boost to compensate pressure losses is provided by a volumetric difference inside the cylinder between the brine and the feedwater (Fig. 11.15).

The key components of the OSMOREC® Energy Recovery Device are as follows:

- An array of three vertical pressure exchange vessels, where energy contained in the high-pressure brine is transferred to feedwater.
- Pistons (one for each vessel), which directly transfer energy from the brine to the feedwater while preventing them from mixing.
- Three-way rotary brine flow distribution valves (one for each vessel), with a unique design allowing highly efficient energy transfer from concentrate to feedwater while eliminating leakage.

High- and low-pressure brine and feedwater check valves, which supply brine and deliver pressurized feedwater to the RO membranes.

11.2.4 Isobarix

Using the same technology as the PX Pressure Exchanger, this device is also made of ceramic and the external vessel is made from titanium. At the time of writing there is still not enough commercial adoption (Fig. 11.16). As per

Fig. 11.15 Osmorec energy recovery device (ERD).

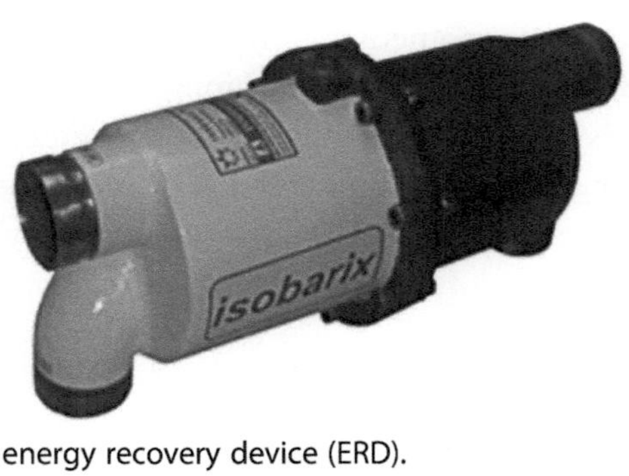

Fig. 11.16 Isobarix energy recovery device (ERD).

the datasheets the efficiency and performance of this device is similar to ERI Pressure Exchanger, but field performance data is not available.

11.2.5 Salino

Also manufactured by KSB, Salino is marketed as an all-in-one RO device. It contains HPP, ERD, and booster pump. The ERD is a rotary isobaric device made of stainless steel alloys, so it is very similar in the ERD portion to the Danfoss iSave (Fig. 11.17). At the time of writing, there are no references published in the KSB official channels. Performance could not be evaluated as the technical info is not available.

Fig. 11.17 Salino energy recovery device (ERD).

11.3 ADVANCES IN ERDs

11.3.1 Advances in Centrifugal ERDs

11.3.1.1 Removable Hydraulic Insert Technology

Traditional Design Methods

Historically, there have been two methods by which the volute is created for a centrifugal pump or centrifugal ERD. The pump and the turbine volutes have been either cast or cast then machined into the pressure containment casing. The casting method is a simple and relatively inexpensive way to achieve the general volute geometry required for the equipment. With a cast volute (Fig. 11.18), there is no extra machining operations performed and the surface roughness is that of the casting method whether it is a sand casting or investment casting. Typically, an investment casting has a smoother surface finish than a sand casting. A cast volute typically provides a low

Fig. 11.18 Cast pump volute.

efficiency product due to inconsistencies in volute geometry, less accurate tolerances, and poor surface finish. The volute cannot be changed or customized once it is cast [4].

With a machined volute, a specific set of machining operations are performed on a raw casting to ensure that the volute geometry is more accurate and the surface finish is better than a raw casting. Since the volute must be machined into the casing, there is some waste associated with this process and the inherent extra cost due to the higher weight of the initial casting component. There is also the labor and overhead cost associated with the machining processes that must be done. The advantage of a machined volute is that it can be customized to the exact operating conditions of the system. Both of these methods are very common in the pumping industry today [4].

The problem with both of these design methods is that once the volute is either cast or machined, it is very difficult to adapt to variations in operating conditions. However, there are two ways to bring the high-pressure pumping section of the plant back to optimal efficiency. The first is to buy all new pumping equipment—an expensive and cumbersome option. Once the new equipment is delivered to the site, the system must be taken offline in order to complete installation. For large systems, the time to install and commission this equipment can take weeks, costing the operator hundreds of thousands of dollars in lost revenue related to downtime. The second option is to re-machine the casings, volutes, impellers, etc. of the existing equipment. While this is a less expensive option, the unit or a full spare will need to be sent back to the manufacturer, thus adding significant cost or downtime. In both options, the negatives outweigh the positives and are rarely seen as viable solutions.

New Design Method

To overcome these obstacles a new design method for pump and centrifugal ERD hydraulics was developed in 2012. The design method allows for rapid customizability and adaptability while also reducing machining and labor time. The machining of the volutes into steel, especially 2205 Duplex Stainless Steel, which common in seawater applications, is a labor intensive and slow process. Duplex steel is a hard material to work with and requires numerous machine passes and multiple, high-strength tools. Every hour of machine time saved contributes to reducing the cost of the product. Removal of all customization from the casing itself was an effective approach. Consequently, the volutes constituted independent, replaceable components within the standard pump and/or turbocharger casings (Fig. 11.19).

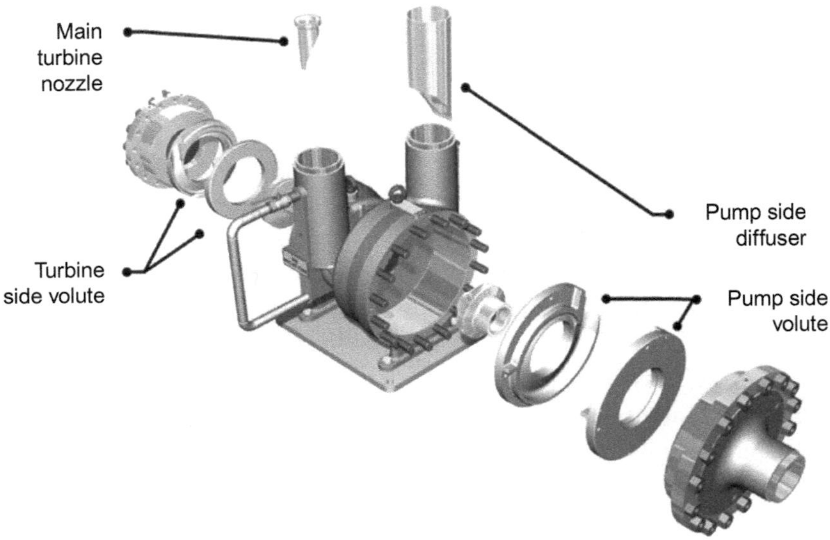

Fig. 11.19 TurboCharger exploded view with removable volutes.

This enabled elimination of custom machining work from the metal casings themselves leaving them to be generically machined and identical every time. The same methodology was used for the nozzles of the turbochargers as well as the pump outlet diffusers for the pumps and turbochargers [4].

Each of the main metal components, including the pump casing and end cap as well as the two turbocharger end caps and the main casing, are generically machined. All of the customization is machined into the volutes, diffusers, and nozzles of the pumps and turbochargers. The volutes are split radially to facilitate both machining and assembly/disassembly.

11.3.2 Advances in isoBaric ERDs

The greatest advances in isobaric devices in recent years have been focused on minimizing the energy consumption, introducing engineered materials of construction, and adding simplicity and reliability to the different designs.

The appearance of rotary isobaric devices with only one moving part has increased the adoption of isobaric technologies all over the world, especially in big projects. Efforts had been put in material science in order to engineer materials highly resistant to wear, unaffected by corrosion, and maintenance free, thus increasing reliability and design life of the devices.

Piston-type isobaric ERDs have also started with the introduction of improvements in order to compete with the newly developed rotary isobaric devices that have captured the biggest portion of the market. The main efforts in piston-type ERDs have been put in to improve the reliability of the valves design and their actuators and in the use of new materials such as fiberglass reinforced polyester (FRP) for the cylinders, thus avoiding problems related to corrosion and welding in the traditional stainless steel cylinders.

As the isobaric technologies offer very high efficiency performance, with peak values of 98%, it could be thought that there is not much room for further improvement. However, as explained by Stover and Andrews [2], even marginal improvements can lead to very significant energy savings in the long-term life cycle cost analysis. Therefore, the additional resources spent on developing these marginal improvements would get a quick payback and many years of additional savings throughout the life of a desalination plant.

The reduction of the isobaric devices "losses" such as mixing, leakage, pressure losses, etc., by improving the hydraulic design and the quality of materials of construction will bring the new generation of piston and rotary isobaric devices.

11.4 SYSTEM DESIGN AND OPERATION

11.4.1 System Design and Operation With Isobaric ERDs

Fig. 11.20 illustrates the flows within a typical SWRO system equipped with an isobaric ERD unit. Filtered seawater is pumped to both the main HPP and the low-pressure inlet of the ERD (B). The pressurized reject brine from the SWRO membranes (G) enters the ERD and pressurized seawater leaves the ERD unit to the booster pump. The booster pump adds the small amount of pressure lost to friction in the ERD unit, the membranes and the associated piping. Fully pressurized seawater from the booster pump merges with the seawater from the HPP. The combined seawater flow is directed to membrane array. Most of the seawater pumped by the HPP (C) becomes permeate (F). A small percentage of the seawater pumped by the HPP leaks through the seals of the ERD and flows to waste at H.

The water going to the drain (H) is the depressurized concentrate, but the HP concentrate flow rate to the ERD cannot be determined by measuring the flow rate to the drain at H. If a flowmeter is placed at H, the measured flow rate will be equal to the ERD unit LP inlet flow rate (B) plus the lubrication flow rate.

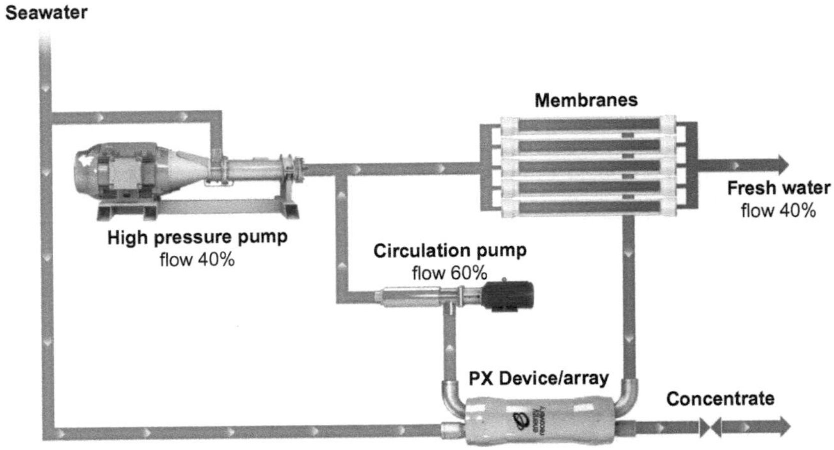

Fig. 11.20 Seawater reverse osmosis (SWRO) system configuration with isobaric energy recovery devices (ERDs).

11.4.1.1 Flow Control

Three flow rates must be considered in a system built around isobaric energy recovery technology—the high-pressure flow rate through the isobaric device, the low pressure flow rate through the isobaric device, and the flow rate through the HPP. The two flow rates through the isobaric unit must remain within the specified limits at all times—during start-up, normal operation, flushing, cleaning, and shutdown. The three flows are entirely independent of each other—changing one does not affect the others. All must be measured and controlled independently.

HPP flow—The HPP flow path is illustrated with a dashed line in Fig. 11.21. The flow rate of pressurized feedwater to the membranes determines the permeate flow rate. Therefore, care must be exercised when specifying the HPP capacity and/or some means to change the pump flow rate must be installed. A variable frequency drive (VFD) can be used or sheaves and belts can be changed on PD pumps to change the flow rate. Permeate flow from the system is usually metered and it can be beneficial to measure the feedwater flow to the HPP. The permeate flow will be less than the feed flow to the HPP by the lubricant flow rate through the isobaric device hydrodynamic bearing.

ERD high-pressure (HP) flow: The ERD unit internals are constructed in such a manner as to isolate LP from HP. It is important to understand that changing the LP flow rate will not affect the HP flow rate or vice versa.

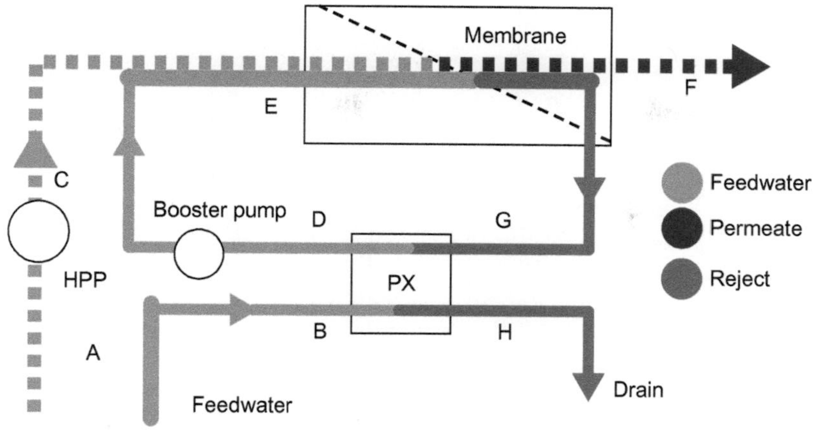

Fig. 11.21 High-pressure pump (HPP) flow path.

With reference to Fig. 11.22 or 11.23, imagine that there are two pieces of straight pipe inside the ERD unit—one carrying the LP flow from "B" to "H" and another one carrying the HP flow from "G" to "D". As the flows are contained in separate pipes, they are isolated. Flow meters must be installed on each flow circuit.

The flow of high-pressure water from the membranes to the isobaric device, then to the booster pump, and back to the membranes can be considered a closed loop. The HP flow circuit is illustrated with a dashed line in Fig. 11.22. Pressure is lost to friction in the membranes, piping, and in the ERD. The booster pump is necessary in most applications to repressurize the flow from the ERD so that it can enter the membrane feed stream.

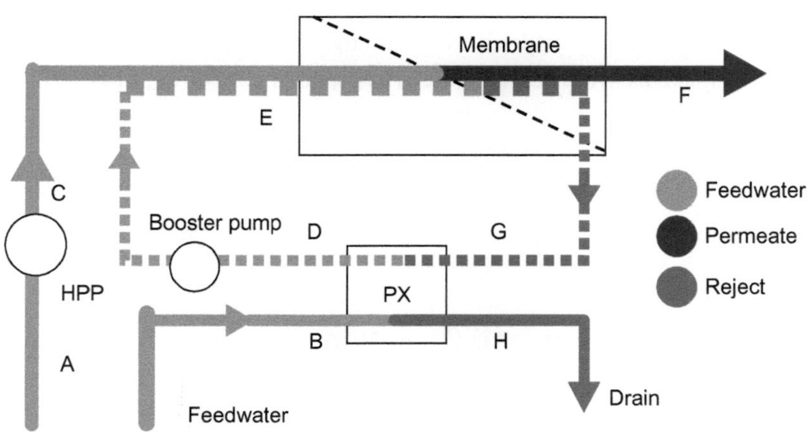

Fig. 11.22 Energy recovery device (ERD) high-pressure flow circuit.

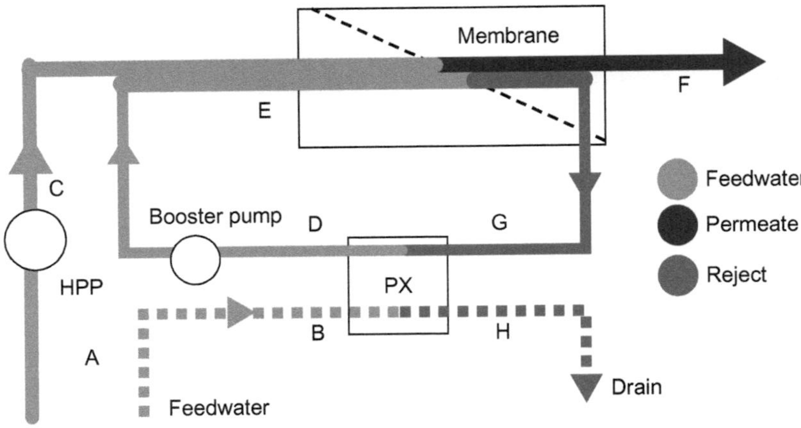

Fig. 11.23 Energy recovery device (ERD) low-pressure flow path.

The booster pump must be sized to make up the pressure losses at the required flow rate. This flow rate is relatively easy to control using a VFD on the booster pump or a control valve downstream from the booster pump. A flow meter is required in this flow circuit.

ERD low-pressure (LP) flow: The flow rate of low-pressure water to the ERD unit and out to the brine waste is usually set to equal the ERD unit HP flow rate. The LP flow path is illustrated in Fig. 11.23 with a dashed line. By changing this flow rate, the operator can slightly alter the salinity of the membrane feed. If the LP feedwater pressure can be controlled and held constant, LP flow control can be accomplished with a high-quality manual valve installed downstream of the ERD and upstream of the brine disposal point. It is not always easy to hold the feedwater pressure constant. Often, the LP feedwater to the ERD and the HPP feedwater come from the same source making it more difficult to maintain a constant ERD LP feedwater pressure, especially when starting the system, during flushing, or when more than one train is being fed from the same feedwater header system. Although not always practical, the best and safest design (as drawn above) is to have the ERD unit LP feedwater come from a dedicated source (pump, tank, etc.) and control the flow with a valve downstream of the ERD unit. A flow meter is required in this flow path.

11.4.2 System Design and Operation With Centrifugal Turbochargers

Fig. 11.24 illustrates how the turbocharger works as a feed pressure booster in an RO system. The turbocharger device is designed to produce a pressure

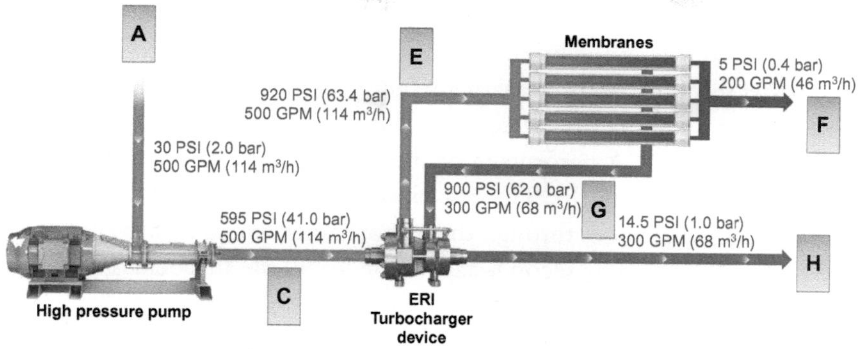

Fig. 11.24 Typical turbocharger installation for seawater reverse osmosis (SWRO).

boost in the RO feed stream (A, C, and E) using the hydraulic energy available in the brine stream (G). The low-pressure feed flow (A) passes through the main feed pump which provides a pressure boost (C–A). The feedwater passes through the turbocharger which provides an additional pressure boost (E–C). The feedwater then enters the membrane pressure vessels (E). A percentage of the feedwater exits the membrane as permeate (F). The rest exits as high-pressure brine (G). The brine passes through the turbocharger which extracts the pressure energy (G–H). The brine leaves the turbocharger at low pressure for disposal (H).

Fig. 11.25 illustrates how the turbocharger works as an interstage pressure booster in a two-stage RO system. The turbocharger device is designed

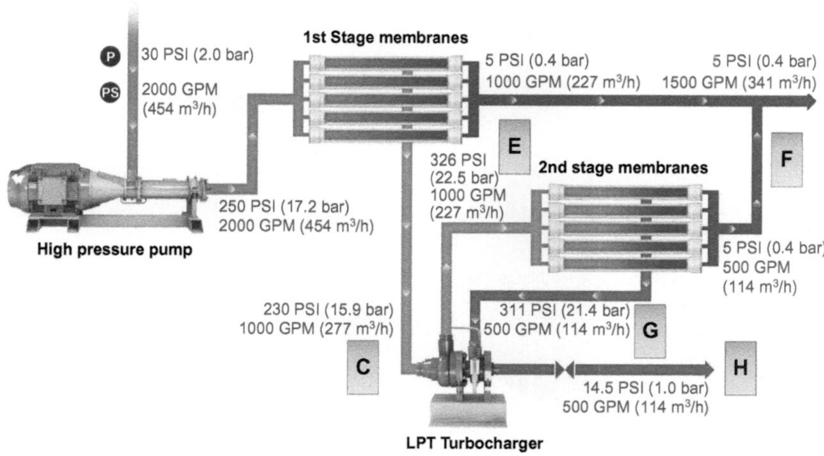

Fig. 11.25 Typical turbocharger installation for two-stage system.

to produce a pressure boost in the RO feed stream (C & E) using the hydraulic energy available in the brine stream (G). The brine water from the first-stage passes through the turbocharger which provides the required interstage pressure boost (E–C). The water then enters the second stage membrane pressure vessels (E). A percentage of the second-stage feedwater exits the membrane as permeate (F). The rest exits as high-pressure brine (G). The second-stage brine passes through the turbocharger which extracts the pressure energy (G–H). The second-stage brine leaves the turbocharger at low pressure for disposal (H).

11.5 TECHNICAL AND COMMERCIAL EVALUATION

The question is thus no longer whether we should use an ERD, but what is the most economical ERD for a specific project. A comprehensive technical and commercial evaluation of ERDs needs to be considered to determine the most suitable ERD for a specific set of project conditions. Many times the initial capital expenditure (CAPEX) is the only factor that is considered when deciding which ERD to select for a given project. This is a fairly common practice but can result in significant economic losses over the useful design life of the facility.

11.5.1 Life-Cycle Cost

The life-cycle cost of a plant is one of the most important factors when planning and specifying equipment. Over the 20–25 year life of a large SWRO facility, the operating cost will tend to outweigh the initial capital investment. As demonstrated in Fig. 11.26, the largest component of the

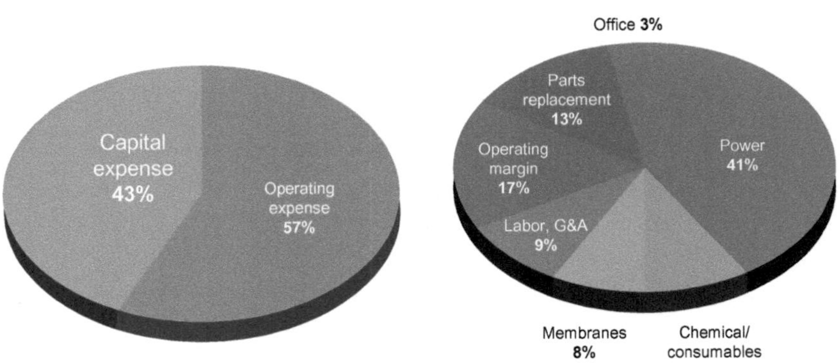

Fig. 11.26 Life-cycle analysis charts.

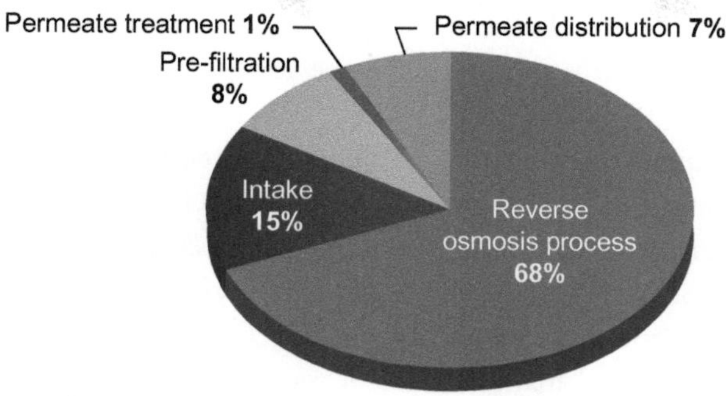

Fig. 11.27 Power breakdown in seawater reverse osmosis (SWRO) plant.

operational expense is the cost of power. Fig. 11.27 shows which components of the system consume the most power in the SWRO plant. The RO process which includes the high-pressure pumping equipment can consume over 20% of the entire plant power and in some cases up to 45%. The required power is a direct result of the efficiency of equipment. High-efficiency centrifugal pumps as well as both isobaric and centrifugal ERDs have become commonplace to help reduce power consumption, greenhouse gas emissions, etc.

The economics of ERD selection can be broken down into two primary categories of capital and operational expenditures. Both, CAPEX and operational expenses (OPEX) have many subsets that can be quantified and carefully analyzed to ensure maximum return on investment [5].

11.5.2 CAPEX

In the RO portion of the process, there are three primary components:
- RO membranes
- main HPP and
- the energy recovery system (ERS)

The cost of the ERDs is roughly 2%–5% of the total CAPEX of the SWRO facility. Again, making a major purchase and decision based on CAPEX alone can be an extremely costly mistake. There are also several items in addition to the ERD cost that need to be considered when evaluating the CAPEX of an ERS [5].

Installation cost: ERD technologies vary tremendously on the amount of installation cost required to meet the manufactures specifications. Piston-type ERDs require additional civil works; have independent PLC and hydraulic systems, and consume varying degrees of floor space (i.e., foot print).

The cost of the ERD racks and manifolds also add cost to each ERD offering. Centrifugal-type ERDs tend to have the lowest footprint and installation requirements.

Auxiliary equipment cost: Isobaric-type ERDs require an additional circulation booster pump while centrifugal-type ERDs do not. Connection types, number, and size of connections, and instrumentation all need to be taken into account during the CAPEX analysis.

Other cost: Depending on the type of ERD, there may be specific cost associated to a specific manufacturer or technology. Pelton-turbine ERDs may require an additional pump and sump system to displace the exhaust brine. Filtration and flushing requirements are several other costs that are predominately ERD manufacture specific but can quickly add cost to a proposed solution.

11.5.3 OPEX

Maintenance: Fewer moving components will reduce the amount of maintenance required. Consider the device spare parts costs to maintain the ERD over its life span. Some ERDs may require specialized tools or shop equipment for routine maintenance, as well as downtime for repairs.

Durability: To ensure the long-term and trouble-free lifetime of the SWRO process and its enabling technology, it is essential to utilize the most advanced and reliable materials of construction. One of the more advanced and unique materials currently in use in SWRO desalination applications is high purity (>99%) aluminum oxide (alumina) ceramics.

Availability: Availability can be defined as the probability that a system or piece of equipment used under the specified conditions operates satisfactorily at any given time.

11.6 CASE STUDIES

11.6.1 Magtaa Case Study

The Magtaa desalination plant was built near Oran in northwest Algeria. It was constructed by Tahlyat Myah Magtaa on a design–build–own–operate–transfer (DBOOT) basis. Algerian Energy Company (AEC) won the Magtaa contract from the Algerian Government in an open tender and subcontracted the project to Hyflux, which has a record of delivering large-scale desalination projects efficiently. Hyflux and Malaysia's Malakoff signed an agreement with AEC to run the plant for 25 years. The AEC and Hyflux also incorporated joint venture company Tahlyat Myah Magtaa.

The Magtaa SWRO desalination system in Algeria was the largest membrane desalination facility in the world when it was commissioned in 2014 with a capacity of 500,000 m^3/day (132 million US gallons per day). This facility utilized the largest Turbocharger ERDs installed in the world at the time, the AT-7800 Turbocharger. The AT turbocharger technology was chosen for many reasons, with one of the most significant advantages of the AT is its ability to potentially save the operators millions of dollars in operating expenses (OPEX) over the life of the plant. These potential savings are directly related to the turbocharger's ability to accommodate at any time, any, and all required hydraulic changes, allowing plant operators to operate continuously at optimal conditions and at a minimal cost. This unique AT design was discussed previously in Section 11.3.1 of this chapter.

There are 25 AT-7800 turbochargers installed at the Magtaa facility. Each turbocharger is designed for over 8500 gpm of feed flow to the RO membranes. The plant began operating in November of 2014 with all 25 trains in use and has been successfully operated since that time. Each train saves approximately 1.6 MW of energy for a total plant savings of 40.0 MW. This results in an annual energy savings of over $25 Million dollars.

11.6.2 Carlsbad Case Study

The Carlsbad SWRO desalination plant, located in the southern California coast, has a designed capacity of 204,412 m^3/day (54 MGD) and was commissioned during the second half of 2015. Poseidon Water is acting as the project developer responsible for securing the water purchase agreement (WAP) with San Diego County Water Authority (SDCWA). A joint venture between Kiewit and JF Shea Construction contracted IDE Technologies as the Engineering, Procurement and Support Services contractor. The privately finance project will produce enough drinking water to serve 300,000 local residents and provide the county with approximately 7% of its total water supply by 2020. IDE technologies is also responsible for the plant's operation and maintenance (O&M) having responsibility for all process and performance guarantees.

The intake and outfall of the SWRO plant will be connected to the existing Encina Power Station Cooling Seawater system. This characteristic avoids additional and complicated civil works that would have created environmental and economic effects in the Agua Hedionda lagoon marine ecosystem, its seafood-farming industry, and recreational areas located around the lagoon.

As the outfall from the desalination plant is also connected to the discharge of the power station cooling system, then the brine water reject from the desalination plant is diluted with seawater before being discharged back to the ocean, thus reducing the effect of increased salinity in the marine life near to the discharge.

After a pretreatment consisting in flocculation and gravity medial filters, the water collected by the intake is ready to pass through the RO process consisting of a two-pass system, having the first pass a single stage and a second pass with up to three stages, known as "cascade" RO process.

The innovation in the first pass process is the use of a "pressure center" concept, where different to conventional designs, there are no pumps and ERSs dedicated to individual RO membranes trains. The Carlsbad pressure center consist of 3 + 1 HPPs feeding a common header that distributes the pressurized seawater to each one of the 14 RO membranes trains; similarly the high-pressure brine reject is collected by a big size common header that directs the concentrate to 7 + 1 energy recovery skids consisting of 18 Pressure Exchangers PX-Q300 each (Fig. 11.28). This arrangement provides enhanced flexibility and reliability as the membranes trains are able to work with any combination of three HPPs and seven PX skids. Furthermore one or more membranes trains, HPPs, and PX systems can be isolated and put offline to reduce the plant production, adjusting the remaining PX systems

Fig. 11.28 PX-Q300 units at Carlsbad Desalination Plant.

Table 11.1 PX-Q300 predicted performance at Carlsbad desalination plant
PX technology performance

PX unitary flow	58.7	m^3/h
Salinity increase at membranes	2.8%	
Volumetric mixing VM	5.4%	
Lubrication flow (LF) per PX array	12.0	m^3/h
LF as % of concentrate flow	1.1%	
HP DP	0.7	bar
LP DP	0.7	bar
RO specific energy[a]	2.00	kWh/m^3
Efficiency	96.7%	

[a]Does not include feedwater supply pump energy consumption.

and HPP's operating points, thus allowing a great downturn capacity of the plant while maintaining an extremely high efficiency in the ERS due to the very flat efficiency curve of the isobaric PX-Q300 devices.

There is no doubt that energy recovery innovations in the SWRO desalination process have been one of the most important factors that have driven down the cost of the desalinated water and have made this technology widely accepted and affordable worldwide for the production of potable water. The rapid growth of desalination and the construction of mega-size plants in many countries around the world could have not been conceived without the achievement of these important milestones in terms of lowering energy consumption driven by the development of the modern isobaric ERDs.

For example, the predicted performance for the PX-Q300 system at the Carlsbad's first large-scale desalination plant in California is illustrated in Table 11.1. There will be eight PX arrays consisting of 18 PX-Q300 units each.

11.7 CONCLUSIONS

The use of ERDs in SWRO desalination was fundamental in the success of this process; it decreased dramatically the operating costs along with the improvements in polyamide membranes and pumping equipment.

Isobaric ERDs are the current "state-of-the-art" technology for recovering energy in SWRO plants.

While the efficiency in energy recovery for isobaric devices is up to 98%, there is still margin for improvement and continuous innovation and R&D

efforts can be done in order to get further reductions in SEC for SWRO desalination plants.

In most of the cases, isobaric technology offers the lowest life-cycle cost and therefore isobaric devices are the preferred technical solution implemented in the modern desalination plants that have been built in the last 10 years. However, the selection of the appropriate ERD sometimes will depend not only on technical factors but also on the project structure (e.g., EPC, BOO, BOT, etc.) and the particularities of the territory where the project will be developed; for instance, in a country with a very low electricity price, this variable could turn the selection in favor of centrifugal ERDs because the OPEX would be less important compared with countries with high electricity tariffs.

REFERENCES

[1] Watereuse Association Desalination Committee. Seawater Desalination Costs White Paper.
[2] Stover R, Andrews W. In: Energy Recovery Devices—Past, Present and Future. Proceedings of the IDA World Congress, Perth, Australia; 2011.
[3] Andrews WT, Laker DS. A twelve-year history of large scale application of work-exchanger energy recovery technology. Desalination 2001;138(20):201–6.
[4] Kadaj E. A novel turbocharger hydraulic design method utilized in the World's Largest Membrane Desalination System. IDA Congress October 2013;2015.
[5] Clemente R. The economics on energy recovery devices. AMTA-AWWA Conference September 2014;2014.

FURTHER READING

[6] Guirguis, Mageed Jean. Energy Recovery Devices in Seawater Reverse Osmosis Desalination Plants with Emphasis on Efficiency and Economical Analysis of Isobaric versus Centrifugal Devices, Graduate Theses and Dissertations, 2011. http://scholarcommons.usf.edu/etd/3135.

CHAPTER 12

Thermodynamic, Exergy, and Thermoeconomic analysis of Multiple Effect Distillation Processes

Pietro Catrini, Andrea Cipollina, Francesco Giacalone, Giorgio Micale, Antonio Piacentino, Alessandro Tamburini
University of Palermo, Palermo, Italy

Nomenclature

a_w water activity
a_s salt activity
C molar concentration (mol/L)
e specific exergy (kJ_{ex}/kg)
$\dot{E}$ exergy flow (kW_{ex})
h specific enthalpy (kJ/kg)
$\tilde{h}$ partial molar enthalpy (kJ/kmol)
$\dot{m}$ exergy destruction rate (kW_{ex})
k^* specific exergy cost (kW_{ex}/kW_{ex})
$\dot{m}$ mass Flow rate (kg/s)
n number of species in a mixture
$\dot{N}$ molar flow rate (mol/s)
p pressure (Pa)
$\dot{P}$ power (kW)
R Universal constant of gases (kJ/(kmol K))
s specific entropy (kJ/(kg K))
$\dot{S}$ partial molar entropy (kJ/kmol)
t temperature (°C)
T absolute temperature (K)
x molar fraction (mol/mol)

Greek Symbols

ϕ Osmotic coefficient (dimensionless)
γ salt activity coefficient (dimensionless)
η efficiency (dimensionless)
μ chemical potential (J/mol)

Renewable Energy Powered Desalination Handbook
https://doi.org/10.1016/B978-0-12-815244-7.00012-X

Subscript

0 related to reference or "dead" state
B related to brine
ch related to chemical exergy
D related to distillate
eff related to effect
el related to electricity
ex related to exergy
in related to inlet
out related to outlet
ph related to preheater
th related to thermal exergy
s related to salt
sol related to solution
w related to water

12.1 INTRODUCTION

Population growth, economic developments, and improvement of living standards result in an increase in water usage during past decades. At the same time, water shortage is a pressing concern in many countries around the world, above all in arid zones as the Middle East and North Africa (MENA). According to the World Health Organization (WHO) statistics [1], about 20% of the world's population lives in countries where water is scarce. This problem is expected to worsen during the next decades.

The desalination industry plays a fundamental role in addressing water scarcity, by supplying freshwater from natural sources as seawater, brackish, and brines. According to the Worldwide Desalting Inventory [2], the total capacity of commissioned desalination plants in the world at 2016 is around 88.6 million m^3/day and 300 million is the number of people around the world who rely on desalinated water for some or all their daily needs. Currently, according to data provided by [2], the vast majority of desalination processes are based on reverse osmosis (RO), multistage flash (MSF), and multieffect distillation (MED). More specifically, as shown in Fig. 12.1, among these technologies, RO has achieved a dominant role during past decades due to improvements in membrane properties, accounting for nearly 65% of the total worldwide installed capacity [2]. The thermal desalination process accounts for nearly 28% of global installed capacity, which is shared 21% by MSF and 7% by MED. The remaining fraction of global desalination capacity is covered by nanofiltration (NF), electrodialysis (ED), and other processes.

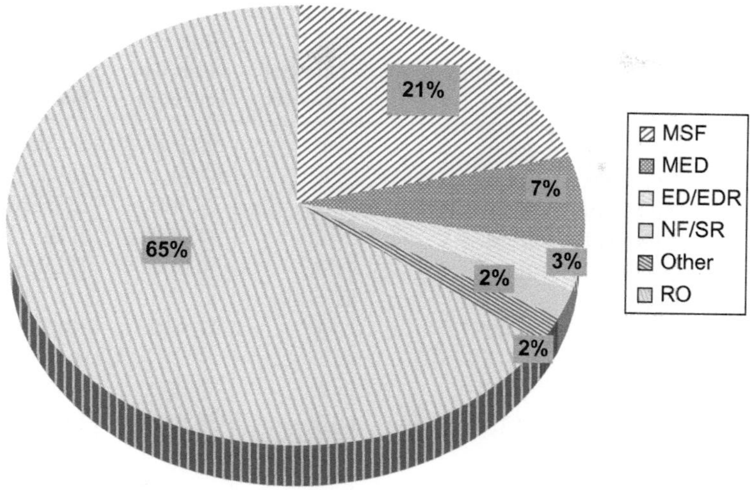

Fig. 12.1 Global desalination capacity by process.

Even though the RO process is leading the worldwide desalination market [3–5], thermal desalination processes are still predominant in *Gulf cooperation council countries* (GCC), that is, Bahrain, Kuwait, Oman, Qatar, Saudi Arabia, and the United Arab Emirates [4,5]. Indeed, in GCC, thermal processes account for about 56% of installed desalination capacity, with the remaining fraction, that is, 44%, being covered by the RO process. The usage of thermal technologies in these countries is mainly due to (i) the poor quality of water sources to be desalted and (ii) the high availability of cogenerative steam from the local thermoelectric plant. Firstly, feedwater supplied from the Arabian Gulf is characterized by high temperature, high salinity, and high impurity, thus increasing the operating cost for the RO process due to membrane replacement and pretreatment processes. In addition, in those countries, the large amount of low-grade thermal energy from thermal power plants allows for a cogeneration operating mode which leads to a more cost-effective and energy-saving desalination process.

From the previous data, it is clear that MED technology is not extensively used from a worldwide point of view. Although the MED process is more efficient from a thermodynamic point of view than MSF [6], its initial spread was limited due to serious scaling and corrosion problems occurring on the external surface of the evaporator bundle. However, during the past decades, improvement in MED technology has contributed to solving the above mentioned technical issues, thus becoming more competitive

Table 12.1 Examples of recent MED installations around the world [7,8]

Country	Location	Commissioning year	Number of units	Unit capacity (m^3/day)	Manufacturer
UAE	Layyah	2006	1	36,368	SIDEM
India	Gujarat	2006	4	8450	IDE
Bahrain	Al Hidd	2007	10	27,300	SIDEM
India	Jamnagar	2008	4	24,000	IDE
Saudi Arabia	Al Jubail	2009	27	30,000	SIDEM
UAE	Fujairah II	2009	12	37,900	SIDEM
Lybia	Zawia	2009	4	20,000	SIDEM
Qatar	Ras Laffan C	2010	10	28,640	SIDEM
China	Tianjin	2010	4	25,000	IDE
India	Gujarat	2012	2	24,000	IDE
Saudi Arabia	Yanbu	2013	2	31,674	SIDEM
India	Gujarat	2013	4	25,000	IDE
India	Jamnagar	2015	3	24,000	IDE
Kuwait	Az Zour	2016	10	49,279	SIDEM

with respect to MSF technology. Nowadays, according to Refs. [7,8], the MED unit capacity usually ranges from 600 to 36,000 m^3/day.

MED processes are highly capital intensive and energy consuming processes. For instance, in Refs. [6,9,10], energy consumption of the MED process is nearly 2–3 times higher than the energy consumption of the RO process. It is clear that lowering energy requirements for the MED process is the current challenge in order to achieve a more competitive, cost-effective, and sustainable process.

In Table 12.1, some recent MED installations around the world during the decade 2006–2016 are shown. One of the largest MED plants is installed in Al Jubail (Saudi Arabia) with a total capacity of 800,000 m^3/day (27 units of 30,000 m^3/day each) [7,8].

12.2 FUNDAMENTALS ON MED TECHNOLOGY

MED is prevalently used for seawater desalination, but it is often applied in the process industries such as sugar, paper and pulp, dairy, textiles, acids and bases, and desalination. It represents a well consolidated technology, which has observed a large spread since the late 1950s.

The MED process consists of a sequence of effects, whose number usually ranges up to 16, where pure water is separated from the feed by means of evaporation. The necessity of operating with a multiple effect configuration is due to the poor performance of a single-effect system [11]. In fact, recovering the thermal energy content of the generated vapor to drive the successive effect allows to avoid wasting of a large amount of energy that would be rejected in a single-effect configuration.

In most MED processes, the *falling film evaporator* is used for heat transfer between the feed seawater to be evaporated and the heat source. In this configuration, seawater is sprayed over the outer surface of heated tubes. Steam produced within each effect is then transferred to the next effect for boiling seawater or the brine coming from the previous effect. Other evaporator configuration (e.g., vertical falling film, submerged tube) have been investigated [6,12].

In order to avoid scaling and reduce corrosion on the outside of the tube bundle, the top brine temperature (TBT) has been limited to 70°C [6,11]. This temperature limit allows for the exploitation of low grade heat, or the integration of vapor compression within the process. Another advantage is the use of alternative construction materials which leads to a decrease in capital cost. More specifically, aluminum alloys are used for the heat transfer tubes, as well as carbon steel epoxy coated shells for the evaporator shells.

Temperature gradually decreases when passing through one effect to the next one, reaching about 30–40°C in the last effect.

Since temperature decreases from the first to the last stage, each effect operates at a decreasing level of pressure in order to allow water evaporation. Also, the amount of distillate generated in each effect is less than the distillate generated in the previous effect due to the increase in the specific latent heat of vaporization when decreasing the effect temperature.

The intake seawater is preheated within the down condenser by condensing a fraction of the distillate generated from the last effect. After that, a fraction of the intake seawater, (i.e., *cooling seawater)*, is discharged into the sea. Cooling seawater aims at removing the excess heat added into the system in the first effect [11]. Conversely, the remaining fraction is further preheated by distillate generated before being supplied to the effects for producing freshwater.

Regarding the pumping unit, the MED process is usually equipped with pumps for seawater intake, distillate product, brine blowdown, and chemical dosing. The feedwater stream is chemically treated and deaerated before being sprayed into the effects [6,11].

12.2.1 MED Process Configurations

MED plants can be classified according to the direction of the feedwater flow and heat flow, respectively, as: *forward feed plants* (FF-MED), *parallel feed plants* (PF-MED), and *backward feed plants* (BF-MED) [6,11,13]. Schemes for each MED configuration are shown in Fig. 12.2A–C. Some details on each configuration plant are provided as follows:

- *Forward feed plants*: as shown in Fig. 12.2A, the feedwater, after being preheated into the down condenser and the series of preheaters, is introduced into the first effect where it is separated into pure water and brine. Then the brine flow is transferred to the next effect where pure water is further separated by means of the heat released from the distillate generated into the previous effect. Obviously, brine concentration increases when passing from the first effect to the last, where the highest concentrated brine is found at the lowest temperature, avoiding scale formation. The FF-MED system is not found on an industrial scale for the desalination industry, due to its complex layout with respect to the parallel feed configuration. Conversely, it is widely used in the sugar and paper industries.
- *Parallel feed plants*: as shown in Fig. 12.2B, the feedwater, after being preheated into the down condenser, is equally distributed among effects. In each effect, pure water is separated from feedwater by means of heat released from the condensing distillate formed in the previous effect. The main advantages of this configuration are its simple layout and the lower risk of scale formation. For these reasons, it is largely used as the base technology for seawater desalination.
- *Backward feed plants*: as shown in Fig. 12.2C, the feedwater is introduced into the last effect, which operates at the lowest pressure and temperature of the system. Then the brine flows from the last effect to the first for further distillate generation. The main advantage of this configuration is that no feedwater preheaters are required, thus lowering the amount of capital investment necessary. Conversely, the increase in the operating pressure across the effects requires the use of pumps, thus increasing operating costs. Also, the highest brine concentration occurs in the first effect, which operates at the highest temperature of the system, thus increasing the risk of scale formation. For this reason, this configuration is not applied to seawater desalination.

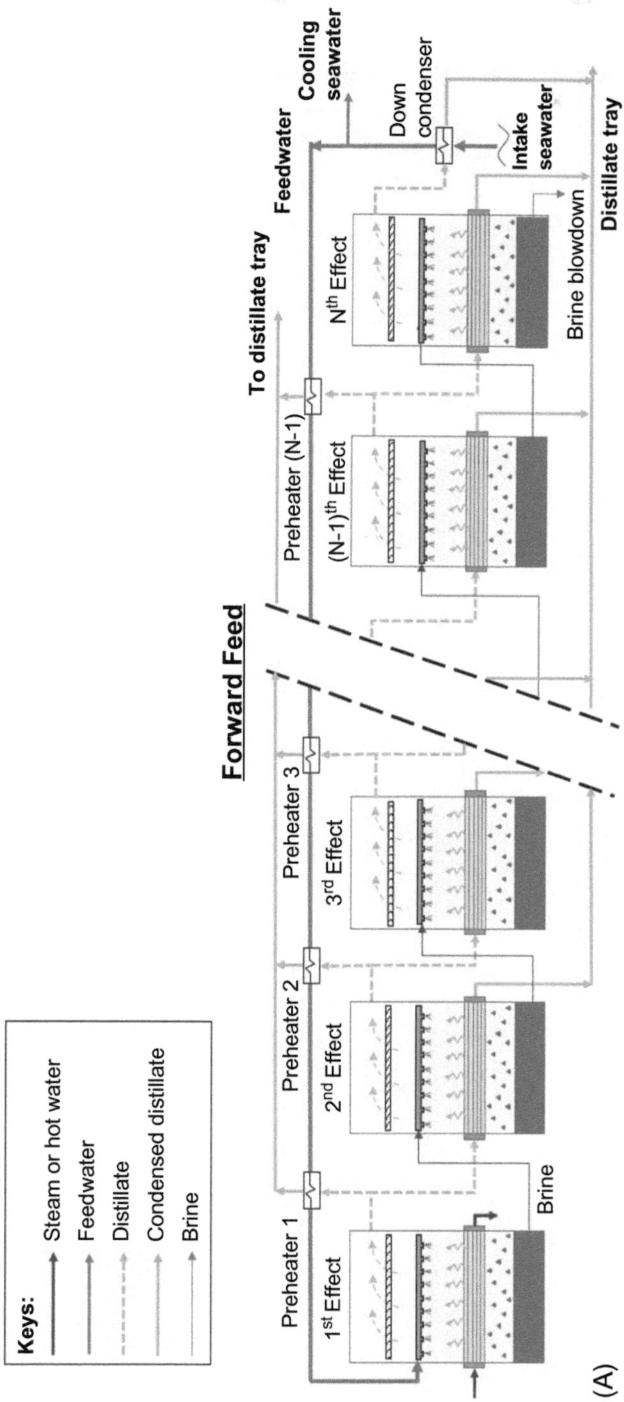

Fig. 12.2 MED process configuration: (A) Forward feed.

(Continued)

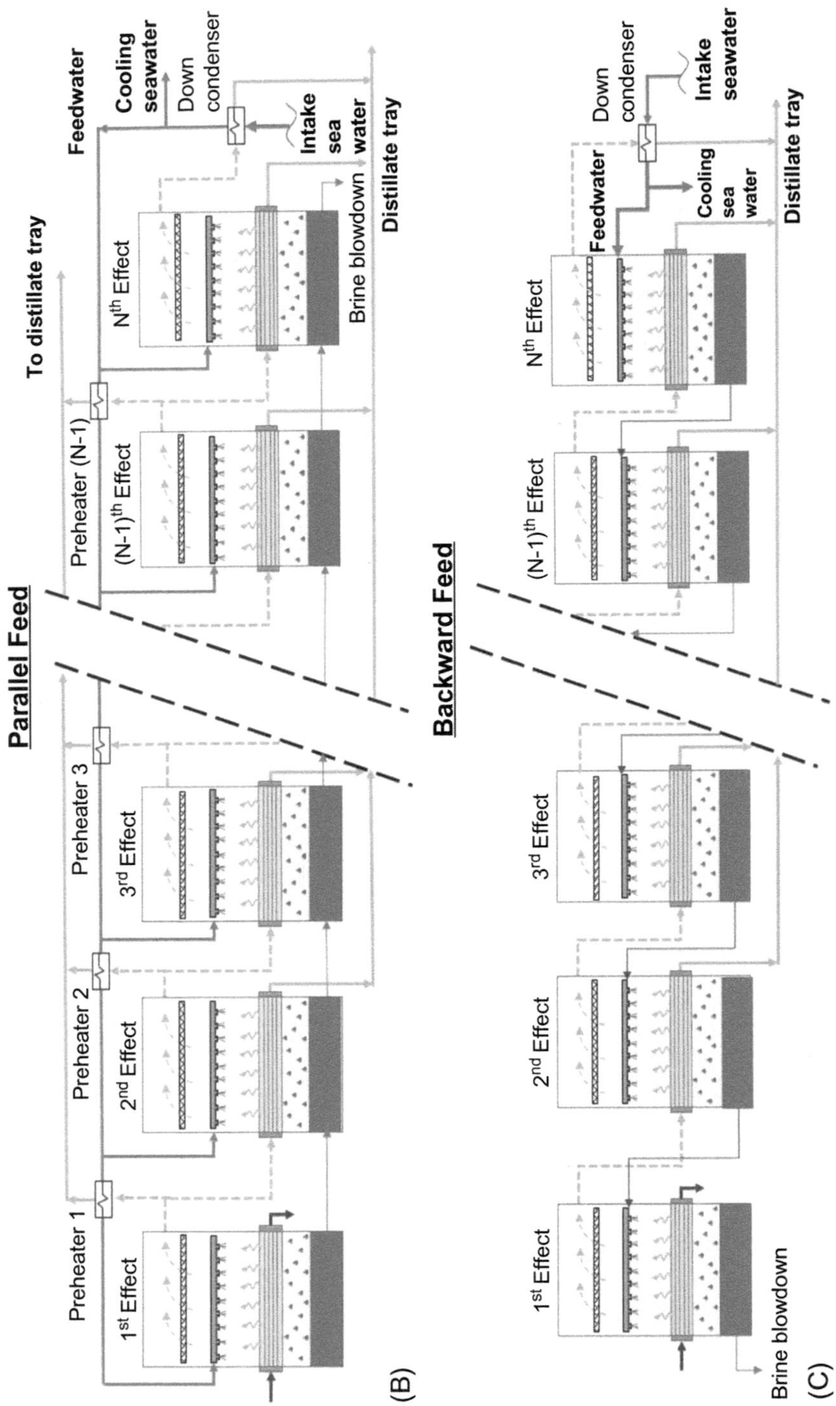

Fig. 12.2, Cont'd (B) Parallel feed. (C) Backward feed.

12.3 ENERGY CONSUMPTION OF THE MED PROCESS

MED processes are usually driven by external heat flow supplied by means of a dedicated combustion of fossil fuel within a boiler or by waste heat available from an industrial process. Typical values of 40–80 kWh_t/m^3 are found in the literature [6,8,11]. Electricity consumed for pumping and auxiliaries ranges between 1.5 and 2.5 kWh_e/m^3.

Following the El-Dessouky approach [11], the following indicators are adopted for assessing the MED process performance.

- *Gain output ratio* (GOR) expresses the amount of thermal energy consumed in the desalination process, and is defined as the ratio of the mass of distillate to the mass of the input steam. As shown in [11], GOR values up to 10 may be obtained for a standard MED process. This indicator heavily depends on the number of effects of the MED system and, in particular, an increase in the number of effects allows for an additional distillate generation and higher of GOR.
- The *specific heat transfer surface area* expresses the area required for the heat transfer process for a unit distillate product. As shown in [11], a typical value of $250\,m^2/(kg_{dist}/s)$ is obtained for a MED process with 12th effects and a TBT equal to 60°C. This indicator is highly sensitive to the number of effects and to the TBT values. More specifically, its values decrease when operating with high-values TBT, due to the higher driving force available for the heat transfer process.
- The *specific cooling water flow rate* indicates the amount of cooling water supplied to the down condenser for condensing the unit mass of distillate generated. This indicator is a measure of the electricity consumed for pumping and treatments. Its values decrease from 9 to 2.7 $(kg_{CW}/s)/(kg_{dist}/s)$ while increasing the number of effects, respectively, from 6 to 12 [11].

Both the *GOR* and the *specific flow rate of cooling* are a measure of MED operating costs, since they are related respectively to the supplied motive steam and to the electricity consumed for pumping the intake seawater. Conversely, the *specific heat transfer surface area* is a measure of capital investment required for process realization.

12.3.1 Enhancement of the MED Process Performance by Means of "Vapor Compression"

In order to enhance the process energy efficiency, MED units are often equipped with a vapor compression system, which compresses part of the distillate vapor generated within the last effect before supplying it to the first effect. The following systems are used for vapor compression: (i) thermal vapor compression (MED–TVC) and (ii) mechanical vapor compression (MED–MVC).

In MED–TVC configuration, a fraction of the steam from the last effect is entrained and compressed to the pressure of the first effect in a steam jet ejector by means of a middle pressure steam. A scheme of the MED–TVC plant is provided in Fig. 12.4. Once compressed, the steam flow is supplied to the first effect where it releases its energy content to the feedwater [6,11,13]. By means of this configuration, the GOR of the plant increases for about 30%–40% with respect to the standard MED process, thus producing 15–16 kg of distillate per kg of motive steam supplied to TVC. Also, the specific electricity consumption is lowered for this configuration with respect to a standard MED process. Indeed, owing to steam extraction from the last effect, less cooling seawater is required to reject heat from the distillate coming from the last effect, thus reducing the electricity consumption for pumping intake seawater. Typical specific electricity consumption values range from 1.5 to 2 kWh_e/m^3. As concerns specific heat transfer surface not sensible variation are observed for a MED–TVC configuration with respect to a standard MED [11]. It is worth stressing that all the recent MED installations showed in Table 12.1 are of the MED–TVC type.

In the MED–MVC system, only electric power is supplied to the process. The MVC process was developed in the early 1980s. A scheme of this system is shown in Fig. 12.3. The necessity of developing this technology was motivated by the need to have an alternative thermal desalination system which uses electric power as a unique energy source. The electrical consumption of such a system ranges from 8 to 15 kWh_e/m^3 [8,14]. Owing to the current limitation in compressors technology, the maximum capacity of MED–MVC units is limited to 5000 m^3/day per unit. [14].

12.3.2 On "Dual Purpose" and "Hybrid Plant" Configuration for the MED Process

One of the main advantages of the MED desalination process is the possibility to be supplied by low temperature heat [14,15]. This feature allows for the exploitation of low grade heat available either from an industrial process (which would be wasted if it was not used) or by condensing steam extracted from power plants. Freshwater production from a cogeneration system allows for a decrease in specific cost and in environmental burdens of freshwater production.

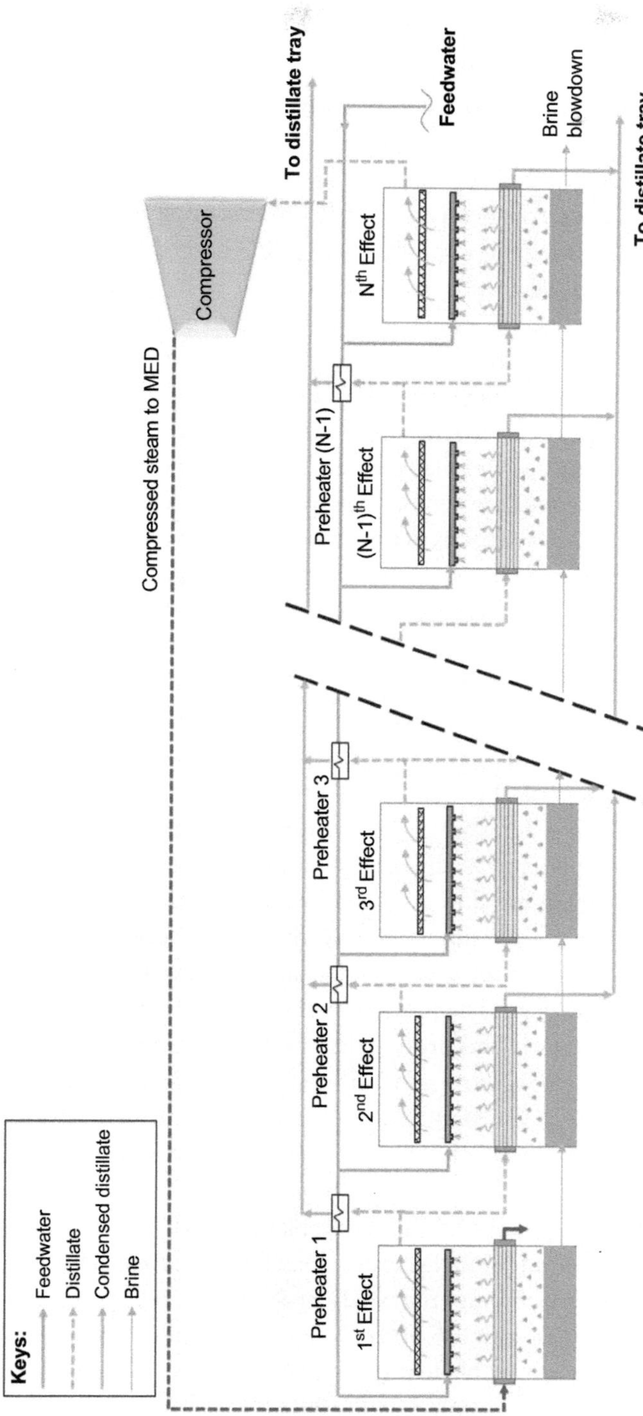

Fig. 12.3 Typical MED–MVC system layout.

According to [14], the typical cost value for a MED process operated in a cogeneration mode is 0.5 US$/$m^3$. Valero et al. [16] carried out a life cycle assessment for the MED process in order to assess environmental impacts of this technology. In particular, by comparing results for a stand-alone configuration with respect to a cogeneration operating mode, it was shown that CO_2 emission reduces respectively by 61% when coupling the MED process with a natural gas fueled combined cycle (CC), and by 55% when considering a coal fueled CC.

Economic benefit for cogeneration of water and electricity may be achieved when cogeneration is supported by national directives as in the case of the European Union Directive [17]. Piacentino et al. [18] carried out a study on eligibility as "high-efficiency" cogeneration of a MED–TVC plant coupled with a steam power plant, where the effects of changes in the steam extraction pressure and the type of fuel supplied were assessed.

One of the main drawbacks of the cogeneration system for electricity and freshwater is the simultaneous water and electricity generation [19], which does not allow for plant flexibility during plant operation in those contexts where seasonal water and power demands are extremely variable and not perfectively matched. In this situation, the system is not run at its maximum efficiency values, thus not achieving profitable energy saving.

One way to address this problem is the use of "hybrid desalination systems" [20] in which thermal and membrane processes are combined. The main advantages of these systems are: (i) reduction in specific energy consumption, (ii) flexibility in operation in the case of highly variable water and electricity demand, (iii) cost saving due to reduction in posttreatment of freshwater, and (iv) possibility to use a unique low intake/outfall construction.

In other words, by means of a "hybrid plant configuration," full load operation for the overall plant is maintained in those periods when power demand decreases significantly. The profitability of a hybrid system such as CC+MED+RO with respect to a stand-alone RO plant is highly sensitive with fuel market prices [21]. Clearly, in GCC, where fossil fuels are extensively subsidized, profitability of a hybrid plant with respect to stand-alone RO is still assured [21].

One of the largest hybrid desalination plants in the world is "Fujairah 2," located in the United Arab Emirates [8]. The plant capacity is equal to 596,500 m^3/day. It combines the MED process with RO, whose capacities are equal to 450,000 and 136,500 m^3/day, respectively. By means of this configuration, the power plant runs at its maximum capacity all year long. The desalination unit is coupled with a 2000 MW power plant. Also, steam for the MED operation is supplied by steam extraction from turbines. Regarding the RO operation, electricity is supplied mainly in the winter season, when electric power demand from the grid is low.

12.4 ENVIRONMENTAL IMPACTS OF THE MED PROCESS

As explained in Refs. [6,22,23], the main environmental burdens of the desalination process can be grouped as follows: (i) greenhouse gases (GHG) and air pollutants emissions due to energy consumptions, (ii) brine disposal back to the sea, (iii) the use of large quantities of seawater for cooling purposes and as feed water, and (iv) construction-related impacts on coastal and nearshore habitats.

As concerns GHG emissions, it is clear that since most of the MED process still relies on the fossil fuel-powered generation system, a large amount of CO_2 is produced for supplying steam to the MED process. In [16], CO_2 emissions for a stand-alone process account for 18–24 kg_{CO2}/m^3 according to the type of fossil fuel used for steam generation.

With respect to brine disposal, it is important to consider the effect of the scale inhibitor, antifoam chemicals, and metals (usually nickel and copper are released from corrosion of the evaporator bundle) which could be harmful to the receiving environment. [22]. Brine temperature is usually 15°C higher than seawater. In order to avoid "thermal pollution," rigorous guidelines provided by national authorities must be followed in order to avoid local increase on seawater temperature. With respect to brine concentration, a predilution of this flow is usually carried out by means of waste water before brine disposal to the seawater [6,22,24].

12.5 ECONOMIC ASPECTS OF THE MED PROCESS

Many studies have been carried out during these years on the economics of desalination processes [23,25–29]. From a keen analysis of the scientific literature, it could be noted that the freshwater unit cost from the MED process differs from case to case. As explained in [6,25,29], differences in cost values can be attributed to several factors which may be summarized as follows: (i) plant capacity and design, (ii) site condition, (iii) feedwater properties, in particular salinity and turbidity, (iv) energy cost, and (v) plant life and amortization. Alardin et al. [29] provided a typical breakdown of the freshwater unit cost from the thermal desalination process. More specifically, the overall energy consumption (i.e., both thermal and electrical) account for 60%–80% of the operating cost. Nowadays, typical specific capital cost values for the MED process range between 900 and 2000 US$/($m^3$/day) [5].

Regarding the freshwater unit cost, Soldatos et al. [25] classified values according to MED plant capacity. In particular, for plant capacity of more than 90,000 m^3/day, the reported cost ranged between 0.52 and 1.01 US$/$m^3$. Conversely, for medium MED capacities of 12,000 to 55,000 m^3/day, the cost varies between 0.95 and 1.95 US$/$m^3$. For plant capacity less than 100 m^3/day, the unit cost ranges between 2 and 8 US$/$m^3$.

Freshwater unit cost higher than 2 US$/$m^3$ is typical of the MED process powered by renewable energy sources as solar thermal (for instance, solar pond/MED or concentrating solar plant (CSP)/MED) and geothermal energy [29–38], whose costs are heavily affected by the high capital cost of the renewable energy technology and the low capacity of the MED system.

12.6 DESCRIPTION OF THE REFERENCE MED PLANT

The examined TVC–MED is an existing plant located in Trapani (Sicily), constructed and started up in 1995. The plant is composed of four identical MED–TVC units with a specific capacity of 9000 m^3/day each, for a total plant capacity of 36,000 m^3/day [6,39–41]. A schematic layout is presented in Fig. 12.4.

The MED–TVC units configuration is parallel-feed type with 12 effects each.

Each effect is a falling-film evaporator with the feedwater sprayed on the external part of the tube bundle. The produced distillate within each effect is condensed into the next one, by releasing its latent energy content for generating further distillate from seawater. Each effect is equipped with a tube bundle counting about 11,000 tubes, 7 m long and with an external diameter of 22 mm [41a].

In each unit, steam is extracted from the 12th effect at a 6-kPa absolute pressure and compressed up to the pressure of the first effect (i.e., 22 kPa), by means of thermal vapor compression. The steam recompressed and de-superheated into the TVC is supplied to the first effect at a temperature around 63°C, in order to minimize scaling problems in the tube bundle. A temperature drop between two consecutive stages of 2–3°C is maintained, and the temperature of the brine exiting from the last effect ranges between 35°C and 38°C. Recorded data are provided in

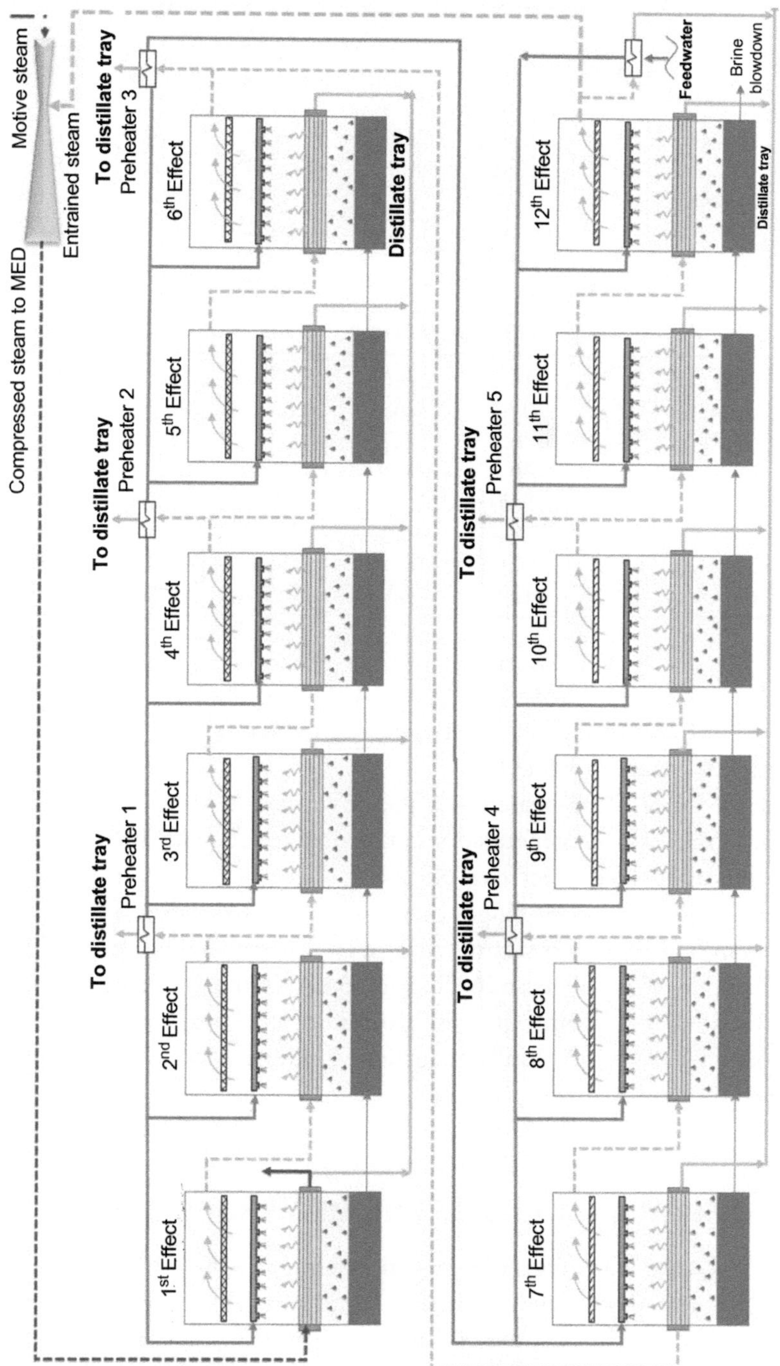

Fig. 12.4 Layout of the reference MED–TVC.

Table 12.2 Thermodynamic data for the reference MED–TVC plant [6,39–41]

		Effect 1	Effect 2	Effect 3	Effect 4	Effect 5	Effect 6
t_{feed}	(°C)	55.0	55.0	53.3	53.3	48.8	48.8
C_{feed}	(mol/L)	0.63	0.63	0.63	0.63	0.63	0.63
$\dot{m}_{feed}$	(kg/s)	26.2	26.2	26.2	26.2	26.2	26.2
t_D	(°C)	61.7	60.3	58	55.8	53.5	51.3
$\dot{m}_D$	(kg/s)	10	9.94	9.8	9.74	9.84	9.08
t_B	(°C)	62.5	61.5	59.4	57.3	55.0	52.9
C_B	(mol/L)	1.021	1.016	1.015	1.014	1.004	1.000
$\dot{m}_B$	(kg/s)	16.2	32.6	49.0	65.3	82.4	99.2
		Effect 7	**Effect 8**	**Effect 9**	**Effect 10**	**Effect 11**	**Effect 12**
t_{feed}	(°C)	44.2	44.2	39.6	39.6	29.6	29.6
C_{feed}	(mol/L)	0.63	0.63	0.63	0.63	0.63	0.63
$\dot{m}_{feed}$	(kg/s)	26.2	26.2	26.2	26.2	26.2	26.2
t_D	(°C)	49	46.7	44.4	42.1	39.8	37.2
$\dot{m}_D$	(kg/s)	9.29	8.20	8.54	7.16	7.63	5.63
t_B	(°C)	50.6	48.4	46.1	43.9	41.5	39.1
C_B	(mol/L)	0.989	0.982	0.968	0.962	0.945	0.934
$\dot{m}_B$	(kg/s)	117.2	134.8	153.8	172.3	192.8	212.7

Table 12.2. For the sake of clarity, the subscript "D" refers to distillate, whereas "B" refers to brine.

The feedwater is preheated into the down condenser by condensing a fraction of the distillate generated into the 12th effect and furthermore feed preheaters are located every two effects.

The *GOR* of the plant ranges between 14 and 16 kg distillate per kg of motive vapor. The motive steam, whose total flow rate is equal to 36–60 t/h, is produced by two boilers at 260°C and 45 bar, with a methane consumption estimated at 5.56 Nm^3/m^3 distillate. Electrical energy consumption is estimated at 2 kWh/m^3 distillate, including pumps and utilities from the intake to the posttreatment section.

12.7 EXERGY ANALYSIS OF THE MED PROCESSES

When analyzing an energy conversion system, it is useful for analysts to adopt rigorous methodologies aimed at identifying the subprocesses or components where the highest margins for improvements are located, due to a poorer performance of energy conversion.

Exergy analysis provides a method for assessing the difference between the energy consumption of a real process and the minimum ideal requirement. The attribute "ideal" is here used to indicate the processes where energy degradation does not occur. By such a rigorous approach, analysts can easily identify the design options and operating conditions (among a set of alternative design and operation scenarios) that achieve the highest overall efficiency, thus leading to a more cost-effective and sustainable system [42–45].

The *exergy* of an energy flow or a stream of matter is the maximum amount of work obtainable when the stream evolves from its thermodynamic state to the thermomechanical and chemical equilibrium with the environment through reversible processes involving mass and heat transfer between the system and the environment [42]. Differently from energy, exergy is destroyed in real processes. Common sources of exergy destruction are heat exchanges across finite temperature differences, uncontrolled mixing of flows at different concentrations or composition and flow friction [42]. For example, in the case of a steam turbine, steam flow rate is supplied in order to generate mechanical work. The maximum amount of work is produced only if no flow friction occurs. This amount of work is equal to the exergy content of the steam flow, decreased by the exergy content of the low-pressure steam exiting the turbine. The difference between the maximum amount of work and the actual work is equal to the exergy destruction (or irreversibility) occurring within the turbine. The more irreversible the process occurring in the turbine, the higher the exergy destruction. Conversely, in the case of an air compressor, if a reversible compression process is considered, a minimum amount of work is required. This quantity is equal to the increase in exergy content of compressed air.

Before applying exergy analysis to the MED process, some details on the exergy of a stream of matter are here given to allow for an easier understanding of the results discussed in the following sections.

12.7.1 Exergy of a Stream of Matter

In order to derive a general expression for the exergy content of a stream flow, the general approach presented in [44] is here adopted. It is useful to consider the scheme shown in Fig. 12.5. A stream flow in a defined thermodynamic state, defined by its temperature T, pressure p, and composition x_i (i.e., the molar fraction of each "ith" species in the mixture), reaches thermomechanical and chemical equilibrium with the environment by

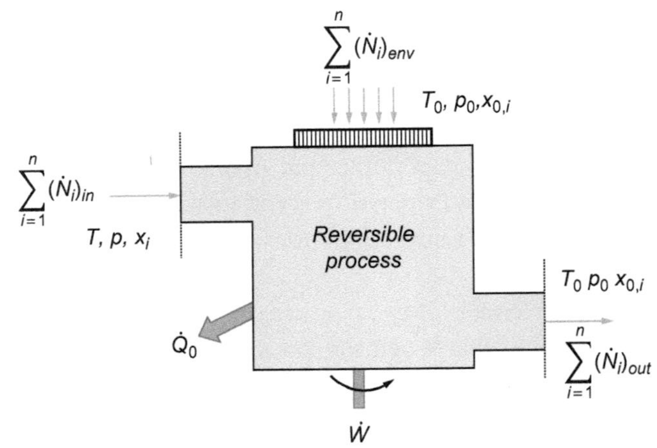

Fig. 12.5 Reversible evolution of a stream of matter to thermodynamic equilibrium with the environment.

exchanging "*reversibly*" mass, heat, and work. The final thermodynamic state of the stream flow is characterized by the same temperature T_0, pressure p_0, and composition x_{i0} (i.e., the molar fraction of the "*i*th" species, at equilibrium) of the environment, commonly named "*dead state.*"

The general expression for the exergy content of the inlet stream is derived by applying to the open system represented in Fig. 12.5 mass, energy and entropy balances. Once introduced the hypothesis of process reversibility, no entropy is generated, thus obtaining the production of the maximum possible amount of work. According to the definition of exergy provided above, the work obtained in this ideal reversible case equals the exergy content of the inlet stream flow and provides the following analytical expression:

$$\dot{E} = \sum_{i=1}^{n}\left[\left(\widetilde{h}_i(T,p,x_i) - T_0\widetilde{s}_i(T,p,x_i)\right) - \left(\widetilde{h}_i(T_0,p_0,x_{i,0}) - T_0\widetilde{s}_i(T_0,p_0,x_{i,0})\right)\right]\cdot \dot{N}_{i,in} \quad (12.1)$$

where N_i, $\widetilde{h}_i$ and $\widetilde{s}_i$, indicate the molar flow of the "ith" species constituting the inlet flow, the partial molar enthalpy and the partial molar entropy, and n is the number of species of the inlet flow [44].

Since the temperature, pressure, and composition of a stream flow change during the evolution from its initial thermodynamic state to the equilibrium with the environment, it is suitable to account separately for

(i) the fractions of the work produced by change in chemical composition, and (ii) those related to changes in temperature and pressure of the stream [46]. From this perspective, the exergy of a stream is usually divided into:

– *Physical exergy*: this quantity indicates the amount of work obtainable when the stream reaches thermal and mechanical equilibrium with the environment, while no change in chemical composition is observed (i.e., not reaching a chemical equilibrium with the surrounding ambient). The final thermodynamic state of the stream flow is characterized by the same temperature and pressure of the environment (T_0, p_0), commonly named "restricted dead state" [42,46]. In Eq. (12.2), expression for a generic stream flow is shown.

$$\dot{E}_{phys} = \sum_{i=1}^{n} \Big[\Big(\widetilde{h}_i(T, p, x_i) - T_0 \widetilde{s}_i(T, p, x_i) \Big) - \Big(\widetilde{h}_i(T_0, p_0, x_i) - T_0 \widetilde{s}_i(T_0, p_0, x_i) \Big) \Big] \dot{N}_{i,in} \quad (12.2)$$

– *Chemical exergy*: this quantity is the maximum amount of work obtainable when the stream, supposedly in physical equilibrium with the restricted dead state, modifies its chemical composition, thus reaching also chemical equilibrium with the environment. The thermodynamic state variables of the stream flow in this case are equal to the ones of the environment, (T_0, p_0, $x_{i,0}$). In Eq. (12.3), expression for the chemical exergy is presented.

$$\dot{E}_{ch} = \sum_{i=1}^{n} \Big[\Big(\widetilde{h}_i(T_0, p_0, x_i) - T_0 \widetilde{s}_i(T_0, p_0, x_i) \Big) - \Big(\widetilde{h}_i(T_0, p_0, x_{i,0}) - T_0 \widetilde{s}_i(T_0, p_0, x_{i,0}) \Big) \Big] \dot{N}_{i,in} = \sum_{i=1}^{n} [(\mu_i(T_0, p_0, x_i) - \mu_i(T_0, p_0, x_{i,0}))] \dot{N}_{i,in} \quad (12.3)$$

Proper selection of dead state variables (i.e., pressure, temperature, and composition), is necessary. With reference to temperature and pressure values, common practice assumes them to be equal to the respective average values of the site where the system is going to operate. As regards the chemical composition of the reference environment, it is common to refer to the most abundant natural element where it appears in the environment (whether it be in the atmosphere, the seawater, or the earth's crust) [46].

As can be seen from Eq. (12.3), the chemical potential calculation for each species in the mixture is implied. Clearly, this means that suitable thermodynamic modeling for solutions is necessary when assessing the chemical exergy content of a stream flow.

12.7.2 Exergy Analysis for the Whole MED Process

Exergy analysis can be applied to MED desalination processes at a whole system level, by referring to Fig. 12.6. As shown in the scheme, according to the disaggregation level adopted, the entire process is in this case treated as a "black box," where only the exergy of energy/matter flows entering and exiting the whole plant is considered.

As previously shown, energy input for an MED process is constituted by the following contributions: (i) heat supplied by means of a "hot stream," commonly in the form of low-pressure steam or, more rarely, hot water produced by a boiler or supplied by an industrial process; (ii) electricity used to drive pumps and auxiliaries. In terms of exergy inputs to the MED process, it must be considered that: (i) the amount of thermal exergy released by the "hot stream" is lower than the heat content of the stream, and it may be calculated by the expression given in Eq. (12.4); (ii) the electricity used to drive pumps and other auxiliaries may be considered as pure exergy.

$$\dot{E}_{\text{hot stream}} = \dot{m}_{\text{hot stream}} \cdot \left[\left(h_{\text{hot stream}} - h_0 \right) - T_0 \cdot \left(s_{\text{hot stream}} - s_0 \right) \right] \quad (12.4)$$

In the above equation, the chemical exergy of the supplied steam or hot water (i.e., hot stream) is not considered, since no variation in its chemical composition is observed.

The exergy output is constituted by the chemical exergy of the brine and freshwater flows exiting the MED process, whose expressions are provided

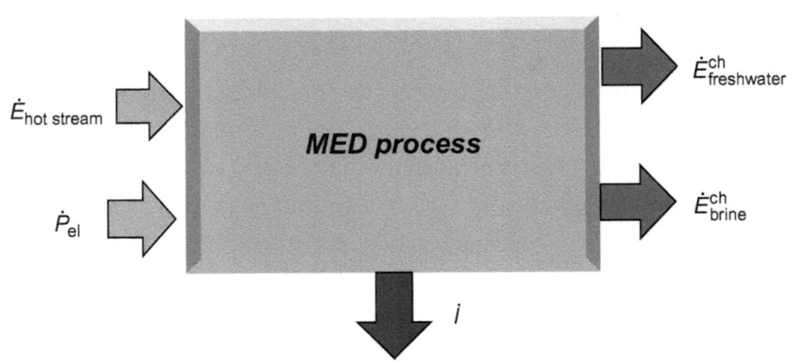

Fig. 12.6 Exergy balance for the whole MED process.

in Eqs. (12.5) and (12.6), respectively. Both expressions are derived from Eq. (12.3), where chemical potentials are explicitly expressed in terms of salt and water activities in each solution and in the reference state. The thermal exergy of these flows is not considered since they are assumed to leave the whole system in thermomechanical equilibrium with the environment.

$$\dot{E}_{ch,fresh} = RT_0\left[\dot{N}_{fresh}\ \ln\left(\frac{1}{a_{w,0}}\right)\right] \tag{12.5}$$

$$\dot{E}_{ch,brine} = RT_0\left[\dot{N}_B\left(x_{s,B}\ \ln\left(\frac{a_{s,B}}{a_{s,0}}\right) + x_{w,B}\ \ln\left(\frac{a_{w,B}}{a_{w,0}}\right)\right)\right] \tag{12.6}$$

It is worth underlining that no exergy flow enters the MED system with the feedwater stream, since it is assumed in thermomechanical and chemical equilibrium with the environment.

By setting the exergy balance to the overall process as shown in Eq. (12.7), exergy destruction due to all the irreversibility occurring within the plant is easily assessed by the following equation:

$$\dot{E}_{\text{hot stream}} + \dot{P}_{\text{el}} - \left(\dot{E}^{\text{ch}}_{\text{brine}} + \dot{E}^{\text{ch}}_{\text{freshwater}}\right) = \dot{I} \tag{12.7}$$

If the unit operated reversibly, no irreversibility would be generated and the minimum amount of exergy input to the system would be required for a specified freshwater production rate. It is trivial that under this ideal condition all the supplied exergy $\left(\dot{E}^{\text{th}}_{\text{hot stream}} + P_{el}\right)$, would be entirely converted into the exergy content of concentrated brine and freshwater.

In order to formulate a synthetic indicator for the thermodynamic performance of the MED process, the exergy efficiency for this unit may be defined as shown in Eq. (12.8). In particular, this 2nd law-based efficiency is calculated as a ratio between the chemical exergy of the freshwater product and the thermal exergy of the supplied hot stream driving the unit.

$$\eta_{\text{ex,MED-TVC}} = \frac{\dot{E}^{\text{ch}}_{\text{freshwater}}}{\dot{E}_{\text{hot stream}} + P_{\text{el}}} \tag{12.8}$$

In Eq. (12.8) the chemical exergy of the exiting brine is not accounted. This choice is justified by the productive scope of MED plants which consists of producing freshwater, without further exploiting the chemical exergy of brine which is usually discarded back to the sea. As observed in [47], only when the exiting brine is supplied to further units (such as a Salinity Gradient Power system that recovers it to produce electricity [48]), its exergy content would represent a useful product of the MED–TVC plant.

12.7.2.1 *Exergy Analysis for the Examined MED–TVC Plant*

After a brief analysis of the fundamentals of exergy analysis and its application to a whole MED process, a more detailed analysis is here presented to assess the exergy performance of the reference MED–TVC system.

The examined MED–TVC plant is constituted by four identical units. For the sake of simplicity, the analysis is carried out for only one of the installed units.

Based on their chemical composition, both seawater and brine streams were properly modeled as water–NaCl solutions. In order to calculate activity coefficients required in Eqs. (12.5) and (12.6), Pitzer's model was adopted, which is outlined in detail in Appendix.

With reference to the environmental conditions to be used when calculating specific exergy of the material streams involved in the MED process, an average ambient temperature $T_0 = 298.15\,\text{K}$ and pressure $p_0 = 101.3\,\text{kPa}$ were considered. Also, a salt concentration of seawater equal to 37,000 ppm was considered as reference for chemical exergy calculations.

The reference MED–TVC is supplied with motive steam produced by pressurized boilers at 260°C and 45 bar. The thermal exergy flow $\left(\dot{E}_{\text{motive,in}} - \dot{E}_{\text{cond to boiler}}\right)$, difference between the exergy of the superheated steam, and the exergy of condensate water recirculated back to the boilers represent the total exergy consumption of the whole MED–TVC process.

The exergy efficiency at a whole plant level is calculated according to Eq. (12.8):

$$\eta_{\text{ex,MED-TVC}} = \frac{\dot{E}^{\text{ch}}_{\text{freshwater}}}{\left(\dot{E}_{\text{motive,in}} - \dot{E}_{\text{cond to boiler}}\right) + \dot{P}_{\text{el}}} = \frac{299.8\,\text{kW}_{\text{ex}}}{(6348\,\text{kW}_{\text{ex}} - 64.42\,\text{kW}_{\text{ex}}) + 750\,\text{kW}_{\text{ex}}} \approx 0.043 \quad (12.9)$$

The low value of exergy efficiency at a whole plant level pinpoints the high incidence of thermodynamic irreversibility occurring along the MED system, when attempting to convert the thermal exergy content of motive steam into chemical exergy of a freshwater product.

12.7.3 Exergy Analysis of the MED System at the Single-Subprocess Level

In order to understand the origin of the low exergy efficiency of the MED–TVC system, exergy analysis may be carried out at a higher disaggregation level, by examining separately the elementary subprocesses involved in the

MED plant. In this way, a complete and detailed picture of exergy flows, irreversibility, and exergy efficiencies of each specific conversion process could be obtained. Looking at the MED–TVC process, the following main subprocesses could be distinguished:

– *TVC of the entrained steam*

As previously mentioned, in the MED–TVC process a fraction of distillate generated within the last effect is entrained and compressed from the pressure level of the last effect to the pressure reigning in the first effect by means of a steam ejector. As clearly explained in [49], the ejector is a static component where low-pressure steam is compressed by means of a mid or high-pressure steam, usually named "motive steam."

The exergetic model adopted for the TVC has been presented in a previous article by the authors [47]. For the sake of clarity, some key aspects are here briefly outlined. Once indicated with $e_{\text{steam to MED}}$, the specific exergy of the compressed steam exiting the TVC (corresponding to the inlet condition to the 1st effect of MED), and with $e_{\text{entrained}}$ the specific exergy of the steam exiting the 12th effect and entrained by the TVC, it follows that:

(1) the exergy consumed by the TVC unit is equal to the exergy variation of the motive steam when expanding from its high-pressure p_{motive} to the pressure of the compressed steam supplied to the 1st effect $p_{\text{steam to MED}}$. Then the amount of exergy consumed is expressed by $\dot{m}_{\text{motive}}(e_{\text{motive}} - e_{\text{steam to MED}})$ as shown in Fig. 12.7;

(2) the increase in the exergy content of entrained steam, due to compression from $p_{\text{entrained}}$ to $p_{\text{steam to MED}}$, is equal to $\dot{m}_{\text{entrained}} \cdot (e_{\text{steam to MED}} - e_{\text{entrained}})$ as shown in Fig. 12.7.

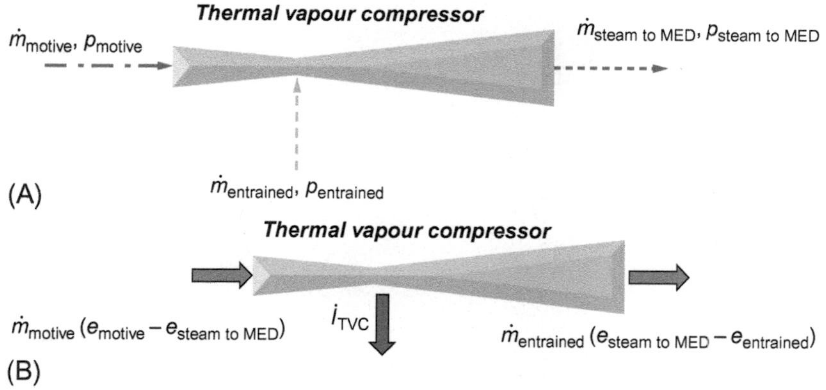

Fig. 12.7 (A) Process scheme of a thermal vapor compressor. (B) Exergy model for a thermal vapor compressor.

In Eqs. (12.10) and (12.11), exergy balance and exergy performance for the TVC are respectively given.

$$\dot{m}_{\text{motive}} \cdot \left(e_{\text{motive}} - e_{\text{steam to MED}}\right) - \dot{m}_{\text{entrained}} \cdot \left(e_{\text{steam to MED}} - e_{\text{entrained}}\right) = \dot{I}_{\text{TVC}} \tag{12.10}$$

$$\eta_{\text{ex,TVC}} = \frac{\dot{m}_{\text{entrained}} \cdot \left(e_{\text{steam to MED}} - e_{\text{entrained}}\right)}{\dot{m}_{\text{motive}}\left(e_{\text{motive}} - e_{\text{steam to MED}}\right)} \tag{12.11}$$

– Feedwater preheating
As shown in Fig. 12.8, in a generic feed preheater the thermal exergy of feedwater is increased (together with its temperature) from $\dot{E}^{\text{th}}_{\text{feed,in}}$ to $\dot{E}^{\text{th}}_{\text{feed,out}}$ by recovering the thermal exergy released by condensing a fraction of the distillate produced within the coupled effect $\dot{E}^{\text{th}}_{\text{D,j to ph}}$. By means of the exergy balance presented in Eq. (12.12), it is possible to assess the amount of exergy destroyed mainly due to the heat transfer across a finite temperature difference between the two streams. The exergy efficiency of a preheater can be calculated by the simple expression provided in Eq. (12.13).

$$\dot{E}^{\text{th}}_{\text{feed,out}} - \dot{E}^{\text{th}}_{\text{feed, int}} + \dot{I}_{\text{ph}} = \dot{E}^{\text{th}}_{\text{D, j to ph}} \tag{12.12}$$

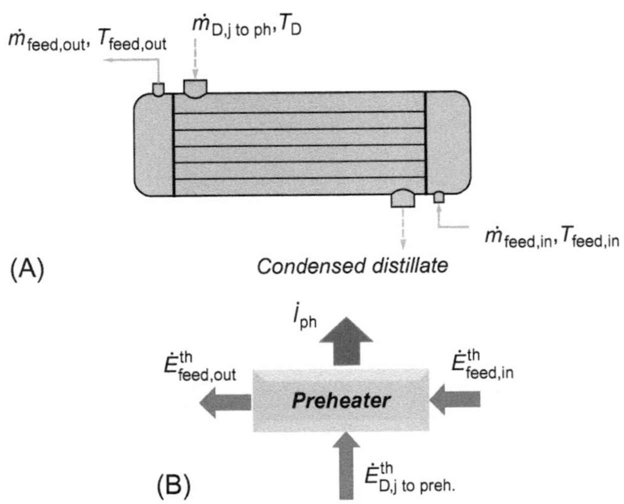

Fig. 12.8 (A) Process scheme of a generic feed preheater. (B) Exergy model for a generic feed preheater.

$$\eta_{ex,ph} = \frac{\dot{E}^{th}_{feed,\,out} - \dot{E}^{th}_{feed,\,int}}{\dot{E}^{th}_{D,\,j\,to\,ph}} \tag{12.13}$$

– Distillate generation within the 1st effect

A schematic representation of the exergy flows entering and exiting the 1st effect is given in Fig. 12.9. The exergy balance for this effect is expressed by Eq. (12.14), whereas the exergy efficiency may be calculated based on Eq. (12.15).

Looking at Fig. 12.9 it may be observed that exergy flows entering the 1st effect are related to the following contributions: (i) the thermal exergy rate $\left(\dot{E}_{steam\ to\ MED} - \dot{E}_{cond\ to\ boiler}\right)$ released by the supplied steam flow (exiting the TVC) which condensates the tube side of the falling film evaporator and (ii) the thermal exergy of feedwater, $\dot{E}^{th}_{feed,1}$.

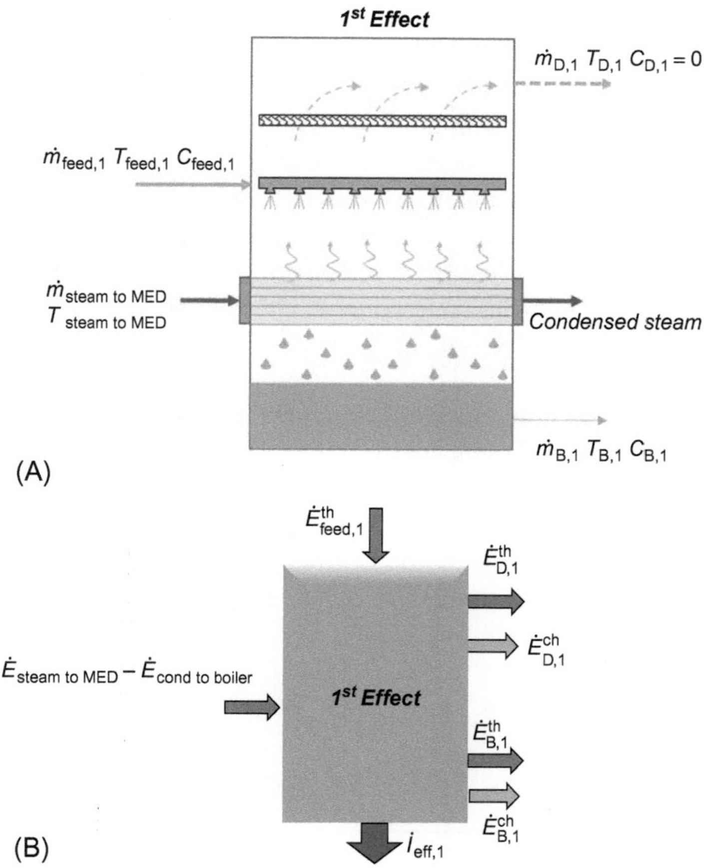

Fig. 12.9 (A) Process scheme of the 1st effect. (B) Exergy model for the 1st effect.

Some additional consideration is needed for the thermal exergy of the distillate produced in the 1st effect, that is, $\dot{E}_{D,1}^{th}$. According to the definition of the exergy efficiency given, for the examined effect, in Eq. (12.15), the thermal exergy content of the distillate is intended as a sort of "nonextracted" (i.e., nonconsumed) fraction of the thermal exergy supplied to the effect, which will be used in successive effects to generate further distillate. This interpretation clarifies the presence of $\dot{E}_{D,1}^{th}$ as a subtractive term at the denominator of the exergy efficiency formula, rather than as a positive term at the numerator (where only the useful exergy product of the effect, that is, the chemical exergy of distillate $\dot{E}_{D,1}^{ch}$ is included).

$$\dot{E}_{feed,1}^{th} + \left(\dot{E}_{\text{steam to MED}} - \dot{E}_{\text{cond to boiler}}\right) - \left(\dot{E}_{D,1}^{th} + \dot{E}_{D,1}^{ch} + \dot{E}_{B,1}^{th} + \dot{E}_{B,1}^{ch}\right) = \dot{I}_{eff,1} \quad (12.14)$$

$$\eta_{ex,\,eff\,1} = \frac{\dot{E}_{D,1}^{ch}}{\dot{E}_{feed,1}^{th} + \left(\dot{E}_{\text{steam to MED}} - \dot{E}_{\text{cond to boiler}}\right) - \dot{E}_{D,1}^{th}} \quad (12.15)$$

– Distillate generation within the generic "jth" effect

As regards a generic "*j*th" effect, for $j = 2$ to N, a simple scheme of exergy flows entering and exiting the effect is presented in Fig. 12.10. Also, the exergy balance and the analytical expression of exergy efficiency are provided in Eqs. (12.16) and (12.17), respectively. For a generic effect, the supplied exergy is constituted by the following contributions: (i) the thermal exergy of feedwater $\dot{E}_{feed,j}^{th}$ (due to the parallel-feed configuration of the examined plant), (ii) the thermal exergy $\dot{E}_{D,\ (j-i)\ to\ j}^{th}$ of the distillate produced at the "$(j-1)$th" effect (excluding the fraction eventually used for feedwater preheating), and (iii) variation of thermal exergy of the brine flow, $\left(\dot{E}_{B,\,(j-1)}^{th} - \dot{E}_{B,j}^{th}\right)$.

The exergy product of the "*j*th" effect is represented by the chemical exergy of the distillate, $\dot{E}_{D,j}^{ch}$. As already explained for the 1st effect, as well as for the generic "*j*th" effect, the thermal exergy of the distillate $\dot{E}_{D,j}^{th}$, is not considered a useful product but as a "nonextracted" fraction of the exergy input.

$$\dot{E}_{feed,j}^{th} + \dot{E}_{D,\ (j-i)\ to\ j}^{th} + \dot{E}_{B,\ (j-i)}^{th} - \left(\dot{E}_{D,j}^{th} + \dot{E}_{D,j}^{ch} + \dot{E}_{B,j}^{th} + \dot{E}_{B,j}^{ch}\right) = \dot{I}_{eff,j} \quad (12.16)$$

$$\eta_{ex,\,eff\,j} = \frac{\dot{E}_{D,j}^{ch}}{\dot{E}_{feed,j}^{th} + \dot{E}_{D,\ (j-i)\ to\ j}^{th} - \dot{E}_{D,j}^{th} + \left(\dot{E}_{B,\ (j-i)}^{th} - \dot{E}_{B,j}^{th}\right)} \quad (12.17)$$

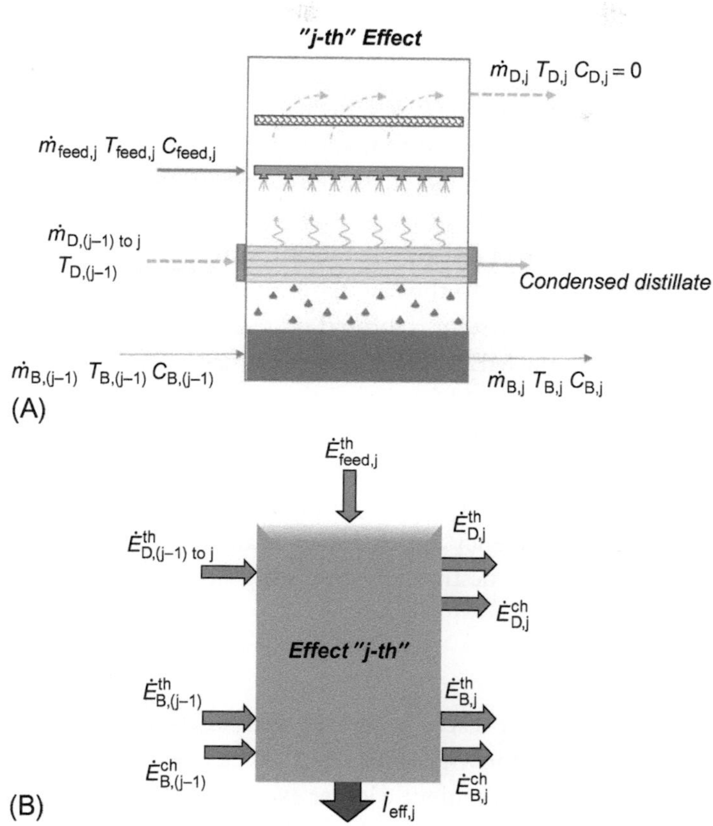

Fig. 12.10 (A) Process scheme of the "jth" preheater. (B) Exergy model for the "jth" preheater.

12.7.3.1 Exergy Analysis for Subprocesses of the Reference MED–TVC

Once defined the general formulation of exergy balances and exergy efficiencies for each of the main subprocesses occurring in an MED unit, a detailed exergy analysis is carried out for the reference MED–TVC, based on its actual operating data examined in detail in Section 12.6.

TVC of Entrained Steam

According to Eq. (12.18), the exergy performance of TVC is equal to 15.64%.

$$\eta_{ex,\,TVC} = \frac{\dot{m}_{entrained} \cdot \left(e_{steam\ to\ MED} - e_{entrained}\right)}{\dot{m}_{motive} \cdot \left(e_{motive} - e_{steam\ to\ MED}\right)} = \frac{709.6}{4537} \cong 0.156 \qquad (12.18)$$

This poor performance highlights that a high exergy destruction rate occurs within this component. Two aspects can be here remarked:

– Despite the low exergy efficiency, the inclusion of the TVC is beneficial in terms of overall exergy performance of the plant, when compared to the case of directly producing the whole amount of low grade steam by methane boilers. In fact, the conversion of chemical exergy of methane into thermal exergy of the supplied steam would achieve an even lower exergy efficiency, due to the low temperature of the required steam;
– Although the TVC system is located upstream the whole energy conversion chain in the MED-TVC system, its very low efficiency does not represent an upper bound for the efficiency of the whole plant, since its layout is not "strictly sequential," due to the presence of exergy recirculating flows (such as the entrained steam). However, as evident from a comparison between Eqs. (12.9) and (12.18), the plant achieves an overall exergy efficiency much lower than the TVC alone, due to the additional exergy destructions occurring along the evaporation processes in each effect.

Feedwater Preheating

Exergy performance for each preheater was assessed on the basis of Eq. (12.13). Results are shown in Fig. 12.11. As can be seen, the down condenser and the preheater 5 achieve a lower exergy efficiency than preheaters 1, 2, 3, and 4. Exergy destruction in the feed preheating process is related to heat transfer across a finite temperature difference between a fraction of the produced distillate which releases its condensation heat and the feedwater being preheated. In the specific case, by analyzing the temperature profile for each preheater, a greater temperature difference is observed for preheaters 5 and down condenser with respect to the others, thus justifying

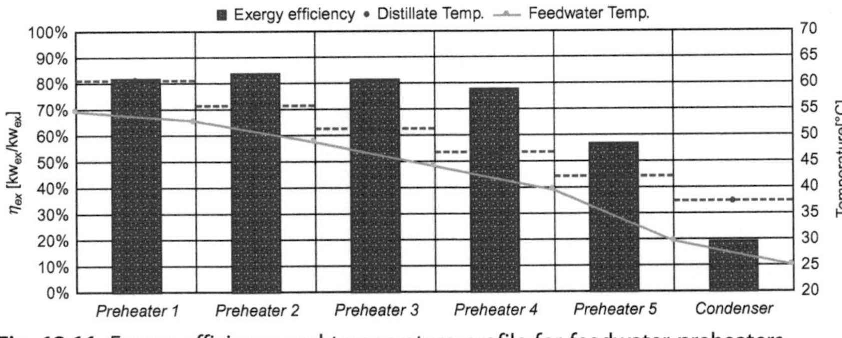

Fig. 12.11 Exergy efficiency and temperature profile for feedwater preheaters.

the lower values of their exergy performance. It can be observed that in most of the preheaters quite high exergy efficiencies are achieved, due to the "tightly" designed temperature profiles between the exchanging fluids.

Exergy Analysis Results at the Whole Effect Level

Exergy flows related to brine, distillate, and feedwater streams for each effect are shown in Table 12.3, also providing the values of specific exergy. It is worth noting that:

- The specific chemical exergy of brine $e_{B,j}^{ch}$ decreases when passing from the first effect to the last: this trend is related to the reduction in brine concentration C_{brine}, as observed from the values provided in Table 12.3. In a PF–MED configuration like the one examined in detail in the present chapter, pure water is not separated from brine exiting the previous effect as occurs in an FF–MED configuration. As a consequence, no increase in salt concentration of the brine is observed when passing from one effect to the next one. Conversely, the chemical exergy flow rate of brine $\dot{E}_{B,j}^{ch}$ increases from the 1st effect to the last, due to the gradual increase in the mass flow rate of brine that overcomes the influence of the aforementioned slight decrease in the specific chemical exergy.
- The specific thermal exergy of brine, $e_{B,j}^{th}$, decreases when passing from the 1st effect to the last, due to the decrease in brine temperature. When combined with the gradual increase in the brine mass flow rate, the above trend leads to a nonmonotone behavior of the thermal exergy of brine flow, which initially increases and, after a maximum value at effects no. 6 and 7, gradually decreases in the successive effects.
- A constant value is obtained for specific chemical exergy of the distillate $e_{D,j}^{ch}$, due to the assumption of a salts-free distillate in all the effects. In terms of exergy flow, a decrease in the chemical exergy of the distillate $\dot{E}_{D,j}^{ch}$ is observed passing from the 1st effect to the last, due to the reduction in the amount of distillate generated at lower temperatures.

In Fig. 12.12, values of exergy efficiency for each effect are shown, calculated based on Eqs. (12.15) and (12.17). As can be seen, the 2nd law-based efficiency ranges between 16% and 22%, indicating that only a small fraction of the thermal exergy used to evaporate feedwater within each effect is converted into chemical exergy of the distillate product.

Main Outcomes of Exergy Analysis

In this section, some brief interpretations of the results of exergy analysis are proposed. Based on the results of the exergy analysis carried out at the

Table 12.3 Exergy analysis results for the examined case

	Brine				Distillate				Feed	
Effect	$e^{th}_{B,j}$ (kJ_{ex}/kg)	$\dot{E}^{th}_{B,j}$ (kW_{ex})	$e^{ch}_{B,j}$ (kJ_{ex}/kg)	$\dot{E}^{ch}_{B,j}$ (kW_{ex})	$e^{th}_{D,j}$ (kJ_{ex}/kg)	$\dot{E}^{th}_{D,j}$ (kW_{ex})	$e^{ch}_{D,j}$ (kJ_{ex}/kg)	$\dot{E}^{ch}_{D,j}$ (kW_{ex})	$e^{th}_{F,j}$ (kJ_{ex}/kg)	$\dot{E}^{th}_{F,j}$ (kW_{ex})
1	9.12	147.7	9.01	7.786	266.6	2666.0	51.48	28.60	5.924	155.2
2	8.65	282.1	8.79	15.29	257.5	2523.8	51.48	28.03	5.924	155.2
3	7.72	378.2	8.72	22.79	242.5	2376.2	51.48	28.03	5.290	138.6
4	6.83	446.3	8.72	30.38	227.8	2065.7	51.48	28.31	5.290	138.6
5	5.92	488.1	8.29	36.46	212.4	1932.4	51.48	26.02	3.777	98.95
6	5.15	510.5	8.15	43.15	197.3	1603.7	51.48	26.88	3.777	98.95
7	4.35	510.2	7.66	47.97	181.4	1487.3	51.48	23.45	2.482	65.02
8	3.65	492.5	7.39	53.23	165.2	1141.7	51.48	24.59	2.482	65.02
9	2.98	459.1	6.86	56.41	148.8	1071.2	51.48	20.59	1.449	37.97
10	2.41	414.6	6.61	60.84	132.1	687.8	51.48	22.02	1.449	37.97
11	1.84	355.4	5.98	61.66	115.2	656.6	51.48	16.30	0.147	3.85
12	1.35	287.8	5.62	63.93	95.8	95.8	51.48	18.02	0.147	3.85

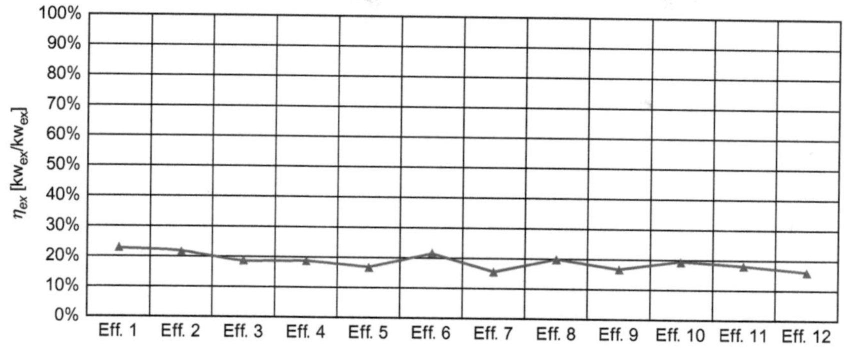

Fig. 12.12 Exergy efficiency of each effect.

subprocesses level, two main criticisms may be identified in the proposed layout:

- The largest contribution to exergy destruction is related to the TVC. Consuming high-pressure motive steam to produce the low-pressure steam supplied to the 1st effect is intrinsically inefficient, even considering a properly designed steam jet ejector. In order to overcome this inherent limitation of the plant, rethinking of the heat supply strategy would be absolutely needed, designing the system to be driven by low-pressure steam produced exploiting heat cascades from a power cycle or a higher temperature industrial process. This need would be even more evident if the boundary of the analyzed system were extended to comprise the additional exergy destructions occurring in the methane boilers.
- The conversion, taking place in the evaporators, of thermal exergy released by condensing vapor into chemical exergy of the distillate is rather inefficient. Possible improvements based on designing the evaporators to operate with a lower temperature difference are limited by both thermodynamic constraints (i.e., related to the Boiling Point Elevation, which consumes part of the driving force available for heat transfer) and cost considerations.

12.8 ON THE EXERGY COST ACCOUNTING METHOD

Exergy analysis provides insights on the location and magnitude of the thermodynamic losses occurring within an energy conversion system. By this tool, however, it is not possible to assess the contribution of each material or energy flow to the total exergy consumption of external resources. In this sense, the use of rigorous and systematic cost accounting techniques provides much more comprehensive results, since it not only reflects the intrinsic exergy efficiency at the component or subprocess level, but also considers the functional interactions among components which depend on the peculiar design of the plant layout. The same component, in fact, even achieving the same exergy efficiency, may contribute differently to the consumption of external resources if located in a different position along the "energy conversion chain" [50,51].

One of the most reliable and comprehensive thermoeconomic approaches for the analysis of complex energy conversion systems is the well-known "Theory of Exergetic Cost" [52,53].

The Exergetic Cost of a material stream or energy flow represents the amount of external exergy consumed by an energy conversion system to obtain it according to a well specified process and sequence of components. Exergetic Costs are calculated based on a rigorous accounting of all the exergy destructions involved along the plant to obtain the examined stream or energy flow. Obviously, the more inefficient the energy conversion process that generates the stream, the higher the amount of irreversibility "charged" on it and, consequently, its exergetic cost (per unit of exergy). By means of this method, insights on the cost formation process are obtained and a clear view of the subprocesses (or the components or the specific inefficiencies) mostly contributing to an increase in the energy consumption of the system may be derived.

The resulting model essentially consists of the following set of equations:

- *Exergy Cost Conservation* (one for each subprocess) that, according to the conservative nature of costs, imposes that the cost of all the exergy resources entering a component equals the cost of the flows exiting the same component [52,53].
- *Auxiliary equations*, formulated based on a number of different principles (exergy extraction, multiple outputs, external assessment) and included whenever in a plant component we have several products (with unknown exergoeconomic costs), thus being insufficient the use of one single "cost balance" equation [54].
- *Allocation of Residues*: in the theory of Thermoeconomics, the term "residues" is used to indicate the exergy flows generated within the process which do not represent either the final products of the plant or useful output to be supplied to other components, thus representing exergy flows that are rejected from the system to the surrounding environment [55,56]. Intuitive examples may be represented by the thermal or chemical exergy of the brine exiting the last effect of a desalination process, or the cooling water exiting from the condenser of a power plant.

From a cost accounting perspective, exergy costs of such flows are not allocated on the component where the exergy is uselessly discarded, but assigned as an additional input cost to the upstream components where the residue has been generated.

12.8.1 Exergy Cost Accounting of the Reference MED–TVC

For the examined MED–TVC plant, the exergy cost equations are presented below in detail for each subprocess. With regard to the adopted symbolism, the unit exergy cost of the "*i*th" component product, indicated as k_i^*, expresses the amount of the external exergy consumed by the plant for producing a unit of exergy of the product itself (being thus measured in kW_{ex}/kW_{ex}).

✓ *Exergy Cost Balance for the TVC*

$$k^*_{\text{motive}}\dot{E}_{\text{motive}} + k^*_{\text{entrained}}\dot{E}_{\text{entrained}} = k^*_{\text{steam to MED}}\dot{E}_{\text{steam to MED}} \tag{12.19}$$

Auxiliary equations for TVC:

$$k^*_{\text{th,D,12}} = k^*_{\text{entrained}} \tag{12.20}$$

The auxiliary equation for the TVC imposes that the thermal exergy cost of entrained vapor is equal to the exergy unit cost of distillate produced in the 12th effect. An external assessment is required for the unit exergy cost of motive steam, since the boundary of the analyzed system does not comprise the methane boiler used to produce steam. As a reasonable assumption, considering an exergy efficiency of the boiler equal to 0.45, k^*_{motive} results equal to $2.22\,kW_{ex}/kW_{ex.}$

✓ *Exergy Cost Balance for the 1st effect*

$$k^*_{\text{steam to MED}}\left(\dot{E}_{\text{steam to MED}} - \dot{E}_{\text{cond to boiler}}\right) + k^*_{\text{th, feed\_1}}\dot{E}^{\text{th}}_{\text{feed\_1}} + \frac{1}{12}k^*_{\text{th,B,12}}\dot{E}^{\text{th}}_{\text{B,12}} = k^*_{\text{ch,D,1}}\dot{E}^{\text{ch}}_{\text{D,1}} + k^*_{\text{th,D\_1}}\dot{E}^{\text{th}}_{\text{D,1}} + k^*_{\text{ch,B,1}}\dot{E}^{\text{ch}}_{\text{B,1}} + k^*_{\text{th,B,1}}\dot{E}^{\text{th}}_{\text{B,1}} \tag{12.21}$$

Auxiliary equations for the 1st effect

$$k^*_{\text{ch,B,1}} = 0 \tag{12.22a}$$

$$k^*_{\text{ch,D,1}} = k^*_{\text{th,D,1}} \tag{12.22b}$$

$$k^*_{\text{th,B,1}} = k^*_{\text{th,D,1}} \tag{12.22c}$$

Exergy balance for the 1st preheater is given in Eq. (12.21). Referring to Fig. 12.9B, all the exergy costs of input flows are summed up and imposed equal to the total cost of the exiting exergy flows.

As can be seen from Eq. (12.21), the thermal exergy of the brine exiting the 12th effect is considered as a residue flow and its exergetic cost, $k^*_{\mathrm{th,B,12}}\dot{E}^{\mathrm{th}}_{\mathrm{B,12}}$, is uniformly reallocated as an additional input cost to all the upstream effects, following the previously mentioned approach proposed in [56] for the allocation of residues.

In Eqs. (12.22a)–(12.22c), auxiliary equations for the first effect are expressed. In particular:

- Eq. (12.22a) imposes that specific cost of chemical brine is null. This condition is coherent with the assumption to consider the chemical exergy of the brine as an unuseful product of each effect (being finally rejected back to sea).
- Eqs. (12.22b) and (12.22c) impose the same unit exergetic cost for the thermal and chemical exergy of the distillate and the thermal exergy of the brine. The rationale of this assumption lies in the fact that, at the proposed plant disaggregation level (with the "effect" modelled as a unique elementary component), it is not possible to distinguish between the subprocesses occurring within the effect and leading to the production of these exergy flows. Then, having these exergy outputs been obtained by a same energy conversion chain, they may be assumed to have implied a same "exergy destruction per unit exergy" and they are consequently assigned a same unit cost.

✓ *Exergy Cost Balance for the "jth" effect*

$$k^*_{th,feed\,j}\dot{E}^{th}_{feed,j} + k^*_{th,D,\int j-1}\dot{E}^{th}_{D,(j-1)\ to\ next}k^*_{ch,B,j-1}\dot{E}^{ch}_{B,j-1} + k^*_{th,B,j-1}\dot{E}^{th}_{B,j-1} + \frac{1}{12}k^*_{th,B,12}\dot{E}^{th}_{B,12} = k^*_{ch,D,j}\dot{E}^{ch}_{D,j} + k^*_{th,D,j}\dot{E}^{th}_{D,j} + k^*_{ch,B,j}\dot{E}^{ch}_{B,j} + k^*_{th,B,j}\dot{E}^{th}_{B,j} \tag{12.23}$$

Auxiliary Equations for the "jth" effect

$$k^*_{ch,B,j} = 0 \tag{12.24a}$$

$$k^*_{ch,D,j} = k^*_{th,D,j} \tag{12.24b}$$

$$k^*_{th,B,j} = k^*_{th,D,j} \tag{12.24c}$$

✓ *Exergy Cost Balance for the generic preheater*

$$k^*_{\mathrm{th,\,feed\,in}}\dot{E}^{\mathrm{th}}_{\mathrm{feed\,in}} + k^*_{\mathrm{th,D}\,j}\dot{E}^{\mathrm{th}}_{\mathrm{D\,to\,preheater}} = k^*_{\mathrm{th,\,feed\,out}}\dot{E}^{\mathrm{th}}_{\mathrm{feed\,out}} \tag{12.25}$$

Referring to Fig. 12.8, Eq. (12.25) expresses that cost balance for the generic preheater.

All the equations presented above constitute a balanced set of linear equations that can be easily solved by a software.

12.8.2 Results of Exergy Cost Accounting for the Examined Plant

Once defined the exergy cost accounting model for the MED–TVC process, it was applied to the examined case and solved by means of the Engineering Equation Solver [57].

In Table 12.4, specific costs of thermal exergy of the motive steam, entrained vapor, and compressed vapor supplied to the first effect of the MED section are shown.

As concerns the specific cost of thermal exergy of the motive steam, the external assessment $k^*_{motive} = 2.22\ kW_{ex}/kW_{ex}$ was used in this analysis. This condition states that motive steam is produced by a boiler whose exergy performance is equal to 0.45 (in particular, $k^*_{motive} = \frac{1}{\eta_{ex,\,boiler}} = \frac{1}{0.45} = 2.22$).

From the exergy cost balance of TVC presented in Eq. (12.18), the specific cost of thermal exergy exiting from TVC and supplied to the first effect $k^*_{steam\ to\ MED}$, results equal to $15.84\ kW_{ex}/kW_{ex}$. This value indicates that the unit exergy of steam supplied to the 1st effect implies the consumption of nearly $15.84\ kW_{ex}$ of external exergy resources. This value is nearly seven times higher than the exergy cost of motive steam, due to the extremely

Table 12.4 Specific exergy cost for the TVC unit

$\dot{E}_{motive}$ (kW_{ex})	6348
k^*_{motive} (kW_{ex}/kW_{ex})	2.22
$\dot{E}_{entrained}$ (kW_{ex})	354.3
$k^*_{entrained}$ (kW_{ex}/kW_{ex})	88.73
$\dot{E}_{steam\ to\ MED}$ (kW_{ex})	2875
$k^*_{steam\ toMED}$ (kW_{ex}/kW_{ex})	15.84

low exergy efficiency and the high exergy destruction by irreversibility generated during TVC of the entrained steam.

The values obtained for the specific exergetic cost and the exergetic cost flow for each effect are presented in Table 12.5.

- The unit cost of the chemical exergy of the distillate $k_{ch,D}^{*}$ increases sharply from 20.05 to 88.73 kW_{ex}/kW_{ex} when passing from the first effect to the last. Analogous results are observed for the specific cost of thermal exergy of the distillate and brine, according to Eqs. (12.24a)–(12.24c). In spite of the parallel configuration adopted to feed salty water to the different effects, the sequential exploitation (along the effects from 1 to 12) of the thermal exergy of compressed steam (supplied to drive the 1st effect) and the high irreversibility occurring in each effect when converting thermal exergy into chemical exergy of the distillate induce a gradual increase in the unit exegetic cost of thermal exergy. Then, in cumulative terms, a unit of chemical exergy of the distillate produced in the 12th effect is "charged" with a higher amount of exergy destruction, compared to the distillate produced in the 1st effect, thus justifying the trend observed in the specific exergetic cost.
- The specific cost of the thermal exergy of the feed k_{feed}^{*}, dramatically increases along the feed preheating line, passing from 69.66 kW_{ex}/kW_{ex} (in the 1st and 2nd effects) to 461.26 $kW_{ex}/kW_{ex,}$ (in the 11th and 12th effects). As previously explained, the thermal exergy of a small fraction of the distillate is consumed as exergy source along the feedwater preheating process. Since $k_{th,D,j}^{*}$ values increase steeply when moving forward along the MED plant (from the 1st to the 12th effect), the preheater 5 (see plant scheme in Fig. 12.11) consumes a much more "precious" or costly exergy input to preheat the feedwater. Then an increase in the specific cost of thermal exergy of the preheated feed is well explained, being also augmented by the lower exergy efficiencies achieved by the last preheaters (see Fig. 12.11).

  The steep increase in specific cost of thermal exergy from 134.91 to 461.26 kW_{ex}/kW_{ex}, observed when passing from the 10th to the 11th effect, is caused by the extremely low exergy efficiency of the down condenser.

12.8.3 On the Possible Exploitation of Additional Thermoeconomic Analyses

In the previous section, exergy cost accounting has been proposed as a method for assessing the cost sustained for freshwater production in terms

Table 12.5 Specific exergy cost and exergy cost profiles along the MED-TVC plant

	Brine		Distillate				Feed	
Effect	$k^*_{th,B,j}$ (kW_{ex}/kW_{ex})	$k^*_{th,B,j}\dot{E}^{th}_{B,j}$ (kW_{ex})	$k^*_{th,D,j}$ (kW_{ex}/kW_{ex})	$k^*_{th,D,j}\dot{E}^{th}_{D,j}$ (kW_{ex})	$k^*_{ch,D,j}$ (kW_{ex}/kW_{ex})	$k^*_{ch,D,j}\dot{E}^{ch}_{D,j}$ (kW_{ex})	k^*_{feed} (kW_{ex}/kW_{ex})	$k^*_{feed}\dot{E}^{th}_{feed}$ (kW_{ex})
1	20.05	2960.2	20.05	53446.6	20.05	573.3	69.66	10810.7
2	23.92	6746.7	23.92	60363.0	23.92	670.3	69.66	10810.7
3	28.34	10718.2	28.34	67336.4	28.34	794.2	74.49	10323.5
4	32.84	14657.4	32.84	74083.1	32.84	92.8	74.49	10323.5
5	37.78	18439.5	37.78	72999.7	37.78	983.1	88.57	8763.8
6	42.32	21602.9	42.32	78495.0	42.32	1137.6	88.57	8763.8
7	48.16	24573.1	48.16	71634.1	48.16	1129.5	107.54	6992.7
8	53.63	26414.3	53.63	76189.2	53.63	1318.9	107.54	6992.7
9	60.16	27623,1	60.16	64445.8	60.16	1238.8	134.91	5122.5
10	67.09	27818.0	67.09	68249.6	67.09	1477.4	134.91	5122.5
11	75.79	26936.9	75.79	49763.5	75.79	1235.4	461.26	1776.9
12	88.73	25538.7	88.73	53532.3	88.73	1598.7	461.26	1776.9

of external exergy resource consumption. The same conceptual procedures could be extended to assess the amount of "monetary resources" needed for operating an MED process or, more in general, any energy conversion system. This well-established method is commonly named in the scientific literature as "Exergoeconomic Cost Accounting" [50,51].

Similar to the procedure described for exergetic cost accounting, Exergoeconomics also assumes that costs (in monetary units) distribute along the plant proportionally to the exergy content of material streams, thus providing a rigorous thermodynamic basis to allocate costs on the different flows. Such an approach can be used for rational pricing of different products in multioutput systems like dual purpose plant or, more generally, polygeneration systems. Exergoeconomic cost quantifies the monetary resources, in terms of capital expenditure (related to equipment purchase and installation) and operating costs (as fuels, operation and maintenance), sustained when the system produces the unit of exergy of its product. Also, in this case, preliminary identification of the *productive history* of each product within the system is necessary for proper cost assessment. In the past decade, some papers in the literature have focused on the exergoeconomic cost accounting of desalination systems. In order to mention just some of the papers reported on this specific topic, in [58], detailed cost accounting for an FF–MED plant was carried out. In [47], Exergoeconomics was proposed as a method for assessing unit cost of freshwater and electricity in a condensing steam cycle coupled to an MED–TVC. In [59], a polygeneration system powered by renewable energy sources for small isolated communities has been investigated by exergoeconomic analysis.

Also, an exergoeconomic method can be easily integrated into optimization procedure in order to support an efficient, cost-effective, and environmental-friendly design and operation of an energy system. Some applications have focused on trigeneration systems [60,61], desalination processes coupled with nuclear power plants [62], or refrigeration plants [63,64].

During the past decades, *Thermoeconomic Diagnosis* has been proposed as a promising approach for the diagnosis of malfunctions occurring within an energy conversion system [65–68]. In this approach the term "malfunctions" is meant to indicate degradation phenomena which are detrimental to the performance of the system, without necessarily implying any interruption in its operation. The method aims at revealing, in a generic off-design operating condition indicated by the distributed control system, which is the "faulty" component within the system and the particular fault (e.g., fouling in heat exchangers, etc.), also quantifying, in case of multiple simultaneous faults, the effects on the overall system performance induced by each specific fault. With this aim, opportune indicators, such as the "malfunction cost" in [66,67], are defined in order to detect the faulty component during the diagnosis procedure.

Thermoeconomic diagnosis has been extensively applied to power plants [68–70], but it is still under investigation to assess its potential application in other fields, such as in refrigeration systems [71–74]. The main obstacles for a reliable application of this technique are: (i) the need for properly filtering the mutual interactions among components, to detect correctly the malfunctioning ones, (ii) the possible distortions induced by the presence of several simultaneous faults, and (iii) the need for a preliminary filtration of the effects of control systems [75,76].

From the above considerations, it is apparent how the research in this field is still open and great efforts will be required in the future to increase the reliability of these techniques and enlarge their possible fields of application.

12.9 CONCLUSIONS

In this chapter, exergy analysis and exergy cost accounting are proposed for the in-depth interpretation of thermodynamic behavior of MED processes. The exergy method allows the analyst to assess in which part of the process the highest exergy destruction occurs. Exergy Cost accounting provides a comprehensive view of the cost formation process, allowing the plant designer or manager to calculate, on a rigorous thermodynamic basis, the exergetic (or economic) amount of external resources consumed to produce each specific material stream or exergy flow. The fundamentals of both methods are thoroughly provided within the chapter.

A 12 effects MED parallel feed coupled with thermal vapor compression has been selected as a case study. The low exergy efficiency of the whole MED process suggests that high exergy destruction occurs when converting thermal exergy of motive steam into chemical exergy of freshwater.

In order to provide a clear interpretation of the reasons for the low exergy efficiency of the MED system, a conveniently high disaggregation level has been adopted, by analyzing separately TVC, feedwater preheating, and distillate generation. As concerns thermal vapor compression, the use of high-pressure steam to produce low-pressure steam supplied to the MED process resulted in being among the main responsible subprocesses for the poor plant performance. This result testifies the great convenience in supplying such low temperature processes, like the examined MED plant, by means of cogenerated heat recovered from power plants or waste heat cascaded from other industrial processes. Also, the conversion of thermal exergy into chemical exergy, which takes place in the effects, resulted not efficient from a thermodynamic viewpoint, mainly due to the finite temperature difference between the heat exchanging fluid, which in its turn is limited by thermodynamic and cost constraints.

By means of exergy cost accounting, the cost formation process of the distillate could be clearly depicted. A substantial increase in specific cost of the freshwater product is observed from the first effect to the last, as a consequence of the irreversibility being cumulated along the quasi-sequential production process.

The applicative example confirmed the potential of exergy and thermoeconomic analysis to provide a comprehensive understanding of plant behavior, allowing the plant designer to identify the possible margins for improvement. Also, it is widely recognized in the literature that the proposed methodologies can be conveniently used during plant operation, to allow for rational pricing of the product (especially in case of dual purpose configurations) and efficient diagnosis of malfunctions.

APPENDIX DETAILS ON THE THERMODYNAMIC MODEL USED FOR NACL SOLUTIONS

As shown, exergy analysis of the desalination process requires an accurate evaluation of thermodynamic properties of solutions. In this work, NaCl solutions were considered as the model for brine and seawater, but similar considerations can be easily extended to other electrolyte solutions. As is well known, the chemical potential of the generic solute MX involved in an electrolyte solution is defined as [77]

$$\mu_{MX} = \mu_{MX}^0 + \nu \cdot R \cdot T \cdot \ln(m_{\pm}\gamma_{\pm}) \tag{A1}$$

In Eq. (A1), μ_{MX}^0 is the chemical potential of the solute in the standard state (i.e., an ideal 1 molal solution), ν is the solute stoichiometric coefficient obtained as the sum of the stoichiometric coefficients for each single ion ($\nu = \nu_+ + \nu -$, equal to 2 in the case of NaCl salt), $m_{\pm}$ the mean molality and $\gamma_{\pm}$ the mean activity coefficient. As regards the mean molality and activity coefficient, they are defined as the geometric mean of the ions properties [77] (for the examined case, Na^+ and Cl^- ions are considered). Similarly, the chemical potential of the solvent (in this case water) is defined as

$$\mu_s = \mu_s^0 + RT \ln(a_s) \tag{A2}$$

in which a_s is the solvent activity (water in the analyzed case) and μ_s^0 is the solvent chemical potential in the standard state. The activity of the solvent is related to the osmotic coefficient ϕ according to Eq. (A3):

$$\varphi = -\frac{1000 \ln(a_s)}{M_w \sum_i \nu_i m_i} \tag{A3}$$

where M_W is the molecular weight of the solvent (g/mol), m_i is the "ith" ion molality in the solution (mol/kg), and ν_i the dissolution stoichiometric coefficients of the "ith" ion. In order to estimate the electrolyte activity coefficient, that is, $\gamma_\pm$, and osmotic coefficient for the solvent, that is, ϕ, Pitzer's model is adopted.

Pitzer's ion interaction model [77] is widely used to describe the behavior of electrolyte solutions. In Pitzer's model, the osmotic and salt activity coefficients are described by semi-empirical equations in which the interactions of all ions in solution are incorporated by means of concentration–independent interaction coefficients. This means that for an arbitrary composition of an aqueous electrolyte mixture, when the interaction coefficients are known, the activities and osmotic coefficients of all ions can be estimated using Pitzer's model.

For the generic electrolyte MX, Pitzer's equations for osmotic and activity coefficients are reported below:

$$\varphi = 1 - |z_M z_X| A_\varphi \frac{I^{0.5}}{1 + bI^{0.5}} + 2m\frac{\nu_M \nu_X}{\nu} B_{MX}^\varphi + 2m^2 \frac{(\nu_M \nu_X)^{3/2}}{\nu} C_{MX}^\varphi \tag{A4}$$

$$\ln(\gamma_\pm) = -|z_M z_X| A_\varphi \left(\frac{I^{0.5}}{1 + bI^{0.5}} + \frac{2}{b} \ln\left(1 + bI^{0.5}\right) \right) + 2m\frac{\nu_M \nu_X}{\nu} B_{MX}^\gamma + 3m^2 \frac{(\nu_M \nu_X)^{3/2}}{\nu} C_{MX}^\varphi \tag{A5}$$

with

$$B_{MX}^\varphi = B_{MX}^{(o)} + B_{MX}^{(1)} \exp\left(-\alpha I^{0.5}\right)$$

$$B_{MX}^\gamma = 2B_{MX}^{(o)} + \frac{2B_{MX}^{(1)}}{\alpha^2 I}\left[1 - \left(1 + \alpha I^{0.5} - \frac{\alpha^2 I}{2}\right) \exp\left(-\alpha I^{0.5}\right)\right] \tag{A6}$$

in which $m_0 = 1\,\text{mol/kg}$ is the standard molality, $b = 1.2\,\text{kg}^{1/2}/\text{mol}^{1/2}$ is an universal parameter; α is a numerical constant equal to 2 for univalent ions; I is the ionic strength.

A_ϕ is the Debye–Huckel parameter for the osmotic coefficient. $B_{MX}^{(0)}$, $B_{MX}^{(1)}$, and C_{MX} are adjustable parameters, called *ion-interaction parameters or virial coefficients*, The virial coefficients are a function of the electrolyte type, temperature, and pressure. For some salts, like NaCl and KCl [78,79], some correlations to evaluate the interaction parameters were found in the literature as a function of temperature and pressure. In the examined case of NaCl solution, reference can be found in [80].

REFERENCES

[1] World Health Organization Report: "Progress on sanitation ad drinking water," Available at: http://www.who.int/en/.

[2] Desaldata, www.DesalData.com.

[3] Mabrouk AN, Fath HES. Technoeconomic study of a novel integrated thermal MSF–MED desalination technology. Desalination 2015;371:115–25.

[4] Shahzad MW, Burhan M, Ang L, Ng KC. Energy-water-environment nexus underpinning future desalination sustainability. Desalination 2017;413:52–64.

[5] Ghaffour N, Missimer T, Amy GL. Technical review and evaluation of the economics of water desalination: current and future challenges for better water supply sustainability. Desalination 2013;309:197–207.

[6] Micale G, Rizzuti L, Cipollina A. Seawater Desalination Conventional and Renewable Energy Processes. Springer; 2009.

[7] SIDEM MED reference list. Available at: www.sidem-desalination.com.

[8] IDE Desalination Plant Reference List. Available at: http://www.ide-tech.com/wp-content/uploads/2016/06/E5643-107-24-stationery_web.pdf.

[9] Al-Karaghouli A, Kazmerski LL. Energy consumption and water production cost of conventional and renewable-energy-powered desalination processes. Renew Sust Energ Rev 2013;24:343–56.

[10] Semiat R. Energy issues in desalination processes. Environ Sci Technol 2008;42:8193–201.

[11] El-Dessouky HT, Ettouney HM. Fundamentals of Salt Water Desalination. 1st ed. Elsevier Science; 2002.

[12] Al-Shammiri M, Safar M. Multi-effect distillation plants: state of the art. Desalination, Vol. 126, pp. 45–99.

[13] Palenzuela P, Alarcón D, Zaragoza G. Concentrating Solar Power and Desalination Plants: Engineering and Economics of Coupling Multi-Effect Distillation and Solar Plants. Springer International Publishing; 2015.

[14] Kronenberg G, Lokiec F. Low-temperature distillation processes in single- and dual-purpose plants. Desalination 2001;136:189–97.

[15] Wu L, Hu Y, Gao C. Optimum design of cogeneration for power and desalination to satisfy the demand of water and power. Desalination 2013;324:111–7.

[16] Raluy RG, Serra L, Uche J, Valero A. Life-cycle assessment of desalination technologies integrated with energy production systems. Desalination 2004;167:445–58.

[17] European Commission. 2011/877/EU: commission Implementing Decision of 19 December 2011 establishing harmonised efficiency reference values for separate production of electricity and heat in application of Directive 2004/8/EC of the European Parliament and of the Council and repealing Commission Decision 2007/74/EC. Off J Eur Union 2011;343:91–6. 23.12.2011.

[18] Tamburini A, Cipollina A, Micale G, Piacentino A. CHP (combined heat and power) retrofit for a large MED-TVC (multiple effect distillation along with thermal vapour compression) desalination plant: high efficiency assessment for different design options under the current legislative EU framework. Energy 2016;115:1548–59.
[19] Cardona E, Piacentino A. Optimal design of cogeneration plants for seawater desalination. Desalination 2004;166(15):411–26.
[20] Hamed OA. Overview of hybrid desalination systems-current status and future prospects. Desalination 2005;186:207–14.
[21] Rensonnet T, Uche J, Serra L. Simulation and thermoeconomic analysis of different configurations of gas turbine (GT)-based dual-purpose power and desalination plants (DPPDP) and hybrid plants (HP). Energy 2007;32:1012–23.
[22] Lattemann S, Höpner T. Environmental impact and impact assessment of seawater desalination. Desalination 2008;220:1–15.
[23] Ophir A, Lokiec F. Advanced MED process for most economical sea water desalination. Desalination 2005;182:187–98.
[24] Sommariva C, Hogg H, Callister K. Environmental impact of seawater desalination: Relations between improvement in efficiency and environmental impact. Desalination 2007;167:439–44.
[25] Karagiannis IC. PG Soldatos Water desalination cost literature: review and assessment. Desalination 2008;233:448–56.
[26] Sambrailo D, Ivic J, Krstulovic A. Economic evaluation of the first desalination plant in Croatia. Desalination 2005;170:339–44.
[27] Reddy KV, Ghaffour N. Overview of the cost of desalinated water and costing methodologies. Desalination 2007;205:340–53.
[28] Al-Sahali M, Ettouney H. Developments in thermal desalination processes: design, energy, and costing aspects. Desalination 2007;214:227–40.
[29] Andrianne J, Alardin F. Thermal and membrane processes economics: selection for seawater desalination. Desalination 2002;153:305–11.
[30] Tzen E, Morris R. Renewable energy sources for desalination. Sol Energy 2003;75:375–9.
[31] Al Hawaj O, Darwish MA. A solar pond assisted multi-effect desalting system. Desalination 1994;99:119–35.
[32] El-Nashar AM, Samad M. The solar desalination plant in Abu Dhabi: 13 years of performance and operation history. Renew Energy 2014;263–74.
[33] Moser M, Trieb F, Fichter T. Potential of concentrating solar power plants for the combined production of water and electricity in MENA countries. J Sustain Develop Energy Water Environ Syst 2013;1:122–40.
[34] Li C, Goswami Y, Stefanakos E. Solar assisted sea water desalination: a review. Renew Sust Energ Rev 2013;19:136–63.
[35] Al-Karaghouli A, Renne D, Kazmerski LL. Solar and wind opportunities for water desalination in the Arab Regions. Renew Sust Energ Rev 2009;13:2397–407.
[36] Barbier E. Geothermal energy technology and current status: an overview. Renew Sust Energ Rev 2002;6:3–65.
[37] Ophir A. Desalination plant using low grade geothermal heat. Desalination 1982;40:125–32.
[38] K. Karytsas, the kimolos geothermal desalination project. In: Proceedings of the international workshop on possibilities of geothermal energy development in the Aegean Islands Region, Milos Island, Greece, 2002, pp. 206–219.
[39] Cipollina A, Micale G, Rizzuti L. A critical assessment of desalination operations in Sicily. Desalination 2005;182:1–12.
[40] A. Cipollina, M. Agnello, A. Piacentino, A. Tamburini, B. Ortega, P. Palenzuela, D. Alarcon, G. Micale A dynamic model for MED-TVC transient operation; Desalination pp. 234-257, 2017.

[41] Temstet C, Canton G, Laborie J, Durante A. A large high-performance MED plant in Sicily. Desalination 1996;105:109–14.
[41a] Available at: http://www.sciencedirect.com/science/article/pii/S0011916416311341-bb0185.
[42] Kotas TJ. The Exergy Method of Thermal Plant Analysis. Butterworths; 1985.
[43] Dinçer I, Rosen MA. Exergy Analysis of Heating, Refrigerating and Air Conditioning: Methods and Applications; 2015.
[44] Bejan A. Advanced Engineering Thermodynamics. John Wiley & Sons; 1997.
[45] Dinçer I, Rosen MA. EXERGY: Energy, Environment and Sustainable Development. Elsevier Science; 2012.
[46] Szargut J. Exergy Method: Technical and Ecological Applications. WIT Press; 2005.
[47] Catrini P, Cipollina A, Micale G, Piacentino A, Tamburini A. Exergy analysis and thermoeconomic cost accounting of a combined heat and power steam cycle integrated with a multi effect distillation-thermal vapour compression desalination plant. Energy Convers Manag 2017;149:950–65.
[48] Cipollina A, Micale G. Sustainable Energy From Salinity Gradients. Woodhead Publishing; 2016.
[49] Hassan AS, Darwish AM. Performance of thermal vapour compression. Desalination 2014;335:41–6.
[50] Valero A, Torres C. Thermoeconomic analysis. In: Frangopoulos CA, editor. Exergy, Energy System Analysis and Optimization, From Encyclopedia of Life Support System (EOLSS). Oxford: EOLSS Publishers; 2004. Available at: http://www.eolss.net.
[51] Torres C. Symbolic thermoeconomic analysis of energy systems. In: Frangopoulos CA, editor. Exergy, Energy System Analysis and Optimization, From Encyclopedia of Life Support System (EOLSS). Oxford: EOLSS Publishers; 2004. Available at: http://www.eolss.net.
[52] Valero A, Lozano MA, Serra L, Torres C. Application of the exergetic cost theory to the CGAM problem. Energy 1994;16:365–81.
[53] Lozano MA, Valero A. Theory of the exergetic cost. Energy 1993;18:939–60.
[54] Usón S, Valero A, Agudelo A. Thermoeconomics and industrial symbiosis. Effect of by-product integration in cost assessment. Energy 2012;45:43–51.
[55] Lozano MA, Valero A. Thermoeconomic Analysis of a Gas Turbine Cogeneration System. ASME Book no. H00874, vol. 30. WAM; 1993. p. 312–20. AES.
[56] Torres C, Valero A, Rangel V, Zaleta A. On the cost formation process of the residues. Energy 2008;33(2):144–52.
[57] Klein SA, et al. Engineering equation solver V9.810. Middleton, WI: Solar Energy Laboratory, F-Chart Software; 2010.
[58] Piacentino A. Application of advanced thermodynamics, thermoeconomics and exergy costing to a multiple effect distillation plant: in-depth analysis of cost formation process. Desalination 2015;371:88–103.
[59] Calise F, Dentice d'Accadia M, Piacentino A. Exergetic and exergoeconomic analysis of a renewable polygeneration system and viability study for small isolated communities. Energy 2015;92:290–307.
[60] Piacentino A, Cardona F. On thermoeconomics of energy systems at variable load conditions: integrated optimization of plant design and operation. Energy Convers Manag 2007;48:2341–55.
[61] Cardona E, Piacentino A. Optimal design of CHCP plants in the civil sector by thermoeconomics. Appl Energy 2007;84:729–48.
[62] Ansari K, Sayyaadi H, Amidpour M. Thermoeconomic optimization of a hybrid pressurized water reactor (PWR) power plant coupled to a multi effect distillation desalination system with thermo-vapor compressor (MED–TVC). Energy 2010;35:1981–96.

[63] Rezayan O, Behbahaninia A. Thermoeconomic optimization and exergy analysis of CO 2/NH 3cascade refrigeration systems. Energy 2011;36:888–95.
[64] Dentice d'Accadia M. Thermoeconomic optimization of a refrigeration plant. Int J Refrig 1998;21:42–54.
[65] Torres C, Valero A, Serra L, Royo J. Structural theory and thermoeconomic diagnosis: part I. On malfunction and dysfunction analysis. Energy Convers Manag 2002; 43(9–12):1503–18.
[66] Valero A, Correas L, Zaleta A, Lazzaretto A, Verda V, Reini M, et al. On the thermoeconomic approach to the diagnosis of energy system malfunctions: part 2. Malfunction definitions and assessment. Energy 2004;29(12–15):1889–907.
[67] Uson S, Valero A. Thermoeconomic Diagnosis of Energy Systems. Zaragoza, Spain: Prensa Universidad de Zaragoza; 2010.
[68] Usón S, Valero A. Thermoeconomic diagnosis for improving the operation of energy intensive systems: comparison of methods. Appl Energy 2011;88:699–711.
[69] Zhang C, Chen S, Zheng C, Lou X. Thermoeconomic diagnosis of coal fired plant. Energy Convers Manag 2007;48:405–19.
[70] Xua J-q, Yang T, Zhou K-y, Shi Yf. Malfunction diagnosis method for the thermal system of a power plant based on thermoeconomic analysis. Energy Sources A Recov Utiliz Environ Effects 2016;38(1):124–32.
[71] Piacentino A, Talamo M. Critical analysis of conventional thermoeconomic approaches to the diagnosis of multiple faults in air conditioning units: capabilities, drawbacks and improvement directions. A case study for an air cooled system with 120 kW capacity. Int J Refrig 2013;36(8):24–44.
[72] Piacentino A, Talamo M. Innovative thermoeconomic diagnosis of multiple faults in air conditioning units: methodological improvements and increased reliability of results. Int J Refrig 2013;36:2343–65.
[73] Piacentino A, Catrini P. Assessing the robustness of thermoeconomic diagnosis of fouled evaporators: sensitivity analysis of exergetic performance of direct expansion coils. Entropy 2016;18:85.
[74] Piacentino A, Catrini P. On thermoeconomic diagnosis of a fouled direct expansion coil: Effects of induced malfunctions on quantitative performance of the diagnostic technique. J Sustain Develop Energy Water Environ Syst 2017;5:177–90.
[75] Verda V, Serra L, Valero A. The effects of the control system on the thermoeconomic diagnosis of a power plant. Energy 2004;29(3):331–59.
[76] Verda V, Baccino G. Thermoeconomic approach for the analysis of control system of energy plants. Energy 2012;41:38–47.
[77] Fahidy TZ. Activity coefficients in electrolyte solutions (second edition), edited by Kenneth S. Pitzer, 1991, 542 + vi pages, CRC Press, Boca Raton, FL; ISBN 0-8493-5415-3. Price: US$ 195.00. Can J Chem Eng 1993;71(3):494.
[78] Rogers PSZ, Pitzer KS. Volumetric properties of aqueous sodium chloride solutions. J PhysChem Ref Data 1982;11(1):15–81.
[79] Archer DG. Thermodynamic propertis of the KCl + H2O system. J Phys Chem Ref Data 1999;28(1):1–16.
[80] Pitzer KS, Pelper JC, Busey RH. Thermodynamic properties of aqueous sodium chloride. J Phys Chem Ref Data 1984;13:1–102.

FURTHER READING

[81] Dawoud M, Al Mulla M. Environmental impacts of seawater desalination: Arabian Gulf case study. Int J Environ Sustain 2012;1:22–37.

CHAPTER 13

Exergy Analysis of Thermal Seawater Desalination—A Case Study

Ibrahim Mustafa, Ahmed Al Ghamdi
Yanbu Industrial College, Yanbu, Saudi Arabia

Nomenclature

C_p specific heat at constant pressure (kJ kg^{-1} K^{-1})
$\hat{h}$ specific enthalpy (kJ kg^{-1})
h_d enthalpy at the dead state (kJ)
M_r mass flow rate of recycled brine (kg s^{-1})
m_d mass flow rate of distillate (kg s^{-1})
m mass flow rate (kg s^{-1})
Q rate of heat transfer (kW)
R ideal gas constant (kJ/kg K)
R' mass of distillate formed per mass of brine (kg kg^{-1})
T_d temperature at the dead state (K)
W rate of shaft work (kW)
X rate of exergy destruction (kW)

Greek letters

λ' latent heat of vaporization (kJ kg^{-1})
ξ exergy associated with mass flow (kJ)

13.1 INTRODUCTION

At present, the worldwide seawater desalination capacity consists of 35%–40% of thermal desalination in the form of multistage flash (MSF), multiple effect desalination (MED), and thermal or mechanical vapor compression (TVC or MVC). These methods are quite energy intensive, as is generally the case with phase change processes. For example, 55% of the total operational and maintenance costs are utilized as energy cost. The energy input to these processes is both thermal or low heat energy and the electrical energy. Thermal energy is supplied by saturated steam to evaporate brine, whereas electrical energy is used to drive pumps. Owing to finite temperature differences across the stages, these processes are thermodynamically inefficient. The irreversibility occurs due to friction, nonequilibrium or thermal

Renewable Energy Powered Desalination Handbook
https://doi.org/10.1016/B978-0-12-815244-7.00013-1

losses, boiling point elevation, and pump inefficiencies. Thus, being energy intensive, these processes are employed mostly in the Arabian Gulf region where a cheap source of energy in the form of oil or gas is abundantly available.

Approximately 70% of desalination plants are located in the Arabian Gulf [1]. Out of these, roughly 88% of desalination comes from thermal methods with MSF being the most widely used [2–4]. The MSF process is preferred because it is insensitive to seasonally varied sea conditions and much less susceptible to scaling and corrosion problems usually encountered in other types of thermal desalinations. However, like other thermal processes, MSF is complex and highly energy intensive. Consequently, a good portion of the thermal desalination research has been directed toward the optimization or reduction in the energy intake of the MSF process. Methods such as redesigning of heat exchangers, efficient process control, reduction in waste heat, and the use of higher efficiency pumps are employed. When all these methods are exhausted or physically restricted, engineers attempt to carry out overall optimization of the process by taking exergy analysis into account.

In recent years, exergy analysis has been used effectively to isolate and improve process plant equipment and parameters that contribute to a lower efficiency of industrial plants [5–10]. In this method, coupled with energy and mass balances, the exergy destruction or the loss of work potential is ascertained across a unit or system. The higher the exergy destruction, the lower is the thermodynamic efficiency of the system. In other words, if the actual exergy decrease is much higher than thermodynamic exergy decrease, then there is always a chance to modify the system design or its operating conditions to improve the thermodynamic efficiency. Thus, units giving higher exergy destructions are indicated and ways are adopted to decrease high exergy losses.

Several articles have been published regarding exergy analysis of MSF desalination [11–42]. Units of various capacities and configurations have been considered such as MSF once through, MSF with recycled brine, MSF coupled with TVC or MVC, MSF with solar energy, and MSF with feed mixing and cooling. In addition to variation in design configurations, these approaches have used different models to calculate thermodynamic properties of seawater, most importantly heat capacity, vapor pressure, density, and boiling point elevation. Simplified models have neglected the salt content and modeled seawater as pure water, whereas little more realistic approaches have opted to model seawater as an ideal mixture of sodium chloride (neglecting other salts) and pure water. However, owing to a strong electrolytic effect, seawater is not an ideal mixture and the model may fail at high salt concentrations. Recently, Sharqawi et al. [28] have developed correlations for seawater properties applicable to a wide range of operating conditions. These correlations reflect true seawater behavior and therefore should be used in all future work.

The results of previous research work on exergy analysis have been summarized in Table 13.1. The summary covers a period of almost 16 years, that is, from 1995 to 2011. Only 11 articles appeared in the literature that focused on determining the exergy loss across an MSF plant. Mostly simplified mathematical models have been used by incorporating constant physical properties, the distillate being salt free, an insignificant contribution of kinetic, potential, and chemical exergies, the distillate product being salt free, and a uniform temperature drop over stages.

It is evident from Table 13.1 that exergy efficiency typically falls in the range of 2.5%–6.5%, with total exergy loss from 60 to 80 kJ/kg product, which is significant and needs improvement. A number of suggestions have been made in the past like increasing the number of stages and top brine temperature (TBT), using higher efficiency pumps, enlarging the specific condensation area, prohibition of scale forming, and lowering steam temperature or combinations thereof. The microthermal analysis of subsystems indicates the heat recovery section (HRS) to be the major contributor toward exergy loss followed by the brine heater.

In this work, exergy analysis of three different configurations of MSF is made and compared with real plant data. The properties of seawater are estimated using recently developed correlations [28]. A noniterative code in MATLAB is written to perform mass, energy, and exergy balances. The details are given in the following sections.

13.2 DESCRIPTION OF THE MSF PLANT, YANBU

As a case study for exergy analysis, the MSF Section of a dual-purpose cogeneration plant located in Yanbu, Saudi Arabia, is selected. A schematic of the plant is shown in Fig. 13.1. The co-generation plant consists of nine gas turbine units and four steam turbine units with a total capacity of 1030.1 MW. The electric generators deliver power at 13.8 kV to the main transformers which step up the voltage to 115 kV and are connected to the 115 kV switch yard. Marafiq operates nine MSF units and one RO with a total installed capacity of 146,160 m^3/day.

Table 13.1 Previous exergy analyses of MSF plants worldwide

MSF features	[42]	[24][a]	[24][b]	[24][c]	[24][d]	[24][e]	[24][f]	[33]	[19]	[39]	[35]
No. of stages (heat recovery)	–	31	14	21	19	13	19	17	–	15	–
No. of stages (heat rejection)	–	3	2	3	3	3	3	3	–	3	–
Total stages	22	34	16	24	22	16	22	20	19	18	28
Top brine temperature (°C)	90.7	115	108	110	90	90	87.8	110	76.5	90	106
Distillate production (kg/s)	272	114	232	232	248	227	123	53	210	416	397
GOR (lb dist./lb steam)	7.8	9.5	6.9	7.0	7.8	6.5	9.5	8.0	–	5.9	8.3
Seawater temperature (°C)	35	29	29	29				27	–	35	25
Seawater salinity (ppm)	46,500	407,500	40,750	40,750	40,750	40,750	40,750	45,000	45,000	–	42,000
Heat exchanger area (m^2)	–	–	–	–	–	–	–	829	–	–	–
Brine blowdown salinity (ppm)	70,093	–	–	–	–	–	–	70,000	–	–	70,000
Brine blowdown temperature (°C)	43.2	–	–	–	–	–	–	40	–	–	40
Feed seawater (kg/s)	2397	–	–	–	–	–	–	–	–	–	1809
Recycled brine (kg/s)	808	–	–	–	–	–	–	–	–	–	
Brine blowdown (kg/s)	536	–	–	–	–	–	–	–	–	–	642
Pumps efficiency (%)	75	–	–	–	–	–	–	–	–	–	
Total exergy loss kW/kg product	60	58	69	79	59	70	55	119.6	40.5	62.5	–
Exergy efficiency %	4.2	5.5	4.8	4.3	6.1	6.0	6.4	2.0	3.6	5.5	2.5

a = Jeddah Phase II, b = Jeddah Phase III, c = Jeddah Phase IV, d = Al-Jubail, e = Al-Khobar, f = Al-Khafji.

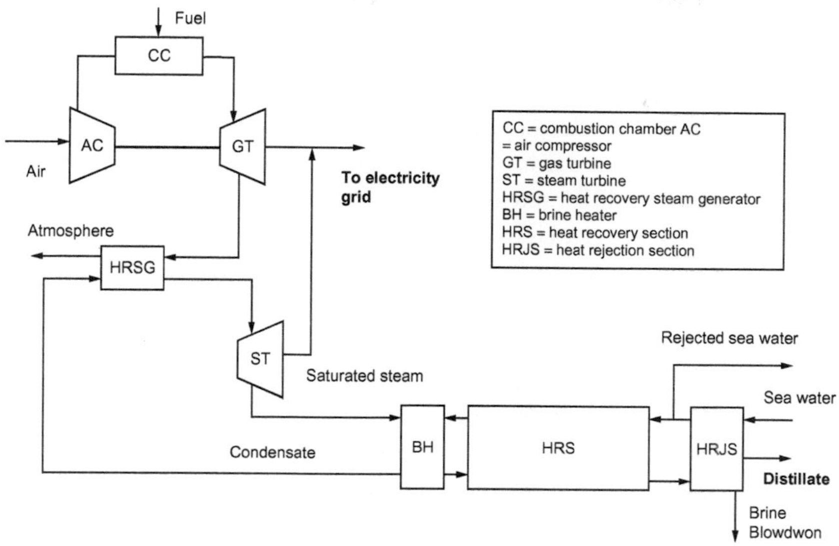

Fig. 13.1 A schematic of a dual purpose cogeneration plant with MSF.

The flow diagram of one MSF unit with 529.75 m^3/h production is shown in Fig. 13.2. The plant consists of 28 flashing stages with 25 stages in the HRS and the remaining in the heat rejection system (HRJS). The unit is designed for an intake seawater temperature of 22.9–35.2°C and a TBT of 110–121°C. The designed seawater supply is 1092.3 kg/s with a performance ratio of 19 kg of distillate per 1000 kcal heat input.

The chemically treated seawater from the Red Sea with 45,000 ppm salinity or 4.5% salt by weight, 22.9–35.2°C temperature, and at

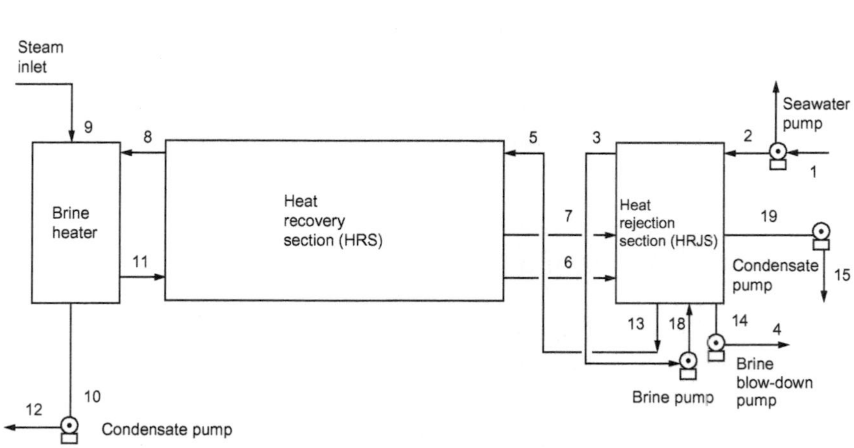

Fig. 13.2 A block flow diagram of one MSF unit at Yanbu.

atmospheric pressure is pumped to the HRJS where its temperature is raised by absorbing latent heat from the condensing vapor on the tubes. In this system, the excess heat added by heating steam in the brine heater is discarded to the environment by circulating seawater. The seawater outflow from the HRJS is divided into two streams. The first stream is recycled back to the last flashing stage where it is mixed as make up brine. The other stream is discharged back to the sea and is called seawater discharge. The main purpose of the flashing stages in the HRJS is therefore to control the brine temperature being discharged to the sea as well as to raise the temperature of the incoming seawater.

The mixed brine from the HRJS is recycled back to the HR Sand it is called recycled brine. This brine at a 31.33°C temperature and salinity of 56,521 ppm enters the HRS where its temperature is increased by absorbing latent heat from the condensing vapors in each stage. There cycled brine leaves the last stage at a temperature of 106.92°C and enters the brine heater where its temperature is increased by condensing steam to 114°C TBT. The high-pressure steam at 125.05°C and 1.99 bar from the electricity generation section is used in the heater as a source of heat. This heated brine then enters the HRS where a gradual reduction in pressure from 1.42 bar to 0.04 bar in each stage causes the vaporization of brine.

A small fraction of steam from the steam turbine is used in the vacuum ejector. The mostly salt free brine vapor in each stage is passed through a demister pad to avoid salt carryover to distilled water and then condensed on the tube surface and collected as desalted water (25 ppm) which is pumped to the storage tank for onward supply. The latent heat lost by the desalted water is gained by the brine flowing inside the heat exchanger tubes. There are 14 heat exchangers arranged across the width of the unit. Each heat exchanger is divided vertically into two sections by stage division plates. Moreover, each stage of the exchanger is 2-pass with a tube bundle containing 18,000 tubes. Important operating conditions are listed in Table 13.2.

13.3 MODEL FORMULATION

To carry out exergy analysis for the purpose of evaluating plant efficiency, a mathematical model can be developed and solved in MATLAB. The major energy consuming equipment such as the HRJS, the HRJ, the brine heater, products and feed pumps are considered. The details of the models are as follows.

Table 13.2 Plant design/operation data

Number of stages (HRJS)	03
Number of stages (HRS)	25
Top brine temperature	114°C
Distillate production	159–200 kg/s
Performance ratio	19 kg distillate/1000 Kcal heat input
Seawater temperature	22.9–35.2°C
Seawater salinity	45,000 ppm
No of heat exchangers	14
Design of heat exchanger	2 pass with tube bundle 18,000 tubes
Brine heater	Shell & Tube
Brine blowdown salinity (ppm)	65,000 ppm
Brine blowdown temperature	32 °C
Feed seawater	1092 kg/s
Recycled brine	1227 kg/s
Brine blowdown	360 kg/s

13.3.1 Model Assumptions

The following assumptions were made to simplify the model [43].

- The distillate product is salt free.
- The specific heat and heat of vaporization are taken at the average temperature of the process.
- The overall heat transfer coefficient is constant and equals to 2.0 kW/m^2 °C.
- The heat loss to the surrounding is negligible.
- The subcooling of the condensate or superheating of the incoming steam to the brine heater is ignored.
- The latent heat of vaporization in each stage is estimated at the average of the TBT and the temperature of the last stage of the HRJS.
- Constant temperature drop per stage of the HRS and HRSJ.
- Noncondensable gases such as air have no effect on the heat transfer process.
- Potential and Kinetic Energies are ignored.
- A mixture of salt and water is an ideal solution so the mixture properties are the average of the component properties.
- The dead state temperature is the temperature of the incoming seawater and the dead state pressure is 1 atm where the system has zero exergy. In addition, NaCl and H_2O are considered at their standard states or most stable states in the incoming seawater to the plant.
- The pumps efficiency is taken as 65%.

13.3.2 Governing Equations

For a given system under consideration, one can generally write mass, energy, and exergy balances for a steady state process as follows:

(i) Mass Balance

$$\text{Overall}: \sum m_i = \sum m_o \tag{13.1}$$

$$\text{Salt Balance}: \sum m_i x_i^s = \sum m_o x_o^s \tag{13.2}$$

where m is the mass and x^s is the mass fraction of the salt.

(ii) Energy Balance

$$\text{Overall}: \sum m_i \hat{h}_i = \sum m_o \hat{h}_o \tag{13.3}$$

where $\hat{h}$ is the specific enthalpy.

The mass of distillate, R, formed in each stage is calculated by equating the latent heat of vapor formed with the decrease in the sensible heat of the brine [43]:

$$R' = C_p \, \Delta T / \lambda' \tag{13.4}$$

where R' is reported in kg of vapor formed per kg of brine and λ' is the average latent heat of vaporization. Then the flow of distillate at each stage and the total flow of distillate formed would be

$$m_d{}^i = M_r \, R'(1 - R')^{(i-1)} \tag{13.5}$$

$$M_d = M_r \left(1 - (1 - R')_n\right) \tag{13.6}$$

where n is the total number of stages in the MSF unit and Mr is the mass flow rate of the recycled brine.

(iii) Exergy Balance

Exergy is defined as the availability of useful energy or maximum work potential of a system at a given state in a specified environment. Therefore, the difference between actual work and exergy indicates the room for improvement in the system's design and/or its operation. Being closed to entropy in nature it is also not a conserved property (no generation of entropy means no destruction of exergy for a reversible process) and depends on both the state of the system and the surroundings. An exergy balance on a control volume therefore yields an extra destruction term '$\dot{X}$' as follows:

$$\sum \left(1 - \frac{T_d}{T_i}\right) \dot{Q}_i - \dot{W} + \sum_{in} \dot{m}\xi - \sum_{out} \dot{m}\xi = \dot{X} \tag{13.7}$$

The first term on the left-hand side (LHS) represents the exergy transfer rate due to heat, the second term represents the exergy transfer rate due to work, and the third and fourth terms represent exergy transfer in and out of the system, respectively, whereas the term on the right-hand side (RHS) represent the rate of exergy destruction during the process. It should be noted that in developing the above equations the heat transfer to a system and the work done by the system have been taken as positive quantities.

The exergy analysis of the MSF, Yanbu plant consists of exergy balances for the brine heater, the HRS, the HRJS and the five pumps while ignoring the steam ejector and ejector condenser because their process temperature and flow rate values are quite low compared with the other systems. For the brine heater the second term on the LHS of the above equation will be dropped, for the pumps the first term on the LHS will be dropped, and for both the HRS and HRJS units both the first and second terms will be dropped by considering the systems to be adiabatic. That is:

$$\text{Heat Recovery System (HRS)}: \quad \sum_{in} \dot{m}\xi - \sum_{out} \dot{m}\xi = \dot{X}_{HRS} \tag{13.8}$$

$$\text{Heat Rejection System (HRJS)}: \quad \sum_{in} \dot{m}\xi - \sum_{out} \dot{m}\xi = \dot{X}_{HRJS} \tag{13.9}$$

$$\text{Brine Heater}: \quad \sum \left(1 - \frac{T_d}{T_i}\right) \dot{Q}_i + \sum_{in} \dot{m}\xi - \sum_{out} \dot{m}\xi = \dot{X} \tag{13.10}$$

$$\text{Pumps}: \quad -\dot{W} + \sum_{in} \dot{m}\xi - \sum_{out} \dot{m}\xi = \dot{X} \tag{13.11}$$

In the MSF plants where nuclear, magnetic, and surface tension effects are not present, the exergy associated with mass that is, 'ξ' is usually taken as the sum of kinetic, potential, physical, mixing, and chemical exergies:

$$\xi = \xi^{ke} + \xi^{pe} + \xi^{ph} + \xi^{ch} \tag{13.12}$$

The changes in kinetic and potential exergies can also be ignored due to the fact that changes in these exergies across a control volume are insignificant. Thus,

$$\xi = \xi^{ph} + \xi^{ch} \tag{13.13}$$

The physical exergy 'ξp' is written as the sum of exergy due to temperature and pressure effects and the exergy due to mixing. The exergy due to temperature and pressure effects is

$$\xi^{ph}_{T,P} = (h - h_d) - (s - s_d)T_d \tag{13.14}$$

whereas the physical exergy due to mixing (ideal solution) is given as

$$\xi^{ph}_{mixing} = \sum x_i \xi^{pure}_i + RT_d \sum x_i \ln(x_i) \tag{13.15}$$

The second term on the RHS indicates exergy loss due to mixing. Upon simplification, the above equations become [44]

$$\xi^{ph}_{mixing} = mT_d R\left(x_s \log\left(\frac{x_s}{x_{o,s}}\right) + x_w \log\left(\frac{x_w}{x_{o,w}}\right)\right) \tag{13.16}$$

where $x_{o,s}$ and $x_{o,w}$ are mole fractions of salt and water in the dead state. The salinity is customarily reported in mass fractions from which the mole fractions of salt (assuming only NACl) and water are calculated using relations

$$x_s = \left(\frac{18}{58.5\left(\frac{1}{x_w} - 1\right) + 18}\right) \tag{13.17}$$

where x_s is the mass fraction of salt and x_w is the mass fraction of water.

Chemical exergy is defined as the exergy of chemical substances with reference to their standard states or most stable states in the environment. The environment is usually taken as species in air, seawater, or lithospheric solids at normal temperature and pressure [44]. The chemical exergy for any species is calculated under standard conditions if affinity of the reaction and exergy of the reactant and products are known. Once the chemical exergy of a species is estimated at standard conditions of T_d and P_d, the chemical exergy other temperatures and pressure can be determined [44]. The chemical exergy of the mixture is calculated by following the rule of mixing:

$$\xi^{ch} = \sum x_i \xi^{ch}_i \tag{13.18}$$

Since there are no reactions involved in the MSF plant and the dead state is assumed to be at the conditions of inlet sea water (T_d=22.9°C/35°C,

$P_d = 1$ atm with NaCl and H_2O as species in their most stable state at a Salinity of 4.5%) and since the chemical compounds do not change their composition (and hence chemical exergies) while passing through the systems, one can safely ignore the chemical exergy contribution in the above equation. Strictly speaking, pressure differences between the dead state and the process condition of a compound do give chemical exergies, but they have also been ignored here [44].

Once exergies of flowing streams are known, the second law of thermodynamic efficiency of an MSF unit can be estimated [42]

$$\eta_{II} = \frac{\text{Minium Exergy Required for Separation}}{\text{Total Exergy in}} \tag{13.19}$$

13.3.3 Properties Estimation

To estimate the exergy of flow streams, the values of several seawater properties are needed. A number of correlations have been developed to estimate the thermodynamic properties of seawater such as enthalpy, entropy, heat capacity, thermal conductivity, density, viscosity, surface tension etc. In this work, developed correlations have recently been used [45] which are given in the Appendix.

13.4 RESULTS AND DISCUSSIONS

A noniterative code in MATLAB was written to solve the above system of equations. To ensure the accurate functioning of the code, checks on the mass and energy balances were incorporated. The program output was also compared with the MSF Yanbu Plant data.

13.4.1 Code Testing and Sensitivity Analysis

In order to test the code's output, the stagewise salinity was calculated and compared with the plant data (Fig. 13.3). The maximum deviation was about 1% at the 14th stage and it was due to the calculation of latent heat of vaporization at the average temperature of the system. This deviation is, however, within tolerance limits. The data match exactly at stage 28 because the brine blowdown salinity is an input to the code and it was taken from the plant.

A further comparison of the values of important process parameters from the code to those obtained from the plant is shown in Table 13.3. As

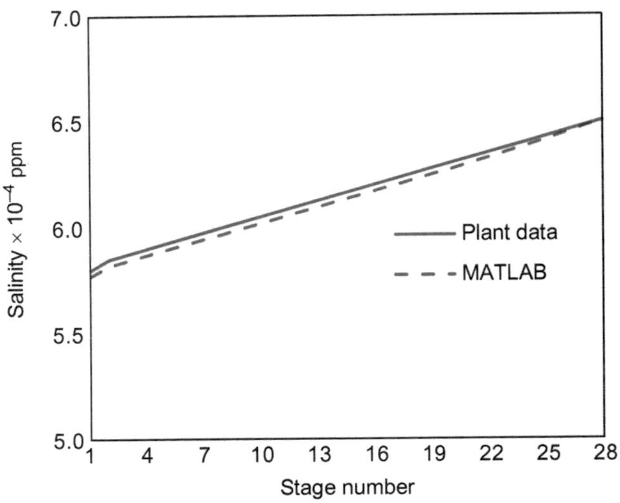

Fig. 13.3 Code output vs plant data for stagewise salinity.

Table 13.3 Comparison of the code output with plant data

No.	Process variable	Plant	Code	% Error
1	Recycled brine (kg/sec)	1228	1162	5.3
2	Brine blow-down (kg/sec)	360	358	0.6
3	Energy consumed at heater (kJ/sec)	34,722	34,902	0.5
4	Performance ratio (kg/1000 kcal)	19	19	0.0
5	Recirculation ratio	7.8	7.7	1.6

indicated, the code output deviation is maximum for the recycled brine flow rate estimation which is about 5%. Again, this is due to the assumption of the latent heat of vaporization being independent of the temperature and evaluated at the average temperature of the MSF plant. The other predictions from the code are quite satisfactory.

To check the robustness of the code, a sensitivity analysis was carried out. The variation of the performance ratio with TBT was estimated. This is shown in Fig. 13.4. As expected the performance ratio increases with the increase in the TBT. The performance ratio has been defined as follows:

$$PR = \frac{\text{Net product water flow (kg/s)} \times 1000}{\text{Energy consumed at B/H (kJ/s)} \times 0.2388} \tag{13.20}$$

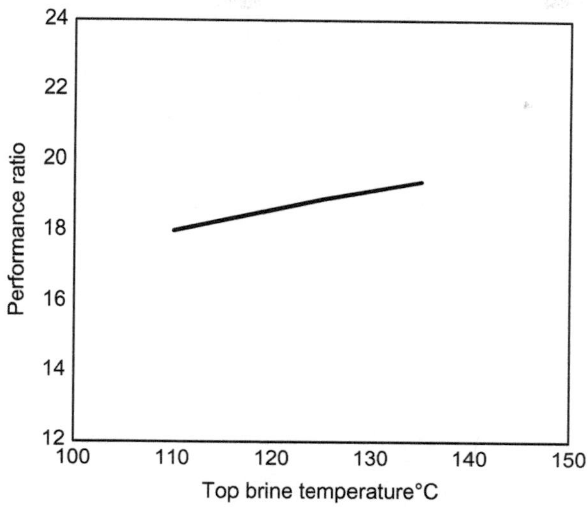

Fig. 13.4 The dependence of PR on top brine temperature (TBT).

13.4.2 Exergy Analysis

An exergy analysis of major process equipment is given in Tables 13.4–13.11. The negative exergy values occur in cases where the salinity is much higher than that of the dead state at moderate to low temperatures. This is because the net input work is required to decrease the salinity of a given stream to the dead salinity, and thus its exergy is negative [42]. The highest exergy destroyed is found in HRS (74.9%) followed by the HRJS (24.2%) and the brine heater (0.9%) as shown in Fig. 13.5.

In order to calculate the second law efficiency of the plant approach used in Ref. [25] was adopted as follows:

$$X_{in} = X_{in,\text{steam}} + X_{in} = 10532.7\,\text{kW} \tag{13.21}$$

$$W_{\text{separation}} = X \; + X_{\text{distillate}} = 388.2\,\text{kW} \tag{13.22}$$

Therefore, the second law of efficiency is estimated from Eq. (13.19) to be 3.22%, which is low but close to previously calculated values in the literature [42]. This indicates that a lot of room exists to improve the design and process conditions of the MSF plant. The simulations are repeated for the case of peak load where the distillate production is about 200 kg/s at a seawater temperature of 35.2°C. A comparison of important process parameters for the two cases is given in Table 13.12. As expected the seawater intake and steam flow rate increase significantly, putting a load on the heater,

Table 13.4 Exergy analysis around brine heater

	Description	Temp (° C)	Pressure (bara)	Mass flow (kg/s)	Salinity (%)	Enthalpy (kW)	Entropy (kW/kg K)	Exergy (kW)
Inlet								
Stream 8	Brine inlet	107	2.74	1228	5.65	526,401	1683	47,441
Stream 9	Steam inlet	125	2.00	15.9	–	43,029	112	9540
Outlet								
Stream 11	Brine outlet to HRS	114	2.44	1228	5.65	561,024	1774	55,933
Stream 10	Condensate to B/H	125	2.24	15.9	–	8406	25	975.5
Exergy destroyed								72.6

Table 13.5 Exergy analysis around heat rejection system (HRJS)

	Description	Temp (° C)	Pressure (bara)	Mass flow (kg/s)	Salinity (%)	Enthalpy (kW)	Entropy (kW/kg K)	Exergy (kW)
Inlet								
Stream 2	Seawater inlet	22.90	4.28	1093	4.5	101,189	401	0.0
Stream 6	Brine from HRS	38.62	0.06	1081	6.42	165,275	629	−3664
Stream 7	Condensate from HRS	29.34	0.04	147	0.0025	23,242	81.84	2958
Stream 18	Brine from stage 26 after deaeration and decarbonation	30.33	0.04	520	4.5	63,779	243.8	9.9
Outlet								
Stream 3	Brine from stage 26	30.33	3.20	1093	4.5	133,981	512	20.8
Stream 13	Recycled brine to HRS	31.33	0.35	1228	5.65	153,807	597	−3145
Stream 14	Brine blowdown	32.06	0.32	360	6.5	45,832	180	−2570
Stream 20	Condensate product	29.34	0.04	159	0.0025	19,864	68	1910
Exergy destroyed								2088

Table 13.6 Exergy analysis around heat recovery system (HRS)

	Description	Temp (°C)	Pressure (bara)	Mass flow (kg/s)	Salinity (%)	Enthalpy (kW)	Entropy (kW/kg K)	Exergy (kW)
Inlet								
Stream 5	Recycled brine	31.33	9.94	1228	5.65	153,895	595	−2561
Stream 11	Brine from heater	114	2.44	1228	5.65	561,024	1774	55,875
Outlet								
Stream 6	Recycled brine to HRJS	38.62	0.06	1081	6.42	165,275	629	−3664
Stream 7	Distillate to HRJS	37	0.06	147	–	23,242	81.84	2958
Stream 8	Seawater to brine heater	107	2.74	1228	5.65	526,401	1683	47,547
Exergy destroyed								6473

Table 13.7 Exergy analysis around seawater feed pump (P-1)

	Description	Temp (°C)	Pressure (bara)	Mass flow (kg/s)	Salinity (%)	Enthalpy (kW)	Exergy (kW)
Inlet							
Stream 1	Seawater inlet	22.9	1.72	1204	4.5	181,232	0
Outlet							
Stream 2	To HRJS	22.9	4.33	1204	4.5	181,525	293
Exergy supplied							293

Table 13.8 Exergy analysis around brine recycle pump (P-2)

	Description	Temp (°C)	Pressure (bara)	Mass flow (kg/s)	Salinity (%)	Enthalpy (kW)	Exergy (kW)
Inlet							
Stream 13	From HRJS	31.33	0.35	1228	5.65	153,808	−3145
Outlet							
Stream 5	To HRS	31.33	10.0	1228	5.65	153,895	−2651
Exergy supplied							494

Table 13.9 Exergy analysis around brine blowdown pump (P-3)

	Description	Temp (°C)	Pressure (bara)	Mass flow (kg/s)	Salinity (%)	Enthalpy (kW)	Exergy (kW)
Inlet							
Stream 14	From HRJS	32.06	0.32	360	6.5	45,833	−1570
Outlet							
Stream 4	To sea	32.06	3.1	360	6.5	46,019	−1478
Exergy supplied							92

Table 13.10 Exergy analysis around brine heater condensate pump (P-4)

	Description	Temp (°C)	Pressure (bara)	Mass flow (kg/s)	Salinity (%)	Enthalpy (kW)	Exergy (kW)
Inlet							
Stream 10	From brine heater	125	2.24	15.7	-	8406	975.5
Outlet							
Stream 12	To boiler house	125	11.26	15.7	-	8421	989.6
Exergy supplied							14.1

Table 13.11 Exergy analysis around distillate pump (P-5)

	Description	Temp (°C)	Pressure (bara)	Mass flow (kg/s)	Salinity (%)	Enthalpy (kW)	Exergy (kW)
Inlet							
Stream 19	From HRJS	32.06	0.32	360.41	0.0025	48,295	357.8
Outlet							
Stream 15	To storage	32.06	3.09	360.41	0.0025	48,395	457.8
Exergy supplied							100

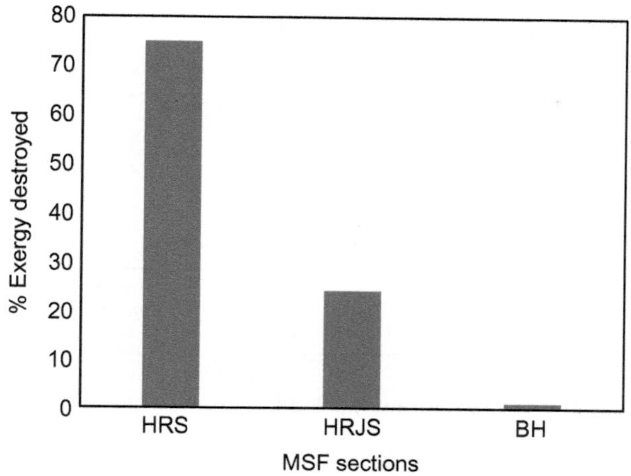

Fig. 13.5 Exergy destruction in multistage flash (MSF) equipment.

Table 13.12 Comparison of process parameters for normal and peak load

Process variable	Normal load	Peak load
Distillate produced (kg/s)	159.2	200.0
Recycled brine (kg/s)	1227.9	1541.7
Brine blowdown (kg/s)	360.4	450.0
Steam flow rate (kg/s)	15.8	19.8
Net seawater intake to HRJS (kg/s)	520.3	649.4
Second law efficiency (%)	3.22	2.84

the seawater feed pump, the recycled brine pump, and the brine blow down pump and thus increasing the exergy input. The second law of efficiency also decreases to 2.84% due to more irreversibility at the peak load that is, high flow rates and high temperatures.

In Fig. 13.6, a simplified exergy flow diagram is shown. It gives a rough estimation of exergy destruction across various equipment of the plant so that the equipments with major exergy destruction are identified.

13.4.2.1 Minimization of Exergy Destruction

The major exergy destruction occurs at the HRS so considerable room for its minimization exists. This can be achieved by analyzing the factors in operation, control, and maintenance that contribute to exergy destruction. This work is related to the operation part of the process.

By using the code, a simulation was performed to check the dependence of exergy destruction on the number of stages in the HRS. This is shown in

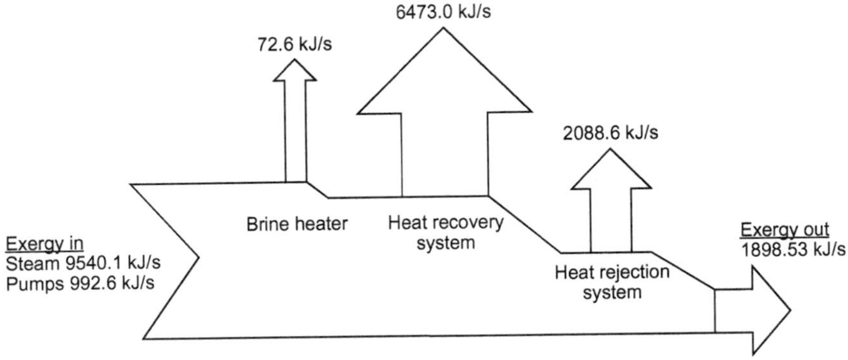

Fig. 13.6 Exergy flow diagram for normal load.

Fig. 13.7. As indicated there is a gradual decrease in exergy destruction (from 74.9% to 69.2%) as the number of stages is increased from 25 to 31. It, therefore, seems feasible to operate the plant with more than 25 stages in the HRS during peak load. The exergy destruction in the HRJS with three stages is almost 20%, which is relatively higher compared to 74.9% in the HRS with 25 stages. Therefore, removing the HRJS and increasing the number of stages in the HRS could be considered. This can be achieved by two types of modifications in the existing plant that is, either converting it once through MSF-OT or MSF-M.

A schematic of MSF-OT is shown in Fig. 13.8. As shown, the HRJS does not exist in such types of MSF units. This simplifies the MSF process

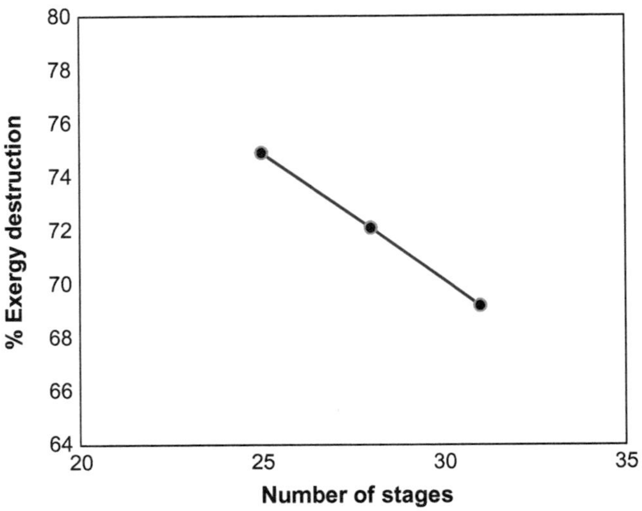

Fig. 13.7 Exergy destruction vs number of stages.

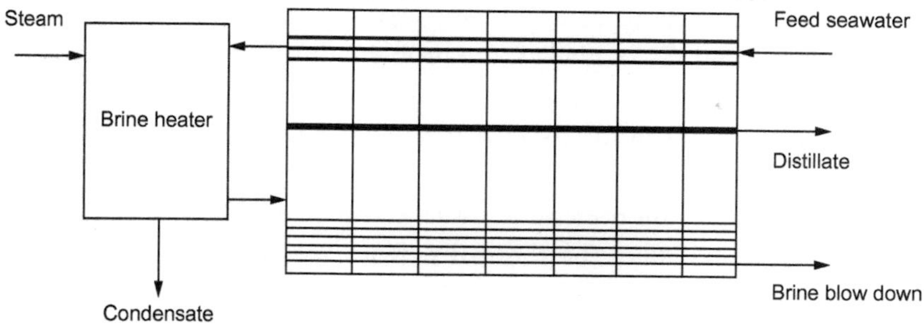

Fig. 13.8 A sketch of MS-OT unit.

operation and maintenance. However, this unit has found limited industrial application in the Gulf because of high temperature variations throughout the year. In winter, when the temperature drops to as low as 20°C, the performance of the unit reduces severely because it does not have any control over the feedwater temperature. A larger temperature gradient in the unit generally calls for greater consumption of steam energy and thus gives a performance ratio as low as 3.0. In addition, a drop in temperature requires the size of the last stages to be increased to compensate for the decrease in vapor density and thus increase in its velocity. At a higher vapor velocity, the entrainment of brine droplets increases. The optimum vapor velocity suggested in the literature is around 4 m/s [43]. Moreover, the feed to product ratio in MSF-OT is usually around 9–10:1, which needs a greater amount of chemicals to be used to ward off scaling, foaming, and corrosion [43].

Another variation of the MSF plant is MSF-M as shown in Fig. 13.9. In this system, the HRJS is replaced with a mixing tank where the part of brine leaving the last stage mixed with the incoming seawater to the plant. The mixing of the brine with seawater is advantageous because it reduces the amount of chemicals used to prevent corrosion and scaling, on the one hand, and to recover heat added to brine in the heater, on the other. This

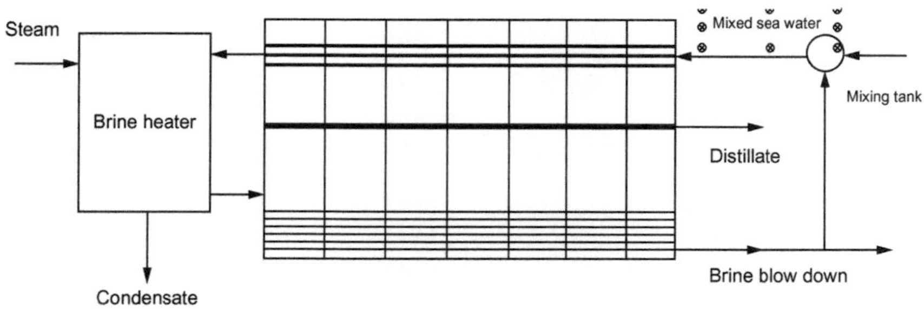

Fig. 13.9 A sketch of MS-M plant.

configuration also helps to control intake seawater temperature which is not available with MS-OT. However, the problem during the mixing of brine and intake seawater is the occurrence of thermal shock due to the temperature difference between the two streams (especially in winter) which promotes decomposition of bicarbonates and subsequent formation of CO_2. If not vented properly, this noncondensable gas might accumulate in the brine stream and severely affect the thermal performance of the unit. With no recycle of brine, this unit reduces to MS-OT.

The existing MSF plant at Yanbu can be easily modified to these configurations. In case of MS-OT, the HRJS could be merged with the HRS system to a single unit of 28 stages. In addition, the brine recycled pump could be removed and the reject seawater pump used to increase the intake seawater flow rate for a given production of the distillate. On the other hand, the modification of existing MSF to MSF-M would be similar, except that a mixing tank would be added and the brine recycled pump diverted to the mixing tank.

To evaluate exergy loss or thermodynamic efficiency across the aforementioned units, the energy and mass balances were adjusted in the MATLAB code. In case of MSF-OT, balances around the HRS and HRJS were done for a combined unit, whereas in case of MSF-M additional balances around the mixing tank were incorporated. The solution procedure and the parameters were identical to the conventional MSF plant. Two cases were simulated to estimate and compare important parameters of all three configurations, that is, the Winter Case (22.9°C)and the Summer Case (35.2°C). In both cases, parameters such as the intake seawater flow rate and its temperature, TBT, and the number of stages were fixed in all the configurations. These values are given in Table 13.11. The simulations were run for these cases and the results are summarized in Table 13.13.

In Table 13.14, all the three configurations are evaluated and compared in the light of important parameters such as performance ratio, thermodynamic efficiency, and distillate production. For both cases, the performance

Table 13.13 Fixed Parameters for MSF configurations for two cases

Parameters	Winter case	Summer case
Seawater flow rate (kg/s)	1092	1373
Sea water temperature (°C)	22.9	35.2
Top brine temperature (°C)	114	114
Number of stages	28	28

Table 13.14 Fixed Parameters for MSF configurations for two cases

	Winter case (22.9°C)			Summer case (35.2°C)		
Parameters	**MS**	**MS-OT**	**MS-M**	**MS**	**MS-OT**	**MS-M**
Performance ratio	19.0	6.3	19.5	19.0	11.2	19.5
Thermodynamic efficiency (%)	3.2	14.8	5.1	2.8	18.5	4.3
Distillate production (kg/s)	147.2	146.8	147.4	200.1	199.6	200.1

ratio of MS-OT is quite low, although it shows very good thermodynamic efficiency compared to MS and MS-M. This can be explained by the fact that a larger amount of steam to produce the same amount of distillate is consumed in reaching the same TBT due to low brine temperature entering the heater. The thermodynamic efficiency or exergy loss is, however, significantly better for MS-OT because of the lower salinity brine blowdown and the removal of the brine recycle pump (see Eq. 13.19). Owing to the very low PR, a larger amount of chemicals requirement and large size stages in the last section of HRSMS-OT is obviously not a viable option.

In comparing MSF and MSF-M, it is found that PR remains the same, whereas thermodynamic efficiency is almost double for MSF-M. This can be attributed to the larger number of stages in the HRS and replacement of the HRJS with a higher exergy efficient mixing tank. The modification of MSF to MSF-M would be simple and feasible and would include the following:

- Brine should be recycled to the mixing tank instead of the last stage of the HRS.
- Flow and temperature control of seawater intake would be removed.
- Preheater tubes of the HRS and HRJS to be joined for continuous flow of seawater.
- Lower capacity pump for intake sea water would be used.

The exergy diagrams for MSF-OT and MSF-M are shown in Figs. 13.10 and 13.11.

13.5 CONCLUSIONS

Exergy analysis is an important tool in controlling energy loss during thermal desalination. It helps determine major contributors to exergy destruction and their readjustment in the process to improve overall thermodynamic

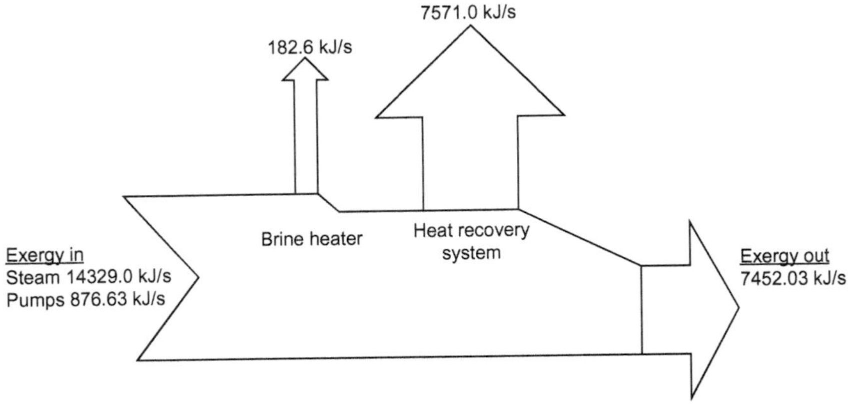

Fig. 13.10 Exergy flow diagram for normal load multistage flash (MSF-OT).

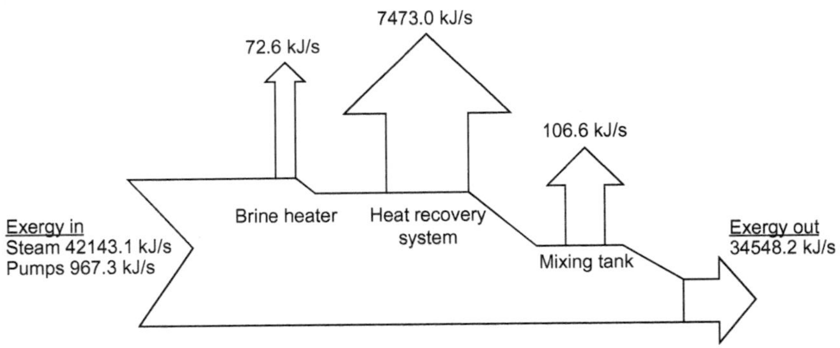

Fig. 13.11 Exergy flow diagram for normal load multistage flash (MSF-M).

efficiency. The method of exergy analysis presented here can be conveniently applied to other types of thermal desalination processes.

It would be fruitful to perform advanced exergy analysis and the exergoeconomics of thermal desalination plants. Whereas classical exergy analysis is used to recognize the causes of equipment inefficiencies on the one hand, advanced exergy analysis focuses on identifying real potential for improvements by splitting exergy loss into avoidable and unavoidable parts on the other. Combining advanced exergy analysis with economic studies is called exergoeconomics, that is, the monetary evaluations of various exergy costs together with electricity costs are obtained by solving the set of equations. The lost costs of each component of the system can also be obtained by this method. Thus, a highly optimized thermal desalination plant can be designed and operated.

A APPENDIX

A.1 Correlation for Seawater Enthalpy

$$h_{sw} = h_w - x_s(a_1 + a_2 x_s + a_3 x_s^2 + a_4 x_s^3 + a_5 T + a_6 T^2 + a_7 T^3 + a_8 x_s T + a_9 x_s{}^2 T + a_{10} x_s T^2 \tag{13.23}$$

where

$a_1 = -2.348 \times 10^4$, $a_2 = 3.152 \times 10^5$, $a_3 = 2.803 \times 10^6$, $a_4 = -1.446 \times 10^7$, $a_5 = 7.826 \times 10^3$

$a_6 = -4.417 \times 10^1$, $a_7 = 2.139 \times 10^{-1}$, $a_8 = -1.991 \times 10^4$, $a_9 = 2.778 \times 10^4$, $a_{10} = 9.728 \times 10^1$

Validity: h_{sw} and h_w in (J/kg K); $10 \leq T \leq 120°C$; $0 \leq x_s \leq 0.12$ kg/kg;

Accuracy: ±0.5% from IAPWS 2008 [19]

where h_{sw} is the specific enthalpy of seawater and h_w is the specific enthalpy of pure water.

A.2 Correlation for Seawater Entropy

$$s_{sw} = s_w - x_s\big(a_1 + a_2 x_s + a_3 x_s^2 + a_4 x_s^3 + a_5 T + a_6 T^2 + a_7 T^3 + a_8 x_s T + a_9 x_s{}^2 T + a_{10} x_s T^2 \tag{13.24}$$

$a_1 = -4.231 \times 10^2$, $a_2 = 1.463 \times 10^4$, $a_3 = -9.88 \times 10^4$, $a_4 = 3.095 \times 10^5$, $a_5 = 2.562 \times 10^1$

$a_6 = -1.443 \times 10^{-1}$, $a_7 = 5.879 \times 10^{-4}$, $a_8 = -6.111 \times 10^1$, $a_9 = 8.041 \times 10^1$, $a_{10} = 3.035 \times 10 - 1$

Validity: s_{sw} and s_w in (J/kgK); $10 \leq T \leq 120°C$; $0 \leq x_s \leq 0.12$ kg/kg

Accuracy: ±0.5% from IAPWS 2008 [21]

The enthalpy and entropy of pure water are evaluated from steam tables.

REFERENCES

[1] Leahy S, Purvis K. Peak Salt: is the desalination dream over for the Gulf States? In: The Guardian. 29th Sept 2016.

[2] www.desaldata.com.

[3] Blank JE, Tusel GF, Nisanc S. The real cost of desalted water and how to reduce it further. Desalination 2007;205:289–311.

[4] Annual Report. Saline Water Conversion Corporation (SWCC). Saudi Arabia.

[5] Ohijeagbon IO, Waheed MA, Jekayinfa SO. Methodology for the physical and chemical exergic analysis of steam boilers. Energy 2013;53:153–4.

[6] Hepbashi A. A key review on exergic analysis and assessment of renewable energy resources for sustainable future. Renew Sustain Rev 2008;12:593–661.

[7] Kotas TJ. The Exergy Method of Thermal Plant Analysis. Butterworths; 1985. ISBN 0408013508.

[8] Aljundi IH. Energy and Exergy analysis of a steam power plant in Jordan. Appl Therm Eng 2009;29(2–3):324–8.
[9] Mistry KH, McGovern RK, Thiel GP, Summers EK, Zubair SM, Lienhard JHV. Entropy generation analysis of desalination technologies. Entropy 2011;13:1829–64.
[10] Aghbashlo M, Mobli H, Rafiee S, Madadlou A. A review on exergy analysis of drying processes and systems. Renewable and Sustainable Energy Reviews 2013;22:1–22.
[11] Al-Karaghouli A, Kazmerski LL. Energy consumption and water production cost of conventional and renewable-energy-powered desalination processes. Renewable and Sustainable Energy Reviews 2013;24:343–56.
[12] Borsani R, Rebagliati S. Fundamentals an costing of MSF desalination plants and comparison with other technologies. Desalination 2005;182:29–37.
[13] Mazozotti M, Rosso M, Beltramini A, Morbidelli M. Dyamic modeling of multistage flash desalination plants. Desalination 2000;127:207–18.
[14] Al-Sahali M, Ettouney H. Developments in thermal desalination processes: Design, energy, and costing aspects. Desalination 2007;214:227–40.
[15] Nafey AS, That HES, Mabrouk AA. 'Thermoeconomic analysis of multistage flash-thermal vapor compression (MSF-TVC) desalination process', Tenth International Water Technology Conference, IWTC10. Egypt: Alexandria; 2006.
[16] Nafey AS, That HES, Mabrouk AA, El-Zeky MA. 'Exergy and thermoeconomic evaluation of MSF desalination plant unsing a new visual package', Ninth International Water Technology Conference, IWTC9. Egypt: Sharm El-Sheikh; 2005.
[17] Catrini, P., Cipollina, A., Micale, G., Piacentino, A., and Tamburini, A., ' Exergy analysis and thermoeconomic cost accounting of a Combined Heat and Power steam cycle integrated with a Multi Effect Distillation Thermal Vapor Compression desalination plant', Energy Conversion and Management (Article in Press).
[18] Fitzsimons L, Corcoran B, Young P, Foley G. Exergy analysis of water purification and desalination: A study of exergy model approaches. Desalination 2015;359:212–24.
[19] Al-Sulaiman FA, Ismail B. Exergy analysis of major recirculating multi-stage flash desalting plants in Saudi Arabia. Desalinatino 1995;103:265–70.
[20] Najafi B, Shirazi A, Aminyavari M, Rinaldi F, Taylor RA. Exergetic, economic and environmental analyses and multi-objective optimization of an SOFC-gas turbine hybrid cycle coupled with an MSF desalination system. Desalination 2014;334:46–59.
[21] Jin C, Chou Q, Jiao D, Shu P. Vapor Compression Flash seawater desalinatin system and its exergy analysis. Desalination 2014;353:75–83.
[22] Ameri M, Ahmedi P, Khanmohammadi S. 'Exergy analysis of a 420 MW combined cycle power plant', Int. J, Energy Res 2008;32:175–83.
[23] ElMoudir W, ElBousiffi M, Al-Hengari S. Process modeling in desalination plant operations. Desalination 2008;222:431–40.
[24] Hamed OA, Al-Sofi MA, Imam M, Mustafa GM, Ba-Mourdouf K, Al-Washmi H. Thermal Performance of Multistage Flash Distillation Plants in Saudi Arabia. Desalination 2000;128:281–92.
[25] Uche J, Serra L, Valero A. Thermoeconomic optimization of a dual purpose power and desalination plant. Desalination 2001;136:147–58.
[26] Baig H, Antar MA, Zubair SM. Performance evaluation of a once-through multi-stage flash distillation system: Impact of brine heater fouling. Energy Conversion and Management 2011;52:1414–25.
[27] Mabrouk AA, Nafey AS, Fath HES. Analysis of a new design of a multi-stage flash-mechanical vapor compression desalination process. Desalination 2007;204:482–500.
[28] Sharqawy MH, Lienhard JH, Zubair SM. On exergy calculations of seawater with applications in desalination systems. Int J Therm Sci 2011;50:187–96.
[29] Gomri R. Energy and exergy analyses of seawater desalination system integrated in a solar heat transformer. Desalination 2009;249:188–96.

[30] Alhazmy MM. Economic and thermal feasibility of multi stage flash desalination plant with brine-feed mixing and cooling. Energy 2014;76:1029–35.
[31] Gude VG, Nirmalakhandan N, Deng S, Maganti A. Desalination at low temperatures: and exergy analysis. Desalin Water Treatment 2012;40(1–3):272–81.
[32] Darwish MA, Yousef FA, Al-Najem NM. Energy consumption and costs with a multi-stage flashing (MSF) desalting system. Desalination 1997;109:285–302.
[33] Nafey AS, Fath HES, Mabrouk AA. Thermo-economic investigation of multi effect evaporation (MEE) and hybrid multi effect evaporation—multi stage flash (MEE-MSF) systems. Desalination 2006;201:241–54.
[34] Mabrouk AA, Nafey AS, Fath HES. Thermoeconomic analysis of some existing desalination processes. Desalination 2007;205:354–73.
[35] Hosseini SR, Amidpour M, Shakaib SE. Thermodynamic and exergy analysis of a combined power and desalination plant. J Energy: Eng Manage 2012;2(4), Winter 1391, 32–9.
[36] Hamed OA, Zamamiri AM, Aly S, Lior N. Thermal performance and exergy analysis of a thermal vapor compression desalination system. Energy Convers Manage 1996;37 (4):379–87.
[37] Hamed OA, Mohammad AK, Imam I, Mustafa GM, Khalid B, Al-Washmi H. In: Thermal performance of multistage flash desalination plants in Saudi Arabia. International Desalination Association World Congress on Desalination and Water Reuse, San Diego, USA; 1999.
[38] Fredrick B. Combined power and process: an exergy approach. London: Mechanical Engineering Publication Limited; 1995.
[39] Gagioli RA, El Sayed YM, Nashar A, Kamaluddin B. Second law efficiency and costing analysis of a combined power and desalination plant. J Energy Resour Technol 1988;110:114–8. ASME.
[40] Ettouney H, El-Dessouky H, Al-Juwayhel F. Performance of the once-through multistage flash desalination process. Proc Inst Mech A, J Power Energy 2002;216:229–41.
[41] Mussati SF, Aguirre PA, Scenna NJ. Improving the efficiency of the MSF once through (MSF-OT) and MSF-mixer (MSF-M) evaporators. Desalination 2004;166:141–51.
[42] Kahraman N, Cengel YA. Exergy analysis of a MSF desalination plant. Energy Convers Manage 2005;46:2625–36.
[43] Dessouky HT, Ettouny HM. Fundamentals of salt water desalination, 1st edn. Elsevier Science; 2002. ISBN 9780444508102.
[44] Sato N. An introduction to chemical thermodynamics to engineers. Elsevier Science and Technology Books; 2004.
[45] Sharqawy MH, Lienhard JH, Zubair SM. Thermophysical properties of seawater: a review of existing correlations and data. Desalin Water Treatment 2010;16:354–80.

CHAPTER 14

Exergy and Technoeconomic Analysis of Solar Thermal Desalination

Gopal N. Tiwari*, Lovedeep Sahota†
*Bag Energy Research Society, Varanasi, India
†Indian Institute of Technology (I.I.T.) Delhi, New Delhi, India

14.1 INTRODUCTION

The performance of solar thermal system is determined from the available useful energy and exergy of the system. Many researchers [1–8] put forth their efforts to optimize the design and operating parameters of the systems on the basis of energy and exergy analysis The consumption or burning of fossil fuels causes environmental degradation (air pollution) via CO_2 emissions. Fossil fuels have been becoming scare with time due its rapid use with growth in population; hence, it is necessary to find other alternatives to conserve fossil fuels. In this perspective, knowledge and awareness about the utilization of renewable energy (RE) sources, that is, solar energy which is an ample source of energy has gained momentum, globally in recent times.

In recent years, the use of exergy analysis in thermal designs (optimization) has been discussed by many researchers in the literature [9–11]. For the performance of thermal system, an execution of exergy analysis is based on the second law of thermodynamics which overcomes the limitations of an energy-based analysis. The exergy destruction is the measure of irreversibility (source of performance loss). Therefore, an exergetic analysis has been carried out for assessing the magnitude of exergy destruction by identifying the location, magnitude, and the source of thermodynamic inefficiencies in a thermal system [12,13]. The advanced exergy analysis splits total exergy destruction into two parts, viz., avoidable and unavoidable exergy destruction. Moreover, it can be divided into endogenous and exogenous exergy destruction. Exergetic analysis predicts the thermodynamic performance of the energy system as well as the efficiency of the system components by precisely quantifying the entropy generation of the components. Dincer [14] reported the important associations between energy and exergy, exergy

Renewable Energy Powered Desalination Handbook
https://doi.org/10.1016/B978-0-12-815244-7.00014-3

and environment, energy and sustainable development, and energy policy making and exergy.

The exergy loss, time, and capital cost-based correlation has been used in exergoeconomic analysis. This correlation clearly suggests that the system should be constructed in order to attain the overall optimal design corresponding to the exergy and economic characteristics. Basically, the exergoeconomic analysis is an exergy-based economic analysis method which can be used for the comparison of two energy conversion systems. In these systems, the identification and calculation of location, magnitude, causes, and costs of thermodynamic inefficiencies have been carried out through exergoeconomic analysis.

The main motive of the analysis is to achieve the optimal design of the system. For this purpose, mutual techniques of scientific disciplines (mainly thermodynamics) with economic disciplines (mainly cost accounting) have often been used. The cost accounting usually considers unit costs on the basis of energy for energy conversion devices. Worldwide, many researchers have recommend the costs are better distributed among the outputs if cost accounting is based on exergy (thermodynamic quantity) [15–30]. One basis for this declaration is that the consistent or reliable measure of economic value is exergy rather than energy; and different methodologies exist (exergoeconomics, thermoeconomics, etc.) for the exergy-based economic analysis.

Moreover, the development of renewable technology brings about the change in global climate but the issue of economic and financial viability of new and smart RESs is the main concern. In the area of technoeconomic analysis (project cost control, profitability analysis, planning, scheduling, optimization of operational research, etc.), engineering judgment and experience are broadly utilized in the application of scientific principles and techniques. From the user point of view, it is essential to make the RESs economically viable for their proper utilization. Better understanding and knowledge of cost analysis can be used for the effective economic analysis of RES with different methods and cash flow diagrams. Following are the important factors for the technoeconomic analysis of RESs:

(i) Initial investment, present cost, and initial cost for the fabrication of RES.
(ii) Annual operational and maintenance cost of RES.
(iii) Annual energy output (thermal energy or electrical energy).
(iv) Repairing cost of RES, if any.
(v) Interest rate, life span, and salvage of the RES.

Apart from these important points, following factors can also be considered for the economic analysis:

(i) Influence of embodied energy of RES on the environment due to CO_2 emission.

(ii) Annual energy utilized for the process of RES, pretreatments, etc.

(iii) CO_2 or carbon credit (CC) earned with the use of RES, etc.

In this chapter, energy and exergy analysis, the basic knowledge of cost analysis, cash-flow diagram, payback time period, etc. are discussed in detail for the effective economic analysis of RES.

14.2 PASSIVE SSSS AND DSSS

Solar distillation is the simplest technique for producing potable water from contaminated saline water by harvesting solar energy. A schematic view of passive SSSS and DSSS is shown in Fig. 14.1A and B, respectively. The body of the solar distiller or solar still is usually made of fiber-reinforced plastic (FRP) with top cover of transparent glass. The orientation of the solar still is one of the important parameters, therefore it should be kept in such a way that it can retrieve maximum solar radiation (SSSS: south facing; and DSSS: east-west orientation in the northern hemisphere).

The side walls and rectangular base of the solar still is painted mat black in order to absorb the maximum solar radiation. A fraction of solar radiation has been reflected from outer surface of the transparent glass cover and water surface; maximum solar radiation penetrates into the solar still cavity and gets absorbed by the basin water and blackened surface (basin liner). The remaining solar radiation reaches the blackened surface where it gets absorb mostly. After the absorption of solar radiation at the blackened surface, stored thermal energy is convected to the basin water mass and elevates its temperature in the basin (greenhouse effect). Thus, it leads to the higher difference in temperatures between basin water (evaporative surface) and inner surface of the top glass cover. Water vapors reaches inner surface of the top glass cover via mechanism of evaporative heat transfer and gets condensed there after releasing its latent heat to the cover. The condensed water or potable water droplets (free from the contaminants such as bacteria, heavy metals, cancer-causing pollutants, pathogens, salts, and other microscopic organism) trickles into the channels provided at the lower ends of the glass cover under gravity [31]. Solar still is more reliable, environmental friendly, and a simple water purifier plant which can be maintained at home site and easy to operate.

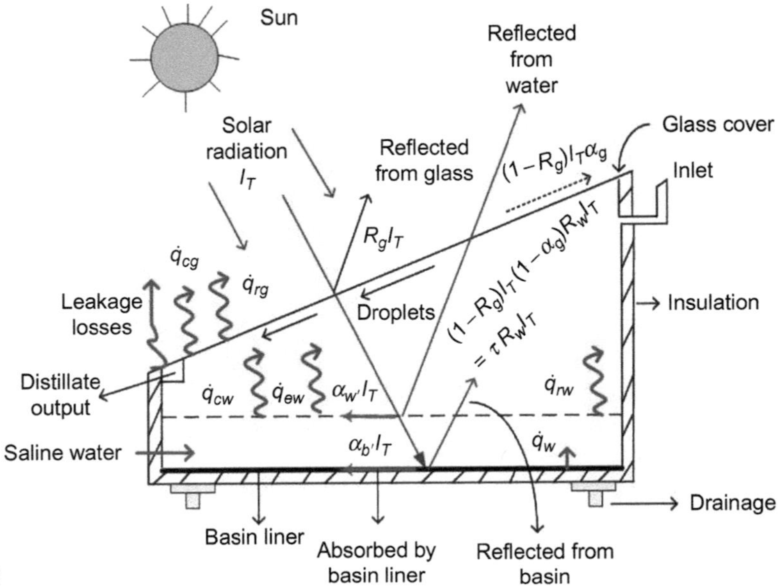

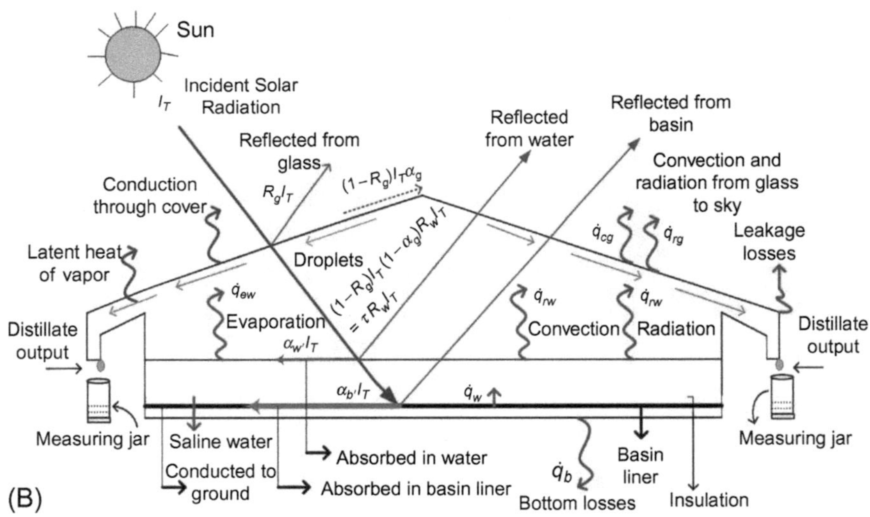

Fig. 14.1 Schematic diagram of (A) single slope solar still (SSSS) and (B) double slope solar still (DSSS).

In the case of DSSS, it is important to mention that when east side acts as a source of solar energy (input) then west side acts as a condensing surface (in morning hours—northern hemisphere) and vice versa; whereas both the glass covers (east and west side) acts as a source during peak sunshine hours.

14.2.1 Energy and Exergy Analysis of Passive SSSS

Following assumptions have been taken into account in writing the energy balance of passive solar still [32]:

(a) All processes in the system are in quasi-steady state.

(b) There is no temperature stratification between the layers of fluid in the system.

(c) The heat capacity of the absorbing material, the insulation (bottom and sides), and the condensing cover is negligible.

(d) The system is vapor leakage proof.

Energy balances of different components of the passive SSSS are given as follows:

(i) Glass cover

$$\alpha_g I_S + h_{1f}\left(T_f - T_{gi}\right) = h_{1g}\left(T_{go} - T_a\right) \tag{14.1}$$

where α_g is the absorptivity of the glass, $I_s(t)$ the solar intensity on the still (W/m^2), $h_{1f}(=h_{ew}+h_{cw}+h_{rw})$ the total internal heat transfer coefficient from fluid (water) to glass (W/m°C); h_{ew}, h_{cw}, h_{rw}, respectively, are the evaporative, convective, and radiative heat transfer coefficients from water surface, $h_{1g}(=h_{rg}+h_{cg})$ the total external heat transfer coefficient from glass to ambient (W/m°C); h_{rg} and h_{cg} are the radiative and convective heat transfer coefficients, respectively, from the glass surface to ambient (W/m°C); T_f the basin water temperature (°C), T_{gi} the temperature of inner surface of the glass cover (°C), T_{go} the temperature of outer surface of the glass cover (°C), and T_a is the ambient temperature (°C).

(ii) Basin liner

$$\alpha_b I_S = h_{b,f}\left(T_b - T_f\right) + h_{ba}\left(T_b - T_a\right)\left(1 + \frac{A_{SS}}{A_S}\right) \tag{14.2}$$

where α_b is the absorptivity of the basin liner, h_{bf} the convective heat transfer coefficient from basin liner to fluid (water) (W/m°C), h_{ba} the heat transfer coefficient from basin to ambient (W/m°C), T_b the temperature of basin liner (°C), and A_{ss} is the area of the sides of solar still (m^2).

The side area A_{SS} being very small compared with the still basin area of solar still (A_S), therefore, the term $\frac{A_{SS}}{A_S}$ can be neglected for lower depth of basin water. Therefore, Eq. (14.2) can be expressed as

$$\alpha_b I_S = h_{b,f}\left(T_b - T_f\right) + h_{ba}\left(T_b - T_a\right) \tag{14.3}$$

(iii) Base fluid (water) mass

$$M_f C_f \frac{dT_f}{dt} = \alpha_f I_S + h_{b,f}(T_b - T_f) - h_{1f}(T_f - T_{gi}) \tag{14.4}$$

where M_f is the basin fluid (water) mass (kg); C_f the specific heat of the fluid (J/kg K); α_f the absorptivity of the basin water; and α_b is the absorptivity of the basin.

The empirical relations for the top loss coefficient are given as follows:

$$h_c = 5.7 + 3.8\,V \quad \text{for} \quad 0 \le V \le 5\,\text{m/s} \tag{14.5}$$

and

$$h_c = 2.8 + 3\,V \quad \text{for} \quad 0 \le V \le 7\,\text{m/s} \tag{14.6}$$

where V is the wind speed (m/s).

Moreover, the relation to evaluate the internal (radiative, convective, evaporative) heat transfer coefficients are expressed as

$$h_{rw} = \sigma \epsilon_{eff} \left[(T_w + 273)^2 + (T_{gi} + 273)^2 \right] (T_w + T_{gi} + 546) \tag{14.7}$$

where $\frac{1}{\epsilon_{eff}} = \frac{1}{\epsilon_w} + \frac{1}{\epsilon_g} - 1$

$$h_{cw} = 0.844 \left[(T_w - T_{gi}) + \frac{(P_w - P_{gi})(T_w + 273)}{(268.9 \times 10^3 - P_w)} \right]^{\frac{1}{3}} \tag{14.8}$$

$$h_{ew} = 16.273 \times 10^{-3} (h_{cw}) \left(\frac{P_w - P_{gi}}{T_w - T_{gi}} \right) \tag{14.9}$$

where

$$P(T) = \exp\left[25.317 - \left(\frac{5144}{273 + T} \right) \right] \tag{14.10}$$

The value of h_{1f} can be determined by considering known values of water and glass temperatures at $t = 0$, that is, $T_f(t=0) = T_{f0}$ and $T_g(t=0) = T_{go}$.

Assuming that $T_{gi} \approx T_{go}$, substitute the values of T_{gi} and T_b from Eqs. (14.1), (14.2), respectively, in Eq. (14.4), one can obtain

$$\frac{dT_f}{dt} + aT_f = f(t) \tag{14.11}$$

where $a = \frac{U_L}{(MC)_f}$; $f(t) = \frac{(\alpha\tau)_{eff} I(t) + U_L T_a}{(MC)_f}$

The unknown terms are given in Appendix 14.2.

The rate of evaporative heat transfer is expressed as

$$\dot{q}_{ef} = h_{ef}\left(T_f - T_g\right) \tag{14.11a}$$

The solution of first-order differential equation (Eq. 14.11) for basin water temperature can be expressed as [32]

$$T_w = \frac{f(t)}{a}\left[1 - e^{-a\Delta t}\right] + T_{w0}e^{-a\Delta t} \tag{14.12}$$

where T_{w0} is the basin water temperature at $t=0$.

Productivity (yield) of the system can be expressed as

$$\dot{m}_{ef} = \frac{\dot{q}_{ef}}{L} \times 3600 = \frac{h_{ef}\left(T_f - T_g\right)}{L} \times 3600\,\text{kg/m}^2\,\text{h} \tag{14.13}$$

where L is the latent heat of vaporization.

The rate of evaporative heat transfer $\left(\dot{q}_{ef}\right)$ can be expressed as

$$\dot{q}_{ef} = h_{ef}\left(T_f - T_g\right) \tag{14.14}$$

Thermal energy and exergy of passive SSSS can be expressed as

$$E_{hourly,\,en} = h_{ef}\left(T_f - T_{gi}\right) \tag{14.15}$$

$$E_{hourly,\,ex} = h_{ef}\left[\left(T_f - T_{gi}\right) - (T_a + 273)ln\left(\frac{T_f + 273}{T_{gi} + 273}\right)\right] \tag{14.16}$$

Input exergy can be calculated using the expression given by Petela [33] as

$$\dot{E}_{ex,\,in} = A_g I(t)\left[1 - \frac{4}{3}\left(\frac{T_a}{T_s}\right) + \frac{1}{3}\left(\frac{T_a}{T_s}\right)^4\right] \tag{14.17}$$

where T_s is the sun temperature (°C).

Further, the instantaneous gain thermal energy efficiency of distillation unit can be determined as follows:

$$\eta_i = \frac{\dot{q}_{ef}}{I(t)} = \frac{h_{ef}\left(T_f - T_g\right)}{I(t)} \tag{14.18}$$

The above equation can be reexpressed by substituting the value of T_f from Eq. (14.12) as given below

$$\eta_i = \left(\frac{h_{ef}h_{1g}}{h_{1f} + h_{1g}}\right)\left(\frac{1}{U_L}\right)\left[(\alpha\tau)_{eff}\left(1 - e^{-a\Delta t}\right) + U_L\frac{\left(T_{f0} - T_a\right)}{I(t)}e^{-a\Delta t}\right] \tag{14.19}$$

The plot of η_i versus $\frac{(T_w - T_a)}{I(t)}$ represents a straight line with $(\alpha\tau)_w$ and $-U_L'$ as the intercept and the slope, respectively.

14.2.2 Energy and Exergy Analysis of Passive DSSS

Following the assumptions given in previous section, the energy balances for different components of passive DSSS (Fig. 14.1B) are expressed below [34].

(i) East side of DSSS

$$\alpha_g I_{SE} A_{gE} + h_{1f,E}\left(T_f - T_{giE}\right)\left(\frac{A_b}{2}\right) - h_{EW}\left(T_{giE} - T_{giW}\right)A_{gE} = \left(\frac{K_g}{L_g}\right)\left(T_{giE} - T_{gOE}\right)A_{gE} \tag{14.20}$$

where α_g is solar energy absorbed by the transparent glass cover, I_{SE} the solar intensity on the east side of the glass cover (W/m^2), A_{gE} the surface area of the glass cover on the east side (m^2), $h_{1f,E}(=h_{ef,E}+h_{rf,E}+h_{cf,E})$ the total internal heat transfer coefficient on the east side (W/m^2°C); $h_{ef,E}$, $h_{rf,E}$, and $h_{cf,E}$ the internal evaporative, radiative, and convective heat transfer coefficient, respectively, on the east side (W/m^2°C); $h_{ef,W}$, $h_{rf,W}$, and $h_{cf,W}$ the internal evaporative, radiative, and convective heat transfer coefficient on the west side (W/m^2°C); $h_{1f,W}(=h_{ef,W}+h_{rf,W}+h_{cf,W})$ the total internal heat transfer coefficient on the west side (W/m^2°C); h_{EW} the internal radiative heat transfer coefficient between the east and the west glass cover (W/m^2°C), K_g the thermal conductivity of glass cover (W/m°C); L_g the thickness of glass cover (m), T_{goE} the temperature of outer surface of the east side glass cover (°C), A_b the area of the rectangular basin (m^2), T_{giE} the temperature of inner surface of the east side glass cover (°C), T_{giW} the temperature of inner surface of the west side glass cover (°C).

$$\left(\frac{K_g}{L_g}\right)\left(T_{giE} - T_{goE}\right)A_{gE} = h_{1gE}\left(T_{goE} - T_a\right)A_{gE} \tag{14.21}$$

where $h_{1gE}(=h_{cg,E}+h_{rg,E})$ is the total external heat transfer coefficient from the east side glass over to ambient (W/m^2°C), $h_{cg,E}$ the external convective heat transfer coefficient from the east side glass over to ambient (W/m^2°C), $h_{rg,E}$ the external radiative heat transfer coefficient from the east side glass over to ambient (W/m^2°C), and T_a is the ambient temperature (°C).

(ii) West side of DSSS

$$\alpha_g I_{SW} A_{gW} + h_{1f,W}\left(T_f - T_{giW}\right)\left(\frac{A_b}{2}\right) - h_{EW}\left(T_{giE} - T_{giW}\right)A_{gW} = \left(\frac{K_g}{L_g}\right)\left(T_{giW} - T_{goW}\right)A_{gW} \tag{14.22}$$

where I_{SW} is the solar intensity on the west side of the glass cover (W/m^2), A_{gW} the surface area of the east side glass cover (m^2), T_f the basin fluid (water) temperature (°C), and T_{goW} is the temperature of outer surface of the west side glass cover (°C).

$$\left(\frac{K_g}{L_g}\right)\left(T_{giW} - T_{gOW}\right)A_{gW} = h_{1gW}\left(T_{goW} - T_a\right)A_{gW} \tag{14.23}$$

where $h_{1f,W} = h_{ef} + h_{cf} + h_{rf}$ and $h_{1gW} = h_{cW} + h_{rW}$

(iii) Basin liner

$$\alpha_b(I_{SE} + I_{SW}) = 2h_{b,f}\left(T_b - T_f\right) + 2h_{ba}(T_b - T_a) \tag{14.24}$$

where $h_{b,f}$ is the heat transfer coefficient from basin liner to fluid (water) (W/m^{2}°C), α_b the absorptivity of the basin liner; T_b the basin temperature (°C); T_a the ambient temperature (°C); T_f the (fluid) water temperature (°C); h_{ba} the heat transfer coefficient from basin liner to ambient air (W/m^{2}°C).

(iv) Base fluid (water) mass

$$M_f C_f \frac{dT_f}{dt} = \alpha_f (I_{SE} + I_{SW})\left(\frac{A_b}{2}\right) + 2h_{b,f}\left(T_b - T_f\right)\left(\frac{A_b}{2}\right) - h_{1f,E}\left(T_f - T_{giE}\right)\left(\frac{A_b}{2}\right) - h_{1f,W}\left(T_f - T_{giW}\right)\left(\frac{A_b}{2}\right) \tag{14.25}$$

where α_f is the solar energy absorbed by the basin fluid (water), C_f the specific heat of water (J/kg K), and M_f is the mass of basin fluid (water) (kg).

On solving Eqs. (14.20)–(14.24), one can obtain

$$T_{giE} = \frac{\left[A + T_f B\right]}{H} \tag{14.26}$$

and

$$T_{giW} = \frac{\left[A' + T_f B'\right]}{H} \tag{14.27}$$

On solving Eq. (14.25) using Eqs. (14.26), (14.27), one can obtain

$$\frac{dT_f}{dt} + aT_f = f(t) \tag{14.28}$$

Again, the solution of above differential equation can be expressed as

$$T_f = \frac{f(t)}{a}\left[1 - e^{-a\Delta t}\right] + T_{f0}e^{-a\Delta t} \tag{14.29}$$

where $a = \left(\frac{A_b}{2H}\right)\left[h_{1f,E}(E_1 + E_2) + h_{1f,W}\left(E_1' + E_2'\right)\right]$

$$f(t) = \left(\frac{A_b}{2H}\right)\{[(K_{1E} + \alpha_{wb})I_{SE}(t) + (K_{1W} + \alpha_{wb})I_{SW}(t)] + T_a(H_1 + H_2 + H_3 + H_4)\}$$

where the unknown terms are given in Appendix 14.2.

Thermal energy and exergy of passive DSSS are expressed as

$$E_{hourly,en} = \left[h_{ef,E}\left(T_f - T_{giE}\right) + h_{ef,W}\left(T_f - T_{giW}\right)\right]\left(\frac{A_b}{2}\right) \tag{14.30}$$

$$E_{hourly,ex} = \left\{h_{ef,E}\left[\left(T_f - T_{giE}\right) - (T_a + 273)ln\left(\frac{T_f + 273}{T_{giE} + 273}\right)\right] + h_{ef,W}\left[\left(T_f - T_{giW}\right) - (T_a + 273)ln\left(\frac{T_f + 273}{T_{giW} + 273}\right)\right]\right\}\left(\frac{A_b}{2}\right) \tag{14.31}$$

Instantaneous gain thermal energy efficiency (η_{th}) of DSSS can be written as

$$\eta_{th} = \frac{\left[h_{ef,E}\left(T_f - T_{giE}\right) + h_{ef,W}\left(T_f - T_{giW}\right)\right]A_b}{A_{gE}I_{SE}(t) + A_{gW}I_{SW}(t)} \tag{14.32}$$

On substituting Eqs. (14.26), (14.27) in above equation, one can obtain

$$\eta_{th} = F'\left[(\alpha\tau)_{eff} + \frac{\left(T_{bf0} - T_a\right)}{I(t)}U_{eff}\right] \tag{14.33}$$

Eq. (14.33) is the characteristic equation of passive DSSS system. The unknowns in above equations are given as

$$(\alpha\tau)_{eff} = \frac{\left\{[(K_{1E} + \alpha_{wb})I_{SE}(t) + (K_{1W} + \alpha_{wb})I_{SW}(t)]\left(1 - e^{-a\Delta t}\right) - \left[K_{1E}'I_{SE}(t) + K_{1W}'I_{SW}(t)\right]\right\}}{A_{gE}I_{SE}(t) + A_{gW}I_{SW}(t)}$$

where the unknown terms are given in Appendix 14.2.

The convective $[h_{cf,E}(T_{bf} - T_{giE})A_b + h_{cf,W}(T_{bf} - T_{giW})A_b]$ and radiative $[h_{rf,E}(T_f - T_{giE})A_b + h_{rf,W}(T_f - T_{giW})A_b]$ thermal losses from the evaporative surface to the inner surface of the top transparent glass cover (east and west sides) are negligible compared with the heat lost on increasing the temperature of water (sensible heat). Therefore, the lost thermal energy efficiency of passive DSSS is obtained as

$$\eta_{L,th} = \frac{M_f C_f (T_f - T_{bf0})}{(A_{gE} I_{SE} + A_{gW} I_{SW})} \tag{14.34}$$

On substituting T_f from Eq. (14.29) with expressions of a and $f(t)$, one can obtain

$$\eta_{L,th} = F'_L \left[(\alpha\tau)_{eff} - U'_{eff} \frac{(T_{f0} - T_a)}{I(t)} \right] \tag{14.35}$$

where $F'_L = \frac{M_f C_f}{H'}; (\alpha\tau)_{eff} = \frac{[(K_{1E} + \alpha_{wb}) I_{SE}(t) + (K_{1W} + \alpha_{wb}) I_{SW}(t)]}{I(t)}; U'_{eff} = H'(e^{-a\Delta t} - 1)$.

For given parameters of the system, the variations η_{th} and $\eta_{L,th}$ with reduced temperature $((T_{f0} - T_a)/I(t))$ gives the instantaneous thermal gain energy efficiency and thermal loss energy efficiency, respectively.

The hourly thermal exergy efficiency of passive DSSS is obtained from the following modified equation:

$$\eta_{hourly,ex} = \left(\frac{1}{0.933 \times A_s \times I_s(t)}\right) \left\{ h_{ef,E} \left[(T_f - T_{giE}) - (T_a + 273) ln\left(\frac{T_f + 273}{T_{giE} + 273}\right) \right] + h_{ef,W} \left[(T_f - T_{giW}) - (T_a + 273) ln\left(\frac{T_f + 273}{T_{giW} + 273}\right) \right] \right\} \left(\frac{A_b}{2}\right) \tag{14.36}$$

The factor 0.933 has been used to convert sun radiation to exergy.

The energy and exergy analysis of conventional single slope (area 1×1 m^2 and inclination 15 degrees) and double slope (area 2×1 m^2 and inclination 15 degrees) solar distillation systems for the climatic conditions of New Delhi (India) have been experimentally studied by Singh et al. [35]. The hourly variation of thermal energy and exergy of both the systems for typical days in the months of January and May is shown in Fig. 14.2.

The higher values of hourly thermal energy and exergy of passive SSSS compared with the passive DSSS is attributed to the fact that SSSS absorbs higher solar radiation for both the months due to its orientation which results in higher temperature in the case of SSSS. Furthermore, the thermal exergy for the month of May is found to be higher for both the systems due to availability of higher solar radiation in the month of May which intern enhances the temperature difference between the evaporative surface and the inner surface of the glass cover. The variation of monthly exergy gain for both the systems is presented in Fig. 14.3.The maximum value of exergy

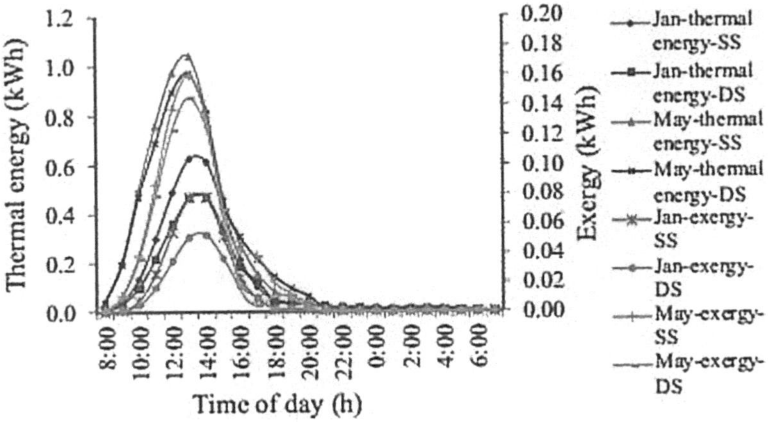

Fig. 14.2 Hourly variation of thermal energy and exergy of passive single slope solar still (SSSS) and double slope solar still (DSSS) for the months of January and May.

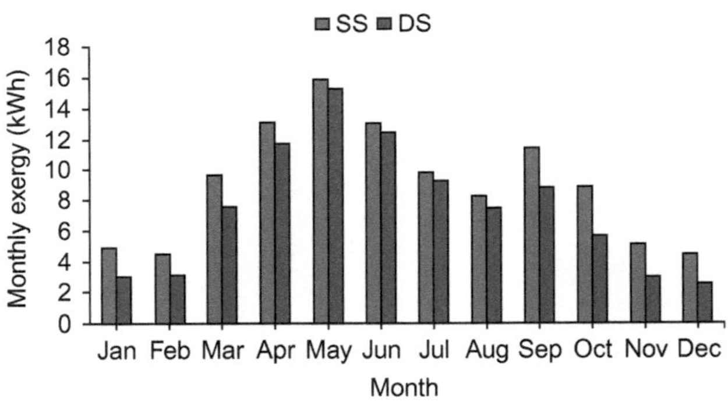

Fig. 14.3 Variation of monthly exergy of passive single slope solar still (SSSS) and double slope solar still (DSSS).

is observed for the month of May for both the systems due to the availability of maximum solar radiation in this month as mentioned earlier whereas the minimum exergy corresponds to the month of December.

The variation of monthly yield obtained from passive SSSS and DSSS for 50 kg basin water mass is depicted in Fig. 14.4. The productivity (yield) has been found to be higher for the month of May for both the systems.

Furthermore, Fig. 14.5 presents the hourly variation of thermal energy efficiency and exergy efficiency of both the systems for the months of January and May. It has been observed that both the thermal energy and exergy efficiency are higher for passive SSSS than DSSS till 2 pm. It is

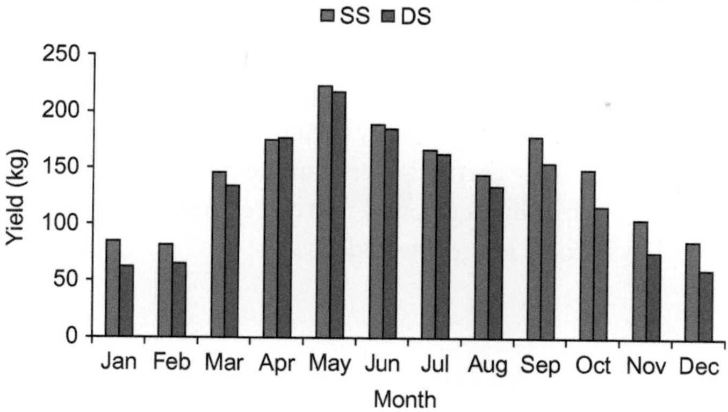

Fig. 14.4 Variation of monthly yield of passive single slope solar still (SSSS) and double slope solar still (DSSS).

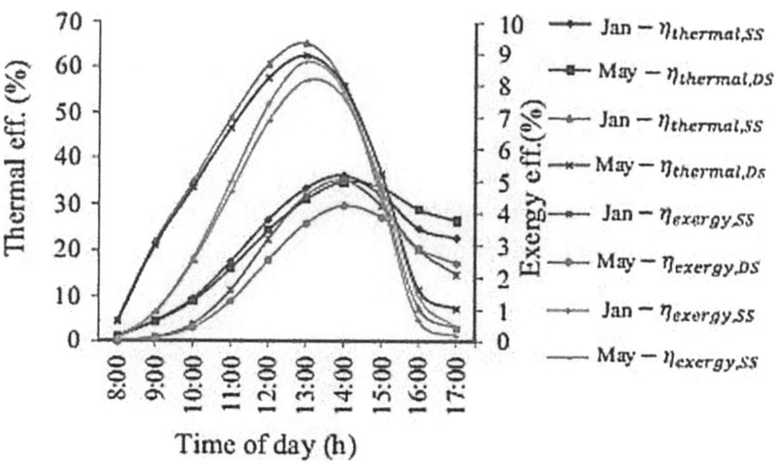

Fig. 14.5 Hourly variation of thermal energy and exergy efficiency of passive single slope solar still (SSSS) and double slope solar still (DSSS) for the months of May and January.

due to the fact that the SSSS receives the maximum solar radiation than the DSSS due to still orientation. After 2 pm, the passive DSSS gives better thermal energy and exergy efficiency due to the orientation of DSSS.

The daily thermal exergy efficiency of passive SSSS and DSSS for the month of May have been found to be 4.49% and 4.35%, respectively; whereas the daily thermal exergy efficiency of passive SSSS and DSSS for the month of January have been found to be 2.91% and 2.50%, respectively. The daily thermal energy efficiency are 45.03% and 42.67% for the SSSS and

DSSS, respectively, for the month of May. On the other hand, the daily thermal energy efficiency 25.12% and 25.01%, respectively, has been found for the SSSS and DSSS for the month of January. Consequently, it has been concluded that the passive SSSS gives better performance than the passive DSSS on the basis of daily thermal exergy and energy efficiency. It happens due to the occurrence of higher hourly thermal exergy and energy efficiency of SSSS than DSSS for a longer period.

14.3 ENERGY AND EXERGY ANALYSIS OF ACTIVE SOAR STILL INTEGRATED WITH N-PHOTOVOLTAIC THERMAL PARTIALLY COVERED FLAT PLAT COLLECTOR (N-PVT-FPC)

In the case of active solar still, the temperature difference between the evaporative surface and inner surfaces of the transparent glass cover can be improved by injecting the external thermal energy into the integrated still basin. For this purpose, the basin water is circulated through the external system either in natural or forced mode of operation reliant upon the requirement. The basin water can be heated either directly (direct feeding of external thermal energy) or indirectly (via heat exchanger). To minimize the thermal losses during circulation, the insulated connecting pipes are used. Systematic diagram of series connected N-photovoltaic thermal partially covered flat plat collectors (PVT-FPC) integrated to SSSS is shown in Fig. 14.6.

Following the assumptions given in Section 14.2.1, the energy balances of different components of the active SSSS are given below.

(a) Inner glass cover

$$\alpha_g' I(t) A_g + h_1 (T_w - T_{gi}) A_g = \frac{K_g}{L_g} (T_{gi} - T_{go}) A_g \tag{14.37}$$

where α_g' is the absorptivity of the glass, $I(t)$ the solar intensity (W/m^2), A_g the area of the glass cover (m^2), $h_1 (= h_{ew} + h_{cw} + h_{rw})$ the total internal heat transfer coefficient from water to glass (W/m°C); h_{ew}, h_{cw}, h_{rw}, respectively, are the evaporative, convective, and radiative heat transfer coefficients from water surface (W/m°C), T_w the basin water temperature (°C), T_{gi} the temperature of inner surface of the glass cover (°C), T_{go} the temperature of outer surface of the glass cover (°C), K_g the thermal conductivity of the glass cover (W/m°C), and L_g is the thickness of the glass cover (m).

(b) Outer glass cover

$$\frac{K_g}{L_g} (T_{gi} - T_{go}) A_g = h_2 (T_{go} - T_a) A_g \tag{14.38}$$

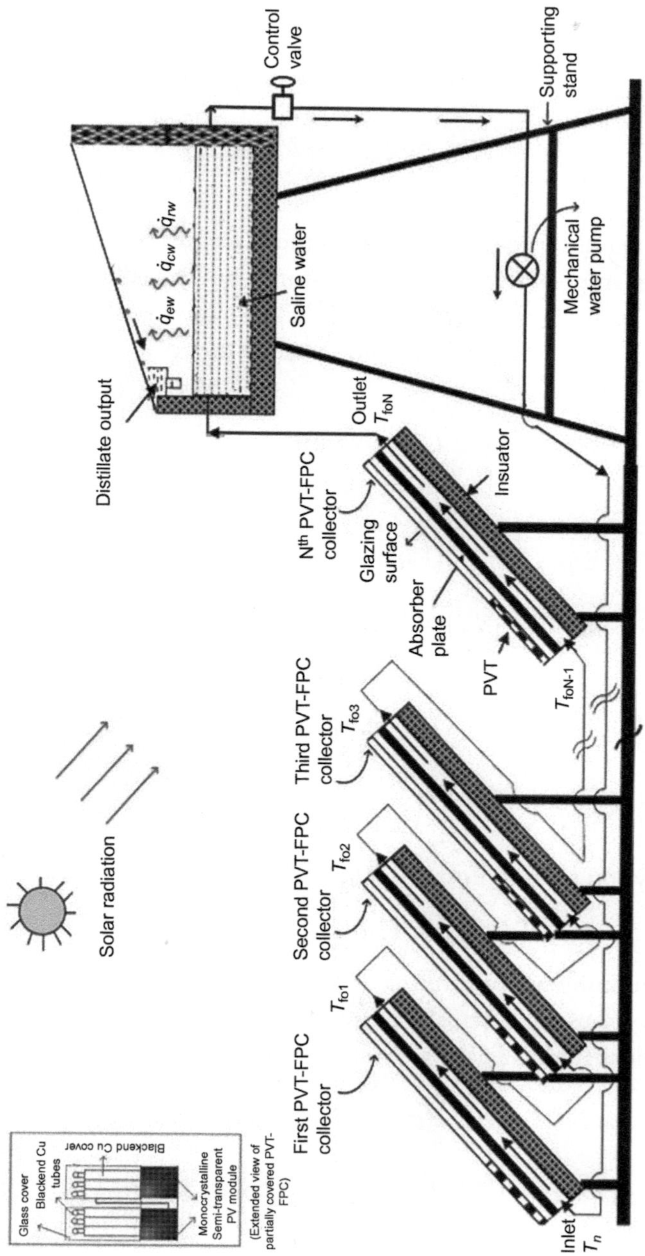

Fig. 14.6 Schematic view of an active single slope solar still (SSSS) coupled with partially covered N-flat plate collectors.

where T_a is the ambient temperature (°C), $h_2(=h_{cg}+h_{rg})$ the total external heat transfer coefficient from glass surface to ambient (W/m^2°C); h_{rg} and h_{cg} are the radiative and convective heat transfer coefficients, respectively, from the glass surface to ambient (W/m°C).

(c) Water mass

$$\alpha'_w I(t)A_b + h_3(T_b - T_w)A_b + \dot{Q}_{uth,N} = M_w C_w \frac{dT_w}{dt} + h_1(T_w - T_{gi})A_b \tag{14.39}$$

where $\alpha_w{}'$ is the absorptivity of water, A_b the basin area of the solar still (m^2), T_b the basin temperature, $h_3(=h_w)$ the natural convective heat transfer coefficient from basin liner to water (W/m^2°C). C_w the specific heat of water, (J/kg K), and M_w is the mass of basin water in the solar still (kg).

(d) Basin liner

$$\alpha'_b I(t)A_b = h_3(T_b - T_w)A_b = h_b(T_b - T_a)(A_b + A_s) \tag{14.40}$$

where A_s is the area of the sides of solar still (m^2) and h_b is the heat transfer coefficient of the basin liner (W/m^2°C).

On solving Eqs. (14.37), (14.38), one can obtain

$$h_1(T_W - T_{gi}) = \left[(T_w - T_a)U_{12} - \alpha''_g I(t)C_1\right] \tag{14.41}$$

$$C_1 = \frac{h_1}{h_1 + U'_{g,a}};\quad U_{12} = \frac{h_1 U'_{g,a}}{h_1 + U'_{g,a}};\quad \alpha''_g = \alpha'_g \frac{A_g}{A_b};$$

$$U_{g,a} = \left(\frac{K_g}{L_g}h_2\right) / \left(\frac{K_g}{L_g} + h_2\right) \text{ and } U'_{g,a} = U_{g,a}\left(\frac{L_g}{K_g}\right)$$

Similarly, on solving Eqs. (14.39), (14.40), one can obtain

$$h_3(T_b - T_w) = C_2\alpha'_b I(t) + U_{3b}(T_a - T_w) \tag{14.42}$$

where the unknowns are given in Appendix 14.2.

The rate of useful thermal energy of series connected partially covered N-photovoltaic thermal–flat plate collectors (N-PVT-FPC is expressed as [36]

$$\dot{Q}_{uth,N} = N(A_c + A_m)\left[(\alpha\tau)_{eff,N} I_c(t) - U_{L,N}(T_{fi} - T_a)\right] \tag{14.43}$$

$$(\alpha\tau)_{eff,N} = \frac{(AF_R(\alpha\tau))_1}{(A_c + A_m)}\left[\frac{1-(1-K_{k,A})^N}{NK_{k,A}}\right];$$

$$U_{L,N} = \frac{(AF_R U_L)_1}{(A_c + A_m)}\left[\frac{1-(1-K_{k,A})^N}{NK_{k,A}}\right];\quad K_{k,A} = \left[\frac{(AF_R U_L)_1}{\dot{m}_f C_f}\right]$$

Therefore, on substituting the values of $\dot{Q}_{uth,N}$, $h_1(T_W - T_{gi})$, and h_3 $(T_b - T_w)$ from Eqs. (14.43), (14.41), (14.42) respectively in Eq. (14.39), one can obtain

$$\frac{dT_w}{dt} + aT_w = f(t) \tag{14.44}$$

where $a = \left[\frac{U + N(A_c + A_m)U_{L,N}}{M_w C_w}\right]$;

$$f(t) = \frac{\left[\alpha_{eff} I(t) + N(A_c + A_m)(\alpha\tau)_{eff,N} I_c(t) + \left(U + N(A_c + A_m)U_{L,N}\right)\right]}{M_w C_w}$$

where the unknowns are given in Appendix 14.2.

The temperature of basin water can be obtained by solving the above first-order differential equation as given below

$$T_w = \frac{f(t)}{a}\left[1 - e^{-a\Delta t}\right] + T_{w0}e^{-a\Delta t} \tag{14.45}$$

Thermal energy and exergy of the active single slope solar distillation system can be obtained from Eqs. (14.15), (14.16), respectively.

Thermal energy and exergy efficiency of the active system can be expressed as

Thermal energy efficiency;

$$\eta_{the,en} = \frac{\dot{m}_{ef} \times L}{[A_b I(t) + N A_c I_c(t)]} \times 100 \tag{14.46}$$

Thermal exergy efficiency;

$$\eta_{the,ex} = \frac{h_{ef}\left[\left(T_f - T_{gi}\right) - (T_a + 273)ln\left(\frac{T_f + 273}{T_{gi} + 273}\right)\right]A_b}{(0.933)[A_b I(t) + N A_c I_c(t)]} \times 100 \tag{14.47}$$

where $I_c(t)$ and $I(t)$ are the total solar radiation falling on the flat plat collector (FPC) and transparent glass cover of the still, respectively, A_c the area of the solar collector (m^2).

Active solar distillation system coupled with two hybrid photovoltaic thermal–flat plate collectors (PVT-FPC) have been experimentally studied by Singh et al. [37] for the climatic conditions of New Delhi (India).

Electrical exergy efficiency can be expressed as

$$\eta^{elc}_{hourly,ex} = \frac{(FF \times V_0 \times I_{sc}) - V_L I_L}{0.933 \times A_m \times I_c} \times 100 \tag{14.48}$$

Overall thermal energy efficiency can be expressed as

$$\eta^{ovr}_{hourly,en} = \left[\frac{\dot{m}_{ef} L}{(A_c I_c + A_s I_s) \times 3600} + \frac{(FF \times V_0 \times I_{sc})}{0.38 \times A_m I_c}\right] \times 100 \qquad (14.49)$$

Overall thermal exergy efficiency can be expressed as

$$\eta^{ovr}_{hourly,ex} = \frac{h_{ef}\left[\left(T_f - T_{gi}\right) - (T_a + 273) ln\left(\frac{T_f + 273}{T_{gi} + 273}\right)\right] A_b + \left[(FF \times V_0 \times I_{sc}) - V_L I_L\right]}{0.933\left[(A_m + A_c) I_c + A_s I_s\right]} \times 100 \qquad (14.50)$$

Fig. 14.7A and B presents the hourly variation of thermal energy efficiency of active solar still for the month of November and February, respectively. It is found to be 11.6% and 11.35% for the month of November and February, respectively. The partially covered portion of the flat plate collector results in low heat gain and this gained heat is further delivered to the basin water mass; this fact is responsible for the low value of daily thermal efficiency. It causes slow rise in the temperature of basin water and hence low rate of evaporation.

The instantaneous efficiency or the characteristic curve of the active hybrid system for both the months is presented in Fig. 14.8A and B, respectively. A fair agreement between the experimental and theoretical results has been observed.

The variation of hourly thermal exergy efficiency for the months of November and February is shown in Fig. 14.9. The daily thermal exergy efficiency is observed to be 3.92% and 3.93% for the month of November and February, respectively.

The variation of hourly electrical exergy efficiency is shown in Fig. 14.10. It is found to be in the range of 2.21%–28.53%; and the daily electrical exergy efficiency has been estimated as 7.24% and 15.36% for the month of November and February, respectively.

Fig. 14.11A and B presents the variation of hourly overall thermal exergy and energy efficiency, respectively, for the months of November and February. The overall thermal exergy efficiency is found to be in the range of 0.03%–25%whereas overall thermal energy efficiency is found to be 1.02%–69.06%. The daily overall exergy and energy efficiency are5.71% and 10.09%, respectively, for the month of November; and 51.28% and 68.13%, respectively, for the month of February.

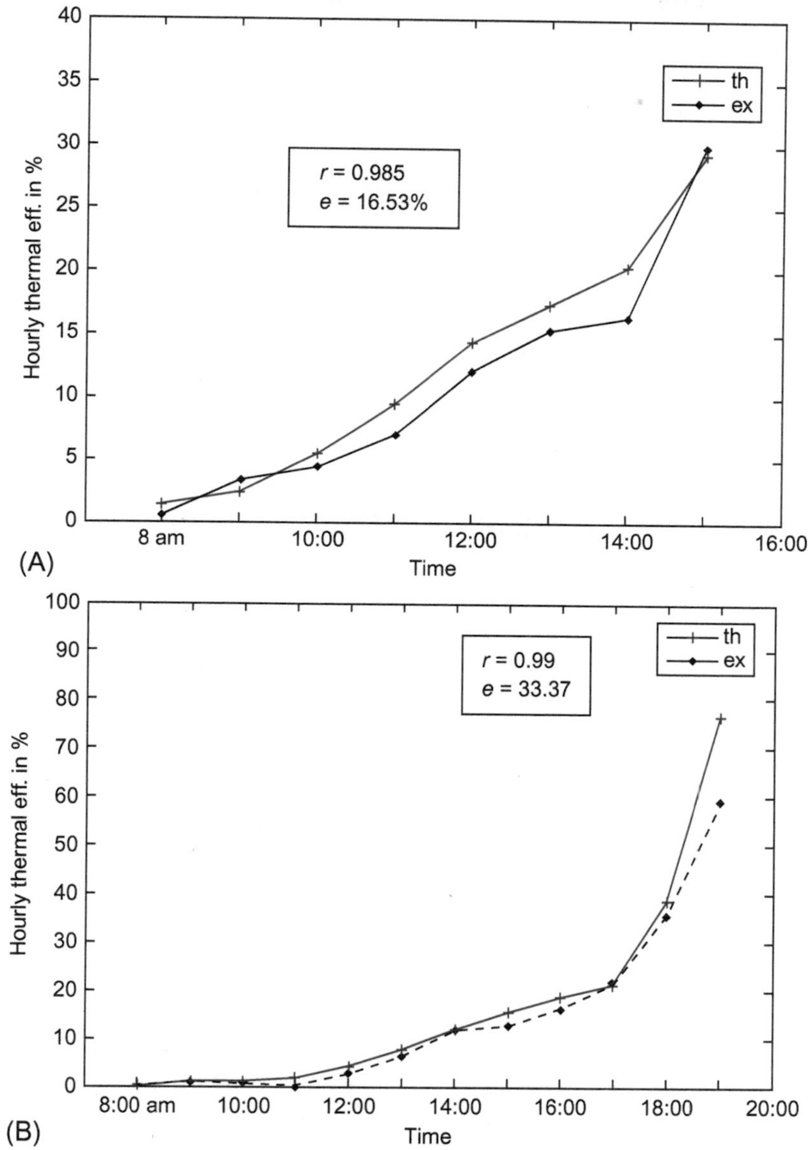

Fig. 14.7 Hourly variation of thermal energy efficiency of active solar still for the month of (A) November and (B) February.

A higher value of overall thermal efficiency is due to the conversion of electrical energy into the equivalent thermal energy obtained by dividing with factor 0.38 (efficiency of conventional thermal power plant); and the

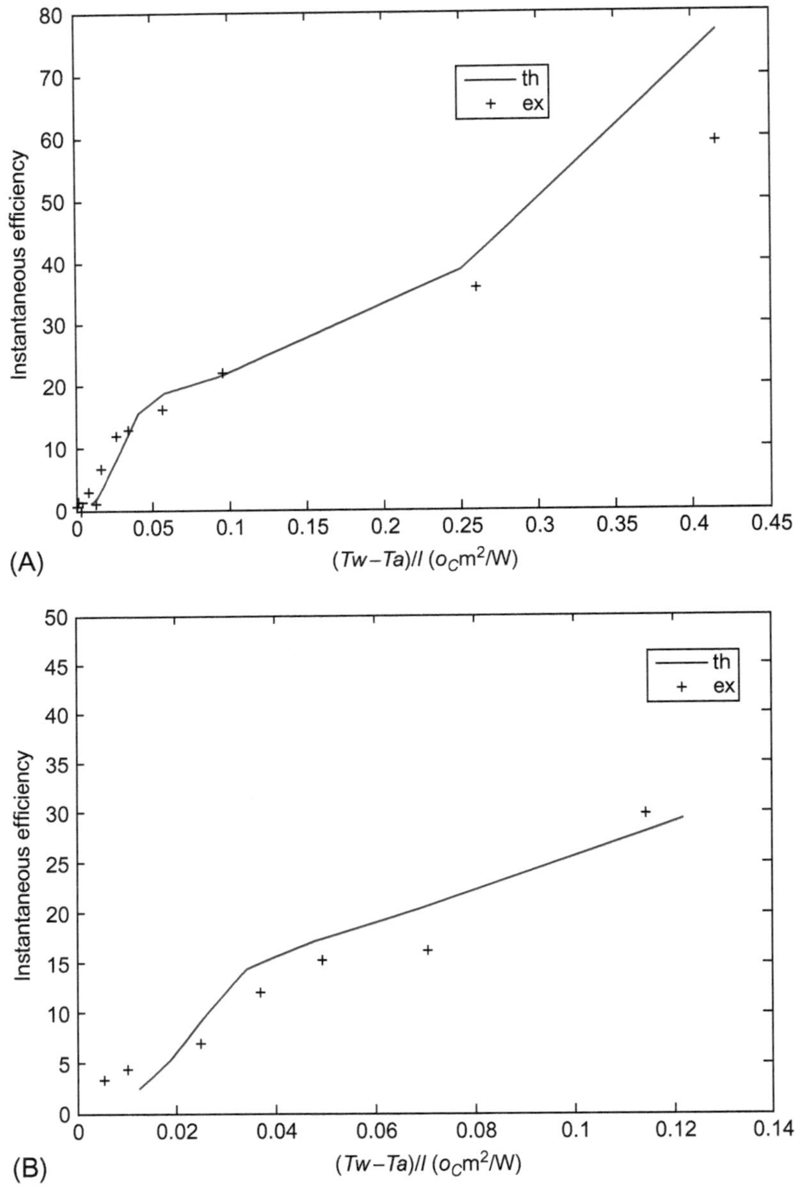

Fig. 14.8 Variation of instantaneous efficiency of the active hybrid system for the month of (A) November and (B) February.

electrical energy and thermal energy provided by the photovoltaic module. Fig. 14.12A and B shows the variation of hourly productivity (yield) obtained from the active hybrid system for the months of November and February, respectively.

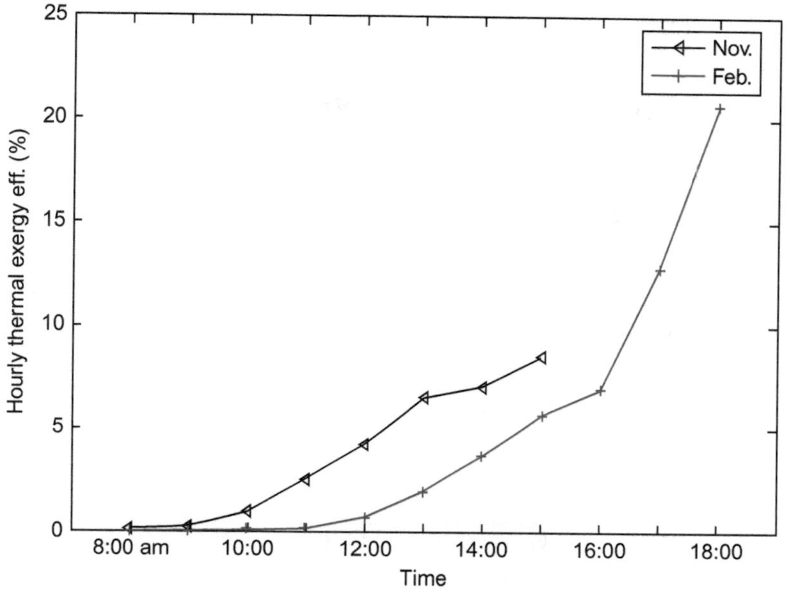

Fig. 14.9 Variation of hourly thermal exergy efficiency for the months of November and February.

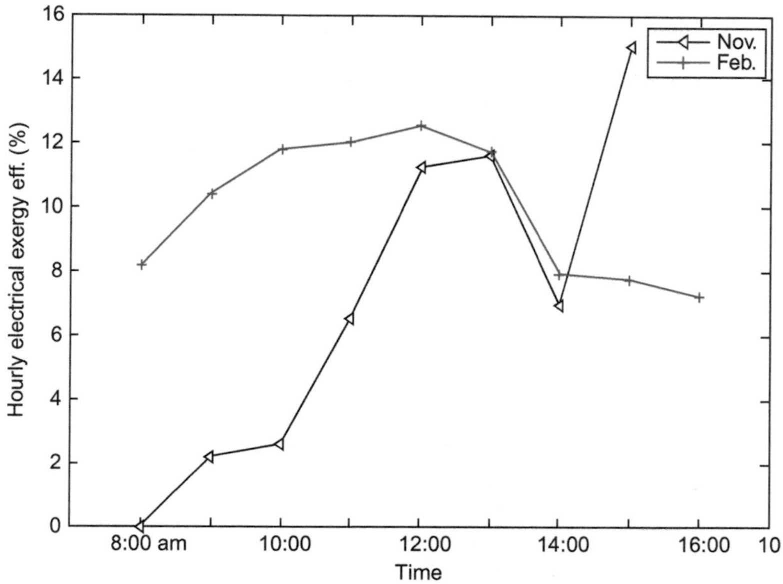

Fig. 14.10 Variation of hourly electrical exergy efficiency for the months of November and February.

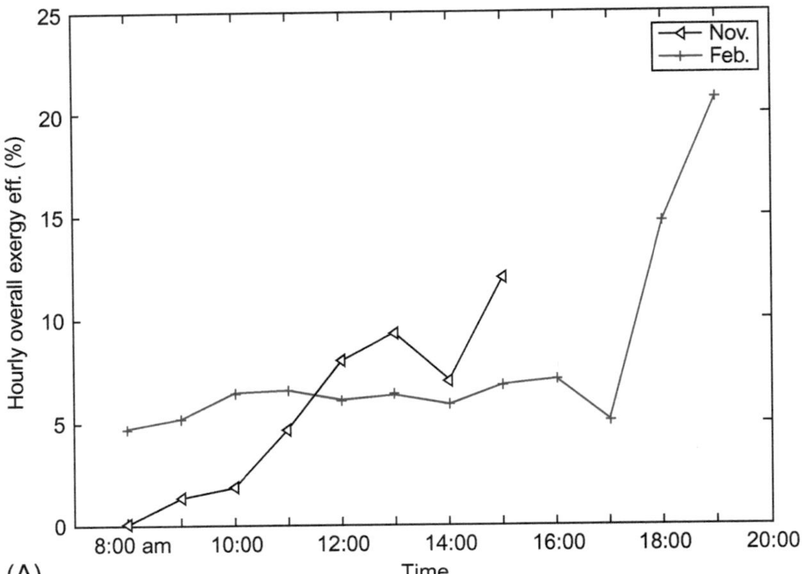

(A)

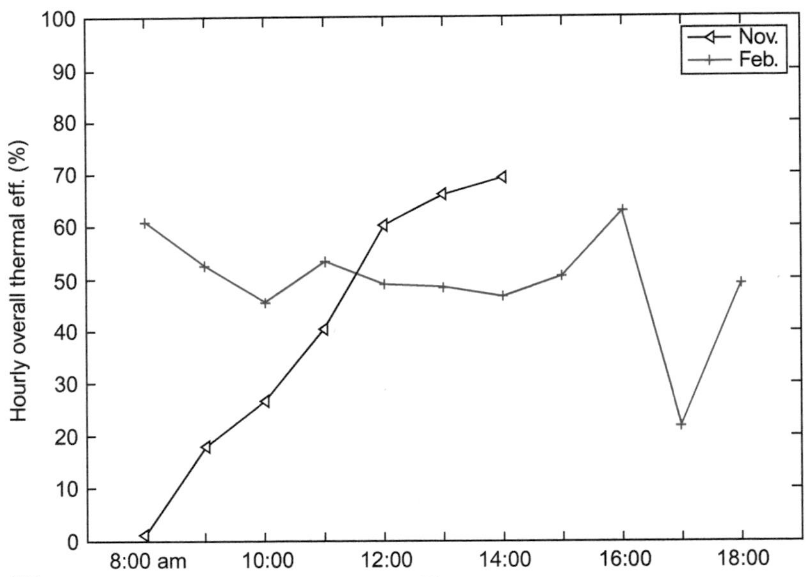

(B)

Fig. 14.11 Hourly variation of overall thermal (A) exergy efficiency and (B) energy efficiency for the months of November and February.

The maximum daily productivity (yield) has been found to be 7.74 kg for the month of May whereas the minimum daily yield 1.67 kg occurs in month of December. Moreover, the monthly yield has been evaluated

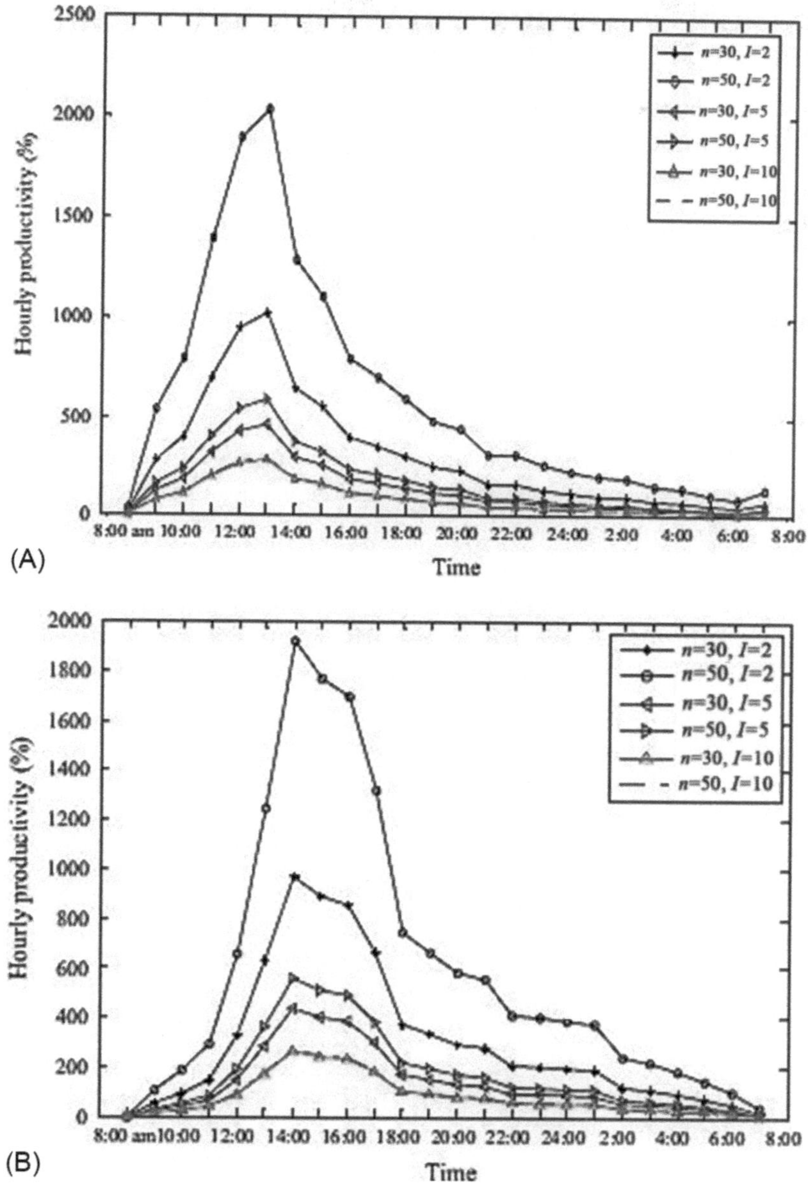

Fig. 14.12 Hourly variation of productivity for the month of (A) November and (B) February.

and its maximum value has been obtained for the month of June (160.44 kg) and its minimum value is found to be 48.43 kg for the month of December. It may be due to solar intensity difference as well as number of clear days obtained in a particular month.

14.4 MODELING FUNDAMENTAL EQUATIONS OF EXERGOECONOMIC ANALYSIS

The exergoeconomic methodology has been discussed below using fundamental equations for appropriate quantities [38].

14.4.1 Mass Balance Equation (MBE)

A mass balance for a quantity in a system can be expressed as

$$\text{Input} + \text{Generation} - \text{Output} - \text{Consumption} = \text{Accumulation} \quad (14.51)$$

where input refers to quantities entering, whereas output refers to quantities exiting through the system boundaries. Accumulation denotes growth (either positive or negative) of the quantity. Within a system, quantities produced and consumed that have been referred correspond to the generation and consumption, respectively.

The MBE can be expressed in integral and differential forms. The differential balance describes the system behavior at a given instant of time. On the other hand, the integral balance expresses the system behaviors between two instants of time, that is, in a particular time interval. Generally, differential balances are applicable on continuous processes, whereas integral balances are applied to the group processes. For the steady-state processes, the accumulation rate term is zero in the differential balance.

In terms of rates of differential terms, Eq. (14.51) (differential form) can be expressed as follows:

$$\begin{aligned}&\text{Input rate} + \text{Generation rate} - \text{Output rate} - \text{Consumption rate}\\&\quad = \text{Accumulation rate}\end{aligned} \quad (14.52)$$

On the other hand, these terms can be expressed in amounts in integral form as follows:

$$\begin{aligned}&\text{Amount input} + \text{Amount generated} - \text{Amount output}\\&\quad - \text{Amount consumed} = \text{Amount accumulated}\end{aligned} \quad (14.53)$$

14.4.2 Thermodynamic Balance Equation (TBE)

Energy obeys the law of conservation whereas exergy is consumed due to irreversibility during the process, hence it is subject to a nonconservation law. Thus, in terms of energy and exergy, the MBE, Eq. (14.51) can be expressed as

$$\text{Energy input} - \text{Energy output} = \text{Energy accumulation} \quad (14.54)$$

and

$$\text{Exergy input} - \text{Exergy output} - \text{Exergy consumption} \\ = \text{Exergy accumulation} \tag{14.55}$$

The output terms in the above two Eqs. (14.54)–(14.55) can be divided into product and waste components as given below:

$$\text{Energy output} = \text{Product energy output} + \text{Waste energy output} \tag{14.56}$$

and

$$\text{Exergy output} = \text{Product exergy output} + \text{Waste exergy output} \tag{14.57}$$

For solar still, the exergy balance equation can be expressed as

$$E_{x,in} - E_{x,out} = E_{x,dest} \tag{14.57a}$$

In above equation, the input exergy [$E_{x,out}$ (W)] and output exergy [$E_{x,in}$ (W)] can be obtained from Eqs. (14.16), (14.17), respectively.

14.4.3 Economic Balance Equation (EBE)

For the cost, which is nonconserved quantity, the MBE, Eq. (14.51) can be expressed as

$$\text{Cost input} + \text{Cost generation} - \text{Cost output} = \text{Cost accumulation} \tag{14.58}$$

In above equation, the cost generation corresponds to the combination of appropriate capital and other costs of the system. Moreover, the input, output, and accumulation cost signifies, respectively, the cost associated with all inputs, outputs, and accumulations for the system. These other costs include the manufacturing and maintenance of the system.

In general, cost generation can be expressed as follows:

$$\text{Cost generation} = \text{Capital cost of equipment} \\ + \text{All other creation and maintenance costs} \tag{14.59}$$

where the other costs include the interest cost and insurance cost. In a differential cost balance, the "cost generation rate" term defines the total cost generation balanced over the working life span of the system. On the other hand, in an integral cost balance, the term "amount of cost created" presents the part of total cost generation accounted between two instants of time under scrutiny.

14.5 IMPORTANT TERMINOLOGY OF EXERGOECONOMIC ANALYSIS

Rosen and Dincer [38] studied two different types of thermodynamic losses with the help of differential forms of the thermodynamic balances as discussed in previous section Eqs. (14.54)–(14.57). For simplicity, the accumulation terms have been assumed to be lossless in the energy and exergy balances. The energy losses can be directly recognized from Eqs. (14.54), (14.56). In the present analysis, the energy loss rate (on the basis of energy) for a system is labeled as L'_{en} for convenience. Hence, from Eq. (14.56), it can be expressed as

$$L'_{en} = \text{Waste exergy output rate} \tag{14.60}$$

Furthermore, on the basis of exergy, the exergy losses can be identified from Eqs. (14.55), (14.57). In general, there are two types of exergy losses, viz.:

(a) waste exergy output (Eq. 14.57)
(b) exergy consumption (Eq. 14.55)

The "waste exergy output" defines the loss associated with exergy released from the system; whereas the "exergy consumption" defines the internal exergy loss due to the occurrence of irreversibility during the process. These combined exergy losses gives the total exergy loss in the system.

Therefore, the exergy-based loss rate, L'_{ex} can be expressed as

$$L'_{ex} = \text{Exergy consumption rate} + \text{Waste exergy output rate} \tag{14.61}$$

Using equations of cost balances (Eqs. 14.58, 14.59), the capital cost (K) is defined here as the part of cost generation attributed to the equipment cost and it can be expressed as

$$K = \text{Capital cost of equipment} \tag{14.62}$$

The capital cost is a single economic term deliberated in the present analysis. From this approach, no benefit has been observed to identify the costs related to inputs as well as cost distributions among outputs. It is predicted that the use of capital cost (Eq. 14.59) is more relevant rather than the entire cost generation term. It is attributed to the fact that the use of the cost generation term considerably raises the complexity of the analysis by adding various other economic details, that is, interest rates; component lifetimes, salvage values, etc.

It can be illustrated that the accumulation terms in Eqs. (14.51)–(14.55), (14.58) are zero for a thermal system functioning normally in a continuous steady state. Consequently, all losses are associated with the loss terms based on energy (L'_{en}) and exergy (L'_{ex}) as discussed earlier. Following equations expresses the energy and exergy loss rates:

$$L'_{en} = \sum_{input} \text{Energy flow rates} - \sum_{products} \text{Energy flow rates} \tag{14.63}$$

$$L'_{ex} = \sum_{input} \text{Exergy flow rates} - \sum_{products} \text{Exergy flow rates} \tag{14.64}$$

Eqs. (14.63), (14.64) are obtained by rearranging Eqs. (14.54)–(14.57) and Eqs. (14.60)–(14.61).

14.6 RATIO OF THERMODYNAMIC LOSS RATE TO CAPITAL COST

A new parameter, "R," is defined as the ratio of thermodynamic loss rate (L') to capital cost (K) as follows:

$$R = \frac{L'}{K} \tag{14.65}$$

On the basis of energy loss rate, it can be written as

$$R_{en} = \frac{L'_{en}}{K} \tag{14.66}$$

and on the basis of exergy loss rate, it can be written as

$$R_{ex} = \frac{L'_{ex}}{K} \tag{14.67}$$

Rosen and Dincer [39,40] considered the following thermodynamic loss rates:

(a) energy (L'_{en})
(b) exergy (L'_{ex})
(c) internal exergy ($L'_{ex,i}$)—consumptions of exergy due to irreversibility of the process inside the system
(d) external exergy ($L'_{ex,e}$)—discarded outputs of exergy through the system boundary

It can be mentioned that $L'_{ex,i} + L'_{ex,e} = L'_{ex}$ for a system; and all energy losses are associated with external waste emissions. Furthermore, for the equipment in a system, capital cost (K) values have been taken into account and other costs are ignored to simplify the economic part of the work. Other costs include the interest and insurance costs. The value of parameter "R" is

Table 14.1 Capital cost and thermodynamic loss data of several devices in a unit of the coal-fired generating station [40]

Device	$K(10^6\$)$	L_{en}'(MW)	L_{ex}'(MW)	R_{en}(W/\$)	R_{ex}(W/\$)
Main condensers (and auxiliaries)	8.69	746	53	85.8	6.10
Preheating devices					
• Case I	16.02	Nil	23	Nil	1.44
• Case II	16.02	11.4	24.2	0.71	1.51
Steam generators	153.95	74	720	0.481	4.68
Turbine generators					
• Case I	63.93	Nil	108	Nil	1.69
• Case II	63.93	775	162	12.1	2.53
• Case III	63.93	1224	304	19.1	4.76
Overall station	274.54	820	916	2.99	3.34

considered on the basis of energy loss rate and total internal and external exergy loss rates. The values of L_{ex}', L_{en}', and K for the unit of the coal-fired electrical generation station are illustrated in Table 14.1. The generation station is classified into following devices viz.: (a) turbine generators (Case I, Case II, and Case III), (b) steam generators, (c) preheating devices (Case I and Case II), (d) condensers, and (e) overall station. The values of R_{en} and R_{ex} for all the devices is listed along with the values of L_{ex}', L_{en}', and K in Table 14.1. Moreover, the variation of thermodynamic loss rate to capital cost ratios for all the devices is presented in Fig. 14.13. It has been observed that the energy loss rate (L_{en}') critically depends on the definition of waste outputs for the turbine generators (three cases) and preheating devices (two cases) whereas exergy loss rate (L_{ex}') is found to be less sensitive.

The value of K is constant for each device for all the proposed cases; and the thermodynamic loss rate to capital cost ratios (R_{en} and R_{ex}) shows the same nature to those exhibited by energy and exergy loss rates (L_{ex}' and L_{en}'). It clarifies that a systematic correlation exists for capital cost and exergy loss but not for capital cost and energy lost. Also, for these devices, the results clearly indicate that the value of R_{ex} stands for a particular value denoted here as R_{ex}^* for these devices. R_{ex}^* does not provide the ideal value of R_{ex} but it can be better understood by considering the wide use of coal-fired electrical generation station. For all the eight devices, the statistical data for these two thermodynamic loss rate to cost ratios (R_{en} and R_{ex}) is presented in Table 14.2. These statistical quantities measure the relative variation in a set of values about the mean and are independent of the scale of

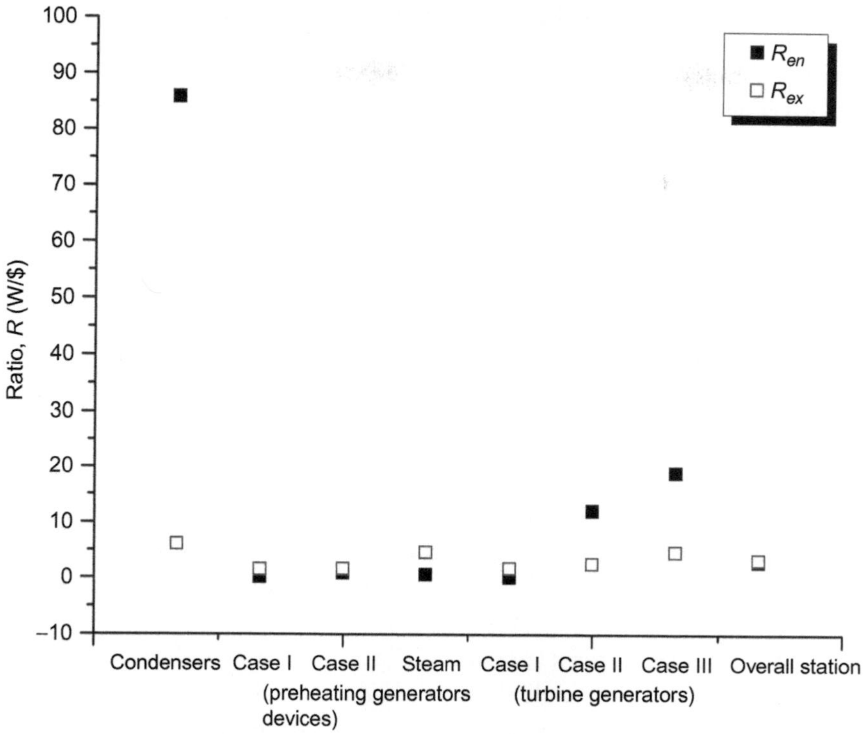

Fig. 14.13 Values of thermodynamic lost rate to capital cost ratio (*R*) for different devices of coal-fired generating station (500 MW) [40].

Table 14.2 Statistical data for "*R*" for different devices in a unit of the coal fired generating station [40]

	On the basis of	
Parameter	**Energy loss**	**Exergy loss**
Minimum, R_{min} (W/$)	Nil	1.44
Maximum, R_{min} (W/$)	85.8	6.10
Mean, $\overline{R}$ (W/ $)	15.1	3.26
Standard deviation, *SD*(*R*) (W/$)	27.4	1.64
Coefficient of variation, *CV*(*R*) (%)	181.5	50.31

measurement; it allows comparison of the variations in several sets of data. Table 14.2 suggest that the relative dispersion in R_{en} and R_{ex} for different devices is large when the ratio is based on energy loss; whereas it is found to be small when ratio is based on exergy loss.

14.7 EXERGOECONOMIC ANALYSIS OF SOLAR STILLS

In the literature, the author studied the exergoeconomic analysis of different thermal systems. In this section, the exergoeconomic analysis of passive and active solar distillation systems has been discussed in detail.:

14.7.1 Exergoeconomic Analysis of Passive SSSS and DSSS

Singh et al. [35] analytically examined the exergoeconomic parameter for passive single slope and double slope solar distillation systems on an annual basis for the climatic conditions of New Delhi (India). Passive SSSS of an area $1 \times 1\ m^2$ and inclination of the transparent glass cover at 15 degrees has been kept in the south to retrieve the maximum solar radiation. On the other hand, passive DSSS (area $2 \times 1\ m^2$ and inclination of the glass cover at 15 degrees) has been kept in the east-west direction for the same purpose. Fig. 14.14 presents the variation of exergoeconomic parameter and productivity with rate of interest for 30-year life span of the proposed systems.

Fig. 14.14 clearly shows that both exergoeconomic parameter as well as annual productivity decreases with increase in interest rate; and it is attributed to the fact that uniform annual cost (UAC) increases with rise in interest rate. Moreover, R_{gex} for single slope is found to be 3.19% higher than DSSS ($n=30$ year; rate of interest 2%) which happens due to higher values of annual exergy gain (17.68%) and UAC (14.18%) in the case of passive SSSS.

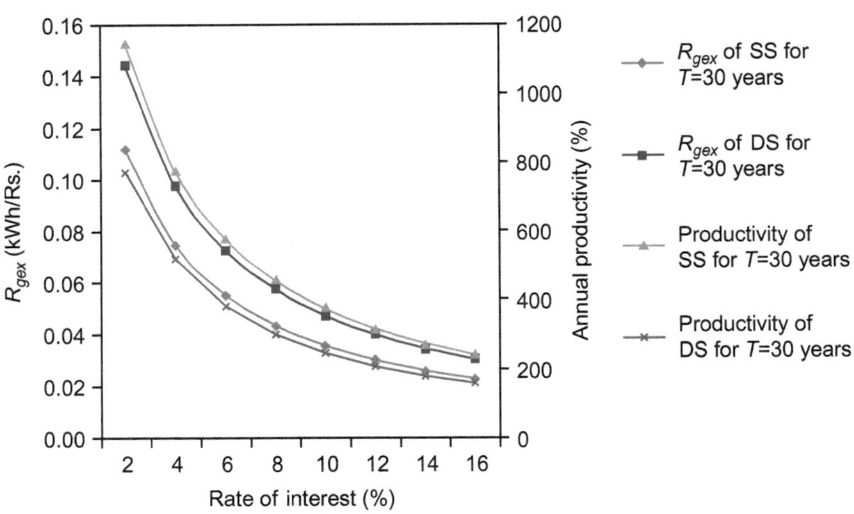

Fig. 14.14 Variation of exergoeconomic parameter (R_{gex}) and annual productivity (η_p) with rate of interest for 30 year life time for passive SSSS and DSSS.

In a similar way, the annual productivity of passive DSSS is found to be higher by 4.07% in comparison to the SSSS for the same parameters (life span and interest rate) and it is attributed to the lower annual yield (11.78%) and UAC (16.52%) of passive DSSS as compared with the SSSS.

14.7.2 Exergoeconomic Analysis of Partially Covered PVT-FPC Active Solar Still

Tiwari et al. [7] experimentally studied the annual performance of partially covered PVT flat plate collector solar distillation system for the climatic conditions of New Delhi (India). The system comprises FRP fabricated SSSS of 1×1 m^2 area with inclination of glass cover at 30 degrees, two flat plate collectors (inclination 45 degrees) each of area 2 m^2, and PV module with maximum power rating of 40 W.

As discussed earlier, the exergoeconomic parameter can be expressed as

$$R_{ex} = \frac{L_{ex,\,ann.}}{UAC} \tag{14.68}$$

where R_{ex} is the exergoeconomic parameter; $L_{ex,ann.}$ the annual exergy loss; and UAC (or K) is the uniform annual cost.

Following Nag [41], exergy loss rate (L_{ex}) for the proposed system can be written as

$$L_{ex} = \left[\left(h_{cwg} + h_{rwg}\right)\left\{\left(T_w - T_g\right) - \left(T_a + 273\right) ln\frac{\left(T_w + 273\right)}{\left(T_g + 273\right)}\right\}\right] + M_w C_w\left[\left(T_{wf} - T_{wi}\right) - \left(T_a + 273\right) ln\frac{\left(T_{wf} + 273\right)}{\left(T_{wi} + 273\right)}\right] \tag{14.69}$$

where h_{cwg} is the convective heat transfer coefficient between water surface and inner surface of the glass cover; h_{rwg} the radiative heat transfer coefficient between water surface and inner surface of the glass cover; T_g the temperature of glass cover; T_w the basin water temperature; T_{wf} the final basin water temperature; and T_{wi} the initial basin water temperature.

The variation of monthly exergy obtained from the system is presented in Fig. 14.15. On multiplying the daily exergy with number of clear days of a typical month, the monthly exergy loss can be estimated. The annual exergy loss (4.38 kWh) is shown in Fig. 14.16.

Moreover, Agrawal and Tiwari [42] performed the experimental and analytical study for the exergoeconomic analysis of series connected glazed hybrid photovoltaic thermal (PVT) module air collector for the climatic

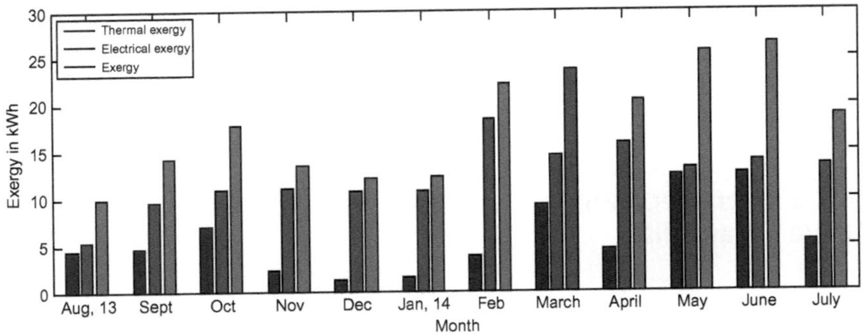

Fig. 14.15 Monthly variations of thermal exergy, electrical exergy, and overall exergy of active solar distillation system.

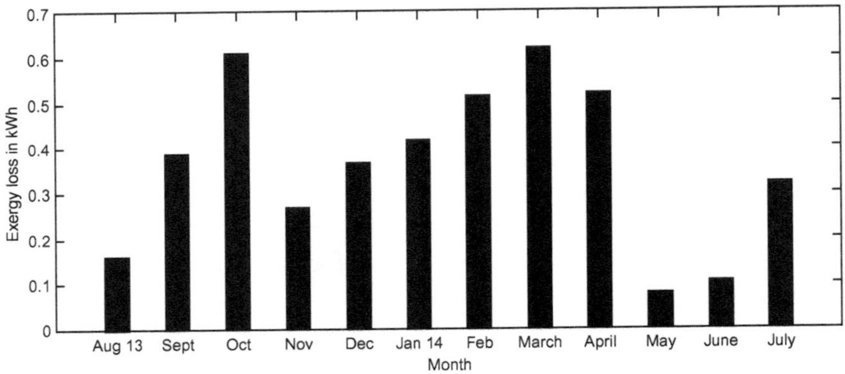

Fig. 14.16 Monthly variation of exergy loss.

conditions of New Delhi (India) (year 2011). They have found good agreements between the experimental and analytical results.

14.8 IMPORTANT TERMINOLOGY OF ECONOMIC ANALYSIS

For the financial estimation of RE technologies, it is essential to compare the combinations of different energy resources with each other for a given end use. Consequently, for such comparisons, it is important to reduce the cash flow at different points in time to an equivalent basis.

14.8.1 Capital Recovery Factor (CRF)

The cash flows cannot be compared directly with each other in the financial evaluation practices. It is due to the fact that cash flows have different values at different points in the time.

Let P be the present amount invested at the beginning ($n=0$) at an interest rate of i per year and S_n be its future value at the end of n years. Then the cash flow can be expressed as

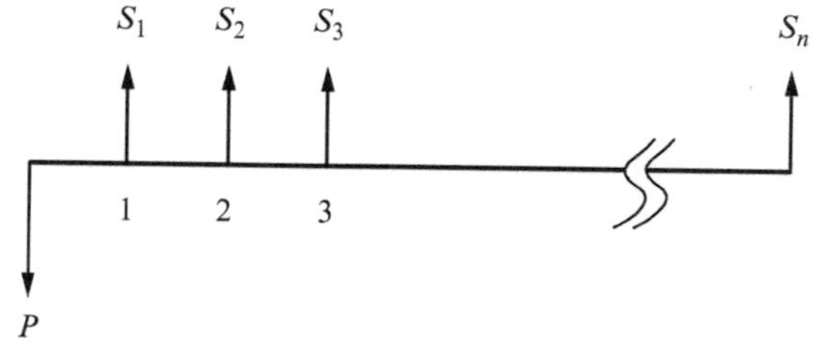

The time value of investment (P) at the end of 1 year ($n=1$) can be written as

$$S_1 = P + iP = (1+i)P$$

Similarly, at the end of second year ($n=2$), it can be written as

$$S_2 = S_1 + iS_1 = P + iP + i(P + iP)$$
$$= P(1+i)^2$$

Similarly, at the end of third and nth year, it can be expressed as

$$S_n = P(1+i)^n$$

For simplicity, above equation can be reexpressed as

$$S = P(1+i)^n \tag{14.70}$$

where $S > P$ for $i > 0$ (compound interest law).

Eq. (14.70) can be simplifies as

$$S = PF_{PS}$$

It defines,

$$\text{Future value} = (\text{Present value}) \times (\text{Future value factor}) \tag{14.71}$$

where F_{PS} can be designated as $F_{PS,i,n}$ and expressed as

$$F_{PS,i,n} = (1+i)^n \tag{14.72}$$

where $F_{PS,i,n}$ is known as the **compound interest factor** or **future value factor**. It gives the future value of the amount by knowing the present value. Thus, the future value is obtained by multiplying the compound interest factor with the present value. For different values of n (from 1 to 20) at different interest rates ($i=0.04\%$, 0.06%, 0.08%, and 0.10%), the values of $F_{PS,i,n}$ are presented in column II of Appendix 14.3.

If 1 year is divided into p number of equal periods, then n becomes np and rate of interest, i become i/p (rate of return per unit period). Using these values, Eq. (14.70) can be expressed as

$$S = P\left(1 + \frac{i}{p}\right)^{np}$$

or

$$S = P\left[\left(1 + \frac{i}{p}\right)^{p}\right]^{n} \tag{14.73}$$

In above equation, the term $\left(1 + \frac{i}{p}\right)^{p}$ can be expressed as

$$\left(1 + \frac{i}{p}\right)^{p} = 1 + \text{effective rate of return (ERR)}$$

or

$$\text{ERR} = \left(1 + \frac{i}{p}\right)^{p} - 1 \tag{14.74}$$

For $p = 1$,

The effective rate of return (ERR) becomes equal to i and remains greater than i if $p > 1$. In the case of simple interest, the future value S after n years can be expressed as

$$S = P(1 + ni) = P + (iP)n \tag{14.75}$$

Eq. (14.70) can be reexpressed as

$$P = S(1 + i)^{-n} \tag{14.76}$$

On evaluating the future amount from Eq. (14.76) against the calendar as present value (at zeroth time), the higher value of future amount (at nth year) will actually reduce, assuming i to be positive. Then the above equation can be expressed as

$$P = SF_{SP}$$

It implies,

$$\text{Present value} = (\text{Future value}) \times (\text{Present value factor}) \tag{14.77}$$

where $F_{SP,i,n}$ is termed as **compound interest factor** or **present value factor**. It is used to convert the future amount into present amount so that present value can be obtained from the product of compound value factor/present value factor and the future value. The column III of Appendix 14.3 gives $F_{SP,i,n}$ for different values of n at different interest rates.

Both the future value ($F_{PS,i,n}$) and present value ($F_{SP,i,n}$) are the compound interest factor but for different purpose and these can be related as

$$F_{PS,i,n} = \frac{1}{F_{SP,i,n}}$$

or

$$F_{PS,i,n} \times F_{SP,i,n} = 1 \tag{14.78}$$

Eqs. (14.70), (14.77) can be used for estimating the future value (S) of initial investment (P) and vice versa, respectively. These can be expressed as

$$S = P(1+i)^n, \text{when moving with the calendar} \tag{14.79a}$$

$$P = S(1+i)^{-n}, \text{when moving against the calendar} \tag{14.79b}$$

In generalized form,

$$A_{t2} = A_{t1}(1+i)^n \tag{14.80}$$

In other words, it can be expressed as

Amount at time t_2 = Amount at time t_1 (compound interest operator)

Eq. (14.80) denotes equation of time value conversion relationship and it concludes that n is positive if estimated using Eq. (14.79a) while it is negative if estimated against the calendar using Eq. (14.79b). The above said method is also referred as life cycle cost analysis (LCCA).

14.8.2 Uniform Annual Cost (UNACOST)

The economic problems are usually solved using the graphical approach. In the diagram, expenses (debits) and receipts (credits) are denoted using vertical lines while the horizontal line represents time. Conventionally, the debits and credits are marked in opposite directions as obvious such that credits are indicated upside while the debits downside.

It is implicit that the unknown value of amount is a time-dependent unit (year). Moreover, it is understood that the amount is equally distributed for n number of years at the end of each year. Thus, it is known as a "uniform end-of-year annual cost" for every year till the period of n years. It is denoted by R and known as UNACOST for a period of n years. Graphically, it is presented below.

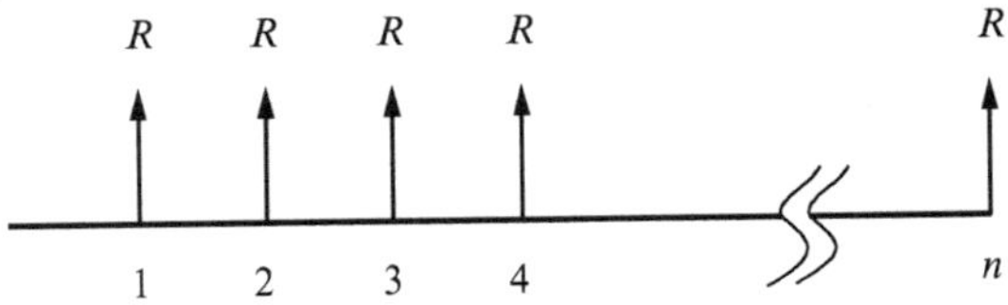

Let P represents single present value at the beginning ($n=0$). Using Eq. (14.77), one can obtain

$$P=R\left[\frac{1}{1+i}+\frac{1}{(1+i)^2}+\frac{1}{(1+i)^3}+\ldots\ldots\ldots\ldots\ldots\ldots+\frac{1}{(1+i)^n}\right] \tag{14.81}$$

or

$$P=R\sum_{1}^{n}\frac{1}{(1+i)^n} \tag{14.82}$$

The summation terms of above geometric series can be expressed as

$$\sum_{1}^{n}\frac{1}{(1+i)^n}=\left(\frac{1}{1+i}\right)\left\{\frac{1-\left(\frac{1}{1+i}\right)^n}{1-\frac{1}{1+i}}\right\}=\frac{(1+i)^n-1}{i(1+i)^n}$$

On substituting this summation term in Eq. (14.82), one can obtain

$$P=R\left[\frac{(1+i)^n-1}{i(1+i)^n}\right]=RF_{RP,i,n}$$

Above equation represents

$$\text{Present value}=(\text{UNACOST})\times(\text{UNACOST present value factor}) \tag{14.83}$$

where

$$F_{RP,i,n}=\frac{(1+i)^n-1}{i(1+i)^n} \tag{14.84}$$

Here $F_{RP,i,n}$ denotes the equal-payment series present value factor or annuity present value factor. The column IV of Appendix 14.3 gives $F_{RP,i,n}$ for different values of n at different interest rates.

Eq. (14.83) can be reexpressed as

$$R = P\left[\frac{i(1+i)^n}{(1+i)^n - 1}\right] \tag{14.85}$$

or

$$R = PF_{PR,i,n}$$

Above equation defines

$$\text{UNACOST} = (\text{Present value}) \times (\text{Capital recovery factor}) \tag{14.86}$$

where

$$F_{PR,i,n} = \frac{i(1+i)^n}{(1+i)^n - 1} \tag{14.87}$$

It is also known as CRF. From Eqs. (14.84), (14.87) respectively, the relation between equal-payment series present value factor and CRF can be attained as

$$F_{RP,i,n} = \frac{1}{F_{PR,i,n}} \tag{14.88}$$

14.8.3 Sinking Fund Factor (SFF)

As discussed earlier, at the end of (n) years, the future value (S) can be distributed into an equal uniform end-of-year annual amount (R). Moreover, it denotes the uniform end-of-year annual amount corresponding to the future value (S).

On substituting Eq. (14.76) in Eq. (14.85), one can obtain

$$R = S(1+i)^{-n}\left[\frac{i(1+i)^n}{(1+i)^n - 1}\right] = S\left[\frac{i}{(1+i)^n - 1}\right]$$

or

$$R = SF_{SR,i,n}$$

Above equation represents

$$\text{UNACOST} = (\text{Future amount}) \times (\text{Sinking fund factor}) \tag{14.89}$$

where

$$F_{SR,i,n} = \left[\frac{i}{(1+i)^n - 1}\right] = SFF \tag{14.90}$$

Corresponding to the salvage value of any system in future, above equation is used to evaluate the uniform end-of-year annual amount. Therefore, Eq. (14.89) can be reexpressed as

$$S = R\left[\frac{(1+i)^n - 1}{i}\right] = RF_{RS,i,n}$$

Above equation implies

$$\text{Future amount} = (\text{UNACOST}) \times (\text{Equal payment series future value factor}) \quad (14.91)$$

where

Equal payment series future value factor,

$$F_{RS,i,n} = \left[\frac{(1+i)^n - 1}{\mathrm{i}}\right] \quad (14.92)$$

Similarly, from Eqs. (14.90), (14.92), the relation between equal-payment series present value factor and SFF can be attained as

$$F_{SR,i,n} = \frac{1}{F_{RS,i,n}} \quad (14.93)$$

A uniform beginning of year annual amount (R_b) can be expressed in terms of P and S as R and R_b, respectively, with following relationship:

$$R = R_b(1+i) \quad (14.94)$$

The estimated values of various conversion factors for given number of years (n) and rate of interest (i) are given in Appendix 14.3.

14.9 CASH FLOW DIAGRAM

The cash flow is the most important part of any business and it is simply the movement of money (in and out) which measures the growth of any healthy business. In other words, it is the single most concern of any economic analysis [2,3]. The cash flow can be understood on time scale diagram known as cash flow diagram. The net cash flow is evaluated as

$$\text{Net cash flow} = \text{Receipts (Credits)} - \text{Expenses (Debits)} \quad (14.95)$$

Eq. (14.95) can be represented by cash flow diagram as follows:

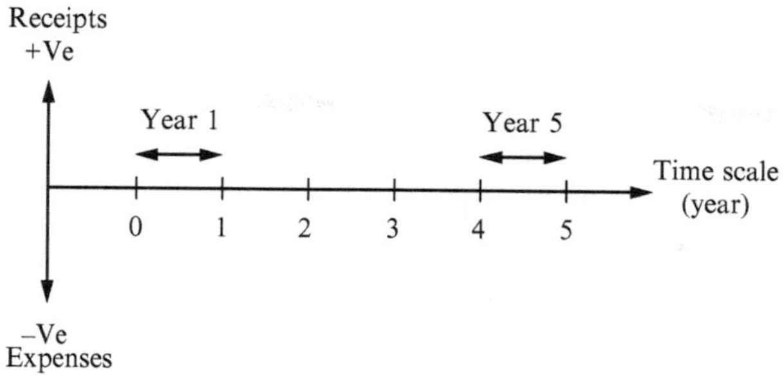

A uniform end-of-year annual amount (R) can be considered at the end of each year in the above cash flow diagram. Other methods such as benefit-cost analysis, cost comparison with equal duration, cost comparison with unequal duration can be used (with cash flow diagram) for the economic analysis of thermal systems [32,36].

The cash flow diagram is used in Example 14.13 for better understanding.

14.10 NET PRESENT VALUE (NPV)

The NPV of the investment is the difference between the present value of the benefits and the costs resulting from an investment. Following points defines the role of an NPV:

(a) A positive NPV implies that the financial position of the investor will be improved by undertaking the project.

(b) A negative NPV would indicate a financial loss.

(c) Zero or null NPV would mean that the present value of all benefits over the useful lifetime is equal to the present value of all the costs [3]. Mathematically,

$$NPV = \sum_{j=0}^{n} \frac{B_j - C_j}{(1+i)^j} \tag{14.96}$$

where B_j is the benefits and C_j is the costs at the end of the period j; n is the useful life of the project and i is the rate of interest.

Eq. (14.96) involves subtraction of the cost from the benefits at any period and then multiplying the result by single payment present worth factor for that period. Eventually, the NPV is determined by algebraically adding the results for all the periods under consideration. It often happens that $(B_j - C_j)$ is constant for all j except for $j=0$. In such a case, Eq. (14.96) can be modified as

$$NPV = B_0 - C_0 + \sum_{j=1}^{n} \frac{B_j - C_j}{(1+i)^j}$$

Since B_0, the benefits in the 0th year, are invariably zero and $B_j - C_j$ is constant ($=B - C$) for $j=1$ to n, then

$$NPV = -C_0 + (B - C)\sum_{j=1}^{n} \frac{1}{(1+i)^n}$$

or

$$NPV = -C_0 + (B - C)\left[\frac{(1+i)^n - 1}{i(1+i)^n}\right] \tag{14.97}$$

where C_0 represents the initial capital investment in the project.

Eqs. (14.96), (14.97) are based on the assumption that the interest rate i remains constant over time. Furthermore, the NPV could be estimated with different rates of interest rate over jth period. Thus, from Eq. (14.96), it can be modified for NPV as

$$\begin{aligned} NPV = (B_0 - C_0) + \frac{B_1 - C_1}{(1+i_1)} + \frac{B_2 - C_2}{(1+i_1)(1+i_2)} + \dots\dots\dots\dots \\ + \frac{B_j - C_j}{(1+i_1)(1+i_2)\dots\dots\dots(1+i_j)} + \dots\dots \\ + \frac{B_n - C_n}{(1+i_1)(1+i_2)\dots\dots\dots(1+i_n)} \end{aligned} \tag{14.98}$$

Eventually, on the basis of NPV method, the acceptance criteria of an investment project is

***NPV*>0, agree to consider the project**

***NPV*=0, remain indifferent**

***NPV*<0, does not agree to consider the project**

A positive NPV represents a positive surplus and the project may be accepted subject to availability of funds, while, a project should not be considered or rejected for negative NPV and the funds can be profitably invested in the other projects. In the case of mutually exclusive alternative investments, the project with highest positive NPV should be preferred.

14.11 EVALUATING ECONOMIC FEASIBILITY OF SOLAR DISTILLATION SYSTEM

Economic or cost analysis of the system includes, viz., the essential initial investment to purchase the systems; administrative costs; and any operational and maintenance costs. Generally, there is no operational cost for the utilities associated with solar energy systems because the customers do not have to pay for the nonconventional source of energy which is required to run the systems. Solar distillation systems offer benefits such as supply of free potable water and reductions in consumption of conventional energy resources (electricity). Apart from this, other benefits may be found in the form of government incentives (tax benefits for capital investment and renewable systems; and a subsidy for RE promotion) offered to promote renewable energies.

The economic feasibility of a solar distillation system can be evaluated mainly on the basis of two points:

(i) Cost per liter (CPL) of the produced potable water.
(ii) Payback period of the investment made for procuring the distillation system.

14.11.1 Cost Per Liter of Distilled/Potable Water

From the UNACOST concept, the net uniform end-of-year annual cost of the solar distillation system for a given initial capital investment can be expressed as

$$UA_{net} = P_s F_{PR,i,n} - P_s F_{PR,i,n} M_s - S_s F_{SR,i,n} \tag{14.99}$$

where UA_{net} is the net uniform end-of-year annual cost of solar distillation system with the values of known parameters, for example, initial capital investment (P_s), interest rate (i), maintenance cost (M_s), life of system (n), salvage value (S_s) after the completion of the system life; $F_{PR,i,n}$ and $F_{SR,i,n}$ represent the CRF and SFF, respectively (discussed earlier).

The first and second terms on the right-hand side of Eq. (14.99) defines the UNACOST corresponding to the initial capital investment (P_s) on a solar distillation system and maintenance cost over this fraction, respectively, while the third term describes the UNACOST of the salvage value of the system after completing a certain life of the system and hence deducted.

The *CPL* of distilled water can be estimated by dividing the UA_{net} of the solar distillation system by annual yield of distilled water (M_Y) of the distillation system. Mathematically, it can be expressed as

$$CPL = \frac{UA_{net}}{M_Y} \quad (14.100)$$

The annual yield (M_Y) can also be evaluated as

$$M_Y = \sum_{m=1}^{12} M_D N_{d,m} \quad (14.101)$$

where M_D is the daily yield per unit area in liters and $N_{d,m}$ is the number of clear days in *m*th month.

14.11.2 Capital Cost of Solar Distillation System (P_s)

To purchase a solar distillation system, the initial capital investment varies according to the design, size, types, materials used for its construction, etc. Therefore, a simple single slope basin-type passive solar distillation system has been considered for the understanding of conceptual calculation of initial capital investment. The initial capital cost of a distillation unit made of glass reinforced plastic (GRP) can be expressed as

$$\begin{aligned} P_s &= P_{GRP} + P_G + P_{cl} + P_{st} + P_{Mi} \\ &= \text{Cost of glass reinforced plastic} + \text{total cost of glass cover} \\ &\quad + \text{cost of clamping unit (frame and clamps)} + \text{cost of stand} \\ &\quad + \text{cost of other necessary miscellaneous items (glass ket, etc.)} \end{aligned} \quad (14.102)$$

In above equation, the cost of different components constituting the distillation unit are in rupees per m^2 of basin area. Similar approach can be implemented to evaluate the initial capital investment for any type and size of the solar distillation system.

14.11.3 Interest Rate (*i*, %)

The possible interest rate to borrow the fund for capital investment could be as follows:

(a) **4%:** A subsidized rate normally offered by government sectors to support the use of RE.
(b) **7%–7.5%:** The rate normally offered by the government banks.
(c) **10%–12%:** The rate offered by private banking sectors.
(d) **12%–16%:** From any other private source.

Thus, any suitable interest rate from 4% to 16% can be considered for economic analysis.

14.11.4 Life of Solar Distillation System (n)

Life of solar distillation system is the total duration (years) in which the system successfully discharges its duty of producing potable water. The climatic conditions to which it is subjected and material used for its construction play a vital role in the life span of the system. The reported durability of the FRP body of the solar distiller is around more than 25 years, whereas the solar distillers made of GRP units are expected to offer 1.5–2 times better durability. Therefore, for any evaluation, the life span of solar distillation system can be taken any value from the range 20–50 years depending on the situation under consideration.

14.11.5 Maintenance Cost (M_s, %)

The following works are the main concerns for the maintenance of solar distillation system under operation:

(i) Consistent filling of brackish water into the distillation unit.
(ii) Removal and collection of potable water from the measuring jar or collection container.
(iii) Proper cleaning of glass cover to receive the maximum solar radiation. Moreover, periodical cleaning of the basin surface is essential to remove the salt deposited.

Lower basin water (contaminated/saline) depth results in higher maintenance cost and requires frequent refilling. The salt concentration of the feed water determines the frequency of basin cleaning or scale removing; therefore, higher frequency means higher maintenance cost of the distillation unit. The maintenance cost is expected to be in the range 8%–16% of the UNACOST corresponding to the initial capital investment. Therefore, the maintenance cost can be estimated by considering its percentage value in this range depending on the conditions prevailing with the operation of the solar distillation system.

14.11.6 Salvage Value (S_s)

After successful completion of the life span of the distillation system, the future amount obtained by selling the scrap of the systems is known as salvage value of the system. Therefore, it is easy to understand that the salvage value solely depends on the quality of the scrap material involved for making a solar distillation system. The components made of iron can give better salvage value after the completion of life of distillation unit if cared well by anticorrosive painting; whereas the concrete, FRP and GRP does not contribute much for the salvage value. The scrap value of iron also increases with the passage of time because the price of iron scrape also increases with time. Hence, the salvage value is calculated by taking the above knowledge into consideration.

14.12 PAYBACK TIME/PAYBACK PERIODS (N_P)

The payback period is the minimum time taken by the system to recover the total investment costs. Basically, the payback period for a system saving energy is computed as the total investment cost divided by the annual revenues from energy saved. In payback analysis, the number of years to payback the investment cost is the measurement unit. In the case of solar distillation system, the importance of annually produced distilled water is the main concern rather than the saving of energy.

Simple payback analysis of solar distillation system takes into account only initial capital costs and the distribution/selling cost of the produced distilled water. Moreover, the understanding of the payback period is used as a rough guide for cost effectiveness. Generally, the solar distillation system is considered to be cost effective if its payback period is significantly less than its expected life span.

The variation of payback period with selling price for different interest rate and at the capital cost of commercially produced solar still is presented in Fig. 14.17. It has been observed that the payback period reduces significantly

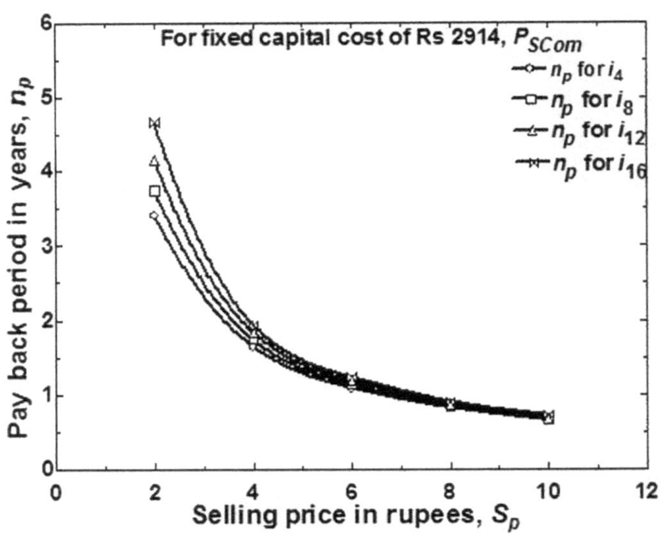

Fig. 14.17 Variation of payback period with selling price for different interest rates (i).

with increase in selling price. It reduces from 4.68 years to the 0.72 years if selling price is increased from Rs. 2 to Rs. 10 per kg even at highest rate interest rate (16%).

14.12.1 Understanding Simple Payback Method

By definition, the payback period is the amount of time (years) in which it recovers the initial investment. Payback deliberates the initial capital costs and the resulting annual cash flows. It is important to mention that the payback method does not account for savings that may continue from a solar distillation system after the coverage of the initial investment.

(a) Payback with equal annual savings

The payback period for the case of equal annual cash flows is obtained by dividing the initial investment by the annual savings. Therefore,

$$\text{Payback period (years)} = \frac{\text{Initial capital cost}}{\text{Annual operating cash flow}}$$

To understand the payback period with equal annual saving, let us consider an example of a battery shop evaluating the purchase of a solar distillation system to quench its need of distilled water. The shop owner analyzes both his current way to purchase distilled water from market and the option of using the solar distillation system. He found that the installation of the system (2 m^2) will cost Rs. 5000 but provide a net annual operational cash flow (savings) of Rs.1825.On dividing the initial capital cost by the net annual operational cash flow, the manager observed that the distillation system will pay for itself in 2.7 years; whereas the life span of distillation system is supposed to be not less than 30 years.

(b) Payback with unequal annual savings

The previous example assumes that the annual cash flow is the same each year. Actually, in reality, there are significant parameters such as interest, maintenance cost, life, salvage value at the life, free distribution of distilled water, selling price of distilled water, etc., that causes to vary cash flows each year. If the annual cash flow differs from year to year, then the payback period can be determined when the accrued cash savings become equal to the initial investment costs.

The payback period (n_p), the number of years necessary to exactly recover the initial capital investment (P_s) is computed by summing the annual cash-flow values and the estimated value of n_p through the relation given below

$$0 = -P_s + \sum_{t=1}^{n} CF_t(F_{SP,i,t}) \tag{14.103}$$

or

$$P_s = \sum_{t=1}^{n_p} \left[CF_t(1+i)^{-n_p}\right] \tag{14.104}$$

where CF_t is net cash flow at the end of each year.

For the same value each year, it can be expressed as

$$P_s = CF(F_{RP,i,n}) \tag{14.105}$$

where $F_{RP,i,n}$ is a factor described by Eq. (14.73) which is multiplied with cash flow (*CF*) and expressed as

$$F_{RP} = \frac{(1+i)^{n_p} - 1}{i(1+i)^{n_p}} \tag{14.106}$$

Above equation can be simplified as

$$n_p = \frac{ln\left(\dfrac{CF}{CF - iP_s}\right)}{ln(1+i)} \tag{14.107}$$

Case I—The distilled water is sold at the price it is produced

In this case the net cash flow is equal to the uniform end-of-year annual cost (UA_{net}) of solar distiller, that is, Eq. (14.107) can be modified as

$$n_p = \frac{ln\left(\dfrac{UA_{net}}{UA_{net} - iP_s}\right)}{ln(1+i)} \tag{14.108}$$

Case II—The distilled water is sold at market price

In this case the cash flow is estimated by selling price and it can be expressed as

$$CF = M_Y S_p \tag{14.109}$$

Above value of *CF* can be used in Eq. (14.107) to find out the payback period in Case II.

Case III—The distilled water is sold at market price considering zero interest rate

In this case, Eq. (14.107) becomes

$$P_s = n_p CF$$

or

$$n_p = \frac{P_s}{CF} \tag{14.110}$$

14.13 SUMMARY

In this chapter, the thermal modeling of conventional single slope and double slope solar still (DSSS); and active solar still integrated with partially covered N-photovoltaic thermal–flat plate collectors (N-PVT-FPC) has been developed in order to derive the analytical expression of basin water temperature. Further, the productivity, energy, and exergy analysis of these solar distillation systems have been carried out. The modeling fundamental equations of exergoeconomic analysis for these systems have been studied including the case study of glazed hybrid PVT module air collector. Moreover, the important terminology of economic analysis has been discussed in detail. The ratio of thermodynamic loss rate to capital cost has been discussed on the basis of energy and exergy for different devices of coal-fired electrical generation station. Also, the life-cycle cost analysis (LCCA) (exergoeconomic and enviroeconomic analysis) and economic feasibility of solar distillation systems have been discussed in detail.

14.14 FUTURE TRENDS

- The performance of solar still with different types of PV technology, and collector assembly can be analyzed; also, the performance with rain water harvesting for a large scale can be studied.
- The effect of different condensing chambers should be analyzed for PVT active slope solar still for fast condensation due to high operating temperature.
- Long-term performance (energy/exergy analysis), LCCA, exergoeconomic, and enviroeconomic analysis of different passive and active solar distillation systems can be studied for different climatic conditions using other working fluids (nanofluids and molten salts via heat exchanger) having better thermo-physical properties as compared to conventional fluid (water).
- The initial cost of above discussed RESs is expected to drop in future. Moreover, there should be a suitable funding policy for improving the economic profitability of RE technology as well as for promoting these RESs at mass-market scale in the preset market structure.

14.15 PROBLEMS AND DESCRIPTIVE QUESTIONS

14.1 Derive an expression for the basin fluid temperature for the following cases:
- **(a)** conventional SSSS
- **(b)** conventional DSSS
- **(c)** active SSSS integrated with N-PVT-FPC

14.2 Develop the characteristic equation of (a) conventional SSSS and (b) active SSSS integrated with N-PVT-FPC.

14.3 Derive an expression for the lost thermal efficiency of passive DSSS.

14.4 Discuss the behaviors of thermal energy and exergy efficiency curve of the active SSSS integrated with N-PVT-FPC.

14.5 Calculate the thermal energy, exergy, instantaneous thermal gain/loss efficiencies, and productivity of the passive DSSS. Given, $L = 2.5 \times 10^6$J/kg, $T_{gi} = 32°C, h_{e,f} = 55$ W/m^{2}°C, $(t) = 860$ W/m^2, $T_a = 20°C$, $(MC)_f = 41,\ 900$ J/°C, $U_L = 9.22$ W/m^{2}°C, $\Delta t = 3600$s, $(\alpha\tau)_{eff} = 0.69$. Other specifications of the system are given below:

Dimensions of passive DSSS		Constants	Numerical value
A_{gE}	$1.025\,m \times 1.025\,m$	σ	5.67×10^{-8} (W/m^2 K^4)
A_{gW}	$1.025\,m \times 1.025\,m$	C_{bf}	4188 (J/kg°C)
A_b	$2\,m \times 1\,m$	α_g	0.05
X	1.5 m	α_b	0.8
θ	30 degrees	α_{bf}	0.6
K_g	0.780 (W/m°C)	$\in_g$	0.95
K_B	0.035 (W/m°C)	where X is the characteristic length, θ is the inclination angle of the glass cover	
L_g	0.004 m		
L_b	0.005 m		

14.6 Calculate the basin fluid temperature of active SSSS integrated with N-PVT-FPC. Use other parameters given below:
Basin area $= 1 \times 1\,m^2$, $L = 2.8 \times 10^6$J/kg, $T_{gi} = 40°C$, $h_{e,f} = 54$ W/m^{2}°C, $I(t) = 850$ W/m^2, $T_a = 22°C$, $(MC)_f = 41,\ 900$ J/°C, $U_L = 10.32$ W/m^{2}°C, $\Delta t = 3600$s, $(\alpha\tau)_{eff} = 0.72$, $A_c = 2m^2$, $I_c = 790$ W/m^2.

14.7 Calculate the thermal energy and exergy; and thermal energy and exergy efficiency for Problem 14.6.

14.8 Discuss the following balance equations in detail:
- **(a)** energy balance equation
- **(b)** thermodynamic balance equation (TBE)
- **(c)** economic balance equation (EBE)

14.9 Explain four main thermodynamic loss rates of a system.

14.10 Discuss the behavior of exergoeconomic parameter of partially covered PVT flat plate collector solar distillation system.

14.11 Explain the following terms in detail:
- **(a)** capital recovery factor (CRF)
- **(b)** uniform annual cost (UNACOST)
- **(c)** sinking fund factor (SFF)

14.12 Discuss the annual cost method and capitalized cost method.

14.13 Explain the concept of NPV for the acceptance/refusal criterion of the project.

14.14 Explain the concept of payback period of solar distillation system in detail.

14.15 Discus the payback period of solar distillation system for the following two cases:
- **(a)** payback period with equal savings
- **(b)** payback period with unequal savings

14.16 Calculate future (F_{ps}) and present (F_{sp}) value factor for a given life of solar distillation system at 10% interest rate and show that $F_{ps}F_{sp}=1$ for each case.

14.17 Calculate the ERR for different value of p at 10% interest rate.

14.18 Calculate capital recovery (F_{RP}) and sinking fund (F_{SR}) factors for different number of years ($n=1, 5, 10, 15, \text{and } 20$) for a given rate of interest $i=0.05\%$, 0.10%, 0.15%, 0.20%.
Also prove that $F_{SR}F_{RS}=1$.

14.19 A hybrid solar dryer purchased for USD $1200 is expected to generate annual revenues of USD $150 and have salvage value of USD $400 at the end of 15 years. If 18% per year required return is imposed on the purchase, compute the payback period. Use cash flow diagram to solve the problem.

14.20 For hybrid solar water heating systems of total amount USD $10,000, a down payment of USD $2000 is requested by a person. For this, the amount of USD $1174 annual end-of-year payment (R) is mandatory for 12 year. Conversely, the person chooses to pay USD $1000 yearly and the remaining payment in the last stage. What will be the balance payment (remaining payment) if money is worth 10% interest.

14.21 A person wants to purchase an energy-efficient equipment and ready to spend the amount of USD $3000 on first, second, third, and fourth year; and also decides to set aside a certain amount in the beginning as well as each year thereafter until the fourth year. If the contribution arrangements have an arithmetical progression by 20% for all years increasing after the first year; then evaluate the initial contribution by a person for the 10%money.

14.22 Besides other routine maintenance costs of USD $ 300 each year, replacement of the wooden duct in the 10th year is predicted to be around USD $ 4000 during its useful life span of 20 years of a BIPVT air-circulating collector with its total cost of USD $35,000. Find the equivalent annual cost of the BIPVT air-circulating collector for 10% interest rate.

APPENDIX 14.1

Example 14.1

Calculate the time taken for the water temperature to rise to 40°C from an initial value of 20°C. Given:

$(\alpha\tau)_{eff}=0.65$, $I(t)=900$ W/m^2, $T_a=8.5$°C, $(MC)_f=41$, 900J/°C, $U_L=9.48$ W/m^{2}°C.

Solution

From Eq. (14.12),

$$\Delta t=-\frac{1}{a}ln\left[\frac{T_f-\frac{f(t)}{a}}{T_{fo}-\frac{f(t)}{a}}\right]$$

where

$$a=\frac{U_L}{(MC)_f}=\frac{9.44}{41,900}=2.26\times10^{-4}\text{ and}$$

$$f(t)=\frac{(0.65\times900)+(9.48\times8.5)}{41900}=0.0159$$

Hence,

$$\Delta t=-\frac{1}{(2.26\times10^{-4})}ln\left[\frac{40-\frac{0.0159}{2.26\times10^{-4}}}{20-\frac{0.0159}{2.26\times10^{-4}}}\right]=2239\text{ s}=37\text{ min}$$

Example 14.2

Estimate the basin fluid (water) temperature, daily yield, energy, and exergy of passive SSSS of area $1 \times 1\ m^2$. Given:

$L=2.5 \times 10^6$ J/kg, $T_{gi}=37°C$, $h_{e,f}=45\ W/m^{2o}C$, $I(t)=700\ W/m^2$, $T_a=15°C$, $(MC)_f=41{,}900$ J/°C, $U_L=9.44\ W/m^{2o}C$, $\Delta t=3600s$, $(\alpha\tau)_{eff}=0.65$.

Solution

From Eq. (14.12)

$$a=\frac{U_L}{(MC)_f}=\frac{9.44}{41{,}900}=2.26\times10^{-4} \text{ and } f(t)=\frac{(\alpha\tau)_{eff}I(t)+U_L T_a}{(MC)_f}$$

$$=\frac{(0.65\times700)+(9.44\times15)}{41{,}900}=0.0142$$

$$T_f=\left(\frac{0.0142}{2.26\times10^{-4}}\right)\left[1-e^{-(2.26\times10^{-4})\times3600}\right]$$

$$+15\times e^{-(2.26\times10^{-4})\times3600}=41.63°C$$

From Eq. (14.15)

$$E_{hourly,en}=h_{ef}\left(T_f-T_{gi}\right)A_b=45\times(41.63-37)=210.15\,W$$

From Eq. (14.16)

$$E_{hourly,ex}=h_{ef}\left[\left(T_f-T_{gi}\right)-(T_a+273)ln\left(\frac{T_f+273}{T_{gi}+273}\right)\right]$$

$$=45\times\left(4.67-(288)ln\left(\frac{314.63}{310}\right)\right)=18.017\,W$$

From Eq. (14.13)

$$\dot{m}_{ef}=\frac{h_{ef}\left(T_f-T_g\right)}{L}\times3600=\left(\frac{45\times(4.67)}{2.5\times10^6}\right)\times3600=0.3026\,kg/h$$

Example 14.3

Estimate the basin fluid (water) temperature, daily yield, energy, and exergy of passive DSSS of area $2 \times 1\ m^2$. Given

$a=3.16\times10^{-4}$;$f(t)=2.01\times10^{-2}$;$h_{ef,E}=48\,W/m^{2o}C$,$h_{ef,W}=54\,W/m^{2o}C$, $T_{giE}=47°C$, $T_{giW}=48°C$;$I_{SE}(t)=750\,W/m^2$,$I_{SW}(t)=650\,W/m^2$, $T_a=22°C$ $(MC)_f=41{,}900\,J/°C$, $\Delta t=3600\,s$, $L=3.1\times10^6\,J/kg$.

Solution

From Eq. (14.29), basin fluid temperature is

$$T_f = \left(\frac{2.01\times10^{-2}}{3.16\times10^{-4}}\right)\left[1-e^{-(3.16\times10^{-4})\times3600}\right] + 22\times e^{-(3.16\times10^{-4})\times3600} = 50.26^{\circ}\mathrm{C}$$

From Eq. (14.30), hourly thermal energy is

$$E_{hourly,en} = h_{ef,E}\left(T_f - T_{giE}\right) + h_{ef,W}\left(T_f - T_{giW}\right)A_b = [48\times(3.26)+54\times(2.26)]\times2 = 0.558\,\mathrm{kW}$$

From Eq. (14.31), hourly thermal exergy is

$$E_{hourly,ex} = h_{ef,E}\left[\left(T_f - T_{giE}\right) - (T_a+273)ln\left(\frac{T_f+273}{T_{giE}+273}\right)\right]\left(\frac{A_b}{2}\right) + h_{ef,W}\left[\left(T_f - T_{giW}\right) - (T_a+273)ln\left(\frac{T_f+273}{T_{giW}+273}\right)\right]\left(\frac{A_b}{2}\right)$$

$$= 48\times\left(3.26-(295)ln\left(\frac{323.26}{320}\right)\right)+54\times\left(2.26-(295)ln\left(\frac{323.26}{321}\right)\right) = 0.0233\,\mathrm{kW}$$

From Eq. (14.13), productivity (yield) of the system is

$$\dot{m}_{ef} = \frac{h_{ef,E}\left(T_f - T_{giE}\right) + h_{ef,W}\left(T_f - T_{giW}\right)A_b}{L}\times3600 = \left(\frac{0.558\times10^3}{3.1\times10^6}\right)\times3600 = 0.648\,\mathrm{kg/h}$$

Example 14.4

Calculate the thermal energy and exergy efficiency of active DSSS integrated with two PVT flat plate collectors. Given

$N=2$, $a=2.96\times10^{-4}$; $f(t)=2.28\times10^{-2}$; $h_{ef,E}=51$ W/m^2°C, $h_{ef,W}=54$ W/m^2°C, $T_{giE}=56$°C, $T_{giW}=58$°C, $I_{SE}(t)=400$W/m^2, $I_{SW}(t)=430$ W/m^2, $T_a=28$°C $(MC)_f=41$, 900 J/°C, $\Delta t=3600$ s, $L=2.80\times10^6$ J/kg, $A_c=1.395$ m^2, $I_c(t)=380$ W/m^2. (Use Eq. 14.45 to find the basin water temperature).

Solution

From Eq. (14.45), basin water temperature is

$$T_f = \left(\frac{2.28 \times 10^{-2}}{2.96 \times 10^{-4}}\right)\left[1 - e^{-(2.96\times10^{-4})\times3600}\right]$$
$$+ 22 \times e^{-(2.96\times10^{-4})\times3600} = 60.13^{\circ}\text{C}$$

From Eq. (14.30), hourly thermal energy is

$$E_{hourly,en} = h_{ef,E}\left(T_f - T_{giE}\right) + h_{ef,W}\left(T_f - T_{giW}\right)A_b$$
$$= [51 \times (4.13) + 54 \times (2.13)] \times 2 = 0.652\,\text{kW}$$

From Eq. (14.31), hourly thermal exergy is

$$E_{hourly,\,ex} = h_{ef,E}\left[\left(T_f - T_{giE}\right) - (T_a + 273)ln\left(\frac{T_f + 273}{T_{giE} + 273}\right)\right]\left(\frac{A_b}{2}\right)$$
$$+ h_{ef,W}\left[\left(T_f - T_{giW}\right) - (T_a + 273)ln\left(\frac{T_f + 273}{T_{giW} + 273}\right)\right]\left(\frac{A_b}{2}\right)$$
$$= 51 \times \left(4.13 - (295)ln\left(\frac{333.13}{329}\right)\right)$$
$$+ 54 \times \left(2.13 - (295)ln\left(\frac{333.13}{331}\right)\right) = 0.029\,\text{kW}$$

From Eq. (14.13), productivity (yield) of the system is

$$\dot{m}_{ef} = \frac{h_{ef,E}\left(T_f - T_{giE}\right) + h_{ef,W}\left(T_f - T_{giW}\right)A_b}{L} \times 3600$$
$$= \left(\frac{0.558 \times 10^3}{2.80 \times 10^6}\right) \times 3600 = 0.838\,\text{kg/h}$$

From, Eq. (14.46), thermal energy efficiency is

$$\eta_{the,en} = \frac{\dot{m}_{ef} \times L}{[A_b I(t) + NA_c I_c(t)]} \times 100$$
$$= \frac{0.838 \times 2.80 \times 10^6}{[(2 \times 415) + 2 \times 1.395 \times 380]} = 34.49\%$$

From, Eq. (14.47), thermal energy efficiency is

$$\eta_{the,ex} = \frac{\left\{h_{ef,E}\left[\left(T_f - T_{gi}\right) - (T_a + 273)ln\left(\frac{T_f + 273}{T_{giE} + 273}\right)\right]A_b + h_{ef,W}\left[\left(T_f - T_{giW}\right) - (T_a + 273)ln\left(\frac{T_f + 273}{T_{giW} + 273}\right)\right]A_b\right\}}{(0.933)\left[A_b I(t) + NA_c I_c(t)\right]} \times 100$$
$$= \frac{29.947}{0.0933 \times [(2 \times 415) + 2 \times 1.395 \times 380]} \times 100 = 1.33\%$$

Example 14.5

What will be the rate of return if Rs. 20, 000 compounds to Rs. 28, 240 in 4 years?

Solution

Present value is $P=20,000$, $n=4$. Therefore, from Eq. (14.70)

$$28,240 = 20,000\,(1+i)^4$$

On simplifying, one can obtain

$$i = 0.09$$

or

$$9\%\ \text{per year}.$$

Example 14.6

For compounded annually at 10% per year, what would be the time period for money to be double?

Solution

Let us consider that the money doubles in n years, then $S=2P$

Using Eq. (14.70), one can obtain

$$2P = P(1+i)^n$$

or

$$2P = P(1+0.10)^n$$

On simplifying, one can obtain

$n=7.3$ years, that is, the money will be doubled in 7.3 years.

Example 14.7

Evaluate the ERR for 10% interest rate; and $p=5$ and 12.

Solution

From Eq. (14.74), one can obtain

Effective rate of return (ERR) $= \left(1+\frac{i}{p}\right)^p - 1$

For $p=5$,

$$\mathrm{ERR} = \left(1 + \frac{0.10}{5}\right)^5 - 1 = 0.104$$

For $p = 12$,

$$\mathrm{ERR} = \left(1 + \frac{0.10}{12}\right)^{12} - 1 = 0.104$$

Example 14.8

What is the rate of interest (i) student will pay if he/she borrows Rs. 2000 and returns Rs. 2100 at the end of 6 months?

Solution

Given $S = 2100$, $P = 2000$, and $n = 6/12$

Therefore, using Eq. (14.75)

$$S = P(1 + ni)$$

or

$$2100 = 2000\left[1 + \left(\frac{6}{12}\right)i\right]$$

On simplifying, one can obtain
$i = 0.10$ or 10%.

Example 14.9

A person borrows Rs. 10, 000 at interest rate of 10% for 4 years and 4 months. Evaluate the money paid by considering compound interest.

Solution

Using Eq. (14.75), the future amount after 4 months can be obtained as

$$S = 10{,}000\left[1 + \left(\frac{4}{12}\right)(0.10)\right] = \mathrm{Rs.}10{,}333$$

Now this obtained value become P for another 4 years. From compound interest equation,

$$S = PF_{PS} = P(1 + i)^n$$
$$S = (10, 333)(1 + 0.10)^4 = \mathrm{Rs.}15, 129$$

Example 14.10

Approximate USD $4000 every year is saved in terms of fuel savings from large capacity water heating system. For a discount rate of 12%, what is the present worth of fuel saving in the 5th, 10th, 15th, 20th, 25th, and 30th years (neglecting the effect of rise in the price of fuel saved).

Solution

Given

$i = 0.12$ and the amount of fuel saving = USD $4000 per year.

For the desired years, the values of the present worth factors and the corresponding present worth of annual fuel saving are evaluated and shown in table below:

Year (*n*)	Present worth factor (*PWF*) $\frac{1}{(1+i)^n}$	Present worth of fuel savings (4000 × *PWF*)
5	$1/(1.2)^5 = 0.5670$	2269.7
10	$1/(1.2)^{10} = 0.3220$	1287.8
15	$1/(1.2)^{15} = 0.1827$	730.7
20	$1/(1.2)^{20} = 0.1037$	414.6
25	$1/(1.2)^{25} = 0.0588$	235.2
30	$1/(1.2)^{30} = 0.0334$	133.5

From the above tabulated values, one can observe that it is rather small for the useful life of the domestic solar water heating system in later years. Therefore, for RES with longer useful life, the present value analysis may not be expressive of its actual effectiveness to the user.

Example 14.11

Derive the relation between R_b, P, and S.

Solution

On putting the value of R from Eq. (14.81) into Eq. (14.91), one can obtain

$$R_b(1+i) = P(F_{PR,i,n})$$

or

$$R_b = \frac{P(F_{PR,i,n})}{(1+i)}$$

Similarly, from Eq. (14.89)

$$R_b = \frac{S(F_{SR,i,n})}{(1+i)}$$

Example 14.12

For a discount rate of 10%, determine the present worth of a large capacity PVT solar water heater for estimated salvage value of 5000 at the end of its useful life time of 20 years.

Solution

From Eq. (14.81),

$$P = S\left[\frac{1}{(1+i)^n}\right]$$

Given
$S =$ USD \$5000, $n = 20$ years, and $i = 10\%$
Hence, the present worth is given by

$$P = 5000\left[\frac{1}{(1+0.1)^{20}}\right] = \text{USD } \$743.22$$

Example 14.13

A person deposits USD \$1000 at the end of the year for 8 years to create a forborne annuity. To purchase hybrid solar water heating system, he desires to withdraw the money at the end of 14 years from now. If the money is worth 10% per year, then find the accumulated value at the end of the 14th year.

Solution

Let the amount available is X at the 14th year (as a receipt). The cash flow diagram for the payment is given below

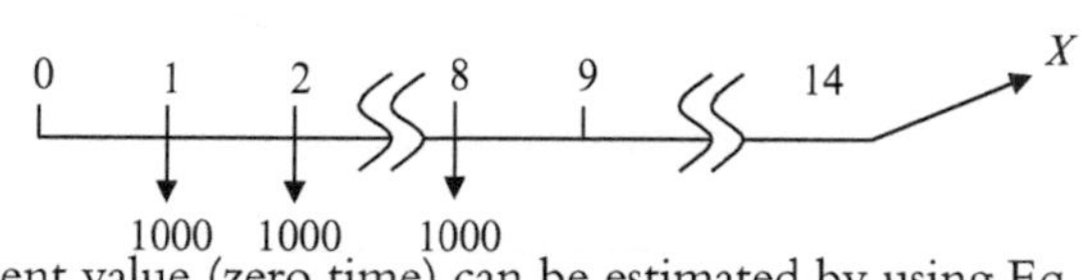

The present value (zero time) can be estimated by using Eq. (14.71) as

$$P = 1000 \times F_{RP,10\%,8} = 1000 \times 5.3349 = \text{USD } \$5334.90$$

On depositing the amount for 14 years, the future value at the end of 14 years can be obtained as

$$S = 5334.90 \times F_{PS,10\%,14} = 5334.90 \times 3.79 = \text{USD } \$20{,}259$$

Moreover, the above cash flow diagram can be drawn by considering USD \$1000 paid for 14 years less USD \$1000 paid as annuity for the last 6 years.

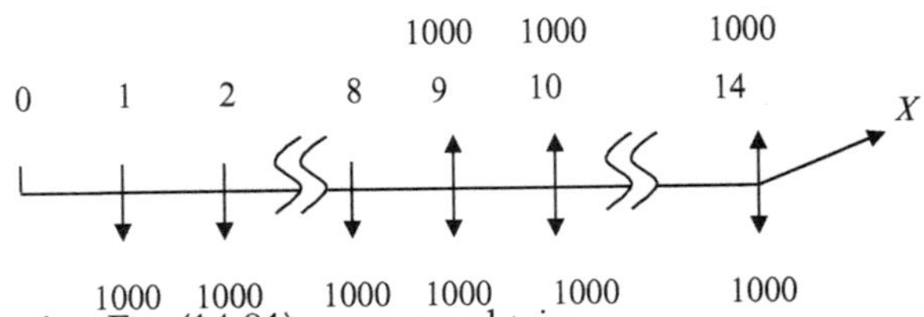

By using Eq. (14.91), one can obtain

$$S = 1000\left[\frac{(1.10)^{14}-1}{0.10} - \frac{(1.10)^{16}-1}{0.10}\right] = 1000 \times (27.9 - 7.71)$$

$$S = \text{USD } \$20,259$$

Example 14.14

The purchase and installation of a PV system for water pumping on the field of a farmer costs USD \$10, 000. The annual saving of diesel of a farmer is estimated to be around USD \$1200; and the annual maintenance cost is expected at USD \$100. If the useful life of the system is 30 years, then what will be the NPV of the investment on the PV system for the 8% interest rate?

Solution

Net annual benefits of using a PV system $= 1200 - 100 = \text{USD } \$1100$

Since the amount of net annual benefits is constant over the useful life of the system as defined in a problem, hence NPV can be obtained from Eq. (14.97) as

$$NPV = -10{,}000 + (1100)\left[\frac{(1+0.08)^{30}-1}{0.08(1+0.08)^{30}}\right] = 2384$$

Thus, the investment in the PV system is financially viable for the farmer.

APPENDIX 14.2

$$(\alpha\tau)_{eff} = a_b\left(\frac{h_f}{h_f + h_b}\right) + a_w + a_g\left(\frac{h_{1f}}{h_{1f} + h_{1g}}\right); \quad U_L = U_b + U_t;$$

$$U_b = \frac{h_{bf}h_b}{h_{bf} + h_b}; \quad U_t = \frac{h_{1f}h_{1g}}{h_{1f} + h_{1g}}$$

$$K_{1E} = \left[h_{1bf,W}\left(\frac{A_b}{2A_{gW}}\right) + h_{EW}\left(1 + \frac{1}{h_{1bf,E}}\right) + U_{gaW}\right]A_{gE}A_{gW}\alpha_g h_{1bf,E}$$

$$K_{1W} = \left[h_{1bf,E}\left(\frac{A_b}{2A_{gE}}\right) + h_{EW}\left(1 + \frac{1}{h_{1bf,W}}\right) + U_{gaE}\right]A_{gE}A_{gW}\alpha_g h_{1bf,W}$$

$$K'_{1E} = \left[h_{1f,W}\left(\frac{A_b}{2}\right) + h_{EW}\left(1 + \frac{h_{1f,W}}{h_{1f,E}}\right)A_{gW} + U_{gaW}A_{gW}\right]A_{gE}\alpha_g h_{1f,E}$$

$$K'_{1W} = \left[h_{1f,E}\left(\frac{A_b}{2}\right) + h_{EW}\left(1 + \frac{h_{1f,E}}{h_{1f,W}}\right)A_{gE} + U_{gaE}A_{gE}\right]A_{gW}\alpha_g h_{1f,W}$$

$$H_1 = \left(\frac{A_b}{2}\right)\left(U_{gaE}A_{gE} + U_{gaW}A_{gW}\right)h_{1bf,E}h_{1bf,W}$$

$$H_2 = A_{gE}A_{gW}h_{EW}\left(U_{gaE}h_{1bf,E} + U_{gaW}h_{1bf,W}\right)$$

$$H_3 = A_{gE}A_{gW}U_{gaE}U_{gaW}\left(h_{1bf,E} + h_{1bf,W}\right)$$

$$H_4 = A_{gE}A_{gW}h_{EW}\left(U_{gaE}h_{1bf,W} + U_{gaW}h_{1bf,E}\right)$$

$$E_1 = U_{gaE}\left[h_{EW} + h_{1bf,W}\left(\frac{A_b}{2A_{gW}}\right)\right]; \quad E_2 = U_{gaW}\left(h_{EW} + U_{gaE}\right)A_{gE}A_{gW}$$

$$E'_1 = U_{gaW}\left[h_{EW} + h_{1bf,E}\left(\frac{A_b}{2A_{gE}}\right)\right]; \quad E'_2 = U_{gaE}\left(h_{EW} + U_{gaW}\right)A_{gE}A_{gW}$$

$$F' = \left(\frac{A_b}{H}\right); \quad U_{eff} = (H')e^{-a\Delta t} = (H_1 + H_2 + H_3 + H_4)e^{-a\Delta t};$$

$$I(t) = A_{gE}I_{SE}(t) + A_{gW}I_{SW}(t)$$

$$\alpha_{wb} = \left(\frac{A_b}{2}\right)\left(\alpha_w + 2\alpha_b U_{ga}\right); \quad C_2 = \frac{h_3}{h_3 + h'_b}; \quad U_{3b} = \frac{h_3 h'_b}{h_3 + h'_b};$$

$$h'_b = h_b\left(1 + \frac{A_s}{A_b}\right)$$

$$\alpha_{eff} = \alpha'_w A_b + C_2\alpha'_b A_b + \alpha''_g C_1; \quad U = U_{3b}A_b + U_{12}A_b$$

$$(\alpha\tau)_{eff,N} = \frac{(AF_R(\alpha\tau))_1}{(A_c + A_m)}\left[\frac{1 - (1 - K_{k,A})^N}{NK_{k,A}}\right];$$

$$U_{L,N} = \frac{(AF_R U_L)_1}{(A_c + A_m)}\left[\frac{1 - (1 - K_{k,A})^N}{NK_{k,A}}\right]; \quad K_{k,A} = \left[\frac{(AF_R U_L)_1}{\dot{m}_f C_f}\right]$$

APPENDIX 14.3

The values of conversion factors [43]

n	F_{PS}	F_{SP}	F_{RP}	F_{PR}	F_{RS}	F_{SR}	F_{PK}
$i = 0.04$							
1	1.04	0.962	0.962	1.04	1	1	26
2	1.082	0.925	1.886	0.53	2.04	0.49	13.255
3	1.125	0.889	2.775	0.36	3.122	0.32	9.009
4	1.17	0.855	3.63	0.275	4.246	0.235	6.887
5	1.217	0.822	4.452	0.225	5.416	0.185	5.616
6	1.265	0.79	5.242	0.191	6.633	0.151	4.769
7	1.316	0.76	6.002	0.167	7.898	0.127	4.165
8	1.369	0.731	6.733	0.149	9.214	0.109	3.713
9	1.423	0.703	7.435	0.134	10.583	0.094	3.362
10	1.48	0.676	8.111	0.123	12.006	0.083	3.082
11	1.539	0.65	8.76	0.114	13.486	0.074	2.854
12	1.601	0.625	9.385	0.107	15.026	0.067	2.664
13	1.665	0.601	9.986	0.1	16.627	0.06	2.504
14	1.732	0.577	10.563	0.095	18.292	0.055	2.367
15	1.801	0.555	11.118	0.09	20.024	0.05	2.249
16	1.873	0.534	11.652	0.086	21.825	0.046	2.146
17	1.948	0.513	12.166	0.082	23.697	0.042	2.055
18	2.026	0.494	12.659	0.079	25.645	0.039	1.975
19	2.107	0.475	13.134	0.076	27.671	0.036	1.903
20	2.191	0.456	13.59	0.074	29.778	0.034	1.84
$i = 0.06$							
1	1.06	0.943	0.943	1.06	1	1	17.667
2	1.124	0.89	1.833	0.545	2.06	0.485	9.091
3	1.191	0.84	2.673	0.374	3.184	0.314	6.235
4	1.262	0.792	3.465	0.289	4.375	0.229	4.81
5	1.338	0.747	4.212	0.237	5.637	0.177	3.957
6	1.419	0.705	4.917	0.203	6.975	0.143	3.389
7	1.504	0.665	5.582	0.179	8.394	0.119	2.986
8	1.594	0.627	6.21	0.161	9.897	0.101	2.684
9	1.689	0.592	6.802	0.147	11.491	0.087	2.45
10	1.791	0.558	7.36	0.136	13.181	0.076	2.264
11	1.898	0.527	7.887	0.127	14.972	0.067	2.113
12	2.012	0.497	8.384	0.119	16.87	0.059	1.988
13	2.133	0.469	8.853	0.113	18.882	0.053	1.883

Continued

n	F_{PS}	F_{SP}	F_{RP}	F_{PR}	F_{RS}	F_{SR}	F_{PK}
14	2.261	0.442	9.295	0.108	21.015	0.048	1.793
15	2.397	0.417	9.712	0.103	23.276	0.043	1.716
16	2.54	0.394	10.106	0.099	25.672	0.039	1.649
17	2.693	0.371	10.477	0.095	28.213	0.035	1.591
18	2.854	0.35	10.828	0.092	30.906	0.032	1.539
19	3.026	0.331	11.158	0.09	33.76	0.03	1.494
20	3.207	0.312	11.47	0.087	36.786	0.027	1.453
$i = 0.08$							
1	1.08	0.926	0.926	1.08	1	1	13.5
2	1.166	0.857	1.783	0.561	2.08	0.481	7.01
3	1.26	0.794	2.577	0.388	3.246	0.308	4.85
4	1.36	0.735	3.312	0.302	4.506	0.222	3.774
5	1.469	0.681	3.993	0.25	5.867	0.17	3.131
6	1.587	0.63	4.623	0.216	7.336	0.136	2.704
7	1.714	0.583	5.206	0.192	8.923	0.112	2.401
8	1.851	0.54	5.747	0.174	10.637	0.094	2.175
9	1.999	0.5	6.247	0.16	12.488	0.08	2.001
10	2.159	0.463	6.71	0.149	14.487	0.069	1.863
11	2.332	0.429	7.139	0.14	16.646	0.06	1.751
12	2.518	0.397	7.536	0.133	18.977	0.053	1.659
13	2.72	0.368	7.904	0.127	21.495	0.047	1.582
14	2.937	0.34	8.244	0.121	24.215	0.041	1.516
15	3.172	0.315	8.559	0.117	27.152	0.037	1.46
16	3.426	0.292	8.851	0.113	30.324	0.033	1.412
17	3.7	0.27	9.122	0.11	33.75	0.03	1.37
18	3.996	0.25	9.372	0.107	37.45	0.027	1.334
19	4.316	0.232	9.604	0.104	41.446	0.024	1.302
20	4.661	0.215	9.818	0.102	45.762	0.022	1.273
$i = 0.10$							
1	1.1	0.909	0.909	1.1	1	1	11
2	1.21	0.826	1.736	0.576	2.1	0.476	5.762
3	1.331	0.751	2.487	0.402	3.31	0.302	4.021
4	1.464	0.683	3.17	0.315	4.641	0.215	3.155
5	1.611	0.621	3.791	0.264	6.105	0.164	2.638
6	1.772	0.564	4.355	0.23	7.716	0.13	2.296
7	1.949	0.513	4.868	0.205	9.487	0.105	2.054
8	2.144	0.467	5.335	0.187	11.436	0.087	1.874
9	2.358	0.424	5.759	0.174	13.579	0.074	1.736

Continued

n	F_{PS}	F_{SP}	F_{RP}	F_{PR}	F_{RS}	F_{SR}	F_{PK}
10	2.594	0.386	6.145	0.163	15.937	0.063	1.627
11	2.853	0.35	6.495	0.154	18.531	0.054	1.54
12	3.138	0.319	6.814	0.147	21.384	0.047	1.468
13	3.452	0.29	7.103	0.141	24.523	0.041	1.408
14	3.797	0.263	7.367	0.136	27.975	0.036	1.357
15	4.177	0.239	7.606	0.131	31.772	0.031	1.315
16	4.595	0.218	7.824	0.128	35.95	0.028	1.278
17	5.054	0.198	8.022	0.125	40.545	0.025	1.247
18	5.56	0.18	8.201	0.122	45.599	0.022	1.219
19	6.116	0.164	8.365	0.12	51.159	0.02	1.195
20	6.728	0.149	8.514	0.117	57.275	0.017	1.175

REFERENCES

[1] Dunkle RV. In: Solar water distillation: the roof type still and a multiple effect diffusion still. International development in heat transfer, ASME Proceedings, part 5. University of Colorado; 1961. p. 895–902.

[2] Cooper PI. The maximum efficiency of single-effect solar stills. Solar Energy 1973;15:205–14.

[3] Dwivedi VK, Tiwari GN. Annual energy and exergy analysis of single and double slope passive solar stills. Appl Sci Res 2008;3:225–41.

[4] Tsilingiris PT. Modeling heat and mass transport phenomena at higher temperatures in solar distillation systems—the Chilton–Colburn analogy. Solar Energy 2010;84: 308–17.

[5] Singh RV, Dev R, Hasan MM, Tiwari GN. In: Comparative energy and exergy analysis of various passive solar distillation systems, solar thermal—application. World renewable energy congress, Linkoping, Sweden; 2011. p. 8–13.

[6] Ranjan KR, Kaushik SC, Panwar NL. Energy and exergy analysis of passive solar distillation systems. Int J Low Carbon Technol Adv Access 2013;18:147–71.

[7] Tiwari GN, Yadav JK, Singh DB, Al-Helal IM, Abdel-Ghaney AM. Exergoeconomic and enviroeconomic analyses of partially covered photovoltaic flat plate collector active solar distillation system. Desalination 2015;367:186–96.

[8] Fitzsimons L, Corcoran B, Young P, Foley G. Exergy analysis of water purification and desalination: a study of exergy model approaches. Desalination 2015;359:212–24.

[9] Cengel YA, Boles MA. Thermodynamics: an engineering approach. 5th ed. Boston: McGraw-Hill; 2006.

[10] Jones JB, Dugan RE. Engineering thermodynamics. Indiana, USA: Prentice Hall; 1996.

[11] Moran MJ, Shapiro HN. Fundamentals of engineering thermodynamics. 7th ed. Wiley, New York, USA.

[12] Flavio G, Segio A, Silvia AN. Thermoeconomic evaluation of a gas turbine cogeneration system. Energy Convers Manage 2000;41:1191–200.

[13] Zhang G, Hua B, Chen Q. Exergoeconomic methodology for analysis and optimization of process systems. Comput Chem Eng 2000;24:613–8.

[14] Dincer I. The role of exergy in energy policy making. Energy Policy 2000;30:137–49.

[15] Evans RB, Crellin GL, Tribus M. Thermoeconomic considerations of sea water demineralization. In: Spiegler KS, editor. Principles of desalination. 2nd ed. New York: Academic; 1980. p. 1–54 [chapter 1].
[16] El-Sayed YM, Tribus M. Strategic use of thermoeconomics for system improvement. In: Gaggioli RA, editor. Efficiency and costing: second law analysis of processes. ACS symposium series, vol. 235. Washington, DC: American Chemical Society; 1983. p. 215–38.
[17] El-Sayed YM, Evans RB. Hermoeconomics and the design of heat systems. Trans ASME J Eng Power 1970;92:27–34.
[18] Gaggioli RA, Wepfer WJ. Exergy economics. Energy Int J 1980;5:823–38.
[19] Reistad GM, Gaggioli RA. Available energy costing. In: Gaggioli RA, editor. Thermodynamics: second law analysis. ACS Symposium Series, vol. 122. Washington, DC: American Chemical Society; 1980. p. 143–59.
[20] Tsatsaronis G, Winhold M. Exergoeconomic analysis and evaluation of energy conversion plants, Parts 1& 2. Energy Int J 1985;10:69–94.
[21] Poredos A, Kitanovski A. Exergy loss as a basis for the price of thermal energy. Energy Convers Manag 2002;43:2163–73.
[22] Rosen MA. The relation between thermodynamic losses and capital costs for a modern coal-fired electrical generation station. In: Computer-aided energy systems analysis. vol. 21. New York: AES, American Society of Mechanical Engineers; 1990. p. 69–78.
[23] Tsatsaronis G. A review of exergoeconomic methodologies. In: Moran MJ, Sciubba E, editors. Second law analysis of thermal systems. New York: American Society of Mechanical Engineers; 1987. p. 81–7.
[24] Moran MJ. Availability analysis: a guide to efficient energy use. revised edition. New York: American Society of Mechanical Engineers; 1989.
[25] Kotas TJ. The exergy method of thermal plant analysis. reprint ed. Malabar, FL: Krieger; 1995.
[26] Edgerton RH. Available energy and environmental economics. Toronto: D.C. Heath; 1982.
[27] Szargut J, Morris DR, Steward FR. Exergy analysis of thermal, chemical and metallurgical processes. New York: Hemisphere; 1988.
[28] Bejan A. Entropy generation through heat and fluid flow. In: Rosen MA, Dincer I, editors. Applied thermal engineering. vol. 23. Toronto: Wiley; 2003. p. 643–58.
[29] El-Sayed V, Gaggioli RA. A critical review of second law costing methods, Parts I and II. Trans ASME J Energy Resour Technol 1989;111:1–15.
[30] Valero A. On the energy costs of present day society. In: Krane RJ, editor. Thermodynamics and the design, analysis and improvement of energy systems. vol. 35. New York: AES—American Society of Mechanical Engineers; 1995. p. 1–7.
[31] Tiwari GN, Sahota L. Review on the energy and economic efficiencies of passive and active solar distillation systems. Desalination 2017;401:151–79.
[32] Tiwari GN. Solar energy: fundamentals, design, modeling and applications. New Delhi, India: Narosa Publishing House; 2004.
[33] Petela R. Exergy of undiluted thermal radiation. Solar Energy 2003;74:469.
[34] Sahota L, Tiwari GN. Effect of nanofluids on the performance of passive double slope solar still: a comparative study using characteristic curve. Desalination 2016;388:9–21.
[35] Singh DB, Tiwari GN, Al-Helal IM, Dwivedi VK, Yadav JK. Effect of energy matrices on life cycle cost analysis of passive solar stills. Solar Energy 2016;134:9–22.
[36] Tiwari GN, Mishra RK. Advanced renewable energy resources. Cambridge, UK: RSC Publishing; 2012.
[37] Singh DB, Yadav JK, Dwivedi VK, Kumar S, Tiwari GN, Al-Helal IM. Experimental studies of active solar still integrated with two hybrid PVT collectors. Solar Energy 2016;130:207–23.

[38] Rosen MA, Dincer I. Exergoeconomic analysis of power plants operating on various fuels. Appl Therm Eng 2003;23:643–58.
[39] Rosen MA. An investigation of the relation between capital costs and selected thermodynamic losses. In: Second law analysis-industrial and environmental applications. vol. 25/HTD-vol. 191. New York: American Society of Mechanical Engineers, AES; 1991. p. 55–62.
[40] Rosen MA, Dincer I. Thermoeconomic analysis of power plants: an application to a coal-fired electrical generating station. Energy Convers Manag 2003;44:1633–51.
[41] Nag PK. Basic and applied thermodynamics. New Delhi, India: Tata McGraw-Hill; 2004. ISBN 0-07-047338-2.
[42] Agrawal S, Tiwari GN. Exergoeconomic analysis of glazed hybrid photovoltaic thermal module air collector. Solar Energy 2012;86:2826–38.
[43] Hossain MA, Woods JL, Bala BK. Optimisation. Renew Energy 2005;30:729–42.

INDEX

Note: Page numbers followed by *f* indicate figures, and *t* indicate tables.

H

T

U

V

W

FSC
www.fsc.org
MIX
Papier | Fördert gute Waldnutzung
FSC® C083411

Kolibri 360 GmbH
Ferdinand-Jühlke-Straße 7
99095 Erfurt, Deutschland
produktsicherheit@kolibri360.de